Evaluating Fault and Cap Rock Seals

Edited by
Peter Boult
John Kaldi

AAPG Hedberg Series, No. 2

Published by
The American Association of Petroleum Geologists
Tulsa, Oklahoma

Copyright © 2005
By the American Association of Petroleum Geologists
All Rights Reserved

ISBN: 0-89181-901-0

AAPG grants permission for a single photocopy of an item from this publication for personal use. Authorization for additional copies of items from this publication for personal or internal use is granted by AAPG provided that the base fee of $3.50 per copy and $.50 per page is paid directly to the Copyright Clearance Center, 222 Rosewood Drive, Danvers, Massachusetts 01923 (phone: 978/750-8400). Fees are subject to change.

AAPG Editor: Ernest A. Mancini
Geoscience Director: James B. Blankenship

This publication is available from:

The AAPG Bookstore
P.O. Box 979
Tulsa, OK U.S.A. 74101-0979
Phone: 1-918-584-2555 or 1-800-364-AAPG (U.S.A. only)
Fax: 1-918-560-2652 or 1-800-898-2274 (U.S.A. only)
E-mail: bookstore@aapg.org
www.aapg.org

Canadian Society of Petroleum Geologists
No. 160, 540 Fifth Avenue S.W.
Calgary, Alberta T2P 0M2
Canada
Phone: 1-403-264-5610
Fax: 1-403-264-5898
E-mail: Jaime.croft@cspg.org
www.cspg.org

Geological Society Publishing House
Unit 7, Brassmill Enterprise Centre
Brassmill Lane, Bath BA13JN
U.K.
Phone: +44-1225-445046
Fax: +44-1225-442836
E-mail: sales@geolsoc.org.uk
www.geolsoc.org.uk

Affiliated East-West Press Private Ltd.
G-1/16 Ansari Road, Darya Gaaj
New Delhi 110 002
India
Phone: +91-11-23279113
Fax: +91-11-23260538
E-mail: affiliat@vsnl.com

The American Association of Petroleum Geologists (AAPG) does not endorse or recommend products or services that may be cited, used, or discussed in AAPG publications or in presentations at events associated with AAPG.

About the Editors

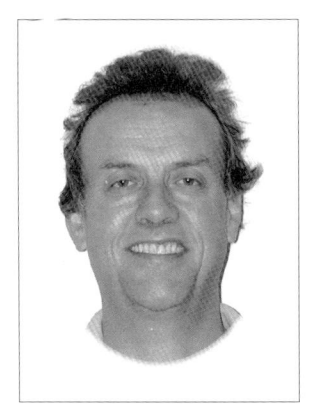

Peter Boult graduated (Geology) from Leeds University in 1976. He has 23 years experience in the oil industry; 6 years experience on the well site in Europe, the Middle East, and Australia; 10 years experience as an academic researcher working in the Cooper, Eromanga, and Papuan Basins; 4 years experience as a company geoscientist working in the Otway and Bass Basins; and 3 years experience as Chief Geologist with the Petroleum Group at Minerals and Energy Resources, South Australia. Boult gained his Ph.D. on hydrocarbon seals in 1996 from the National Centre for Petroleum Geology and Geophysics (Australia). He has continued to be actively involved in seals research through his position as visiting researcher at the Australian School of Petroleum in Adelaide.

John Kaldi received a Ph.D. from Cambridge University, England, in 1980. He spent 16 years working in the upstream sector of the industry (Shell, ARCO, VICO) and since 1998 he has been in academia. First, as the Director of the Australian National Centre for Petroleum Geology and Geophysics in Adelaide, and, for the last year, as Head of School, Australian School of Petroleum, University of Adelaide. Kaldi is a member and active participant in a number of professional organizations, including the American Association of Petroleum Geologists (AAPG), Indonesian Petroleum Association (IPA), Society of Petroleum Engineers (SPE), Society of Economic Paleontologists and Mineralogists (SEPM), International Association of Sedimentologists (IAS), and Petroleum Exploration Society of Australia (PESA). Kaldi's specialties include carbonate sedimentology and diagenesis, seal evaluation, reservoir geology, and multi-disciplinary studies. He was an organizer of two AAPG Hedberg Converences on Seals, has served as Book Review Editor for JSP, Associate Editor for the CSPG Bulletin, and has been Technical Program Chair for the IPA. He has been a member of the AAPG Grants-in-Aid Committees and is presently a member of the House of Delegates. He is Program Manager for the Australian Petroleum Cooperative Research Centre's Program on Seal Evaluation. Kaldi received AAPG's Special Commendation Award for Significant Contributions to Petroleum Geology in 1997.

Acknowledgments

AAPG wishes to thank the following
for their generous contributions to

Evaluating Fault and Cap Rock Seals

Statoil ASA
Statoil Rotvoll

Contributions are applied toward the production cost of publication,
thus directly reducing the book's purchase price
and making the volume available to a larger readership.

Reviewer Acknowledgments

The chapters in this book were reviewed rigorously by multiple experts, and we appreciate their candid comments and constructive suggestions, all of which improved the quality of this publication.

The reviewers were: Andrew Aplin, Wayne Bailey, Peter Bretan, Alton Brown, Jamie Burgess, David Castillo, Conrad Childs, Ben Clenell, Gary Couples, Andrew Davids, David N. Dewhurst, Terry Engelder, Quentin Fisher, Domenque Grauls, Christian Hermanrud, Richard R. Hillis, Susan Hippler, Suzanne Hunt, Simon Lang, Steve R. Larter, Leslie Leith, Dave Lowry, Helen Lweis, Scott Mildren, Isabelle Moretti, Hegge Nordgård Bolås, Geoff O'Brien, Signe Ottesen, Elveina Paraschivoiu, Stephen Priest, Mark Smith, Richard Suttill, Paul Theologou, Barbara Tilly, Neil Tupper, James Undershchultz, John Walsh, Robert Weedon.

Table of Contents

About the Editors iii

Acknowledgments v

Reviewer Acknowledgments vii

Foreword xiii

Chapter 1
Fluid Flow, Pore Pressure, Wettability, and Leakage in Mudstone Cap Rocks 1
Andrew C. Aplin and Steve R. Larter

Chapter 2
Seal Failure Related to Basin-scale Processes 13
Christian Hermanrud, Hege M. Nordgård Bolås, and Gunn M. G. Teige

Chapter 3
Potential New Method for Paleostress Estimation by Combining Three-dimensional Fault Restoration and Fault Slip Inversion Techniques: First Test on the Skua Field, Timor Sea 23
A. P. Gartrell and M. Lisk

Chapter 4
Fault Healing and Fault Sealing in Impure Sandstones 37
David N. Dewhurst, Peter J. Boult, Richard M. Jones, and Stuart A. Barclay

Chapter 5
A Regional Analysis of Fault Reactivation and Seal Integrity Based on Geomechanical Modeling: An Example from the Bight Basin, Australia 57
S. D. Reynolds, E. Paraschivoiu, R. R. Hillis, and G. W. O'Brien

Chapter 6
FAST: A New Technique for Geomechanical Assessment of the Risk of Reactivation-related Breach of Fault Seals 73
Scott D. Mildren, Richard R. Hillis, David N. Dewhurst, Paul J. Lyon, Jeremy J. Meyer, and Peter J. Boult

Chapter 7
Seals: The Role of Geomechanics 87
Gary D. Couples

Chapter 8
The Influence of Stress Regimes on Hydrocarbon Leakage 109
Hege M. Nordgård Bolås, Christian Hermanrud, and Gunn M. G. Teige

Chapter 9
Investigating the Effect of Varying Fault Geometry and Transmissibility on Recovery:
Using a New Workflow for Structural Uncertainty Modeling in a Clastic Reservoir 125
Signe Ottesen, Chris Townsend, and Kjersti Marie Øverland

Chapter 10
Quantifying the Impact of Fault Modeling Parameters on Production Forecasting
for Clastic Reservoirs 137
Guillaume Lescoffit and Chris Townsend

Chapter 11
Using Buoyancy Pressure Profiles to Assess Uncertainty in Fault Seal Calibration 151
Peter Bretan and Graham Yielding

Chapter 12
Evaluation of Late Cap Rock Failure and Hydrocarbon Trapping Using a Linked Pressure
and Stress Simulator 163
A. E. Lothe, H. Borge, and Ø. Sylta

Chapter 13
Sealing by Shale Gouge and Subsequent Seal Breach by Reactivation: A Case Study of
the Zema Prospect, Otway Basin 179
Paul J. Lyon, Peter J. Boult, Richard R. Hillis, and Scott D. Mildren

Chapter 14
Distinct-element Stress Modeling in the Penola Trough, Otway Basin, South Australia 199
Suzanne P. Hunt and Peter J. Boult

Chapter 15
Sedimentology and Petrophysical Character of Cretaceous Marine Shale Sequences in
Foreland Basins—Potential Seismic Response Issues 215
W. R. Almon, Wm. C. Dawson, F. G. Ethridge, E. Rietsch, S. J. Sutton, and B. Castelblanco-Torres

Chapter 16
Using Gas Chimneys in Seal Integrity Analysis: A Discussion Based on Case Histories 237
Roar Heggland

Chapter 17
Formation Fluids in Faulted Aquifers: Examples From the Foothills of Western Canada
and the North West Shelf of Australia 247
J. R. Underschultz, C. J. Otto, and R. Bartlett

Chapter 18
Economic Evaluation of Prospects with a Top Seal Risk 261
David C. Lowry

Foreword

Three AAPG Hedberg research conferences on seals have been held in the past 30 yr (Denver, 1983; Crested Butte, 1993; and more recently, Barossa Valley, South Australia, 2002). Each conference represented a quantum leap in the understanding and methodology of the subject of seals. This volume is a compendium of the proceedings of the Barossa meeting. The key driver for this meeting was the recognition that knowledge of risk (in the estimation of sealing capacity and fault-seal potential) is important in making judgements at the exploration, appraisal, and development stages of the petroleum business. In addition, incorporating seal risk in the overall assessment of hydrocarbons in place can affect decisions to drill prospects, the location of appraisal and development wells, as well as reserve estimation. Improved methods to estimate seal capacity and fault integrity can lead to savings in well costs, improved recoveries through optimum placement of wells, and greater certainty of meeting contractual requirements through improved estimates of hydrocarbon in place.

The 2002 meeting was the first ever Hedberg held in Australia, and the venue, a glorious setting in the heart of the famous South Australia wine-growing region, may have had something to do with the attendance of 85 delegates, most of whom traveled halfway around the world to be there. The meeting consisted of 53 presentations over a period of two-and-a-half days with robust debate taking up almost half of the time. This volume is the result of the papers presented, debated, revised, and finally submitted to AAPG as part of a thematic state-of-the-art publication on seals.

The volume contains 18 chapters that reflect the spectrum of presentations at the conference. The knowledge imparted by these chapters will be a window on the state of knowledge at this juncture of time. It will be a lasting tribute to the efforts of the individuals and the synergy of the group, as a whole, that was established at the conference.

We thank the companies, academic institutions, and government organizations for their investment in support given to researchers to do their research, the provision of data, the sponsorship of students to attend the conference, and, of course, for sending delegates halfway around the world. We further acknowledge the AAPG for its foresight and commitment to this volume.

The plethora of new science presented at the Barossa meeting was obviously an evolutionary outgrowth of previous seals conferences. Threads connecting to the Crested Butte conference included the question of the importance of wettability in seals. At Crested Butte, the long held assumption that all rocks are water wet was questioned. Leith et al. (1993) showed that oil had penetrated seals in some North Sea traps, but whether this was by the process of hydraulic fracturing or capillary leakage through a water-wet seal was not resolved. Over the last 10 yr, very little work on seal wettability has been carried out or at least documented. **Aplin and Larter's** chapter in this volume addresses some of the issues raised in 1993. Their conclusions are that *"hydrophilic organic compounds in reservoirs, followed by diffusion into cap rock pores, may create oil-wet pathways into cap rocks and drive leakage."* This means that not only do petrophysicists and reservoir engineers have to consider wettability when estimating the saturation and flow properties of reservoirs, but explorationists also have to consider the phenomenon when risking entrapment. **Aplin and Larter's** chapter not only probes the wettability question but also describes a methodology for determining, directly from logs, properties that control the flow of hydrocarbons through muddy seals that may undergo wettability change. They do this by investigating the key relationships between (1) porosity and effective stress, (2) porosity and permeability, and (3) porosity and threshold capillary entry pressure. The authors conclude that these properties are *"strongly influenced by the grain-size distribution or clay content of the sediments."* These properties can be calibrated against cores and constrained by cuttings data.

Although **Aplin and Larter** do not consider the role of leakage via hydraulic fracturing of clay-rich cap rocks, **Hermanrud et al.** do. In their chapter, they address *"three sets of such subsurface processes, and how they impact on hydrocarbon leakage: sediment compaction, fluid mobility in a two-phase (water-plus-hydrocarbons) reservoir, and relationships between stress and rock failure."* They describe the conditions for sediment compaction and remind us that stress-insensitive chemical compaction becomes increasingly important above 60°C. They state, conclusively, that *"All pressure compartments leak fluids as long as [chemical] compaction occurs."* **Hermanrud et al.** argue that the relative permeability of seals and reservoirs to water never reaches zero on a geological timescale. This, they say, holds true despite the fact that water in the reservoir may be at a conventional irreducible saturation. Furthermore, they tell us that the *"failure to recognize the mobility of this water may result in too-pessimistic seal capacity estimates."* They also describe the conditions for the creation of structural permeability that will allow the bleed-off of overpressure and the leakage of hydrocarbons.

The abundance of papers in the conference and in this volume containing elements of geomechanics attests to the importance of this area of investigation. Some of the presentations and papers also demonstrate that as a discipline, the investigation of hydrocarbon flow through seals has much to gain from other industries such as mining and construction. Whether leakage occurs through the cap rock as a result of perturbations in the stress field caused by fault discordance as suggested by **Hunt and Boult** or, as **Mildren et al.** suggest, the faults themselves are the mechanism for leakage as they dilate under the influence of the stress field, is still debatable. It would serve any explorationist well to keep a foot in both camps because both may, to some extent, be correct.

Two chapters are present in the volume that introduce, as **Gartrell and Lisk** put it, a *"new layer of complexity to fault reactivation and associated hydrocarbon leakage."* **Gartrell and Lisk's** chapter tests a new methodology for assessing paleostress regimes, and they conclude that, despite the potential errors inherent in the methodology they use, their estimation of the paleostress tensor at the southern end of the Timor Sea is *"not in conflict with structural histories formulated by more traditional methods."* For the prediction of the paleostress tensor, they bring together, for the first time, a three-dimensional fault restoration technique, which is commonly performed on seismic data sets to verify fault and horizon interpretation, and a fault slip inversion technique, which had previously only been used in outcrop studies. They admit that this is very early work, and one data point does not mean the method is proven, but all such empirical methods have to start somewhere. Moreover, they conclude that without such an approach, paleostress estimations based on simple geometric analysis (such as fault orientations) are possibly prone to errors.

Dewhurst et al. however, definitely take the deterministic approach to fault-seal analysis. They examine the detail of the fault rock itself in terms of its response to geomechanical laboratory testing and relate results to petrological properties. Significantly, they find that in two study areas, phyllosilicate framework fault rocks are *"stronger than their host reservoirs as a direct result of syn- and postdeformational physical and diagenetic processes."* They point out that with the trend toward using fault reactivation and consequent seal assessment methods derived from theoretical geomechanics, their observations of *"strength recovery through healing"* means that faults cannot be considered as being purely cohesionless. Thus, algorithms, such as dilation and slip tendency, which do not incorporate rock strength, may be misleading when assessing fault-seal risk. Furthermore, in a regional context, as **Hunt and Boult** point out, the regeneration of fault strength (via diagenesis) influences stress distribution. Within a regional top seal, this may result in fracturing (creation of pervasive structural permeability) and the concomitant loss of hydrocarbons.

Reynolds et al's chapter is based on the application of geomechanical theory, but in this case, they explain what can be done to assess faults for leakage potential on a basinwide scale when little or no data are available for depth conversion. Their approach is a sensitivity analysis of fault assessment based on a geomechanical estimation of the relative risk of reactivation. They assume that risking reactivation of faults in Late Cretaceous sediments of the Great Australian Bight is equivalent to risking leakage. This chapter, along with those of **Mildren et al.** and **Lyon et al.**, assumes that during reactivation, the rate of strain and the rheology of the fault rocks are such that when they interact, reactivation occurs by brittle failure, and dilation occurs to allow leakage. Although this may not be true for the more prolific hydrocarbon-producing basins of the world, where stress regimes are more stable and rocks are younger and weaker, it would appear to be the case for Australia, which, as described by **Reynolds et al.**, currently has and historically has had a highly variable stress tensor in space and time relative to its continental margins.

Although not included in this volume, Fisher presented key data at the 2002 Hedberg Conference that helped clarify the debate on whether faults dilate or leak as they move (reactivate). Fisher described the controls on the geomechanical properties of sediments as they undergo burial and compaction and heating and diagenesis. Fisher illustrated the control of grain size, geothermal gradient, and burial rates on the depth of the ductile-to-brittle transition (Figure 1), which may also coincide with the top of the seismogenic zone. Although a useful guide for considering the impact of faults on production, Fisher's data, like many other petrological studies, are largely confined to reservoir sandstones, because very little data are available from seal lithologies.

Couples' chapter, "The role of geomechanics," contributes to the debate associated with the assumption that faults that reactivate exhibit dilation. This chapter revisits some of the fundamental physical definitions of what a seal is and how geomechanics can describe its behavior. **Couples** provides an interesting definition of seals as *"power transmission barriers,"* which are places *"where the rate of achievable energy dissipation (flow) is less than the rate at which potential energy can be imposed or renewed."* The keyword here is rate, and this will vary depending on whether one is dealing with the relatively high rates of stress change as experienced by Australia compared to the more stable stress regimes such as on the passive margins of the Atlantic. **Couples'** chapter provides insights to understanding "the big picture" of geomechanics.

Couples also warns us about some of the pitfalls of upscaling geomechanical laboratory measurements to large-scale models. He points out that *"a scale dependency to the identification of localized vs. delocalized deformation is present,"* which is *"basically the same issue that arises in the way that many people use the terms brittle and ductile."* This is followed by a discussion of poroplasticity and its relation to volume loss or gain (dilation) during deformation. Although in many respects, this chapter is aimed at the geomechanically inclined, **Couples** is still concerned with getting his message across to the general practitioner. This message is that an understanding of *"poro-plasticity provides the critical link to allow us to make realistic geomechanical predictions about seals."* He also provides some interesting new insights into everyday observations, such as *"we can interpret dilational deformations as representing material that is 'too solid' to be able to undergo the imposed distortions, so extra porosity must be created to 'weaken' the material enough to allow cataclastic flow."* However, the main strength of this chapter is that he works through an example (without using mathematics) where he demonstrates how geomechanics explains most of what we know about the operation of seals and overpressured rocks. In addition, he describes how geomechanical modeling can be used to help understand the phenomena, such as the formation of shale gouge, the non-Andersonian

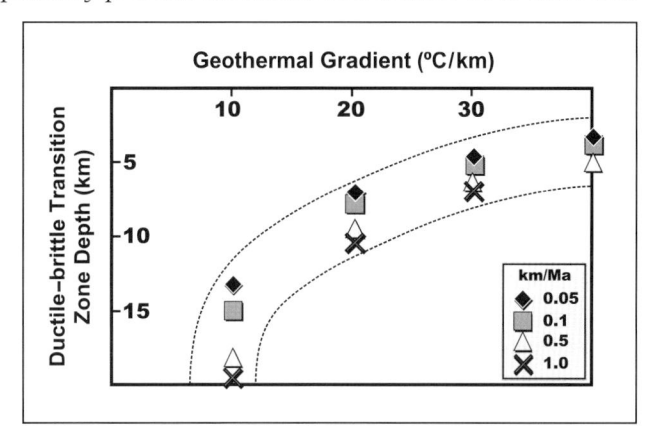

FIGURE 1. Plot of depth of the ductile-brittle transition zone for a medium-grained sandstone against geothermal gradient at variable burial rates.

behavior of stress around faults, and the prediction of petrophysical properties for input into migration simulation models.

In the previous Hedberg conferences, there was considerable interest in pressure compartments, and the term "pressure seal" was in common usage. From this, the term retention capacity (being the difference between the least principal stress and the pore pressure) was born, because seals were assumed to only fail either by hydraulic fracturing or capillary leakage (Watts, 1987). Thus, in water-wet shales with extremely high capillary entry pressures, high retention capacities were thought to be synonymous with good pressure seals. However, in this volume, **Nordgård Bolås et al.** show that in their study area, where differential stress is high, shear failure instead of tensile hydraulic fracturing occurs, and seals with high retention capacities may be breached. Shear fracturing is caused by very recent increases in differential stress attributable to ice loading and unloading and the advance of sedimentary wedges. Furthermore, their analysis indicates that in their case study, shear failure and associated leakage reduce pore pressure and, therefore, increase retention capacity.

Shear-induced failure and mixed-mode failure are mechanisms that have long been thought to be responsible for the creation of structural permeability through cap rocks in and around Australia (Hillis, 1998) because of the recent collision of that continent with New Guinea and Indonesia. In this volume, **Nordgård Bolås et al.** show that mechanisms other than continental collision and a resultant, highly variable stress tensor can cause rapid increases in differential stress, leading to shear failure. Those working in deep offshore basins where the postcharge advance of sedimentary wedges has occurred, should consider the findings of **Nordgård Bolås et al.** Nevertheless, situations in low differential stress environments where retention capacity (i.e., failure by tensile hydraulic fracturing) is still a good indicator of seal adequacy remain. Knowing where to apply this relationship is the main problem. Fortunately, this can be solved by use of the delta-P parameter or FAST approach, as described by **Mildren et al.** Delta-P is a measure of the increase in pore pressure that would be required to create fractures by either shear or tensile failure. The methodology considers not only the stress field and the orientation of planes of weakness and faults in that field, but also the strength of the fault rocks. The easiest way to picture this (**Mildren et al.**'s figure 2) is in terms of the distance between poles to planes in a Mohr circle (representing stress state) and the Coulomb failure envelope (representing rock strength) of the rock that is perceived as the weak link in the sealing system. Whether this is a fault that cuts the seal or the seal itself depends on their relative strengths.

What of initial entrapment in fault-related traps? As previously mentioned, structural uncertainty has certainly been reduced in recent years by rapid advances in technology. Whether it is worth focusing on this kind of uncertainty at the exploration stage is still debatable, given all the other uncertainties of a petroleum system that may still need to be addressed. This lack of concern is reflected in the absence of papers presented on this subject at the conference. However, after a significant discovery has been made, the quantification of structural uncertainty becomes of greater significance. This is long overdue because, as **Ottesen and Townsend** point out, *"the statistical treatment of tectonic heterogeneities in reservoir simulation is not as rigorous as that of their sedimentary counterparts."* However, the increasing tendency toward exploration and discovery of structurally complex traps demands that this aspect be risked in a similarly rigorous fashion. Their chapter describes a workflow for incorporating structural uncertainty into a geological model that can be used in a reservoir simulator. **Ottesen and Townsend** cite fault density and fault-relative permeability as the most influential structural factors that bear on ultimate recovery.

Lescoffit and Townsend add to the contribution of **Ottesen and Townsend** by investigating the relative importance and effect of structural uncertainty on production predicted from models built into a reservoir simulator. To simplify their experiment, they did not vary fault location or geometry and only considered fault connectivity, fault displacement, and parameters related to displacement that affect transmissibility (e.g., fault thickness and gouge-smear prediction). Their models are based on a typical North Sea tilted fault block, and the authors conclude that whereas the sequence-stratigraphic models still dominate, this is not true for all simulation cases. Of the structural parameters that they did vary, the fault-seal algorithm and the subseismic fault pattern used had greatest impact on production prediction.

Bretan and Yielding have done considerable work in the past toward empirically constraining fault-seal capacity in the static case and transmissibility in the production case. Their chapter in this volume very succinctly describes the use of buoyancy pressure profiles for calibrating and incorporating uncertainty in the *V*clay parameter, which, in turn, is used along with fault throw to generate fault algorithm parameters, such as shale gouge ratio (SGR) and shale smear factor. The use of the SGR factor is demonstrated in papers by **Lyon et al.** and **Ottesen et al. Lyon et al.** use it to help understand the original hydrocarbon-trapping mechanism in the Zema structure in the Otway Basin, which has subsequently undergone leakage because of shear failure either along or associated with the major bounding fault. **Ottesen et al.** use SGR in combination with a clay smear algorithm to determine fault transmissibility.

Lothe et al. use the methodologies learned from reservoir simulations and apply these to the province-scale linked pressure and stress simulator. Their chapter describes a validated, time-stepping model for predicting overpressure and hydrocarbon leakage from linked pressure compartments. This, in turn, is based on the interaction of across-fault leakage and hydraulic fracturing of the cap rock in each compartment. Through an ingenious method of altering permeability in faults and modeling stress through time in the cap rocks, they create multiple realizations of leakage. The authors compare their results with actual pressure data in wells and zero in on the most likely hydrocarbon migration route across an area. This, of course, is potentially invaluable in pinpointing prospective areas to reduce exploration uncertainty.

Lothe et al. bring faults and cap rock-sealing mechanisms together in a province-scale linked pressure and stress simulator. **Hunt and Boult,** however, bring these aspects together in an overall assessment of trap integrity. In the Otway Basin where **Dewhurst et al.** recognize that some faults are stronger than surrounding rocks, **Hunt and Boult** have observed stress perturbations that corroborate this observation and use discrete element stress modeling to investigate this phenomena. They base their model on data from multiple fault-bound traps, some of which have leaked. They assign three different fault sets and realistic geomechanical parameters of cap rocks to run their model forward until zones of high and low differential stress appear. They conclude that a positive observable relationship exists between zones of modeled high shear stress and trap leakage, and that in this case where observations indicate that faults are stronger than the surrounding rock, it is fracturing and the development of structural permeability in the cap rock that has caused leakage.

Lyon et al.'s chapter is a case study on the Zema structure, which occurs in the data set used by **Hunt and Boult.** This structure contains a well-documented paleo-oil-gas column in the Otway Basin. A core from the seal was analyzed to determine the original trapping mechanism, and **Lyon et al.** attribute the weak link to shale gouge that developed on the main bounding fault. They then consider the mechanism for failure. The seal over the Zema structure provides the rare opportunity to investigate structural permeability in the form of a deviation on the SP log, which correlates with an interpretation of a fault on the dipmeter log. An in-depth structural interpretation of the Zema structure provides an insight into how this structural permeability developed. The observed structural permeability is not associated with the main bounding fault but appears to be linked to younger faults, which formed under the current stress regime that propagate through the top seal and into the reservoir near its crest.

Almon et al.'s chapter is one of few dealing exclusively with top seal, reflecting the change in the industry's focus since the previous Hedberg conference. This chapter continues to push the knowledge boundary on top seal risk reduction by describing a methodology for seismic-based predrill evaluation of top seal capacity. The authors describe the facies variation seen in outcrop and in shallow drill holes of the Lewis Shale in sufficient petrological detail to enable their separation using discriminant function analysis. They then measured the petrophysical properties for each facies and note that where calcareous laminated shale (from highstand systems tracts) lies directly above organic laminated shales (from transgressive systems tracts), a strong seismic reflection and an associated amplitude-vs.-offset anomaly is generated. These seismic responses could be confused with reservoired hydrocarbons.

Heggeland's chapter is a reminder of how far seismic display and interpretation have come in the last decade. He shows examples of mobile hydrocarbons in gas chimneys from the North Sea, Gulf of Mexico, Nigeria, and the Caspian Sea and classifies them into two types. Type 1 chimneys are chimneys associated with faults. These commonly have circular cross sections with diameter in the order of 100 m (330 ft) and create pockmarks where they erupt on the seafloor. They are interpreted to have a high flux rate and, depending on fault location, can significantly increase exploration risk. Type 2 chimneys are not associated with faults, and their lateral extent can be on the order of several hundred meters. These are interpreted to have a low flux rate and have proven to be good indicators of commercial hydrocarbons.

Undershultz et al.'s chapter encourages the incorporation of hydrodynamics in seal studies. They comment that hydrodynamics is a commonly underused methodology for assessing fault-sealing characteristics by stating that *"hydrodynamic analysis of aquifers cut by faults can be used as an indirect indicator of the fault zone hydraulic properties."* They then use *"case studies from the foothills of western Canada and the North West Shelf of Australia to define a workflow for hydrodynamic analysis in faulted strata"* and provide us with a checklist for identifying which faults, or parts of faults, are sealing to aquifer flow. Faults that show medium to strong aquifer flow, either in an upfault or across-fault sense, are unlikely to be sealing to hydrocarbons.

Lowry makes a very important contribution by discussing various prospect-risking methodologies and how to risk seal, especially top seal, and how this should be incorporated into expected monetary value investment calculations. He extols the virtues of valuing prospects by varying the chance of success with the reserve distribution and states that *"If seal capacity is a continuously varying function of column height, using a single Fill-Factor becomes a blunt instrument likely to lead to a distorted evaluation."* Being a realist, **Lowry** does not wish to overcomplicate the risking procedure and provides a hint as to when the more complicated method, which he has proposed, should be used by stating that *"A common early warning sign is when a group of explorationists focused on risking a prospect start to ask 'What are we risking?'"*

Clearly, not all leakage through top seals is caused by buoyancy pressure of the hydrocarbon phase exceeding threshold capillary entry pressures, perhaps aided by a change of wettability, and much debate still exists over the role of interconnected fractures as a mechanism for leakage. A marked contrast between the Hedberg conference on seals in 1993 and that of 2002 was the increase in the number of presentations on seals related to fault traps in 2002. This probably reflects the maturation of the industry in terms of other disciplines especially geophysical interpretation, which has increased the certainty of structural modeling that allows previously problematic fault-related traps to be viable targets. The interaction of seal rocks with the Earth's paleo- and contemporary stress field is inevitably more complicated in faulted terrain, where discordance abounds, than in gently undulating plays that rely on four-way dip closures. The importance of incorporating the risk element into all such studies is a marked improvement and brings such work on seals directly in congruence with the bottom-line of industry needs.

Altogether, we believe that this volume is a valid representation of the quality of the 2002 Barossa Valley Hedberg conference on seals, and we hope you learn as much from these chapters as we have from editing them. We look forward to meeting you at the next AAPG Hedberg conference on seals, wherever that may be.

Peter Boult and John Kaldi
Conference co-convenors and editors

REFERENCES CITED

Hillis, R. R., 1998, Mechanisms of dynamic seal failure in the Timor Sea and central North Sea, *in* P. G. Purcell and R. R. Purcell, eds., The sedimentary basins of Western Australia 2: Proceedings of the Petroleum Exploration Society of Australia Symposium, Perth, Western Australia: p. 313–324.

Leith, T. L., I. Karstadd, J. Connan, J. Pierron, and G. Caillet, 1993, Recognition of cap rock leakage in the Snorre field, Norwegian North Sea: Marine and Petroleum Geology, v. 10, p. 29–41.

ns, 2005, Fluid flow, pore pressure, wettability, and
leakage in mudstone cap rocks, *in* P. Boult and J. Kaldi, eds., Evaluating fault
and cap rock seals: AAPG Hedberg Series, no. 2, p. 1–12.

Fluid Flow, Pore Pressure, Wettability, and Leakage in Mudstone Cap Rocks

Andrew C. Aplin
School of Civil Engineering and Geosciences, University of Newcastle, Newcastle upon Tyne, United Kingdom

Steve R. Larter[1]
School of Civil Engineering and Geosciences, University of Newcastle, Newcastle upon Tyne, United Kingdom

ABSTRACT

This chapter considers some of the issues surrounding the modeling of one- and two-phase fluid flow in mudstones. For single-phase flow, key relationships include those between porosity and (1) effective stress, (2) permeability, and (3) capillary breakthrough pressure. All three relationships are strongly influenced by the grain-size distribution or clay fraction of mudstones, but a quantitative description is currently only available for the porosity-effective stress relationship. The importance of lithology or clay fraction as a control on the key flow properties of mudstones indicates the practical significance of estimating clay fraction directly from geophysical logs. This chapter illustrates how artificial neural networks can be used to perform this task.

Having considered some of the basic flow properties of mudstones, the second part of the chapter discusses aspects of two-phase flow through mudstone pore systems. Rates, mechanisms, and pathways of petroleum leakage through mudstone pore systems remain poorly constrained. In this chapter, field and experimental data is combined with theoretical arguments to suggest that once a water-wet cap rock is breached, the leak path will become more oil wet as a result of sorption of hydrophilic and ultimately hydrophobic organic compounds onto mineral surfaces. Oil-water partition of hydrophilic organic compounds in reservoirs, followed by diffusion into cap rock pores may even create oil-wet pathways into cap rocks and permit leakage. In these cases, cap rocks simply retard the vertical migration of petroleum, and column height is a function of the rates of petroleum supply and loss. Modeling the rate of loss of petroleum requires a better understanding of mudstone relative permeability.

---

[1]*Present address:* Department of Geology and Geophysics, University of Calgary, Calgary, Alberta, Canada.

Copyright ©2005 by The American Association of Petroleum Geologists.
DOI:10.1306/1060752H23158

INTRODUCTION

Compaction, absolute and relative permeability, pore pressure, and seal capacity are the inextricably linked properties of mudstone cap rocks that must be accurately described in two-phase fluid-flow models of petroleum systems. The three key relationships required for any Darcy-based flow models are those between (1) porosity (or void ratio, e) and effective stress (σ'), (2) flow rate and hydraulic gradient (Darcy's law), and (3) permeability and porosity. When modeling petroleum systems involving two- or three-phase flow, additional data requirements include (1) the relative permeability of the porous medium as a function of phase saturation and (2) the definition of a threshold capillary pressure that must be exceeded in a water-wet porous medium to allow passage of a nonwetting phase into and across a defined rock unit. Dynamic models in which rocks deform and lose porosity under the influence of increasing burial loads require a quantitative description of the way that these parameters evolve through time as a function of, for example, porosity.

This chapter is divided into two main sections. In the first part of this chapter, some of the progress that has been made in providing the quantitative descriptions of the relationships required to model fluid flow in mudstones is considered. The chapter is purposefully limited to a consideration of flow through the matrix pore structure and does not deal with the potentially catastrophic flow that may occur through fractures or faults. The chapter shows that the relationships between (1) porosity and effective stress, (2) porosity and permeability, and (3) porosity and threshold capillary entry pressure in mudstones are strongly influenced by grain size, and describes how the grain size or clay fraction of mudstones may be rapidly and pragmatically estimated from downhole petrophysical logs. In this way, it becomes possible to rapidly estimate mudstone pore pressure and to populate basin models directly from downhole log inputs.

Having established some of the main controls on the basic flow properties of mudstones, the second part of the chapter focuses on issues related to the leakage or transmission of petroleum through mudstone pore systems. This is an area of petroleum geoscience where considerable uncertainty remains. Conventional descriptions of the system involve a water-wet cap rock pore system that will not transmit any petroleum until the buoyant force exerted by a petroleum column exceeds a critical capillary entry pressure dictated by the dimensions of the cap rock's pore-throat system. Flow is then generally assumed to occur across the cap rock according to a version of Darcy's Law, which is modified to consider two-phase flow. This common model raises several issues that are currently only partially resolved and that are discussed in this chapter. In particular, the following should be answered: (1) How do we define a critical or threshold capillary entry pressure of a heterogeneous cap rock sequence? (2) What is the initial wetting state of a mudstone pore system? (3) Is the wetting state of a mudstone pore system altered by leakage of petroleum, and how does this affect the way we assess seal capacity? (4) How do we define relative permeability curves for mudstones? (5) Where leakage has occurred, how can we estimate the amount of leaked petroleum?

COMPACTION AND FLUID FLOW

Porosity-effective stress

Mudstone compaction has commonly been described as a purely mechanical process in which porosity is lost in response to increasing effective stress. These concepts are rooted in soil mechanics (Terzaghi, 1943; Skempton, 1970; Burland, 1990) and have since been used to describe the higher stress, longer time, and higher temperature regimes relevant to sedimentary basins (e.g., Smith, 1971; Rieke and Chilingarian, 1974; Ungerer et al., 1990; Schneider et al., 1993; Jones, 1994; Audet, 1996; Karig and Ask, 2003). Because compaction represents a volumetric reduction of porosity, relationships between porosity and effective stress should strictly be couched in terms of mean effective stress. However, whereas vertical stress can generally be estimated quite accurately either directly from density logs or from well-established regional trends, estimates of minimum stress rely on high-quality leak-off test data; intermediate stresses are even less well constrained. Most commonly and pragmatically, therefore, porosity is related to vertical effective stress. In circumstances where the maximum principal stress is vertical, this will generally be acceptable. In circumstances where the maximum principal stress is not vertical, or where concerns about the magnitude of coupling between pore pressure and stress exist (e.g., Harrold et al., 1999), a more complex analysis may be required.

In soil mechanics, the relationship between porosity and effective stress is described as

$$e = e_{100} - \beta \ln\left(\frac{\sigma'_v}{100}\right) \quad (1)$$

$$\sigma'_v = \sigma_v - u \quad (2)$$

$$e = \frac{\phi}{1 - \phi} \quad (3)$$

In these equations, ϕ is porosity, e_{100} is the void ratio at 100 kPa effective stress, and β is the slope of

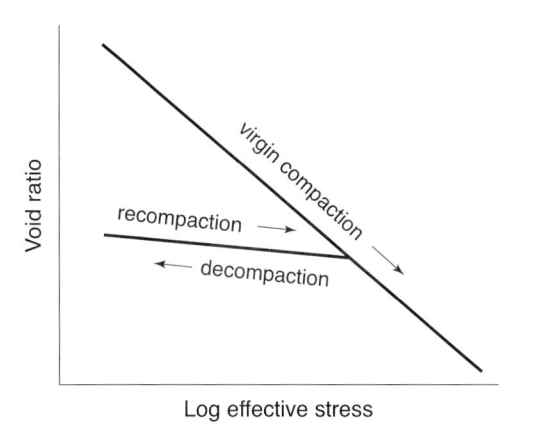

FIGURE 1. Soil mechanics-based description of the one-dimensional, mechanical compaction of fine-grained clastic sediments.

the linear relation between void ratio and the natural logarithm of vertical effective stress. Effective stress (σ'_v) is defined as the difference between total stress (σ_v) and pore fluid pressure (u). The form of equation 1 is such that void ratio is a linear function of the logarithmic value of effective stress. This is shown in Figure 1, which indicates the virgin compression line along which initial compaction occurs. Figure 1 also shows that because mudstone deformation is a predominantly plastic process, the major part of the deformation is not recovered when sediments are unloaded. If an unloaded mudstone is reloaded, the sediment returns to the virgin compaction line, after which, further increases in effective stress drives the mudstone along the virgin line. The porosity of a mudstone is thus an indication of the maximum effective stress to which it has been subjected.

It is also important to note that equations 1 and 2 can be used to estimate pore pressure from mudstone porosity, if the compaction is on the virgin compression line and the values of the compression coefficients in equation 1 are known (Alixant and Desbrandes, 1991). This is useful because the low permeability of mudstones means that pore pressures cannot be measured using conventional measurement techniques.

Quantitative use of the porosity-effective stress equation as a means of (1) describing mudstone compaction and (2) evaluating pore pressure from mudstone porosity data requires us to know the numerical values of the two coefficients β and e_{100}. The work of Skempton (1944, 1970) and Burland (1990) showed that β and e_{100} are strongly correlated both with each other and with the sediment's void ratio at liquid limit (e_L). Although e_L is not a geologically useful term, it is strongly correlated to sediment clay fraction, where clay fraction is defined as the percentage of particles less than 2 µm in diameter (Skempton, 1944). Recently, Yang and Aplin (2004) have used data from both the soil mechanics literature and from geological samples to construct the relationship between clay fraction and compression coefficients. Using a large data set covering an effective stress range of 0.8–40 MPa collected from in-situ, geologically compacted, fine-grained clastic sediments from the North Sea and Gulf of Mexico, they constructed the following relationships:

$$e_{100} = 0.3024 + 1.6867 \text{ clay} + 1.9505 \text{ clay}^2 \quad (4)$$

$$\beta = 0.0407 + 0.2479 \text{ clay} + 0.3684 \text{ clay}^2 \quad (5)$$

Substitution of these coefficients in equation 1 allows one to describe mudstone compaction directly as a function of lithology. Three curves for mudstones with differing clay fractions are shown in Figure 2, illustrating the wide range of compaction behavior. In general, silty mudstones are deposited with lower porosities than more clay-rich materials but are less compressible. At high effective stresses, the porosity of mechanically compacted mudstones thus converge (Figure 2).

Pore Pressure

The low permeability of mudstones precludes direct determination of pore pressure by conventional techniques such as repeat formation tests. For this reason, and because of the safety and cost issues related to drilling through overpressured, mud-rich sequences, a strong drive has existed for many years to estimate

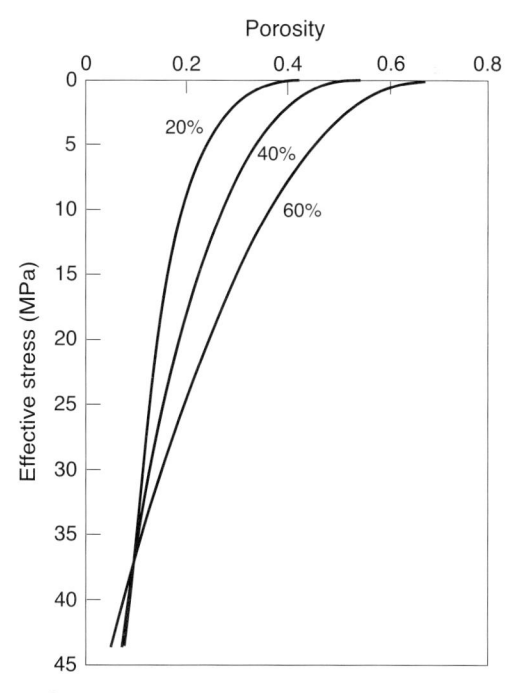

FIGURE 2. Porosity-vertical effective stress trends modified from Yang and Aplin (2004) for geologically compacted mudstones.

mudstone pore pressure indirectly from porosity. Older approaches based on primary mechanical compaction, such as those pioneered by Hottman and Johnson (1965) and Eaton (1975) have proved to be useful but suffer in two ways: first, they assume that all mudstones compact according to a single relationship; and second, their use requires the certain identification of hydrostatically (normally) pressured, shallow-buried sediments. Combining porosity data with an explicit effective stress relationship circumvents these difficulties in that it is based on an explicit physical principle and does not require local calibration to shallow and putatively hydrostatically pressured sections (Alixant and Desbrandes, 1991; Yang and Aplin, 2004). Using this approach, the basic compaction equation 1 can be rearranged such that effective stress can be evaluated if the void ratio (e) is known, and the compression coefficients e_{100} and β can be estimated from the clay content of the mudstone:

$$\sigma' = 100 \times \exp\left(\frac{e_{100} - e}{\beta}\right) \quad (6)$$

Pore pressure is then the difference between total (vertical) stress and (vertical) effective stress. Examination of Figure 2 shows that estimates of effective stress and, thus, pore pressure could easily be 10 MPa in error if mudstone lithology is not considered.

Any quantitative approach to mudstone compaction using the Terzaghi approach should consider its several limitations. We have already mentioned that the model assumes that porosity loss is driven by increases in vertical effective stress instead of mean effective stress, but that this is unlikely to be a serious issue in many geological settings. Second, the model describes the porosity-effective stress relationship as an exponential relationship. This is convenient, and although it appears to work at stresses as much as 40 MPa (Yang and Aplin, 2004), other relationships may generate a more accurate description. Third, the relationships described here are only valid for mechanical compaction. At higher temperatures (above approximately 80–100°C), the recrystallization of clay minerals such as smectite plus the dissolution and remobilization of carbonates and opaline silica change the microfabric of mudstones (Ho et al., 1999; Aplin et al., 2003; Charpentier et al., 2003) and may lead to a loss of porosity that is independent of effective stress (e.g., Bjørlykke, 1999; Nadeau et al., 2002). Chemical compaction occurs extensively in both carbonates and sandstones (e.g., Garrison and Kennedy, 1977; Tada and Siever, 1989; Bjørkum, 1996), although its importance in clastic mudstones is less well constrained.

The fourth and perhaps most important limitation of the porosity-effective stress relationship is that the inelastic nature of mudstones means that it can only be used to estimate the maximum effective stress to which a mudstone has been subjected. Unloading because of uplift or the lateral transfer of pore pressure to basin highs along sands connected deeper in the basin (Yardley and Swarbrick, 2000), for example, will lead to a change in effective stress but essentially no change in porosity. Although this leads to an underestimate of present-day pore pressure using tools based on porosities evaluated from wireline or seismic data, it does give some insight into pore-pressure histories, for example, where pore pressures have increased as a result of recent, differential subsidence in the center of basins.

Porosity-Permeability

Hydraulic conductivity is the coefficient K (LT^{-1}) that relates the rate of fluid flow q (L^3T^{-1}) through a cross-sectional A (L^2) to the imposed hydraulic gradient in Darcy's Law:

$$q = KiA \quad (7)$$

Permeability (k; L^2) is related to hydraulic conductivity through the viscosity (η) and unit weight of water (ρ_w):

$$k = K(\eta/\rho_w) \quad (8)$$

Reviews by Neuzil (1994) and Dewhurst et al. (1999a) indicate that the hydraulic conductivity of mudstones ranges over eight orders of magnitude (10^{-8}–10^{-16} ms^{-1}) and by three orders of magnitude at a single porosity. As stated by Darcy's Law, fluid flux is directly proportional to permeability, so that a permeability uncertainty of three orders of magnitude is unacceptable for modeling purposes; the difference, for example, in dissipating overpressure over a period of 1 million or 1 billion yr.

One difficulty in modeling variations in permeability is the sparse database that exists for well-characterized mudstones. These were summarized by Dewhurst et al. (1999a), and very few new data have subsequently appeared in the literature (but see Hildenbrand et al., 2002). Much of the variation in the permeability of reasonably homogeneous mudstones at a given porosity can be explained by variations in grain size (Dewhurst et al., 1998, 1999b; Yang and Aplin, 1998). For example, Dewhurst et al. (1998, 1999b) measured hydraulic conductivities of the London Clay that varied by close to three orders of magnitude at a single porosity, quite close to the total range suggested by Neuzil (1994) for mudstone cores. The London Clay samples had clay fractions ranging from 27 to 65%, representing close to the full range measured in other studies (20–85%; e.g., Aplin et al., 1995).

Although we currently lack a quantitative model that describes the relationship between porosity and permeability as a function of clay fraction, it seems likely that a fairly strong relationship exists. At porosities for which we have permeability data for lithologically characterized mudstones (20–40%), permeability can be predicted to within an order of magnitude if the clay fraction of the sample is known. Although this is, by no means, perfect, it represents a two-orders-of-magnitude improvement over the full range. At lower porosities, the poroperm relationship as a function of clay content is still poorly established.

Relative Permeability

For two- or three-phase flow, relative permeability is the dimensionless variable that describes the ratio of the effective permeability with respect to a specific phase to the intrinsic permeability (k) of the porous medium. Relative permeability thus represents a semiempirical extension of Darcy's Law. The relative permeabilities of oil and water can be represented as k_{ro} and k_{rw} and have values between 0 and 1:

$$k_{ro} = \frac{k_o}{k} \text{ and } k_{rw} = \frac{k_w}{k}$$

An extensive literature exists describing the relative permeabilities of oil, gas, and water of reservoir lithologies with permeabilities above 1 md (e.g., Aziz and Settari, 1979; De Marsily, 1986). In contrast, very few relative permeability data have been reported for nonreservoir lithologies, not least because of the extreme difficulty of making measurements on samples that have intrinsic permeabilities on the order of 0.1–10 nd. In the absence of data for true mudstones, Okui and Waples (1993) and Okui et al. (1998) collected data from a range of clastic lithologies that varied in grain size from sands to silts. The data showed that with decreasing grain size, residual water saturation increases, flow of oil commences at increasingly small oil saturations, and the crossover point at which oil permeability becomes greater than water permeability occurs at increasingly low oil saturations. Extrapolating to finer grained lithologies, Okui et al. (1998) proposed the mudstone relative permeability curves shown in Figure 3, suggesting that water permeability may decrease to zero at water saturations of 70–80%, that flow of oil may occur at oil saturations of less than 10%, and that the effective permeability of oil may be substantial at oil saturations of greater than or equal to 20%. Although directly measured relative permeability curves for mudstones do not yet exist, the best supporting evidence for Okui and Waples (1993) ideas come from Hildenbrand et al. (2002) experimental de-

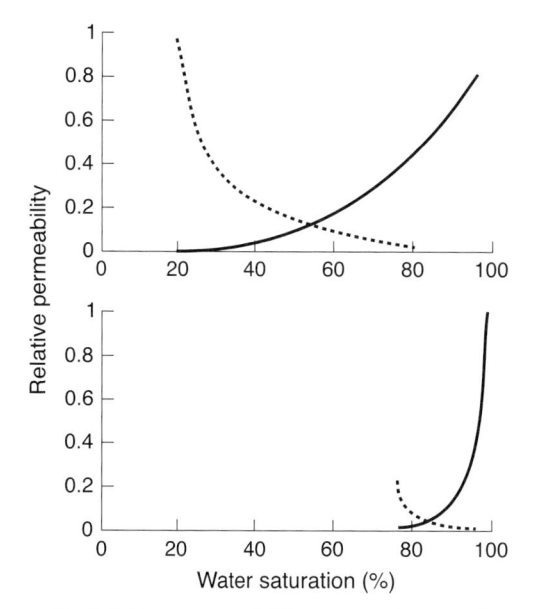

FIGURE 3. Relative permeability curves for sandstones (upper) and mudstones (lower) modified from Okui and Waples (1993). Solid lines and dashed lines represent water and oil, respectively.

terminations of both absolute and gas-phase effective permeabilities in a series of mudstones with porosities of 10–25%. Data in Hildenbrand et al. (2002) suggest that connected gas stringers may form at gas saturations significantly below 1%, and that at higher saturations, relative permeability data range to as much as 1, with many data between 0.1 and 0.4.

Compared with sands, therefore, flow of petroleum through mudstone pore systems occurs at low saturations of the nonwetting phase. Substantial flow of petroleum at low saturations is consistent with both experimental data (e.g., Schowalter, 1979; Hildenbrand et al., 2002) and theoretical work based on the percolation theory (Hirsch and Thompson, 1995; Sahimi, 1995), which predicts that whereas saturations of 10–20% are required to form continuous filaments of petroleum across centimeter-scale volumes of homogeneous reservoir lithologies, saturations are scale dependent and decrease with increasing rock volumes. They are also qualitatively consistent with the permeability model developed by Yang and Aplin (1998), which estimates permeability from the pore-size distribution, calculating the contribution of individual pores to the total permeability. Because permeability is a function of the square of the pore diameter (e.g., Scheidegger, 1974), large pores contribute a disproportionately large fraction of the total permeability. Compared with sandstones, the relative importance of large pores is enhanced in mudstones by their relatively broad pore-size distributions, especially in less compacted material (Dewhurst et al., 1998; Yang and Aplin, 1998).

Threshold Capillary Entry Pressure

The threshold capillary pressure (CP_t) of a cap rock may be defined as the pressure that must be overcome to build a continuous filament of the nonwetting phase (petroleum) across a defined rock unit, thus allowing flow of the nonwetting phase. The pore size (or more properly, the pore-throat size) and, thus, the CP_t are generally assessed from mercury injection data, and a reasonably large public database of mudstone pore-size distributions now exists (Borst, 1982; Katsube et al., 1991; Katsube and Best, 1992; Katsube and Williamson, 1994; Kaldi and Atkinson, 1997; Schlomer and Krooss, 1997; Sneider et al., 1997; Dewhurst et al., 1998, 1999b; Yang and Aplin, 1998; Hildenbrand et al., 2002). Although the porosity of the samples analyzed in these studies is commonly reported, the lithology of the materials is not often quantified, so that we are still missing the framework with which to quantify the way in which CP_t evolves as a dual function of porosity and lithology. This problem is exacerbated by the difficulty of defining a CP_t from mercury injection data, reflected in the various estimation methods, including measurement of the 90th percentile on the cumulative pore-size distribution (e.g., Schowalter 1979; Schlomer and Krooss, 1997) and the pressure corresponding to the inflection point where, on a graph of pressure vs. saturation, the curve becomes convex upward (e.g., Katz and Thompson, 1987). Whereas this is close to the pressure at which a significant volume of the pore space becomes saturated with the nonwetting phase, flow of the nonwetting phase may occur at much lower saturations, perhaps below 1% (Hildenbrand et al., 2002). These methods are subjective and strictly work only for materials with unimodal pore-size distributions. For more complex materials, for example, core samples with bi- or trimodal pore-size distributions (Dewhurst et al., 1998) and sequences of lithologically heterogeneous mudstones (O'Brien and Slatt, 1990), it is much harder to define a CP_t. Definition of the connected weak point in a complex and heterogeneous mudstone cap rock system awaits methods to define the sedimentology of mudstone depositional systems.

Despite the difficulties involved in the precise definition of a CP_t from either mercury injection data or by laboratory flow experiments, data sets have been compiled from which correlations have been drawn between absolute permeability and some measure of CP_t (Ibrahim et al., 1970; Schowalter, 1979; Hildenbrand et al., 2002). Although the slope and intercept of the regression lines vary according to the data sets used, the correlations are quite strong, as would be expected, because both parameters strongly relate to the diameter of connected pore throats. It follows that because the permeability of mudstones at a given porosity is strongly influenced by lithology or grain-size distribution, the CP_t must also be a joint function of porosity and lithology. This is qualitatively clear in data sets such as those published by Dewhurst et al. (1998, 1999b) and Yang and Aplin (1998) and implicit in the comments by Krushin (1997) that mudstones with higher quartz contents (i.e., siltier) tend to display lower CP_t. Equally, Kaldi and Atkinson (1997) showed that the siltier mudstones deposited in higher energy environments in the upper Oligocene Talar Akar formation, offshore northwest Java, had relatively low CP_t.

Assessing Mudstone Lithology from Petrophysical Logs

It is clear that both the porosity-permeability and porosity-effective stress relationships of mudstones are strongly influenced by lithology, which can be expressed simply but robustly by clay fraction: the percentage of particles that have a diameter of less than 2 μm. It is also clear, although less well constrained currently, that the critical capillary entry pressure of mudstones is influenced, at a given porosity, by grain size. Estimation of mudstones' clay fraction thus yields a promising way of constraining the void ratio-effective stress, permeability-void ratio, and porosity-capillary entry pressure relationships, which are required both for basin modeling and pore-pressure evaluation.

Although clay fraction can be measured in the laboratory, a measurement-centered, particle-based approach to mudstone properties is unrealistic on a large scale. From a practical perspective, an approach using wire-line logs is thus appealing. Previous workers have used either the gamma-ray log or a combination of the neutron and density logs to assess the clay mineral content of mudstones (see Doveton, 1994; Hearst et al., 2000). Developed largely for the log analysis of sandstones, the conventional approaches using gamma or neutron-density do not extrapolate easily to mudstones. Because the relationship between log response and clay fraction is quite complex and nonlinear, Yang et al. (2004) developed an artificial neural network (ANN) technique to estimate mudstone clay fraction from wireline data. The ANNs were trained, using a data set of around 530 analyzed mudstone samples, to estimate the clay fraction (and also grain density and total organic carbon) of mudstones from standard wire-line log data (gamma, resistivity, sonic, density, caliper). An example of the output from the program, which incorporates the ANNs as part of a procedure to evaluate mudstone pore pressure and permeability directly from wire-line logs, is shown in Figure 4.

Once clay content has been established from wire-line log data, pore pressure can be evaluated using equation 6. The clay fractions determined from wireline data allow the compression coefficients to be calculated and, thus, the effective stress and pore pressure to be

FIGURE 4. Estimation of pore pressure in a west African well from wire-line logs using the lithology defined porosity-effective approach encapsulated in equation 6 in the text. Laboratory-measured clay fractions are shown as open circles. Porosity estimates (dashed line, central panel) were made from a combination of the density and sonic logs. The modeled normal compaction curve (solid line, central panel) represents the porosity the sediment would have if it was normally pressured. Depths are below kelly bushing (KB), and the depth between KB and the seabed is 771 m (2529 ft). On the right panel, the dashed line is the lithostatic pressure, the solid line is the hydrostatic pressure, and pore pressures determined with the Repeat Formation Tester are shown as circles. Figure modified from Yang et al. (2004).

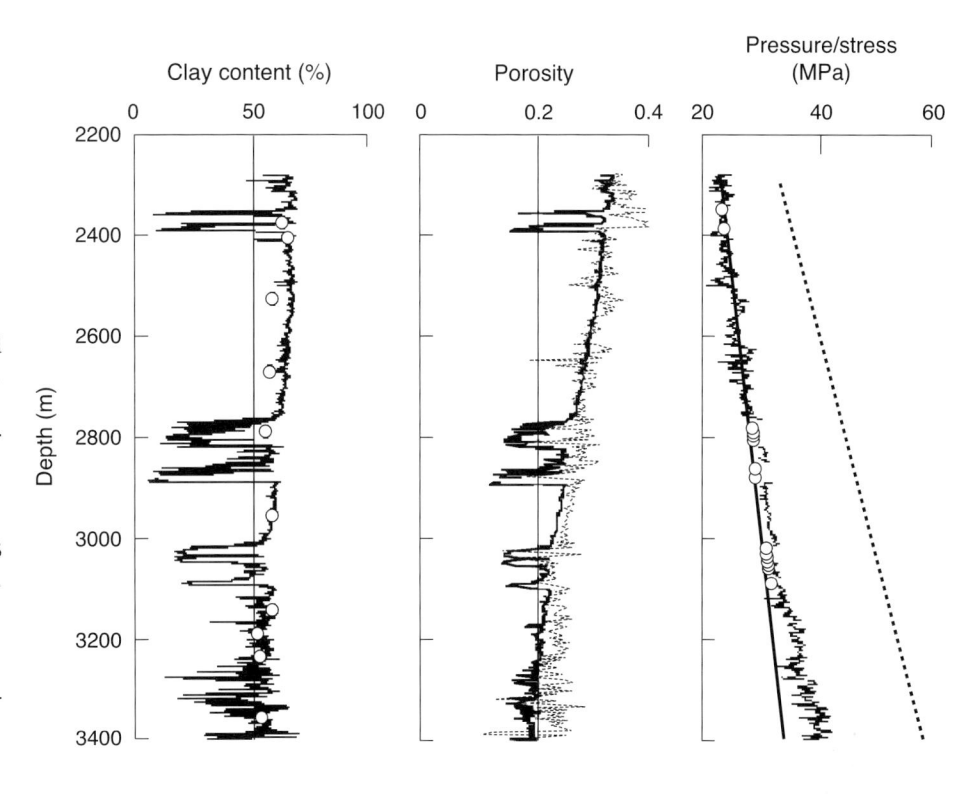

evaluated (Figure 4). Furthermore, once coefficients relating (1) porosity to permeability and (2) CP$_t$ to porosity are known as a function of clay fraction, the same approach can be used to estimate those parameters directly from wireline data. These data can then be fed into fluid-flow models that use either (or both) Darcy or (and) percolation models to describe flow (Figure 5).

PETROLEUM LEAKAGE

The ideas presented thus far provide a basic framework within which to model fluid flow through mudstones. From the petroleum exploration point of view, fluid flow through mudstones critically manifests itself as the leakage of petroleum through cap rocks, or,

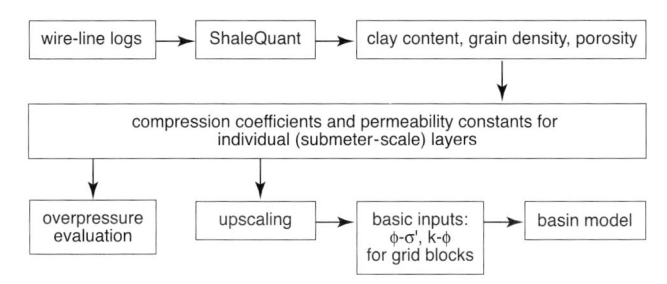

FIGURE 5. Workflow for estimating pore pressure and for obtaining key input data for flow models directly from wire-line logs. ShaleQuant is an ANN model that estimates mudstone clay fraction from wireline data.

from a different perspective, as vertical migration to stratigraphically higher prospects. Indeed, much of the world's petroleum has migrated vertically through large thicknesses of fine-grained sediments, although migration rates, mechanisms, and flow paths are poorly constrained (e.g., faults, fractures, and pores) and, thus, subject to considerable debate. The rest of this chapter considers the flow of petroleum through the bulk pore system of fine-grained sediments, concentrating, in particular, on geochemical evidence that, first, suggests that the wettability of cap rock pore systems may be altered once the cap rock has been breached and, second, helps to define the volume of petroleum that has migrated through the cap rock. The ideas are based on theoretical, experimental, and field evidence that all show that low-molecular-weight petroleum nonhydrocarbons, such as phenols (hydrophilic organic oxygen compounds) and carbazole derivatives (more hydrophobic nitrogen compounds), partition between oil, water, and rock according to rules that suggest substantial adsorption onto mineral surfaces and the distinct possibility that the wetting state will be altered from water- to oil- or mixed-wet (Larter and Aplin, 1995; Larter et al., 1996; Bennett and Larter, 1997; Taylor et al., 1997; van Duin and Larter, 2001).

Field Data: Snorre

Snorre is a large oil field in the Tampen Spur area of the Norwegian North Sea from which petroleum has leaked several hundred meters into the cap rock

(Caillet, 1993; Leith et al., 1993; Leith and Fallick, 1997). Snorre contains approximately 300 m (1000 ft) of undersaturated oil in the Triassic and Jurassic Lunde and Statfjord formations and is successively capped by the Lower Cretaceous Cromer Knoll Group (as much as 20 m [66 ft]) and as much as 600 m (2000 ft) of Upper Cretaceous Shetland Group and Tertiary Rogaland Group mudstones. Leakage through the Snorre cap rock is well documented geochemically (Leith et al., 1993; Leith and Fallick, 1997), with liquid petroleum having migrated vertically to as much as 600 m (2000 ft) through the cap rock succession, not only at the crest of the reservoir but across a broad front throughout most of the reservoir area (Leith et al., 1993; Bond, 2001).

Neither the timing nor the mechanisms of leakage in Snorre are perfectly constrained. Caillet (1993) argued that the currently substantial overpressure in the reservoir (14 MPa) may be high enough to cause hydrofracturing of the cap rock, whereas Bond (2001) suggests that sufficient column height exists in the reservoir to drive capillary failure. From a geochemical standpoint, the key point is that, whereas the hydrocarbons in the cap rock and reservoir are very similar and suggest a common origin (Leith and Fallick, 1997), the oil in the cap rock is enriched in both low- and high-molecular-weight nonhydrocarbons compared to the reservoired oil (Bond, 2001). We suggest here that the high concentrations of nonhydrocarbons in cap rock oils have implications for the wetting state of the pore system and serve as a marker for the volume of the oil that has migrated through the cap rock.

Among the many nonhydrocarbons that occur in oil, benzocarbazoles (aromatic nitrogen compounds) have been studied in some detail, because they sorb strongly to minerals and have thus been used to monitor the extent to which migrating oil has interacted with mineral surfaces (Larter and Aplin, 1995; Larter et al., 1996; Terken and Frewin, 2000). Loss of benzocarbazoles along migration pathways lowers their abundance in reservoired oils but should correspondingly increase their concentration in core extracts from migration pathways. High concentrations of benzocarbazoles in core extracts from cap rocks would thus be evidence for substantial migration through the cap rock pore system.

Figure 6 shows the total concentration of two nonhydrocarbons, benzo[a]carbazole plus benzo[c]carbazole, in solvent-extracted organic matter from the Snorre cap rock. Also plotted are equivalent data for core extracts from the reservoir and an unstained cap rock from the Shetland Group overlying the Troll field, which has not leaked oil. Core extracts from the reservoir have a mean benzocarbazole concentration of 2.3 parts per million (ppm), whereas those from the Troll cap rock, representing benzocarbazole values from an unbreached cap rock, are close to 4 ppm. In contrast, many of

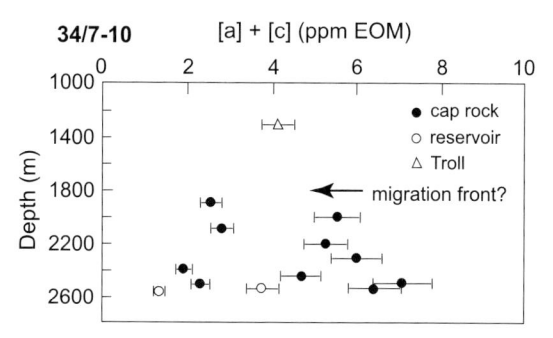

FIGURE 6. Benzo[a]carbazole plus benzo[c]carbazole data from core extracts taken from the Snorre field cap rock (circles) and reservoir (squares) plus the Troll Field cap rock (triangles). Concentrations are expressed in parts per million of the extractable organic matter (EOM). Data for Troll represent a background value for a cap rock that is similar to the Snorre cap rock but which has not suffered oil migration. Concentrations above those of the reservoir oil suggest that greater than 1 volume of oil has migrated through the pore system.

the extracts in the cap rock have values between 5 and 7 ppm. The most logical explanation for this enrichment is that more than one volume of oil has passed through the pore space, with partial or complete sorptive loss of benzocarbazoles from the migrating oil to mineral surfaces. In other parts of the cap rock, benzocarbazole concentrations are similar to or lower than those in the reservoir oil, suggesting loss of benzocarbazoles by sorption elsewhere on the leakage pathway.

Experimental Data

High concentrations of benzocarbazoles in Snorre cap rocks point to sorption onto mineral surfaces and the possibility that the sorption of polar molecules may result in changes to the wetting state of the cap rock pore system. However, benzocarbazoles are hydrophobic, begging the question of how they and other hydrophobic moieties become sorbed to minerals in water-wet pore systems. We suggest that the key compounds involved in sorption and, thus, the alteration of wetting state are actually small, hydrophilic molecules such as phenols. Strong evidence for this comes from a core flood experiment carried out under realistic subsurface conditions on a clay rich siltstone with a microdarcy-range permeability (Larter et al., 2000). Results showed that many nonhydrocarbons were removed on timescales of months from high API gravity oil during its passage through the initially water-saturated rocks. Small hydrophilic molecules such as phenols were rapidly removed and strongly retained in the cores (Figure 7). Comparison of the molecular composition of the oil exiting the core with that injected into the core indicated that predominantly low-molecular-weight polar compounds were removed, and

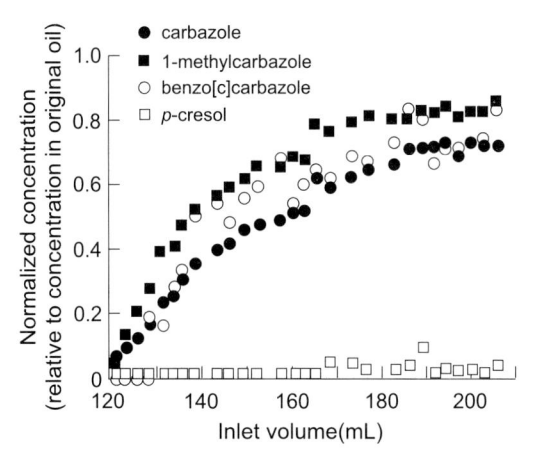

FIGURE 7. Variation in concentrations of selected, representative nonhydrocarbons in oil samples collected at the outlet of a siltstone core during a core-flood experiment in which, first, brine, and then, 41° API oil were flowed through the core (modified after Larter et al., 2000). Concentration data have been normalized to values obtained for the original (input) oil. Whereas both carbazoles and phenols are retained by the core, phenols (as exemplified here by *p*-cresol) show much stronger depletion than carbazole compounds, being strongly retained and essentially removed from the migrating oil in the core. The first oil was eluted from the core after about 120 mL of oil had been injected, the horizontal scale indicating the volume of oil injected. After the experiment, the core inlet showed a marked oil-wet character.

that many of these are multifunctional compounds with mixed nitrogen and oxygen (and sulfur) functionality. Furthermore, environmental scanning electron microscopy indicated that at the core inlet, the initially water-wet core had become strongly oil wet during the experiment, the oil wetness of the core decreasing toward the outlet of the core. These data strongly support the idea that the hydrophilic organic molecules such as alkyl phenols partition first into water and then onto mineral surfaces, altering the wetting state of the pore system and promoting the sorption of more hydrophobic compounds such as benzocarbazoles.

Theoretical Considerations

Theoretical considerations support the experimental and field data pointing to the sorption of polar organic molecules to mineral surfaces. Van Duin and Larter (2001) report a series of molecular dynamics chemical computational simulations of the phase behavior of organic-water mixtures in the presence of quartz and calcite surfaces. When changing the polarity of the organic phase in the simulations from a charge-neutral cyclohexane phase via apolar carbazole and polar phenol to highly polar acetic acid, distinct changes in the organic phase-water-mineral phase behavior were observed. Cyclohexane and carbazole form discrete or-

ganic phases separated from the mineral surfaces by a water film. In contrast, phenol and acetic acid show sufficient water solubility to penetrate through these films to compete with the water molecules for mineral surface adsorption sites. This is consistent with the rapid removal of phenols and other small hydrophilic moieties observed during the siltstone core flood (Larter et al., 2000). Further molecular dynamics simulations with preadsorbed phenol compounds on a calcite surface demonstrate that these small polar molecules can have a profound impact on surface wettability, making the mineral surface accessible even to completely nonpolar compounds, including high-molecular-weight hydrocarbons. This provides strong support for a two-stage process producing wettability changes in oil-water-mineral systems, commencing with small polar species partitioning from oil to water (Larter and Aplin, 1994, 1995; Taylor et al., 1997; Bennett and Larter, 1997) and sorbing to an initially water-wet mineral surface. Sorption results in a more hydrophobic mineral surface, after which water film rupture occurs and leads to the sorption of more apolar compounds (e.g., petroleum hydrocarbons) onto and around these polar surfactants. Mineral surfaces in siltstones and, by inference from the modeling and field observations, cap rocks can become oil wet on a timescale of months.

Wettability: Summary

Field and laboratory studies of fine-grained rocks clearly indicate that polar petroleum compounds are removed onto mineral surfaces, despite the initial presence of water films. Computational chemistry confirms that low-molecular-weight polar compounds, of which phenols are probably just one class, readily penetrate mineral water films to change surface properties, such that hydrocarbons can wet the mineral surfaces. Examination of siltstone cores flooded with brine, then oil, indicates that sorption of polar compounds appears to correlate with changes from an initially water-wet state to a more oil-wet state. The high abundance of benzocarbazoles in core extracts from the Snorre cap rock suggests that similar processes occur in natural systems, so that initial breaching of the cap rock pore network would result in a more oil-wet pore system.

We conclude that there is reasonable doubt that cap rocks in contact with oil would remain water wet. In this scenario, cap rocks then represent low-permeability chokes to petroleum systems instead of capillary seals. Failure could occur by a creep-seep mechanism by which low-molecular-weight hydrophilic oil components partition into cap rock porewaters and change wettability, allowing limited Darcy flow. This is then repeated until the seal is breached, and leakage by continuous Darcy flow occurs. In this case, mudstone cap rocks do not act as permanent seals but simply retard the inexorable flow of petroleum to the basin surface.

SUMMARY AND CONCLUSIONS

In this chapter, we have reviewed some of the issues relating to the understanding and modeling fluid flow in fine-grained sediments. From the practical standpoint of requiring well-constrained input data for flow models, it is apparent that for clastic mudstones, the relationships between (1) porosity and effective stress, (2) porosity and permeability, and (3) porosity and threshold capillary pressure are strongly influenced by the grain-size distribution or clay fraction of the sediments. Whereas coefficients that define mechanical compaction using soil mechanics (Terzaghi) theory have been reported as a function of lithology, quantitative descriptions of the porosity-permeability and porosity-threshold CEP relationships are still required.

Because lithology or clay fraction exerts a strong influence on the key flow properties of mudstones, a method to quickly determine clay fraction from geophysical logs is of practical importance. Among various approaches, we have used an ANN-based program to determine clay fraction directly from wireline data and then to estimate pore pressure using a mechanical compaction model.

Leakage or two-phase flow in cap rock pore systems remains poorly constrained. Accurate and agreed methods for estimating threshold capillary pressures are lacking both for homogeneous mudstones and especially for heterogeneous sequences, for which we are unable to define the critical leak path for petroleum. Field and experimental data combine with theoretical arguments to suggest that once a water-wet cap rock is breached, the leak path will become more oil wet as a result of sorption of hydrophilic and, ultimately, hydrophobic organic compounds onto mineral surfaces. Once this has happened, cap rocks simply retard instead of halt the vertical migration of petroleum. Indeed, oil-water partition of hydrophilic organic compounds in reservoirs, followed by diffusion into cap rock pores, may create oil-wet pathways into cap rocks and drive leakage even in the absence of any pressure gradient to drive Darcy flow. In this case, the rate of loss of petroleum through the cap rock is a function of its relative permeability.

ACKNOWLEDGMENTS

The ideas in this chapter were developed over several years with support from multiple sources: a Royal Society Research Fellowship, a European Union Training and Mobility of Researchers fellowship, a Natural Environment Research Council Realising Our Potential Award, the Polaris consortium (Exxon and Statoil), the GeoPOP consortium (Agip, Amerada Hess, Amoco, ARCO, BP, Chevron, Conoco, Elf, Enterprise, Japan National Oil Corporation, Mobil, Norsk Hydro, Phillips, Statoil, and Total), the Caprocks consortium (BP, ConocoPhillips, ENI, ExxonMobil, Norsk Hydro, Statoil, Total, and Unocal), the European Union Fourth Framework Hydrocarbon Reservoir Programme, and direct support from BP and Norsk Hydro. Additional support to SRC was provided by the Natural Sciences and Engineering Research Council of Canada and the Alberta Ingenuity Fund. We gratefully acknowledge the constructive comments of the reviewers Ben Clennell, Les Leith, and Peter Boult.

REFERENCES CITED

Alixant, J. L., and R. Desbrandes, 1991, Explicit pore pressure evaluation: Concept and application: Society of Petroleum Engineers Paper 19336, p. 50–56.

Aplin, A. C., Y. L. Yang, and S. Hansen, 1995, Assessment of β, the compression coefficient of mudstones and its relationship with detailed lithology: Marine and Petroleum Geology, v. 12, p. 955–963.

Aplin, A. C., I. F. Matenaar, and B. A. van der Pluijm, 2003, Influence of mechanical compaction and chemical diagenesis on the microfabric and fluid flow properties of Gulf of Mexico mudstones: Journal of Geochemical Exploration, v. 78–79, p. 449–451.

Audet, D. M., 1996, Compaction and overpressuring in Pleistocene sediments on the Louisiana Shelf, Gulf of Mexico: Marine and Petroleum Geology, v. 13, p. 467–474.

Aziz, K., and A. Settari, 1979, Petroleum reservoir simulation: London, Applied Science, 475 p.

Bennett, B., and S. R. Larter, 1997, Partition behaviour of alkylphenols in crude oil/brine systems under subsurface conditions: Geochimica et Cosmochimica Acta, v. 61, p. 4393–4402.

Bjørkum, P. A., 1996, How important is pressure in causing dissolution of quartz in sandstones?: Journal of Sedimentary Research, v. 66, p. 147–154.

Bjørlykke, K., 1999, Principal aspects of compaction and fluid flow in mudstones, in A. C. Aplin, A. J. Fleet, and J. H. S. Macquaker, eds., Muds and mudstones: Physical and fluid-flow properties: Geological Society (London) Special Publication 158, p. 22–43.

Bond, K. J., 2001, Mudstone cap rocks as vertical migration pathways: Case studies from the Norwegian sector of the North Sea: Ph.D. Thesis, University of Newcastle upon Tyne, Newcastle upon Tyne, 469 p.

Borst, R. L., 1982, Some effects of compaction and geological time on the pore parameters of argillaceous rocks: Sedimentology, v. 29, p. 291–298.

Burland, J. B., 1990, On the compressibility and shear strength of natural clays: Geotechnique, v. 40, p. 329–378.

Caillet, G., 1993, The cap rock of the Snorre field, Norway: A possible leakage by hydraulic fracturing: Marine and Petroleum Geology, v. 10, p. 42–50.

Charpentier, D., R. H. Worden, C. G. Dillon, and A. C. Aplin, 2003, Fabric development and the smectite to illite transition in Gulf of Mexico mudstones: An image analysis approach: Journal of Geochemical Exploration, v. 78–79, p. 459–463.

De Marsily, G., 1986, Quantitative hydrogeology for engineers: New York, Academic Press, 440 p.

Dewhurst, D. N., A. C. Aplin, J. P. Sarda, and Y. L. Yang, 1998, Compaction-driven evolution of poroperm in natural mudstones: An experimental study: Journal of Geophysical Research, v. 103, p. 651–661.

Dewhurst, D. N., Y. L. Yang, and A. C. Aplin, 1999a, Permeability and fluid flow in natural mudstones, in A. C. Aplin, A. J. Fleet, and J. H. S. Macquaker, eds., Muds and mudstones: Physical and fluid-flow properties: Geological Society (London) Special Publication 158, p. 22–43.

Dewhurst, D. N., A. C. Aplin, and J. P. Sarda, 1999b, Influence of clay fraction on pore-scale properties and hydraulic conductivity of experimentally compacted mudstones: Journal of Geophysical Research, v. 104, p. 29,261–29,274.

Doveton, J. H., 1994, Geologic log analysis using computer methods: AAPG Computer Applications in Geology, v. 2, 169 p.

Eaton, B. A., 1975, The equation for geopressure prediction from well logs: Society of Petroleum Engineers Paper 5544, 11 p.

Garrison, R. E., and W. J. Kennedy, 1977, Origin of solution seams and flaser structures in the Upper Cretaceous chalks of southern England: Sedimentary Geology, v. 19, p. 107–137.

Harrold, T. W. D., R. E. Swarbrick, and N. R. Goulty, 1999, Pore pressure estimation from mudstone porosities in Tertiary basins, southeast Asia: AAPG Bulletin, v. 83, p. 1057–1067.

Hearst, J. R., P. H. Nelson, and F. L. Paillet, 2000, Well logging for physical properties, 2d ed.: Chichester, John Wiley & Sons, 492 p.

Hildenbrand, A., S. Schlömer, and B. M. Krooss, 2002, Gas breakthrough experiments on fine-grained sedimentary rocks: Geofluids, v. 2, p. 3–23.

Hirsch, L. M., and A. H. Thompson, 1995, Minimum saturations and buoyancy in secondary migration: AAPG Bulletin, v. 79, p. 696–710.

Ho, N. C., D. R. Peacor, and B. A. van der Pluijm, 1999, Preferred orientation of phyllosilicates in Gulf Coast mudstones and relation to the smectite-to-illite transition: Clays and Clay Minerals, v. 47, p. 495–504.

Hottman, C. E., and R. K. Johnson, 1965, Estimation of formation pressures from log-derived shale properties: Journal of Petroleum Technology, v. 17, June 1965, p. 717–722.

Ibrahim, M. A., M. R. Tek, and D. L. Katz, 1970, Threshold pressure in gas storage: Michigan, Pipeline Research Committee, American Gas Association at the University of Michigan, 309 p.

Jones, M. E., 1994, Mechanical principles of sediment deformation, in A. Maltman, ed., The geological deformation of sediments: London, Chapman and Hall, p. 37–71.

Kaldi, J. G., and C. D. Atkinson, 1997, Evaluating seal potential: Example from the Talang Akar Formation, offshore northwest Java, Indonesia, in R. C. Surdam, ed., Seals, traps and the petroleum system: AAPG Memoir 67, p. 85–101.

Karig, D. E., and M. V. S. Ask, 2003, Geological perspectives on consolidation of clay-rich marine sediments: Journal of Geophysical Research, v. 108 (B4), art. no. 2197, doi: 10.1029/2001JB000652.

Katsube, T. J., and M. E. Best, 1992, Pore structure of shales from the Beaufort-Mackenzie basin, Northwest Territories: Current Research, Part D: Geological Survey of Canada Paper 92-1E, p. 157–162.

Katsube, T. J., and M. A. Williamson, 1994, Effects of diagenesis on clay nanopore structure and implications for seal capacity: Clay Minerals, v. 29, p. 451–461.

Katsube, T. J., B. S. Mudford, and M. E. Best, 1991, Petrophysical characteristics of shales from the Scotian Shelf: Geophysics, v. 56, p. 1681–1689.

Katz, A. J., and A. H. Thompson, 1987, Prediction of rock electrical conductivity from mercury injection measurements: Journal of Geophysical Research, v. 92, p. 599–607.

Krushin, J. T., 1997, Seal capacity of nonsmectite shale, in R. C. Surdam, ed., Seals, traps and the petroleum system: AAPG Memoir 67, p. 31–48.

Larter, S. R., and A. C. Aplin, 1994, Production applications of reservoir geochemistry: a current and long-term view: Society of Petroleum Engineers Paper 28375, p. 105–113.

Larter, S. R., and A. C. Aplin, 1995, Reservoir geochemistry: Methods, applications and opportunities, in J. M. Cubitt and W. A. England, eds., The geochemistry of reservoirs: Geological Society (London) Special Publication 86, p. 5–32.

Larter, S., B. Bowler, E. Clarke, C. Wilson, B. Moffatt, B. Bennett, G. Yardley, and D. Carruthers, 2000, An experimental investigation of geochromatography during secondary migration of petroleum performed under subsurface conditions with a real rock: Geochemical Transactions, http://www.rsc.org/is/journals/current/geochem/geopub.htm (accessed January 4, 2005).

Larter, S. R., et al., 1996, Molecular indicators of secondary oil migration distances: Nature, v. 383, p. 593–597.

Leith, T. L., and A. E. Fallick, 1997, Organic geochemistry of cap-rock hydrocarbons, in R. C. Surdam, ed., Seals, traps and the petroleum system: AAPG Memoir 67, p. 115–134.

Leith, T. L., I. Kaarstad, J. Connan, J. Pierron, and G. Caillet, 1993, Recognition of cap rock leakage in the Snorre field, Norwegian North Sea: Marine and Petroleum Geology, v. 10, p. 29–41.

Nadeau, P. H., D. R. Peacor, J. Yan, and S. Hillier, 2002, I-S precipitation in pore space as the cause of geopressuring in Mesozoic mudstones, Egersund Basin, Norwegian continental shelf: American Mineralogist, v. 87, p. 1580–1589.

Neuzil, C. E., 1994, How permeable are clays and shales?: Water Resources Research, v. 30, p. 145–150.

O'Brien, N. R., and R. Slatt, 1990, Argillaceous rock atlas: New York, Springer-Verlag, 141 p.

Okui, A., and D. Waples, 1993, Relative permeabilities and hydrocarbon expulsion from source rocks, in A. G. Doré, J. H. Augustson, C. Hermanrud, D. J. Stewart, and Ø. Sylta, eds., Basin modelling: Advances and applications: Norwegian Petroleum Society Special Publication 3, p. 293–301.

Okui, A., R. M. Siebert, and H. Matsubayashi, 1998, Simulation of oil expulsion by 1-D and 2-D basin modeling—Saturation threshold and relative permeabilities of source rocks, in S. J. Duppenbecker and J. E. Iliffe, eds., Basin modelling: Practice and progress: Geological Society (London) Special Publication 141, p. 45–72.

Rieke, H. H., and G. V. Chilingarian, 1974, Compaction of argillaceous sediments: Developments in Sedimentology, v. 16, 424 p.

Sahimi, M., 1995, Flow and transport in porous media and fractured rock: Weinheim, VCH, 482 p.

Scheidegger, A. E., 1974, The physics of flow through porous media, 3d ed.: Toronto, University of Toronto Press, 353 p.

Schlömer, S., and B. M. Krooss, 1997, Experimental characterisation of the hydrocarbon sealing efficiency of cap rocks: Marine and Petroleum Geology, v. 14, p. 565–580.

Schneider, F., J. Burrus, and S. Wolf, 1993, Modelling overpressures by effective-stress/porosity relationships in low-permeability rocks: Empirical artifice of physical reality?, in A. G. Doré, et al., eds., Basin modelling: Advances and applications: Norwegian Petroleum Society Special Publication 3, p. 333–341.

Schowalter, T. T., 1979, Mechanics of secondary hydrocarbon migration and entrapment: AAPG Bulletin, v. 63, p. 723–760.

Skempton, A. W., 1944, Notes on the compressibility of clay: Quarterly Journal of the Geological Society (London), v. 100, p. 119–135.

Skempton, A. W., 1970, The consolidation of clays by gravitational compaction: Quarterly Journal of the Geological Society (London), v. 125, p. 373–411.

Smith, J. E., 1971, The dynamics of shale compaction and evolution of pore fluid pressure: Mathematical Geology, v. 3, p. 239–263.

Sneider, R. M., J. S. Sneider, G. W. Bolger, and J. W. Neasham, 1997, Comparison of seal capacity determinations: Conventional cores vs. cuttings, in R. C. Surdam, ed., Seals, traps and the petroleum system: AAPG Memoir 67, p. 1–12.

Tada, R., and R. Siever, 1989, Pressure solution during diagenesis: Annual Reviews in Earth and Planetary Science, v. 17, p. 89–118.

Taylor, P. N., S. R. Larter, M. Jones, J. D. Dale, and I. Horstad, 1997, The effect of oil-water-rock partitioning on the occurrence of alkylphenols in petroleum systems: Geochimica et Cosmochimica Acta, v. 61, p. 1899–1910.

Terken, J. M. J., and N. L. Frewin, 2000, The Dhahaban petroleum system of Oman: AAPG Bulletin, v. 84, p. 523–544.

Terzaghi, K., 1943, Theoretical soil mechanics: New York, John Wiley, 510 p.

Ungerer, P., J. Burrus, B. Doligez, J. Y. Chenet, and F. Bessis, 1990, Basin evaluation by integrated two-dimensional modeling of heat transfer, fluid-flow, hydrocarbon generation and migration: AAPG Bulletin, v. 74, p. 309–355.

Van Duin, A. C. T., and S. R. Larter, 2001, A computational chemical study of penetration and displacement of water films near mineral surfaces: Geochemical Transactions, http://www.rsc.org/is/journals/current/geochem/geopub.htm (accessed January 4, 2005).

Yang, Y. L., and A. C. Aplin, 1998, Influence of lithology and effective stress on the pore size distribution and modelled permeability of some mudstones from the Norwegian margin: Marine and Petroleum Geology, v. 15, p. 163–175.

Yang, Y. L., and A. C. Aplin, 2004, Definition and practical application of mudstone porosity-effective stress relationships: Petroleum Geoscience, v. 10, p. 153–162.

Yang, Y. L., A. C. Aplin, and S. R. Larter, 2004, Quantitative assessment of mudstone lithology using geophysical wireline logs: Application of artificial neural networks: Petroleum Geoscience, v. 10, p. 141–151.

Yardley, G. S., and R. E. Swarbrick, 2000, Lateral transfer: A source of additional overpressure?: Marine and Petroleum Geology, v. 17, p. 523–537.

Seal Failure Related to Basin-scale Processes

Christian Hermanrud
Statoil ASA, Trondheim, Norway

Hege M. Nordgård Bolås
Statoil ASA, Trondheim, Norway

Gunn M. G. Teige
Statoil ASA, Trondheim, Norway

ABSTRACT

The leakage of trapped petroleum is a major concern in hydrocarbon exploration and has led to a large number of exploration failures. Changes in stress state, fluid pressure, and cap rock permeability may all result in a loss of trapped hydrocarbons. Such changes may result from several different subsurface processes.

This chapter describes an examination of important processes that control compaction, fluid flow in reservoirs, fault reactivation, and their influence on leakage from hydrocarbon reservoirs. It was concluded that seal-failure analysis is seldom based on more than a few of the operating processes and therefore does not reach its full potential. Especially, the effects of overpressures on sediment compaction and hydrocarbon leakage seem to have been oversimplified and commonly overstated. The conditions for hydrofracturing and the corresponding loss of hydrocarbons from structural crests are also commonly considered too superficially in sealing analyses.

It was concluded that inadequate leakage assessments can result from neglect of some subsurface processes that influence stress, pore pressure, and hydrocarbon permeability in the seal. Seal integrity predictions can be improved if thorough analyses of the relevant subsurface processes routinely precede the sealing analyses.

INTRODUCTION

Hydrocarbon leakage has been blamed for the failure of a large number of exploration wells. As a result, significant attention has been given to the analyses of the conditions for hydrocarbon entrapment and subsequent seal failure. The basic principles for hydrocarbon movement and entrapment, which are based on hydrodynamics and capillary-flow resistance, were established by Hubbert (1940, 1953) and Berg (1975).

Copyright ©2005 by The American Association of Petroleum Geologists.
DOI:10.1306/1060753H23159

Several authors later focused on the application of these concepts to hydrocarbon column height predictions by proposing exploration guidelines and classification schemes (Schowalter, 1979; Du Rouchet, 1984; Watts, 1987; Grauls and Baleix, 1994; Heum, 1996). None of these authors aimed to cover the underlying processes that eventually influence the most critical parameters for sealing analysis: compaction, stress, pore pressure, and the relationships between oil and water mobility. Secor (1965) and Mandl and Harkness (1987) described the conditions for hydraulic fracturing and thereby fluid leakage in overpressured rocks. Finkbeiner et al. (1998) extended the considerations needed for the assessment of overpressure-induced leakage by simultaneously addressing the conditions for both fracturing and cap rock failure. Although these papers significantly added to the general knowledge of hydrofracturing, they did not elaborate on the underlying processes that result in changes of stress, pore pressures, or the responses of trapped hydrocarbons to various fracturing modes.

This chapter addresses three sets of such subsurface processes and how they impact hydrocarbon leakage: sediment compaction, fluid mobility in a two-phase (water-plus-hydrocarbon) reservoir, and the relationships between stress and rock failure. Most emphasis is put on the conditions for sediment compaction, because compaction processes also influence the relationships between fluid leakage and the other topics that are covered.

COMPACTION AND FLUID OVERPRESSURES

Compaction of sediments under the influence of overburden weight has been a known geologic phenomenon for a long time. According to Rieke and Chilingarian (1974), compaction as a mechanism for porosity reduction was postulated in the 17th century, and Sorby (1908) presented data that demonstrated inverse relationships between porosity and sediment age. Investigations of compaction and compaction mechanisms quickly became a major interest to geoscientists (Hedberg, 1926; Athy, 1930; Nevin, 1931). The basic principles of mechanical compaction of soils were firmly established by Terzaghi (1925) and Biot (1941).

Although experimental compaction of rocks results in porosity reduction of the involved samples, such compaction does not create lithified rocks from loose sediments. As burial proceeds, mud is transformed to shale, loose sand is transformed to sandstone, and carbonate grains are transformed to limestone. These transformations are caused by chemical reactions (diagenesis), at times involving the formation of new mineral types, but at other times just involving the dissolution of minerals at some locations and the precipitation of the same mineral type somewhere else.

Rocks experience significant changes in both temperature and stress as they subside. Burial leads to diagenesis and sediment compaction, and this demonstrates that changes in stress, temperature, or both have led to these rock alterations over geologic time. However, the relative importance of time, temperature, and stress may vary with sediment type and geologic history. The following paragraphs discuss the evidence for different compaction processes in sandstones, mudstones, and carbonates.

Sandstones

Chuhan et al. (2002) demonstrated in laboratory experiments that loose sands compact under the influence of increased effective stress (total stress minus pore pressure), the degree of compaction depending on the grain-size distribution. Their results showed that mechanical porosity reduction may proceed to a burial depth of 2.5 km (1.6 mi), where the porosities are in the 25–33% range. These experimental results probably describe a lower limit for mechanical sand compaction in nature, because cementation in some cases could restrict the ability of sand to mechanically compact at even shallower burial depths. Nevertheless, these investigations still demonstrate that mechanical compaction is an important process in unconsolidated sediments.

Sandstone diagenesis leads to porosity reduction with increased burial. Such porosity reductions have been proposed to be controlled by effective stress magnitude by some authors (Weyl, 1959; Angevine and Turcotte, 1983; Dewers and Ortoleva, 1990), whereas others suggest that under certain conditions, the rate-limiting step in quartz cementation is temperature controlled (Heald, 1955; Bjørkum, 1996; Walderhaug, 1996; Bjørkum et al.,1998a; Bjørlykke, 1999). Temperature-controlled diagenetic porosity reduction may be caused by the dissolution of quartz in contact with mica and subsequent temperature-controlled silica precipitation at adjacent mineral surfaces (Figure 1). These processes have been reported to begin to replace mechanical compaction at temperatures above 60°C (140°F) (Bjørkum and Nadeau, 1998).

Fluid overpressures will retard sediment compaction if the sandstone compaction is limited by the effective stress. Overpressured rocks will have higher porosities than normally pressured rocks under such circumstances. However, elevated sandstone porosities may also be caused by other factors, such as the coating of the grain surfaces by clay (Heald and Larese, 1974; Ehrenberg, 1993) or by microcrystalline quartz (Aase et al., 1996). Thus, the coexistence of high porosities and

FIGURE 1. Sandstone compaction by thermally driven processes. Adapted from Bjørkum et al. (1998b).

overpressures in sandstones does not necessarily imply a causal relationship between the two. Some observations of sandstone porosity variations with overpressure have found little (Ramm and Bjørlykke, 1994; Hermanrud et al., 1998) or no relationship between the two (Walderhaug, 1994). These results are consistent with the theoretical work by Oelkers et al. (1996), which concluded that modeled temperature-controlled diagenetic porosity reduction closely reproduced petrographic observations.

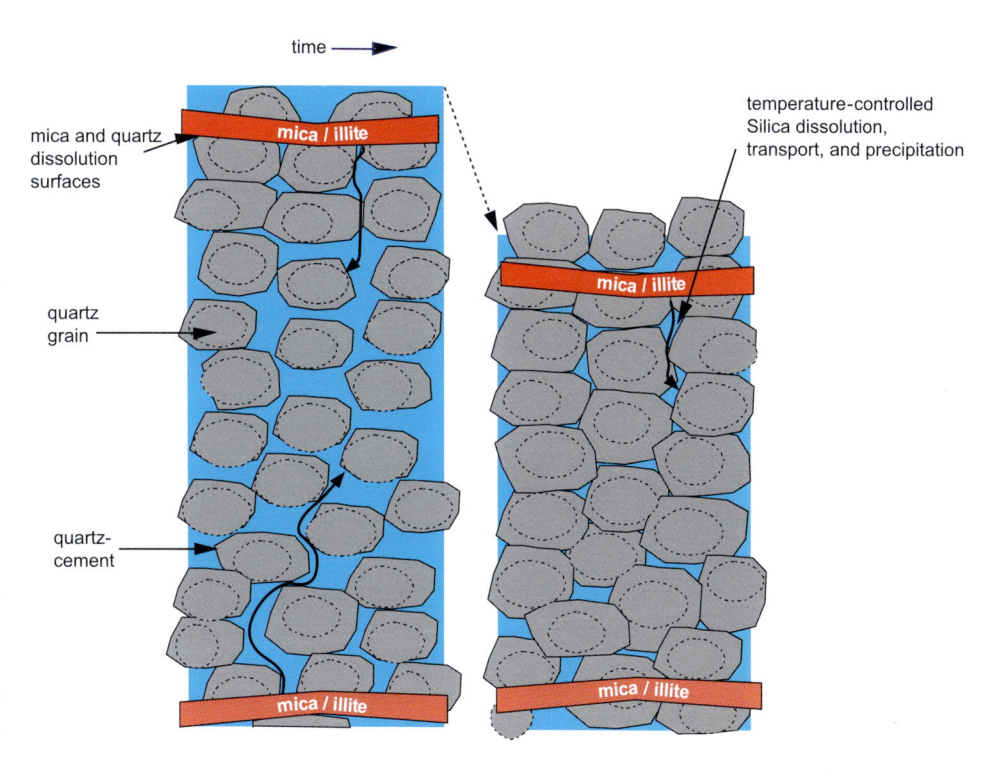

In summary, both mechanical and chemical processes appear to be responsible for the destruction of sandstone porosity during burial. Conditions exist where these chemical processes are largely stress insensitive. Geological analyses based on stress-sensitive compaction as the only mode of porosity reduction may thus lead to erroneous results (Nordgård Bolås et al., 2004).

Mudstones

The compaction of muds and shales may likewise have both mechanical and chemical components. Skempton (1970) presented data from laboratory experiments that documented that increased effective stress acted to reduce the porosity of a large variety of argillaceous sediments. His results were based on experiments with loosely consolidated sediments and with stresses equivalent to 300-m (1000-ft) overburden rocks or less. Dewhurst et al. (1998) performed similar experiments and increased the applied stress to 33 MPa (equivalent to the stress in a rock buried to 3.3 km [2 mi] depth with a pore-pressure gradient of 1.3 g/cm^3). These authors noted, however, that the resulting clay compaction was much lower for these laboratory-compacted samples (porosity of 25%) than what they had observed for those compacted over geological timescales with similar mineralogy and effective stress (porosity of 5–15%).

Muller (1967) suggested that porosity reduction in muds buried below 500 m (1600 ft) is only controlled by grain size, clay mineral composition, and temperature (all stress-independent variables). Mechanical compaction, as well as stress-sensitive chemical compaction, was not considered to be important below this depth of burial. Overpressures will not coincide with elevated porosity under circumstances where the suggestions of Muller (1967) apply.

Overpressures in argillaceous rocks can be caused by several different processes. One popular explanation is that low permeabilities have retarded fluid escape and thereby prevented a rock from attaining the porosity it would have if the rock was normally pressured. This phenomenon has been referred to as compaction disequilibrium (Magara, 1975), nonequilibrium compaction (Plumley, 1980), and disequilibrium compaction (Swarbrick and Osborne, 1998). This process inherently assumes that mudstone compaction is largely controlled by the effective stress, as was the case in the laboratory experiments of Skempton (1970). Overpressured rocks will have higher porosities than normally pressured but otherwise similar rocks under such circumstances.

As in the case for sandstones, associations between high porosity and overpressures in argillaceous rocks have been reported. Notably, overpressured shaly strata are commonly recognized by high porosities and/or high sonic traveltimes (Hottman and Johnson, 1965). As smectite-rich strata have a higher water content than kaolinite-rich strata at the same effective stress (Rieke and Chilingarian, 1974), high subsurface porosities (inferred from density and sonic logs) may be diagnostic of smectic layers. Because smectic layers commonly lead to the preservation of overpressures in underlying permeable rocks, associations between overpressures and high porosities develop. However, as with sandstones, associations between overpressured

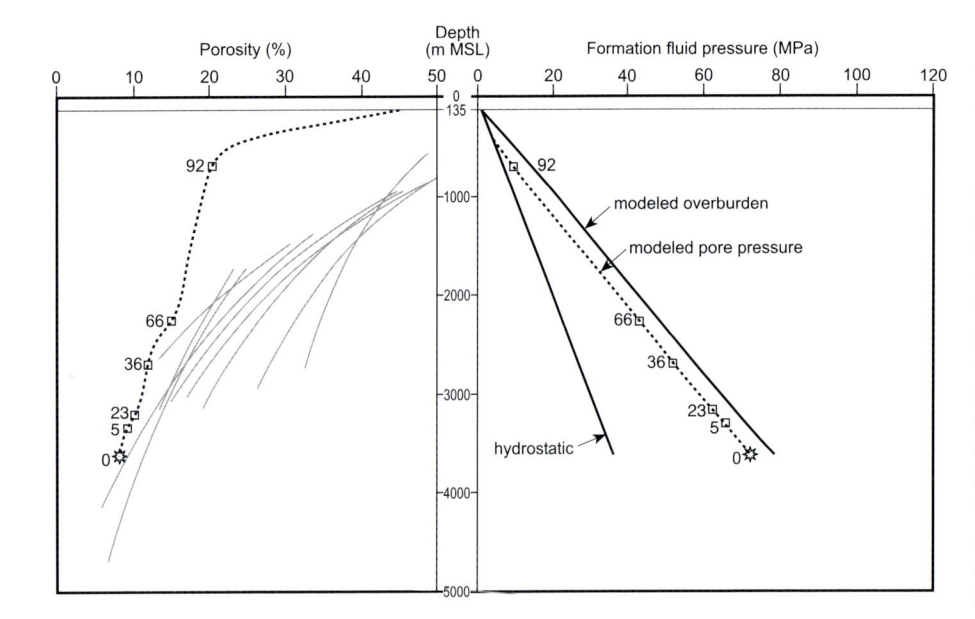

FIGURE 2. Modeled historic development of porosity and pore pressure based solely on stress-sensitive compaction, in the overpressured, low-porosity Cromer Knoll Group in well 34/10-35. The numbers indicate the time (in Ma) when the Cromer Knoll Group reached the various depths during burial history. Solid lines denote porosity-vs.-depth relationships for several different North Sea stratigraphic units (modified from Teige et al., 1999). Unrealistic low paleoporosities (dashed line, left figure) and very shallow pore-pressure buildup (dashed line, right figure) are results from the modeling and demonstrate that the origin of the high fluid pressures in this well are stress-insensitive processes. Adapted from Nordgård Bolås et al. (2004). MSL = mean sea level.

rocks and elevated porosities do not necessarily imply a causal relationship.

Bradley (1975) noted that overburden stress and sealing cap rocks alone cannot cause high overpressures and suggested that temperature changes were the principal factor for the origin of abnormal fluid pressures. The inadequacy of overburden stress and low cap rock permeabilities, as explanations of the generation of overpressures close to the lithostatic pressure, were further supported by Bowers (1994) and Kooi (1997). Nordgård Bolås et al. (2004) used numerical basin modeling to demonstrate that overburden stress and low cap rock permeabilities could result in overpressures close to the lithostatic stress in deeply buried, low-porosity shales, but only through unrealistic permeability- and porosity-vs.-depth relationships in the past (Figure 2).

Teige et al. (1999) compiled porosity and porepressure data from nine different Oligocene to Early Cretaceous shaly stratigraphic units in the Norwegian North Sea, ranging from depths of 1 to 4.5 km (0.6 to 2.8 mi). They found no relationship between the magnitude of overpressures and elevated porosity in any of the investigated mudstones (Figure 3). The overpressures in the investigated rocks were therefore not caused by stress-sensitive compaction. Similar studies in younger and possibly less consolidated rocks have apparently not been reported. Thus, stress-sensitive compaction cannot be dismissed in shallower sediments in this area nor can it be dismissed at burial depths below 1 km (0.6 mi) in other areas.

As for sandstones, both stress-sensitive and stress-insensitive processes influence the porosity reduction of argillaceous rocks during burial. The exact conditions when stress-insensitive processes become dominant have not been thoroughly documented. Basin analysis should honor this uncertainty and should not be solely based on stress-sensitive compaction, unless this is in situations where stress-insensitive compaction can be safely disregarded.

Limestones

Limestone lithification and compaction depend strongly on pore-fluid chemistry. Some limestones lithify more or less at the surface (beach rocks), whereas others survive several kilometers of burial without cementation (Leonard and Munns, 1987). Several carbonate

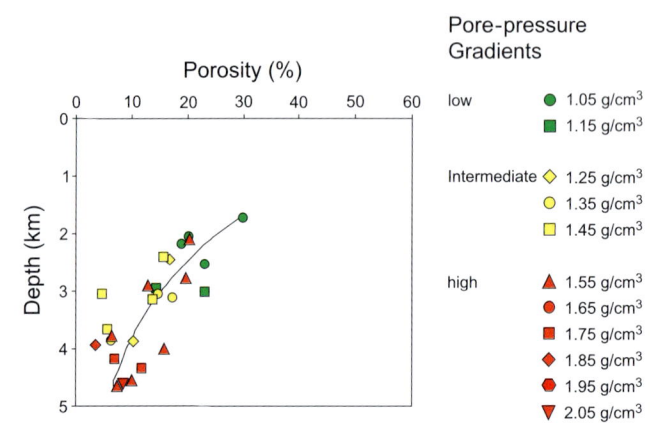

FIGURE 3. Average porosities in the Cretaceous Asgard Formation, Norwegian North Sea, based on density log measurements in 26 wells. Note the absence of a correlation between fluid overpressure and elevated porosity. The dotted line is a standard porosity-vs.-depth curve for the area (Hansen, 1996). Adapted from Teige et al. (1999).

mineral transforms, such as the gypsum to anhydrite transition and dolomitization processes, are known to be temperature controlled. Thus, carbonate porosity reduction is influenced by stress-insensitive processes under certain conditions.

High porosities (as much as 45%) have been reported in several overpressured chalk fields in the Norwegian North Sea (Spencer et al., 1987). Significant reservoir compaction and subsidence during pressure depletion associated with production of these fields testify that stress-sensitive compaction is important in the depth (2–3 km; 1.2–1.8 mi) and temperature (80–120°C [176–248°F]) ranges of these fields. However, modeling performed by H. M. Nordgård Bolås and C. Hermanrud (unpublished data) demonstrate that stress-insensitive compaction is required to explain the existence of high fluid pressures in low-porosity chalks that are observed at burial depths below 4 km (2.5 mi).

It thus appears that both stress-sensitive and stress-insensitive processes influence porosity reduction in limestones and clastic rocks alike, and both processes may operate in reservoirs as well as in cap rocks. Generally, stress-sensitive compaction appears to be most important at shallow burial depths, whereas stress-insensitive compaction takes over at greater depths and temperatures.

Implications for Seal Failure

Processes that control leakage from hydrocarbon-filled reservoirs are closely linked to those that control sediment compaction. Rocks that undergo stress-sensitive compaction will not compact further in the hypothetical case when the pressure compartment is completely sealed off. At this stage, subsidence may proceed, but because the effective stress is not further increased, the pore pressures stay constant.

However, this is not the case when stress-insensitive compaction is operating. In this case, porosity reduces irrespective of pore pressure, and fluids must leave the compartment. The lower the permeability of the rocks that delineate the compartment, the more the overpressures will increase. If the permeability is insufficient to allow the fluids to escape even at high overpressures, then fracturing or fault reactivation will occur.

Thus, the important factors to consider in sealing analysis, when stress-insensitive compaction is operating, are not whether a compartment leaks fluids but where and how it leaks (Figure 4). All pressure compartments leak fluids as long as stress-insensitive compaction occurs. The location of a compartment's outlet depends primarily on whether the leakage occurs in the pore network or whether it occurs along faults or (micro)fractures.

Sealing analyses commonly only address single parts of a pressure compartment's boundary, such as analyses of individual faults or sealing rocks (top and bottom seals). This approach is, however, not appropriate if stress-insensitive compaction occurs. Under such circumstances, sealing analysis should concentrate on identifying the compartment's weakest boundaries where leakage must occur instead of addressing single fault or cap rock properties.

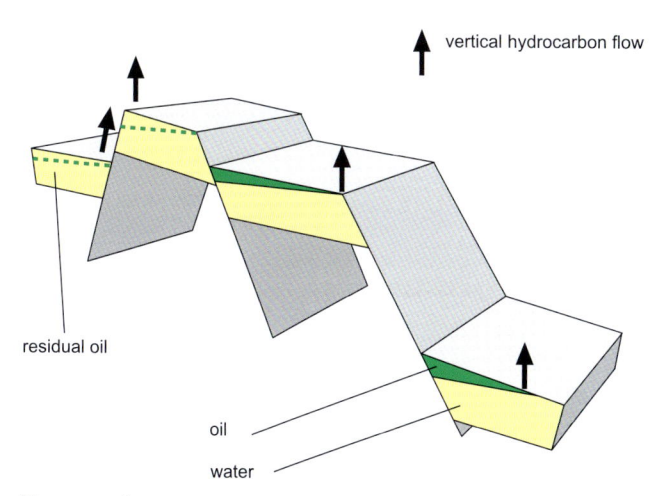

FIGURE 4. All pressure compartments leak as porosity is reduced irrespective of high fluid pressures. Identification of the weakest boundary is the key challenge for seal analysis in such cases.

LEAKAGE OF WATER AND HYDROCARBONS THROUGH THE CAP ROCK

Both hydrocarbons and water may migrate through sealing rocks from a pressure compartment via the pore network of the cap rocks. However, the processes that control hydrocarbon retention below a seal are different from those that control the water movement through reservoirs and cap rocks.

Water flux is controlled by the intrinsic permeability and hydraulic gradient in rocks according to Darcy's Law (Hubbert, 1940). Water will therefore leak in all directions toward lower hydraulic heads (overpressures). The water flux from downflank positions (below the hydrocarbon-water contact) will not be influenced by the presence of hydrocarbons.

Reservoired hydrocarbons will flow through the pore network of a water-wet seal only if the capillary entry pressure (Pc_e) has been overcome by the buoyancy of the underlying hydrocarbons (Berg, 1975) (equation 1):

$$Pc_e < (\rho_w - \rho_{hc}) g h \quad (1)$$

where

$$Pc = 2\gamma \cos \theta / r \quad (2)$$

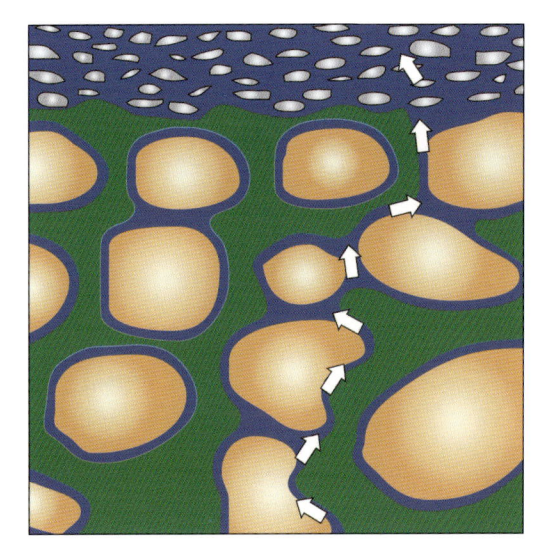

FIGURE 5. Water flow from the aquifer through the wetting-phase water at the top of the reservoir and further into the cap rock. Comparatively high permeabilities for this water, as calculated from the experiments of Teige et al. (2005), prevent high water pressures from pushing the oil through the seal.

through a water-wet, oil-saturated core (Figure 6). This experiment demonstrated that water could in fact flow through a core sample and further through a membrane, whereas the oil was retained in the core plug by this same membrane. The pressure difference between the oil and water phases was set to 0.5 MPa in the laboratory experiment, which is equivalent to the pressure differential that would result from the buoyancy of a 150-m (500-ft) oil column. The permeability of the moving water was calculated to 0.8×10^{-3} md. Higher pressure differences will reduce the mobility of the wetting-phase fluid. The results from these laboratory experiments are in agreement with the conclusions of Bjørkum et al. (1998b, 1999), who also suggested that the permeability of water in the hydrocarbon column of an oil reservoir was significantly higher than the cap rock permeability, and that this water could therefore flow through water-wet reservoirs that are overlain by water-wet cap rocks.

The flow of wetting-phase water through hydrocarbon columns will significantly impact the conditions for the trapping of hydrocarbons. First, such flow

Here, Pc_e = capillary entry pressure of the seal; ρ_{hc} is hydrocarbon density; ρ_w is water density; g is the acceleration of gravity; h is the hydrocarbon column height; γ is the hydrocarbon-water interfacial tension; θ is the contact angle between water and hydrocarbons; and r is pore-throat radius of the seal. Small pore throats (r) result in large capillary entry pressures, with large sealing potential for hydrocarbons.

All hydrocarbon reservoirs were water filled in the past, before hydrocarbon charge commenced. As hydrocarbons filled the traps, most of the water was displaced. However, residual water saturations remained in the reservoirs, mainly confined in acute corners of the pores but also as a water film around the sand grains in water-wet reservoirs. The mobility of this water is reduced with increasing oil columns, because the water will be confined to progressively smaller parts of the pore network.

Experimental work by Dullien et al. (1986) demonstrated that this wetting-phase (commonly termed "irreducible") water was still mobile at capillary pressures (= oil minus water pressures) as high as 0.1 MPa (15 psi), and the water saturation was progressively reduced as long as the pressure of the (nonwetting) oil phase was increased. The fact that this water is mobile suggests that the irreducible water can also flow through a hydrocarbon reservoir and through a seal, whereas the reservoired hydrocarbons stay behind, because the capillary entry pressure has not been overcome (Figure 5).

Teige et al. (2005) set up a laboratory experiment to investigate the possibility of such water movement

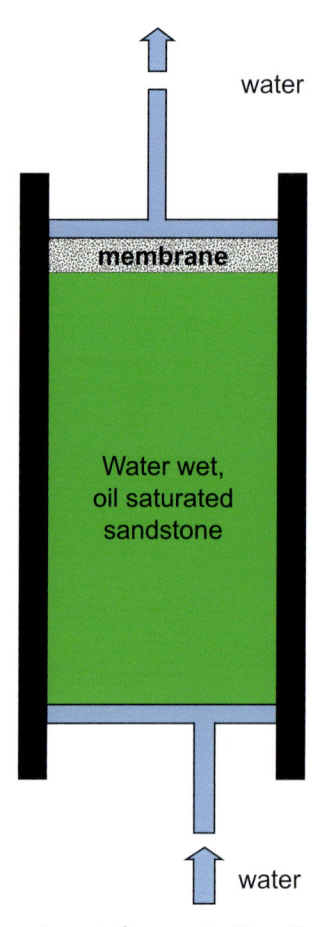

FIGURE 6. Experiment demonstrating that wetting-phase water can move through an oil-saturated core plug. Adapted from Teige et al. (2005).

implies that a pressure compartment can still preserve its hydrocarbons, even if fluids are leaking from an area that includes its shallowest position. Such leakage may explain why several overpressured North Sea oil fields are associated with broad seismic chimneys (Gullfaks, Gullfaks South, Hild, e.g., Nordgård Bolås and Hermanrud, 2003; Nordgård Bolås et al., 2005). Second, as suggested by Bjørkum et al. (1998b), the difference in water pressure between a reservoir and the overlying cap rock should not be added to the hydrocarbon buoyancy term of equation 1, as long as this water pressure is transmitted to the irreducible water at the top of the reservoir through an (almost) hydrostatic pressure gradient. Such a pressure gradient will be present when the wetting-phase fluid has permeabilities comparable to those calculated by Bjørkum et al. (1999) and to the experimental results of Teige et al. (2005). Furthermore, the introduction of fluid pressures to equation 1, but not necessarily as suggested by Clayton and Hay (1994) and Corcoran and Doré (2002), would be appropriate if the permeability of the wetting-phase water was comparable to that of a tight cap rock.

Because calculated seal capacity will be significantly reduced by the addition of a water pressure difference term in equation 1, estimates of wetting-phase water permeability and cap rock wettability should be included in evaluations of membrane sealing capacity. Failure to recognize the mobility of this water may result in too-pessimistic seal capacity estimates.

CONDITIONS FOR FRACTURING AND FAULT REACTIVATION

Fracturing and fault reactivation commonly result in seal breaching. The ability to estimate the likelihood for such events is therefore an important aspect of seal evaluation.

Fracturing and faulting are initiated when the stress state is such that the failure condition for the rock is met. Such stress and failure conditions are conveniently described in Mohr's diagrams (Figure 7). Rock failure is initiated when Mohr's circle touches the failure envelope. This can happen either because the pore pressure is increased (circle moves to the left) or because the stress anisotropy is increased (circle is enlarged).

Figure 7 displays two different fracture criteria, both of which are commonly applied. The existence of a linear (Mohr–Coulomb) failure envelope that intersects the origin would preclude tensile failure, because the circle will always intersect this failure envelope before the least principal stress (left intersection between circle and X-axis) becomes zero. This envelope describes a situation where cohesionless fractures exist (the rock has no tensile strength). The Mohr-Coulomb

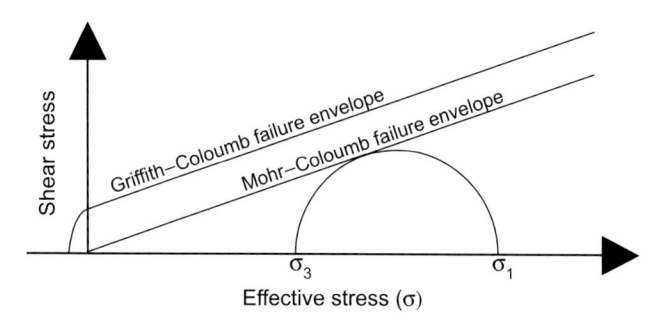

FIGURE 7. Mohr's circle with linear (Mohr–Coulomb) and curved (Griffith–Coulomb) fracturing criteria. The former will always lead to the prediction of shear failure as a fracturing mechanism, whereas the latter can lead to the prediction of tensile failure when the stress anisotropy is small. σ_1 is the largest effective stress (stress minus pore pressure), and σ_3 is the least.

fracture criteria may also be well suited for the description of failure where none of the principal effective stress components are small. This failure criterion was applied by Barton et al. (1995), Castillo et al. (1998), and Finkbeiner et al. (1997), all of whom suggested that shear failure was the dominating mode of failure.

The other fracture criterion of Figure 7 describes intact rocks with a cohesive strength. This criterion was suggested by Secor (1965) and was based on experiments by Griffith (1921, 1925) on the conditions for tensile failure in glass. Applications of this failure criterion to rocks in a basin scale are performed by assuming similar fracturing characteristics of the two media (rocks and glass), because the basin-scale tensile strength of rocks cannot be measured. This Griffith–Coulomb criterion suggests that tensile failure occurs as the fluid pressure increases in overpressured rocks with little stress anisotropy. Studies that suggest that the retention capacity (least principal stress minus pore pressure) can reach zero before fracturing is initiated (Gaarenstroom et al., 1993; Grauls and Baleix, 1994) appear to base their conclusions on this failure envelope. However, fracturing could be initiated at both positive and negative effective stress in rocks with nonzero tensile strengths. Only in special cases will fracturing be initiated at the instance when the tensile stress reaches zero.

The mode of failure can significantly influence the sealing characteristics of an area. Differences in sealing properties between the North Viking Graben and the Halten Terrace areas, both offshore Norway, have been attributed to different fracture modes between the areas by Hermanrud and Nordgård Bolås (2002), Nordgård Bolås and Hermanrud (2003), and Nordgård Bolås et al. (2005). These studies suggest that areas that have experienced large stress anisotropies because of flexuring have been subject to shear failure, with extensive hydrocarbon leakage as a consequence. Areas that have experienced less stress anisotropy seem to have avoided shear failure and have retained appreciable hydrocarbon volumes.

Whereas shear failure preferentially occurs along dipping or vertical surfaces oriented approximately 60° to the minimum, tensile fracturing preferentially occurs along vertical or horizontal surfaces oriented normal to the minimum stress direction (Hobbs et al., 1976). The mode of failure thus also determines which fault orientation is likely to slip first and therefore be expected to control the depth to the fluid contacts in reservoirs that have experienced recent fault reactivation. Hydrofracturing, which is expected to be initiated at the shallowest position on a structure, will only occur if the conditions for shear failure have not been met at lower pore pressures. Evaluation of possible hydrofracturing of a structure is thus also intimately linked to the fracture criterion of the rock. Despite this, analyses of fracture criteria have not yet become common practice in sealing analysis.

SUMMARY AND CONCLUSIONS

Whereas seal failure is known to be controlled by stress and pore pressures, analyses of the processes that result in stress and pore-pressure evolution in sedimentary basins are seldom addressed in any depth in seal analyses. This negligence may result in drawing inferior conclusions.

Porosity reduction during burial is commonly dominated by thermal processes, which proceed despite high fluid overpressures. As a result, pressure compartments leak fluids irrespective of the properties of the rocks that seal them. Sealing analysis should include analyses of all the boundaries of the compartments instead of the current practice of only addressing the sealing capacity of individual faults or sealing rocks.

Water may be transported through an oil column in the irreducible wetting-phase water and further through the cap rock. When this happens, hydrocarbon preservation is favored, and failure to recognize this process will result in very pessimistic evaluations of membrane seals. Reservoir overpressures will many times not be negative for hydrocarbon preservation under such conditions unless the conditions for fracturing are met.

The condition for the onset of rock fracturing is controlled by the stress state of the rock and its mechanical properties. Although both the orientation of leaky faults and the stress level required for fracture initiation are determined by these factors, analyses of these are commonly addressed superficially in seal evaluation. The result may well be overlooked exploration opportunities and failed exploration wells.

ACKNOWLEDGMENTS

This chapter is based on results from Statoil projects, and Statoil's permission to publish the results is gratefully acknowledged. S. Larter, D. Lowry, and P. Boult are thanked for constructive reviews of earlier versions of this manuscript. E. Storsteen is thanked for graphical support, and S. Clark is thanked for his improvements to the English text.

REFERENCES CITED

Aase, N. E., P. A. Bjørkum, and P. H. Nadeau, 1996, The effect of grain-coating microquartz on preservation of reservoir porosity: AAPG Bulletin, v. 80, p. 1654–1673.

Angevine, C. L., and D. L. Turcotte, 1983, Porosity reduction by pressure solution: A theoretical model for quartz arenites: Geological Society of America Bulletin, v. 94, p. 1129–1134.

Athy, L. F., 1930, Density, porosity and compaction of sedimentary rocks: AAPG Bulletin, v. 14, no. 1, p. 1–24.

Barton, C. A., M. D. Zoback, and D. Moos, 1995, Fluid flow along potentially active faults in crystalline rock: Geology, v. 23, no. 8, p. 683–686.

Berg, R. R., 1975, Capillary pressure in stratigraphic traps: AAPG Bulletin, v. 59, p. 939–956.

Biot, M. A., 1941, General theory of three-dimensional consolidation: Journal of Applied Physics, v. 12, p. 155–164.

Bjørkum, P. A., 1996, How important is pressure in causing dissolution of quartz in sandstones?: Journal of Sedimentary Research, v. A66, p. 147–154.

Bjørkum, P. A., and P. Nadeau, 1998, Temperature controlled porosity/permeability reduction, fluid migration, and petroleum exploration in sedimentary basins: Australian Petroleum Production and Exploration Association Journal, v. 38, p. 453–465.

Bjørkum, P. A., O. Walderhaug, and P. Nadeau, 1998a, Physical constraints on hydrocarbon leakage and trapping revisited: Petroleum Geoscience, v. 4, p. 237–239.

Bjørkum, P. A., E. H. Oelkers, P. H. Nadeau, O. Walderhaug, and W. M. Murphy, 1998b, Porosity prediction in quartzose sandstones as a function of time, temperature, depth, stylolite frequency, and hydrocarbon saturation: AAPG Bulletin, v. 82, p. 637–648.

Bjørkum, P. A., O. Walderhaug, and P. Nadeau, 1999, Reply to discussion of the paper "Physical constraints on hydrocarbon leakage and trapping revisited: Petroleum Geoscience, v. 4, p. 237–239," by S. Rodgers: Petroleum Geoscience, v. 5, p. 422–423.

Bjørlykke, K., 1999, An overview of factors controlling rate of compaction, fluid generation and flow in sedimentary basins, in B. Jamtveit and P. Meakin, eds., Growth, dissolution and pattern formation in geosystems: Dordrecht, The Netherlands, Kluwer Academic Publishers, p. 381–404.

Bowers, G. L., 1994, Pore pressure estimation from

velocity data: Accounting for mechanisms besides undercompaction: International Association of Drilling Contractors–Society of Petroleum Engineers Drilling Conference, SPE paper 27488, p. 515–530.

Bradley, J. S., 1975, Abnormal formation pressure: AAPG Bulletin, v. 59, no. 6, p. 957–973.

Castillo, D. A., R. R. Hillis, K. Asquith, and M. Fischer, 1998, State of the stress in the Timor Sea area, based on deep wellbore observations and frictional failure criteria: Application to fault-trap integrity, in P. G. Purcell and R. G. Purcell, eds., The sedimentary basins of Western Australia 2: Proceedings of Petroleum Exploration Society of Australia Symposium, Perth, Western Australia, 1998: p. 325–341.

Chuhan, F. A., A. Kjeldstad, K. Bjørlykke, and K. Høeg, 2002, Porosity loss in sand by grain crushing-experimental evidence and relevance to reservoir quality: Marine and Petroleum Geology, v. 19, p. 39–53.

Clayton, C. J., and S. Hay, 1994, Gas migration mechanisms from accumulation to surface: Bulletin of the Geological Society, Denmark, v. 41, p. 12–23.

Corcoran, D. V., and A. G. Doré, 2002, Top seal assessment in exhumed basin settings— Some insights from Atlantic margin and borderland basins, in A. G. Koestler and R. Hunsdale, eds., Hydrocarbon seal quantification: Norwegian Petroleum Society Special Publication 11, p. 89–107.

Dewers, T., and P. Ortoleva, 1990, A coupled reaction/transport/mechanical model for intergranular pressure solution, stylolites, and differential compaction and cementation in clean sandstones: Geochimica et Cosmochimica Acta, v. 54, p. 1609–1625.

Dewhurst, D. N., A. Aplin, J.-P. Sarda, and Y. Yang, 1998, Compaction-driven evolution of porosity and permeability in natural mudstones: An experimental study: Journal of Geophysical Research, v. 103, no. B1, p. 651–661.

Du Rouchet, J., 1984, Migration in fracture networks; an alternative explanation to the supply of the "giant" tar accumulations in Alberta, Canada: Journal of Petroleum Geology, v. 7, p. 381–402.

Dullien, F. A. L., F. S. Y. Lay, and I. F. Macdonald, 1986, Hydraulic continuity of residual wetting phase in porous media: Journal of Colloidal and Interface Science, v. 109, no. 1, p. 201–218.

Ehrenberg, S. N., 1993, Preservation of anomalously high porosity in deeply buried sandstones by grain coating chlorite: Examples from the Norwegian continental shelf: AAPG Bulletin, v. 77, p. 1260–1286.

Finkbeiner, T., C. A. Barton, and M. D. Zoback, 1997, Relationships among in-situ stress, fractures and faults, and fluid flow: Monterey Formation, Santa Maria basin, California: AAPG Bulletin, v. 81, no. 12, p. 1975–1999.

Finkbeiner, T., M. Zoback, B. B. Stump, and P. B. Flemings, 1998, In situ stress, pore pressure, and hydrocarbon migration in the South Eugene Island field, Gulf of Mexico: Overpressures in petroleum exploration, in A. Mitchell and D. Grauls, eds., Workshop Proceedings, Pau, France, April 1998: Bulletin Centre Research Elf Exploration and Production Memoir 22, p. 103–110.

Gaarenstroom, L., R. A. J. Tromp, M. C. de Jong, and A. M. Brandenburg, 1993, Overpressures in the Central North Sea: Implications for trap integrity and drilling safety, in J. R. Parker, ed., Petroleum geology of northwest Europe: Proceedings of the 4th Conference, The Geological Society (London), p. 1305–1313.

Grauls, D. J., and J. M. Baleix, 1994, Role of overpressures and in situ stress in fault-controlled hydrocarbon migration: A case study: Marine and Petroleum Geology, v. 11, no. 6, p. 734–742.

Griffith, A. A., 1921, The phenomena of rupture and flow in solids: Philosophical Transactions of the Royal Society of London, Series A, v. 221, p. 163–198.

Griffith, A. A., 1925, The theory of rupture, in C. B. Biezeno and J. M. Burgers, eds., Proceedings of the First International Congress for Applied Mechanics, 1924, Delft, The Netherlands, Tech. Boekhandel en Drukkerij J. Waltman Jr., p. 55–63.

Hansen, S., 1996, A compaction trend for Cretaceous and Tertiary shales on the Norwegian shelf based on sonic transit times: Petroleum Geoscience, v. 2, p. 159–166.

Heald, M. T., 1955, Stylolites in sandstones: Journal of Geology, v. 63, p. 101–114.

Heald, M. T., and R. E. Larese, 1974, Influence of coatings on quartz cementation: Journal of Sedimentary Petrology, v. 44, p. 1269–1274.

Hedberg, H. D., 1926, The effect of gravitational compaction on the structure of sedimentary rocks: AAPG Bulletin, v. 10, no. 10, p. 1035–1072.

Hermanrud, C., and H. M. Nordgård Bolås, 2002, Leakage from overpressured hydrocarbon reservoirs at Haltenbanken and in the northern North Sea, in A. G. Koestler and R. Hunsdale, eds., Hydrocarbon seal quantification: Norwegian Petroleum Society Special Publication 11, p. 221–231.

Hermanrud, C., L. Wensaas, G. M. G. Teige, H. M. Nordgård Bolås, S. Hansen, and E. Vik, 1998, Shale porosities from well logs on Haltenbanken (offshore mid-Norway) show no influence of overpressure, in B. E. Law, G. F. Ulmishek, and V. I. Slavin, eds., Abnormal pressures in hydrocarbon environments: AAPG Memoir 70, p. 65–85.

Heum, O. R., 1996, A fluid dynamic classification of hydrocarbon entrapment: Petroleum Geoscience, v. 2, p. 145–158.

Hobbs, B. E., W. D. Means, and P. F. Williams, 1976, An outline of structural geology: New York, John Wiley and Sons, 571 p.

Hottman, C. E., and R. K. Johnson, 1965, Estimation of formation pressures from log-derived shale properties: Journal of Petroleum Technology, v. 17, p. 717–722.

Hubbert, M. K., 1940, The theory of ground-water motion: Journal of Geology, v. 48, p. 785–944.

Hubbert, M. K., 1953, Entrapment of petroleum under hydrodynamic conditions: AAPG Bulletin, v. 37, p. 1956–2026.

Kooi, H., 1997, Insufficiency of compaction disequilibrium as the sole cause of high fluid pressures in pre-Cenozoic sediments: Basin Research, v. 9, p. 227–241.

Leonard, R. C., and J. W. Munns, 1987, Vallhall, in A. M. Spencer, E. Holter, C. J. Campbell, S. H. Hanslien, P. H Nelson, E. Nysæther, and E. G. Ormaasen, eds., Geology of the Norwegian oil and gas fields: London, Graham and Trotman, p. 153–164.

Magara, K., 1975, Re-evaluation of montmorillonite dehydration as a cause of abnormal pressure and hydrocarbon migration: AAPG Bulletin, v. 59, no. 2, p. 292–302.

Mandl, G., and R. M. Harkness, 1987, Hydrocarbon migration by hydraulic fracturing, in M. E. Jones and R. M. F. Preston, eds., Deformation of sediments and sedimentary rocks: Geological Society (London) Special Publication 29, p. 39–53.

Muller, G., 1967, Diagenesis in argillaceous sediments, in G. Larsen and G. V. Chilingar, eds., Diagenesis in sediments: Amsterdam, Elsevier, p. 127–177.

Nevin, C. M., 1931, Principles of structural geology: New York, Wiley, 303 p.

Nordgård Bolås, H. M., and C. Hermanrud, 2003, Hydrocarbon leakage processes and retention capacities offshore Norway: Petroleum Geoscience, v. 9, p. 321–332.

Nordgård Bolås, H. M., C. Hermanrud, and G. M. G. Teige, 2004, The validity of disequilibrium compaction in North Sea shales; constraints from basin modelling: AAPG Bulletin, v. 88, no. 2, p. 1–19.

Nordgård Bolås, H. M., C. Hermanrud, and G. M. G. Teige, 2005, The influence of stress regimes on hydrocarbon leakage in P. Boult and J. Kaldi, eds., Evaluating fault and cap rock seals: AAPG Hedberg Series, no. 2, p. 109–123.

Oelkers, E. H., P. A. Bjørkum, and W. M. Murphy, 1996, A petrographic and computational investigation of quartz cementation and porosity reduction in North Sea sandstones: American Journal of Science, v. 296, p. 420–452.

Plumley, W. J., 1980, Abnormally high fluid pressure: Survey of some basic principles: AAPG Bulletin, v. 64, no. 3, p. 414–430.

Ramm, M., and K. Bjørlykke, 1994, Porosity/depth trends in reservoir sandstones: Assessing the quantitative effects of varying pore-pressure, temperature history and mineralogy: Norwegian shelf data: Clay Minerals, v. 29, p. 475–490.

Rieke, H. H., III, and G. V. Chilingarian, 1974, Compaction of argillaceous sediments: Developments in Sedimentology, v. 16, 424 p.

Schowalter, T. T., 1979, Mechanics of secondary hydrocarbon migration and entrapment: AAPG Bulletin, v. 63, p. 723–760.

Secor, D. T., 1965, Role of fluid pressure in jointing: American Journal of Science, v. 263, p. 633–646.

Skempton, A. W., 1970, The consolidation of clays by gravitational compaction: Quarterly Journal of the Geological Society (London), v. 125, no. 3, p. 373–411.

Sorby, H. C., 1908, On the application of quantitative methods to the study and the history of rocks: Quarterly Journal of the Geological Society (London), v. 64, p. 171–232.

Spencer, A. M., E. Holter, C. J. Campbell, S. H. Hanslien, P. H. H. Nelson, E. Nysæther, and E. G. Ormaasen, 1987, Geology of the Norwegian oil and gas fields: London, Graham and Trotman, 493 p.

Swarbrick, R. E., and M. J. Osborne, 1998, Mechanisms that generate abnormal pressures: An overview, in B. E. Law, G. F. Ulmishek, and V. I. Slavin, eds., Abnormal pressures in hydrocarbon environments: AAPG Memoir 70, p. 13–34.

Teige, G. M. G., C. Hermanrud, L. Wensaas, and H. M. Nordgård Bolås, 1999, The lack of relationship between overpressure and porosity in North Sea and Haltenbanken shales: Marine and Petroleum Geology, v. 16, no. 4, p. 321–335.

Teige, G. M. G., C. Hermanrud, W. H. Thomas, O. B. Wilson, and H. M. Nordgård Bolås, 2005, Capillary resistance and trapping of hydrocarbons: A laboratory experiment: Petroleum Geoscience, v. 11, no. 2, p. 125–129.

Terzaghi, K., 1925, Erdbaumechanik auf bodenphysiklischer Grundlage: Leipzig, Deuticke, 399 p.

Walderhaug, O., 1994, Temperatures of quartz cementation in Jurassic sandstones from the Norwegian continental shelf—Evidence from fluid inclusions: Journal of Sedimentary Research, v. A64, p. 311–323.

Walderhaug, O., 1996, Kinetic modelling of quartz cementation and porosity loss in deeply buried sandstone reservoirs: AAPG Bulletin, v. 80, p. 731–745.

Watts, N. L., 1987, Theoretical aspects of cap-rock and fault seals for single- and two-phase columns: Marine and Petroleum Geology, v. 4, p. 274–307.

Weyl, P. K., 1959, Pressure solution and the force of crystallization—A phenomenological theory: Journal of Geophysical Research, v. 64, p. 2001–2025.

Potential New Method for Paleostress Estimation by Combining Three-dimensional Fault Restoration and Fault Slip Inversion Techniques: First Test on the Skua Field, Timor Sea

A. P. Gartrell

Division of Petroleum Resources, Commonwealth Scientific and Industrial Research Organization, Perth, Western Australia, Australia

M. Lisk

Division of Petroleum Resources, Commonwealth Scientific and Industrial Research Organization, Perth, Western Australia, Australia

ABSTRACT

This pilot study indicates that estimating paleostress orientations and magnitudes from seismic data, through analysis of fault slip data obtained using three-dimensional restoration techniques, is possible, and the results generated are consistent with regional observations. The results suggest that the stress regime responsible for late Miocene fault activity in the vicinity of the Skua oil field in the Timor Sea differs from the present-day stress regime. An extensional stress regime, having the maximum principal stress axis (σ_1) oriented vertically, the intermediate principal stress axis (σ_2) oriented approximately east–west, the minimum principal stress (σ_3) oriented approximately north–south, and a stress ratio (R) of about 0.3, was calculated for the late Miocene. In contrast, measurements of the present-day stress field indicate a transtensional stress regime in which σ_2 is vertical, σ_1 is horizontal and trends east-northeast–west-southwest, σ_3 trends north-northwest–south-southeast, and R = 0.8. Estimation of the magnitudes of the principal stresses indicate that the differential stress operating in the late Miocene was similar to the present, but that greater mean stress in the present-day stress state results in a lowering of reactivation risk with time. These results are consistent with regional observations of widespread late Tertiary extensional faulting, with decreasing fault activity to the present day. The work also suggests that the majority of hydrocarbon leakage associated with fault reactivation in this region is less likely to be associated with the present-day stress regime than with the paleostress regime.

INTRODUCTION

Fault reactivation related to ongoing convergence and collision of the Australian continent with the Banda Island arc is postulated to be responsible for the common occurrence of breached hydrocarbon accumulations in the Timor Sea (e.g., Woods, 1992; Shuster et al., 1998; O'Brien et al., 1999; de Ruig et al., 2000). Consequently, fault reactivation represents the principal risk for hydrocarbon exploration in the area. Analysis of the in-situ stress tensor has been used in previous studies in an attempt to quantify fault reactivation risk and to explain patterns of leakage in the Timor Sea (Mildren et al., 1994; Hillis, 1998; Castillo et al., 2000; de Ruig et al., 2000). These studies consider that in a population of faults and fractures, those that are critically stressed with respect to the in-situ stress tensor are more likely to act as conduits for fluid transmission (Jones et al., 2002). However, if the orientation or magnitude of the local stress tensor has changed significantly since the onset of hydrocarbon charge, then a trap assessed as secure in the present day stress regime may have actually failed during an earlier period (de Ruig et al., 2000). Conversely, if the trap was breached because of fault reactivation in a paleostress regime, then a subsequent change in the stress regime may lead to resealing of the trap so that it can be recharged, although evidence for the previous reactivation event may be observed.

Convergence and collision at Australia's northern plate margin has been a complex and progressive process, and therefore, the probability that changes in the stress field have occurred over time seems high. Initial collision between the Australian continental plate and the Philippine plate occurred in the vicinity of New Guinea Island about 20–25 Ma (Etheridge et al., 1991; Hall, 1996). Subsequent convergence and collision between the Australian continental plate and the Banda Arc occurred in the late Tertiary (Woods, 1994) at high angles to the collisional front at New Guinea. Two main stages of late Tertiary collision have been recognized on Timor Island (Charlton et al., 1991; Woods 1994). The first stage occurred in the late Miocene (~8 Ma) when the transitional Australian continental crust reached the subduction system. The second stage occurred when true continental crust entered the subduction system in the middle Pliocene (~3 Ma; Charlton et al., 1991).

The timing of the deformation events observed on Timor Island corresponds to widespread late Tertiary fault (predominantly extensional) activity and increased rates of subsidence commonly observed in the Timor Sea region, and hence, a causal relationship has been assumed (e.g., Woods, 1992; O'Brien and Woods, 1995; O'Brien et al., 1998). However, the mechanism that allowed extensional deformation in the Timor Sea to occur in conjunction with plate collision remains controversial (cf. Bradley and Kidd, 1991; Woods, 1994). O'Brien et al. (1999) suggest that the regional stress regime changed significantly at approximately 2.5 Ma, when volcanism and thrusting on Timor stopped, and convergence between Timor and the Australian mainland apparently ceased. This event coincides with a notable decrease in fault displacements to the present day in the Timor Sea (e.g., Meyer et al., 2002).

A more continuous record of the region's stress history is clearly needed if quantitative and reliable models for predicting leakage of hydrocarbons caused by fault reactivation are to be developed. Information about paleostress regimes can be gained in several ways, including recognition of common structural geometries or families (e.g., conjugate faults, wrench systems, folds, stylolites, and tension gashes), tectonic history, mineral alignments (e.g., analysis of calcite mechanical twins; Laurent et al., 1981; Lacombe et al., 1990), plate boundary force reconstruction (e.g., Coblentz et al., 1998), and inversion of fault slip data (e.g., Angelier, 1990). Of these possible approaches, fault slip inversion techniques arguably offer the most quantitative and widely tested approach for estimating paleostress conditions. Although somewhat controversial, this technique has been applied and tested in many studies from around the world where faults are observed in outcrop, and the results in the overwhelming number of cases have been shown to be consistent with independent geological data (e.g., Letouzey, 1986; Oncken, 1988; Mercier et al., 1991; Angelier, 1994; Delvaux et al., 1995; Hibsch et al., 1995; Heeremans et al., 1996; Saintot and Angelier, 2002).

Traditionally, fault slip data has been obtained by the physical measurement of fault planes and lineations (e.g., slickensides). The problem with using fault slip inversion techniques in an offshore petroleum exploration context is that the faults are located in the subsurface, where seismic imaging and drill coring are the most common methods for obtaining fault information. Fault lineations are below seismic resolution and are not often sampled in a practical way by coring the faults. However, definition of footwall cutoff, hangingwall cutoff, and fault plane geometries is possible where good-quality three-dimensional (3-D) seismic data

FIGURE 1. (A) Structural elements of the Vulcan Subbasin. (B) Schematic cross section through the Rowan, Skua, and Swift fault blocks (modified from Fittall and Cowley, 1992).

exist. In this case, 3-D structural restoration of a fault by matching footwall and hanging-wall cutoffs can be used to establish fault displacements and, hence, slip directions, if variations in the strike of a fault occur.

In this chapter, we propose a new methodology that combines previously described 3-D structural restoration (used to verify fault and horizon interpretation) and fault slip inversion techniques (used for estimating paleostress conditions from outcrop data) to estimate paleostress conditions from 3-D seismic data. The combined methodology is tried for the first time on data from the Skua oil field, southern Timor Sea (Figure 1). The primary aim is to describe this paleostress estimation procedure and secondarily to take a first look at the validity and implications of the results of this pilot study by making comparisons with regional structural observations and hydrocarbon leakage patterns. The work is presented with the acceptance that further testing of the methods is required, particularly in relation to the quantification of errors.

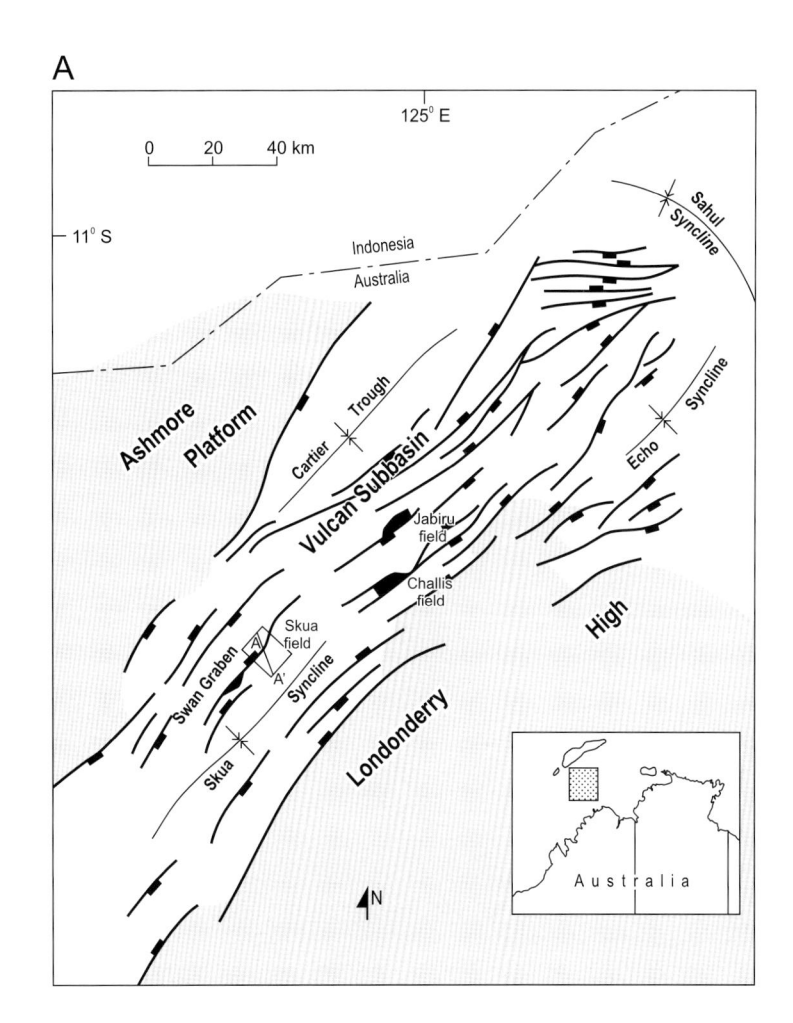

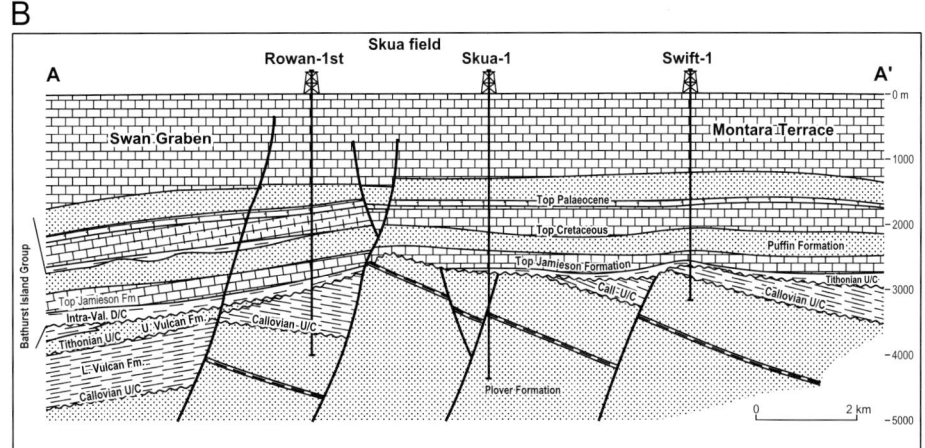

PALAEOSTRESS RECONSTRUCTION USING FAULT DATA

Newly formed (neoformed) conjugate faults that developed in a single stress regime in an isotropic medium theoretically show predictable geometric relationships with the principal stress axes (Anderson, 1951).

σ_1 bisects the acute angle between conjugate faults; σ_2 corresponds to the intersection direction of the faults; and σ_3 bisects the obtuse angle between the faults. Additionally, one of the principal stress axes is commonly vertical because of the effects of gravity and the inability of the Earth's free surface to maintain shear stresses. Therefore, as an example, where a pair of conjugate normal faults are observed and an Andersonian tress system is assumed, a paleostress regime can be determined in which σ_1 was vertical, σ_3 was perpendicular to the strike of the fault, and σ_2 was parallel to the strike of the fault. However, more complicated systems of neoformed faults are possible, including perpendicular and orthorhombic fault systems (e.g., Reches, 1983), for which these simple geometric relationships mentioned above do not hold.

Simple geometric analysis is also not applicable to reactivation of preexisting faults (and other weakness surfaces), which may have any orientation with respect to subsequent stress fields. In these more complicated cases, one approach to obtaining paleostress information is to make use of the hypothesis that slip on a fault plane occurs in the direction of the maximum resolved shear stress (Figure 2A). This hypothesis, proposed by Wallace (1951) and followed by the work of Bott (1959), allows the slip direction on a fault plane to be predicted for any given stress tensor.

The inverse problem involves obtaining the mean stress tensor from the orientations and senses of displacement of numerous fault slips. Several numerical and graphical methods have been developed to search for the reduced stress tensor that best fits the measured fault slip data. The reader is directed to Angelier (1994), wherein detailed coverage of the theory, mathematics, and the development history of using fault slip data to determine paleostress conditions are provided. In summary, inversion of fault slip data can be used to derive a reduced paleostress tensor that describes the orientation of the principal stresses and the stress ratio, $R = (\sigma_2 - \sigma_3)/(\sigma_1 - \sigma_3)$. The full stress tensor (as opposed to the reduced stress tensor) includes the magnitudes of the principal stresses. The value R defines the shape of the stress ellipsoid and can be used to characterize the paleostress regime (e.g., extensional, compressional, strike-slip, transtensional, and transpressional) and to estimate paleostress magnitudes (see below).

A computer program (Delvaux, 1993) was used to perform the inversion required for this study. The procedure optimizes (based on a least squares best fit) the four variables of the reduced stress tensor (σ_1, σ_2, σ_3, R) by successive rotation of a test tensor around each of its axes and by testing different values of R (Figure 2B). The amplitude of the angular misfit between calculated and measured slips are progressively reduced until the tensor is stabilized (Delvaux et al., 1995).

Another aspect of the fault slip inversion technique that we investigate is its use for estimating the full stress tensor, which includes the stress magnitude. A dimensionless Mohr diagram can be constructed to illustrate the relationship between the principal stresses once the stress ratio is determined from the inversion process. Additional knowledge, such as friction laws, failure criteria, and the depth of overburden, allow the position and size of the Mohr circle to be fixed to reconstruct the complete stress tensor.

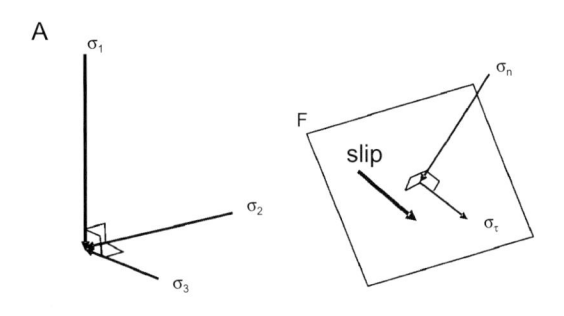

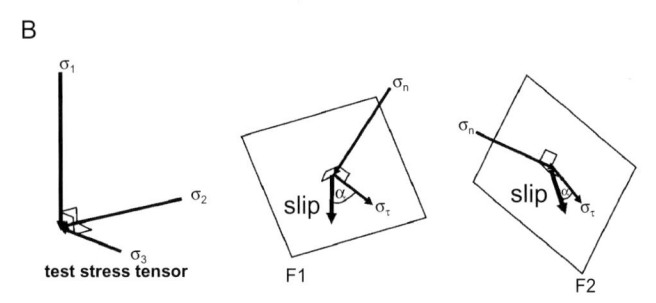

FIGURE 2. Relationship between stress state and slip on a preexisting fault. (A) Preexisting weakness plane F activated as a fault in a given stress state. The orientation of the stress vector σ depends on both the attitude of the weakness plane F and the principal stresses. The stress vector can be thought of as two perpendicular stress components, the normal stress σ_n, and the shear stress σ_τ. According to the Wallace–Bott hypothesis, the direction of slip of the fault should be parallel to σ_τ. (B) Application of a test stress tensor to two preexisting fault planes. The procedure used minimizes the angular misfit (α) between measured and predicted slip vectors for the fault data set. Note that at least four separate items of fault slip data are required to obtain the reduced stress tensor.

FAULT SLIP DATA FROM 3-D RESTORATION OF FAULTS

We follow a two-step procedure similar to that described and tested by Rouby et al. (2000) to restore faulted surfaces (Figure 3A–C). The first step is the unfolding stage, which is designed to remove the effects of ductile deformation and bed rotations, such as that associated with roll-over, drag, fault tips, and relays. Both the footwall and hanging-wall surfaces are restored to a specified level using a given unfolding mechanism (Figure 3A, B). Rouby et al. (2000) tested a series of unfolding mechanisms, including flexural slip, homogeneous inclined shear, and heterogeneous inclined shear. They found that for the data set used in their study, a heterogeneous inclined shear mechanism gave the most favorable restoration result. However, these authors also

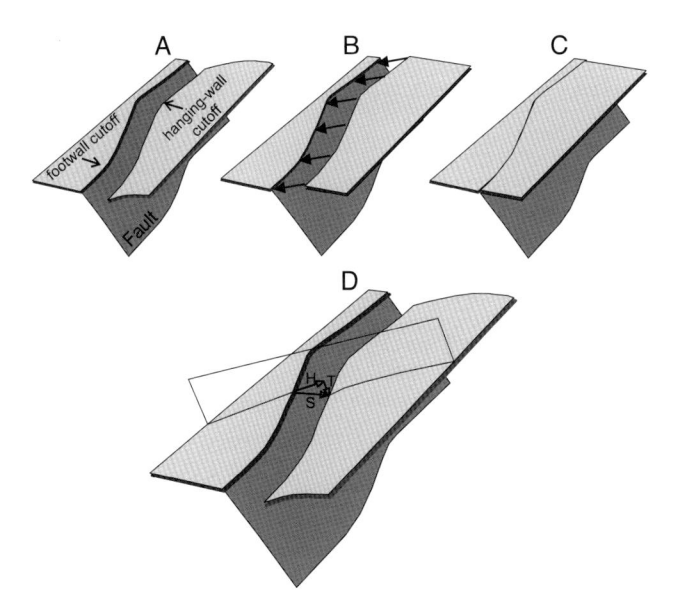

FIGURE 3. Fault slip direction from 3-D restoration. (A) A fault with footwall and hanging-wall cutoffs defined for a given horizon. (B) Horizon flattened using flexural slip unfolding algorithm. (C) Footwall and hanging-wall cutoffs fitted to obtain the best fit heave vector. (D) Slip vector obtained by calculating the intersection between the fault surface and a vertical plane containing the best fit heave vector.

suggest that the choice of unfolding mechanism makes only a small difference to the derived 3-D displacement field for extensional terranes because of the relatively small amount of deformation accommodated by folding compared to that accommodated by faulting. We have used a flexural slip unfolding mechanism for the purposes of this study.

The second part of the restoration procedure involves rigid body translation, which restores the remaining horizontal separation between the footwall and hanging-wall surfaces. Fault gaps are close to a direction that minimizes residual fault gaps and overlaps to obtain the best fit heave direction (Figure 3B, C). This method assumes rigid body motion of the fault blocks (after unfolding), an assumption that is well demonstrated by goodness of fit in several case studies (Rouby et al., 1996). The fault gaps can be closed manually ("by eye") or using an automated block-fitting procedure incorporated in the restoration software (3DMove®). The automated procedure employs simulated annealing, a Monte Carlo approach for minimizing multivariate functions, resulting in a statistical minimum area of misfit between fault blocks. All faults in this study were closed using the automated procedure.

From here, the fault slip at a given location is obtained by determining the pitch of a vertical plane containing the heave direction in the fault plane (Figure 3D). The slip direction data can then be combined with the fault plane information at a given location as input for the inversion approach discussed above. Therefore, if the timing of movement on the fault can be determined, it is possible to estimate the mean paleostress field for that time.

LIMITATIONS OF THE APPROACH

Several limitations, assumptions, and sources of error arise using the paleostress estimation methods described because of the nature of the basic data set (seismic) and the fact that both the structural restoration and inversion techniques are approximations of real geological processes. Our approach to this point of the study has been to minimize these errors where possible and then compare the outcomes of the various procedures with geological observations as a basic test. However, future work will attempt to develop strategies to quantify these errors (see the discussion section above).

The accuracy of the interpreted geometry of the fault surface and horizon cutoffs is vital for the quality of the restoration and subsequent paleostress inversion results. In this study, we have attempted to minimize interpretational errors by carefully picking each fault of interest on every seismic inline (25-m [82-ft] spacing) in the 3-D seismic volume with the aid of a 3-D visualization software. A variance cube (with a 64-ms calculation window and a 3 × 3 neighbor definition) was calculated and used to supplement interpretation of the fault geometries from the 3-D migrated seismic data cube. Horizon cutoffs were then generated by either extrapolating the horizon to the fault or removing sections of the horizon that pass through the fault.

A key assumption for the structural restoration part of the procedure is that deformation associated with a fault can be restored using a two-step process of unfolding followed by rigid block translation. Some error will be present in this approach, because faults do not move in two steps. In addition, the use of the flexural slip mechanism for unfolding will inevitably lead to some errors during restoration, particularly in areas such as fault tips and relay ramps where ductile deformation will occur, and fold axes may be noncylindrical and/or nonparallel to other fold axes on the surface. However, considering the relatively small folding component associated with the extensional system described, we believe such errors will be small. Further problems occur at fault tips during the block-fitting stage of the procedure, where the separation between fault blocks decreases to zero. To counter this problem, we remove the tips of the faults prior to block fitting.

The restoration method also assumes that the current-day separation of footwall and hanging-wall cutoffs along the fault occurred in one slip direction and is not the result of the addition of a series of nonparallel movements. The problem raised by this assumption might be solved by a more detailed fault analysis study across a wider area. This may help to identify multiple slip events, and structural restoration could then be applied to investigate slip directions for each identified event.

A key assumption for the stress inversion technique is that faulting occurred in a homogeneous stress field, and that faults in the data set have moved independently of each other. The faults studied here are in close proximity to each other, and therefore, it is possible that the faults have been influenced by local stress perturbations associated with nearby faults. We attempt to test this effect by comparing the results of the inversion with regional observations. In addition, future work on other seismic data sets from the region should help to clarify whether these effects are significant. Slip displacements at fault jogs may not necessarily be in response to regional stresses but instead may act to maintain slip compatibility along the greater fault plane. For this reason, fault slip measurements are not taken from discrete jogs in the faults.

EXPLORATION AND GEOLOGICAL SETTING OF THE SKUA FIELD

Located at the southern end of the Timor Sea between the Londonderry High and the Ashmore Platform, the Vulcan Subbasin has a complex architecture comprising southwest–northeast and east-northeast–west-southwest-trending, asymmetrical, en echelon subgrabens flanked by intragraben horsts and terraces (Figure 1A) (Pattillo and Nicholls, 1990). The Skua field is situated in one of a series of Late Jurassic tilted fault blocks on the southeastern flank of the Swan Graben (Figure 1B) (Osborne, 1990; Pattillo and Nicholls, 1990; Woods, 1994).

Hydrocarbons in the Skua field are reservoired in Early Jurassic sandstones of the Plover Formation. Top and lateral seal is provided by Early Cretaceous calcareous shales of the Bathurst Island Group, which were deposited during thermal sag and passive-margin development following continental breakup. As with many traps in the region, a combination of factors, including underfilling of the structure, evidence for fault reactivation, the presence of a paleo-oil column, and the presence and location of direct leakage indicators, such as hydrocarbon-related diagenitic zones (HRDZs) and water column seepage anomalies, provides strong evidence of seal breach because of fault reactivation (O'Brien and Woods, 1995; O'Brien et al., 1996; Gartrell et al., 2002). The field is covered by the HV11 3-D seismic survey, which provides full-fold seismic coverage of the area, with a 25-m (82-ft) line spacing, a 12.5-m (41-ft) trace spacing, and data recorded to 5 s (two-way traveltime) at a 4-ms sample interval (Fittall and Cowley, 1992). The quality of the 3-D seismic data is generally good, particularly in the postrift sequence. Notable exceptions include significant degradation of the seismic signal because of strong velocity variations associated with the HRDZs located above parts of the Skua field itself. Well control for the study was provided by Skua wells 1–9 (Skua 1-Arco Australian Ltd.; Skua 2–9-BHP Petroleum) and the nearby Rowan 1-BHP Petroleum and Swift 1-BHP Petroleum wells (Figure 4).

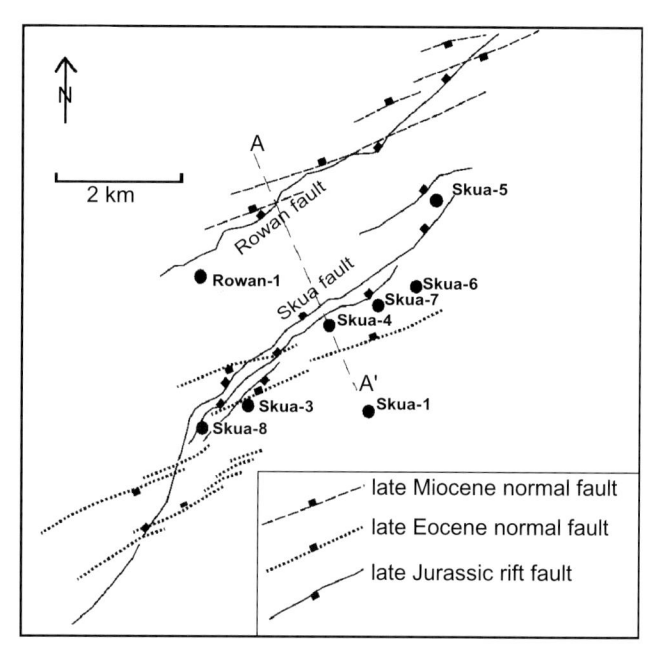

FIGURE 4. Postrift fault pattern above the Rowan and Skua fault blocks. Postrift faults developed above the Rowan fault were active in the late Miocene. The age of postrift faulting above the Skua fault block predates the late Miocene.

Postrift deformation events during the Late Cretaceous, early Tertiary, and late Tertiary have been recognized in the vicinity of the Skua field (Gartrell et al., 2002). Late Tertiary deformation, which is of interest for this study, was accommodated primarily on the Rowan fault located about 3 km (1.8 mi) to the west of the Skua field (Figures 4–6). Postrift faults formed above the Rowan fault clearly terminate upward, slightly above the late Miocene seismic horizon, and subtle sedimentary growth across the faults indicates that reactivation of the Rowan fault occurred at about the late Miocene (Figure 5). However, it is difficult to determine a precise age for this fault activity because of poor age differentiation available in the upper sections of the wells drilled in the area. In contrast, postrift faults located above the Skua fault block formed during the

FIGURE 5. Seismic section (cosine of a phase attribute display) from the northern end of the Skua field. The postrift fault that developed above the Rowan fault crosscuts the late Miocene seismic reflector, indicating late Miocene fault activity. Location of seismic section is indicated as AA' in Figure 4.

late Eocene to early Miocene, as indicated by both subtle sedimentary growth and termination levels, prior to the collisional events at Timor (Figure 5) (Gartrell et al., 2002).

A direct continuation of the Late Jurassic segment of the Rowan fault displaces the Early Cretaceous (base seal) seismic horizon, which was deposited after rift-phase extension ceased (Figure 5). Considerable variation in strike direction is observable along the length of the Rowan fault at this level (Figure 4). A set of left-stepping, en echelon normal faults are observed in the postrift sequence above the Rowan fault (Figures 4–6). In some areas, these faults appear to link downward with the Rowan fault directly, whereas in other areas, the postrift faults detach in the Early Cretaceous shales above the Rowan fault. It is assumed that the displacement observed at the base seal level is associated with the late Miocene movement observed higher in the sedimentary sequence.

FAULT SLIP ANALYSIS OF THE ROWAN FAULT AREA

Displacement on the Rowan fault at the base seal level after backstripping, restoration to late Miocene, and unfolding is shown in Figure 7a. Automatic block fitting of the footwall and hanging-wall cutoffs indicates that the optimal heave direction for restoring the fault at this level trends 175–355° (Figure 7b). Orientations of the fault plane were then measured at eight locations along the length of the fault to characterize slip directions on the fault as described earlier (Table 1).

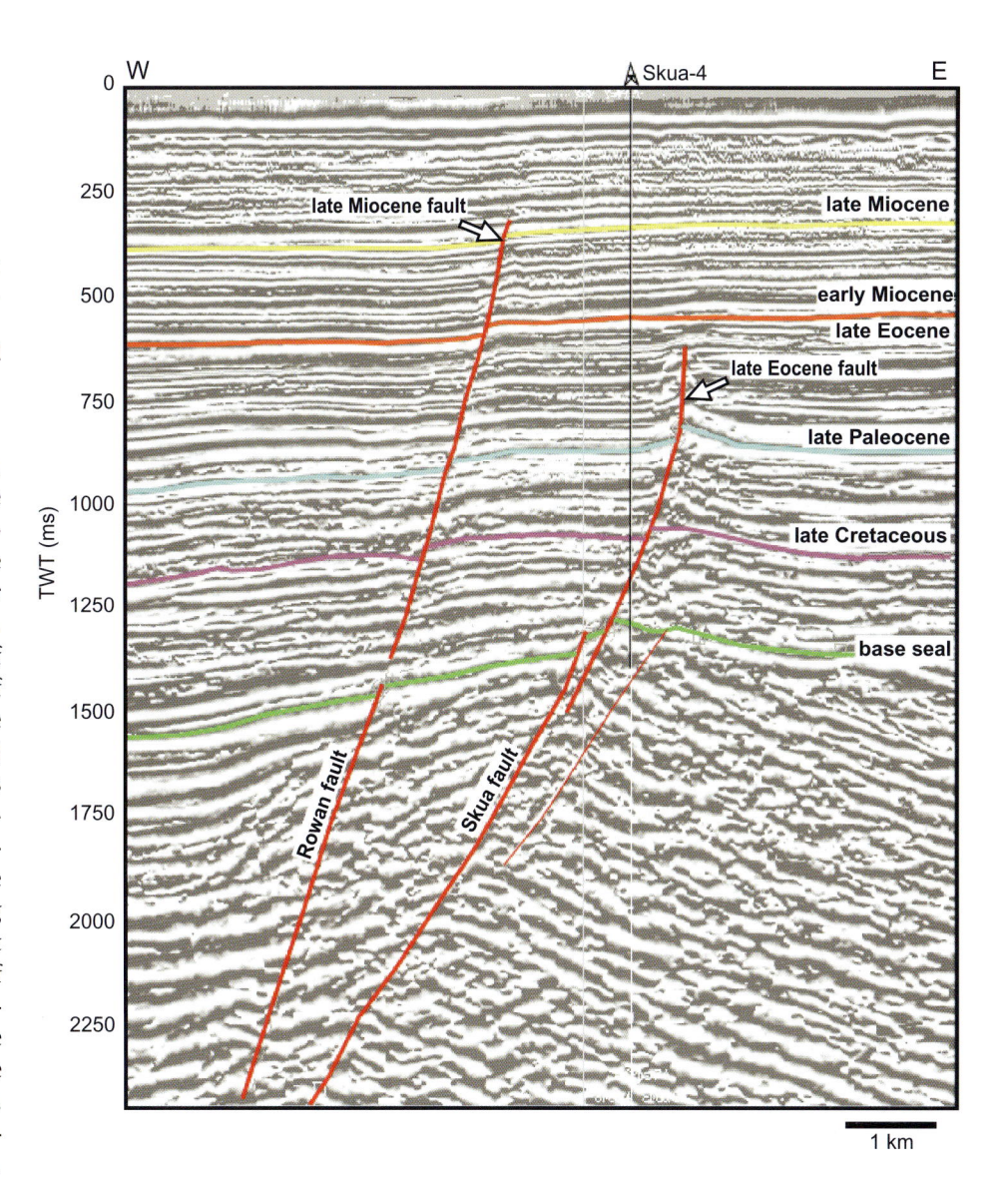

The same process was applied to five of the overlying postrift faults at the early Miocene level. Automatic block fitting on these faults indicates that oblique slip also occurred on these faults, with their transport direction oriented close to north–south (Figure 7c-l). In total, 21 fault slip data pairs were collected from 6 faults and used for the inversion (Table 1).

Inversion of Fault Slip Data

The procedure of inversion that we used (Delvaux, 1993) indicates that a pure, extensive ($R = 0.28$) stress regime, in which σ_1 was subvertical (78:128°) and σ_2 and σ_3 were both subhorizontal, trending approximately east–west (9:264°), and almost north–south (9:355°), respectively, best fits the late Miocene fault slip data obtained in the vicinity of the Rowan fault. An average angular misfit error of 5° was calculated (Figure 8).

A Mohr diagram was used to estimate the magnitudes of the principal stresses during the late Miocene

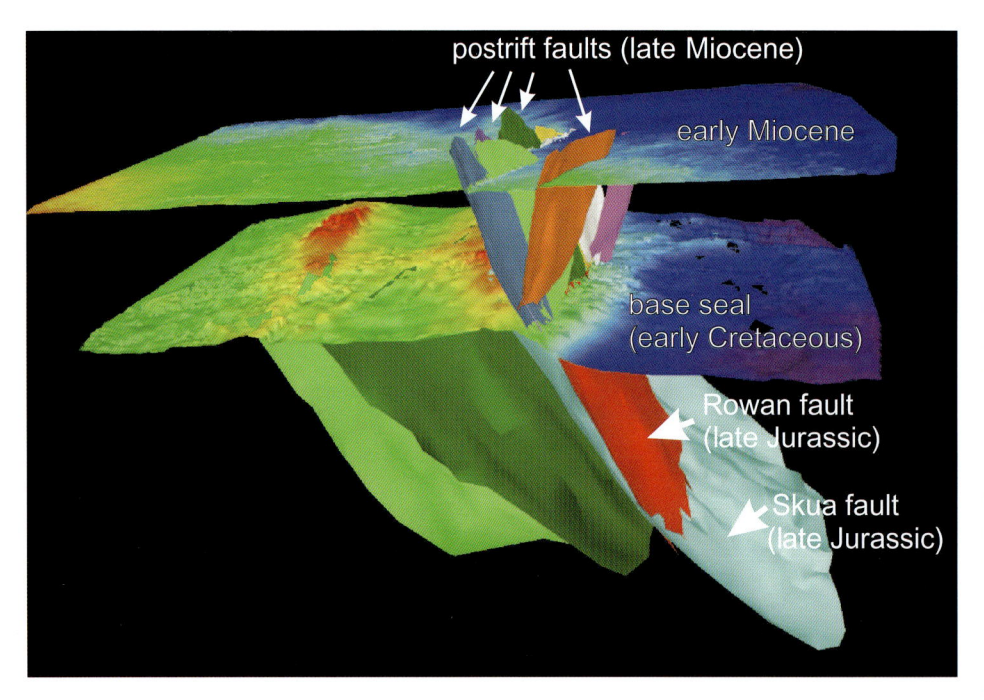

30 Gartrell and Lisk

FIGURE 6. Three-dimensional visualization showing the relationship between the Rowan fault and associated postrift faults.

reactivation event (Figure 9). The vertical stress, σ_v, was estimated at locations on the Rowan fault at the base seal level after it was restored to the late Miocene. The relationship determined in Mildren et al. (2002) for calculating vertical stress at a given depth yields σ_v ~ 47 MPa for a typical depth (to the top of the fault) of about 2200 m (7200 ft). Assuming a pore pressure of 22 MPa (i.e., hydrostatic pressure gradient) at this depth, the effective vertical stress can be estimated at about 25 MPa ($\sigma_v - P_p$). This value is used to constrain the position of σ_1, which the inversion shows is equivalent to σ_v, on the horizontal axis of the Mohr diagram (Figure 9B). An angle of internal friction of 20°, based on the spread of the slip data, and a cohesion value of 5 MPa were adopted to construct a failure envelope (Figure 9C). Streit (1999) suggests that cohesive strength ranges between 4 and 14 MPa for lithified gouges and cataclasites in general, and previously, Mildren et al. (2002) assumed a fault rock cohesive strength of 5 MPa for the Timor Sea. This value was used here to allow a direct comparison between the paleostress estimates and the previous work by Mildren et al. (2002) on the present-day stress system based on borehole data. The size of the Mohr diagram was

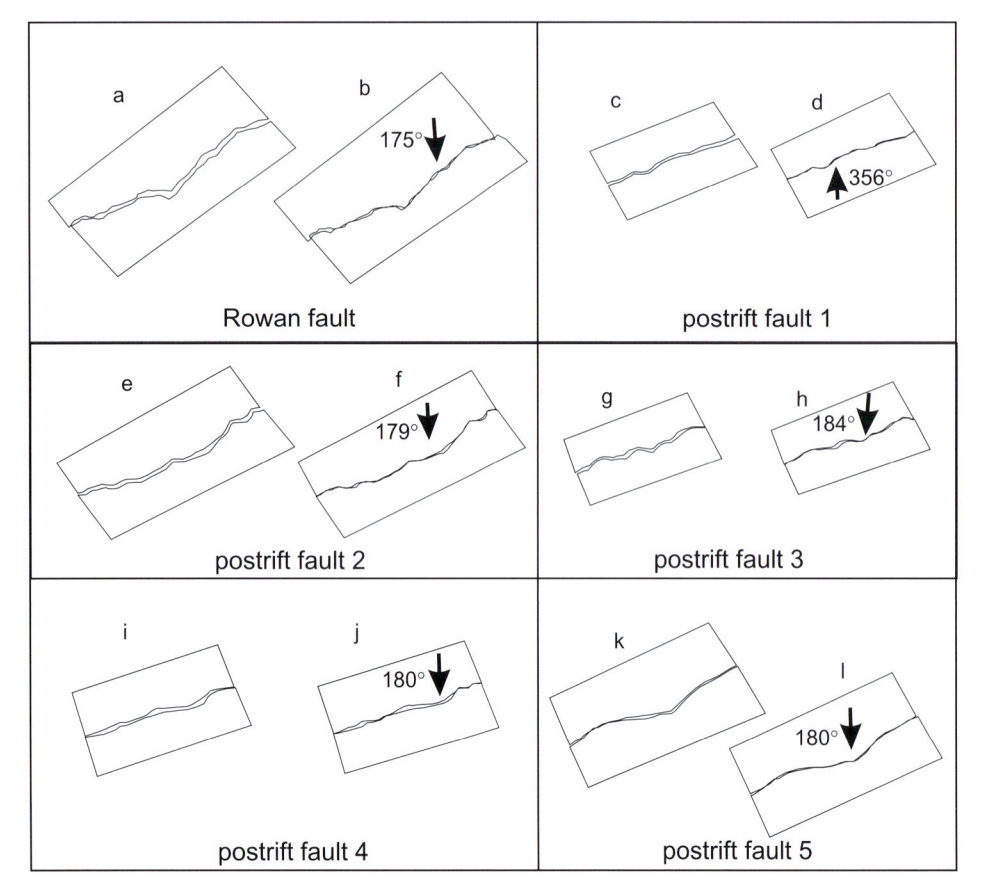

FIGURE 7. Matching footwall and hanging-wall cutoffs from late Miocene faults after restoration to the late Miocene and flexural slip unfolding of the horizon. (a) Fault gap at Early Cretaceous (base seal) level across the Rowan fault prior to block fitting. (b) Fault gap after block fitting with a best fit heave direction toward 175°. (c–l) Block-fitting results for postrift faults 1–5 at the early Miocene level. Angle indicates best fit heave direction. Arrow shown on the hanging-wall block.

TABLE 1. Fault slip data used for the tensor inversion. Orientation data given in dip and dip direction. PRF1–5 = postrift faults 1–5; RF = Rowan fault; ND = normal dextral slip; Neo. = neoformed fault; React. = reactivated fault.

Data Point	Fault Identification	Fault Plane (Dip and Dip Direction)	Slip (Dip and Dip Direction)	Sense	Type
1	PRF1	59:154	57:176	ND	Neo.
2	PRF1	63:166	63:176	ND	Neo.
3	PRF3	65:342	63:003	ND	Neo.
4	PRF3	53:342	51:004	ND	Neo.
5	PRF3	61:344	59:004	ND	Neo.
6	PRF2	74:312	69:355	ND	Neo.
7	PRF2	69:326	65:359	ND	Neo.
8	PRF2	65:332	62:358	ND	Neo.
9	PRF4	72:343	71:359	ND	Neo.
10	PRF4	68:343	67:359	ND	Neo.
11	PRF4	70:339	69:001	ND	Neo.
12	PRF5	68:323	63:359	ND	Neo.
13	PRF5	67:333	65:360	ND	Neo.
14	RF	56:312	47:355	ND	React.
15	RF	51:321	46:355	ND	React.
16	RF	48:327	44:355	ND	React.
17	RF	47:333	45:355	ND	React.
18	RF	44:326	*40:355*	ND	React.
19	RF	58:323	54:355	ND	React.
20	RF	41:309	31:355	ND	React.
21	RF	50:321	44:355	ND	React.

then adjusted until all measured slip planes fall above the failure line, as all the planes plotted on the Mohr diagram (black dots) represent failure planes (Figure 9C). In this way, values of 8 and 1 MPa are estimated for σ_2 and σ_3, respectively.

DISCUSSION

Comparison with Contemporary Stress Regime and Regional Reactivation History

Bearing in mind that this is only a pilot study prior to a more rigorous investigation whereby the limitations of the method will be further tested, we tentatively put forward the following comparison between the interpreted contemporary stress regime and our computed paleostress regime. Contemporary stress measurements in the Timor Sea region, based on borehole information, generally indicate strike-slip or mixed strike-slip and extensional (transtensional) mode stress regimes (depending on locality) in which σ_2 is vertical and σ_1 is horizontal and oriented approximately 55–65° (Hillis,

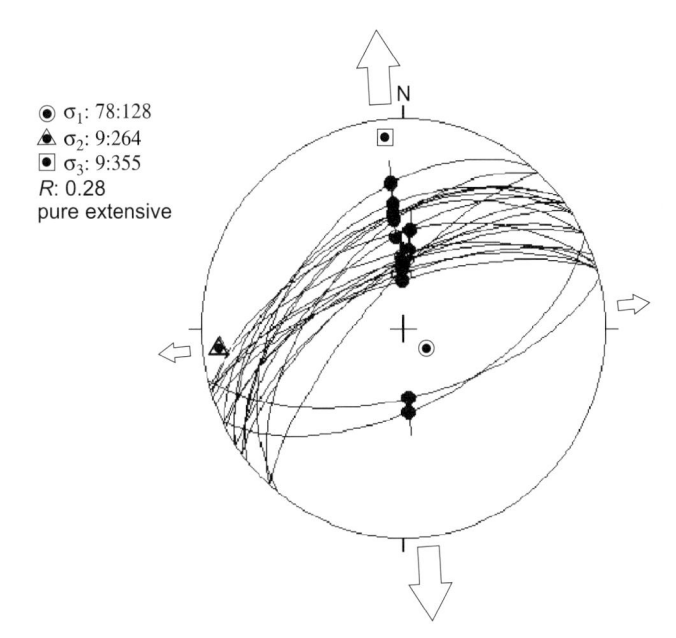

● σ_1: 78:128
▲ σ_2: 9:264
■ σ_3: 9:355
R: 0.28
pure extensive

FIGURE 8. Results of the fault slip inversion plotted on a stereonet (Schmidt lower hemisphere projection). Fault plane and slip lines used for the inversion are listed in Table 1.

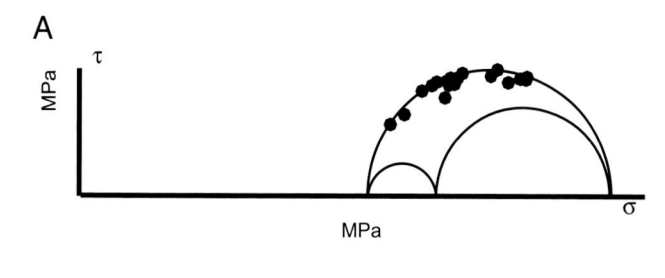

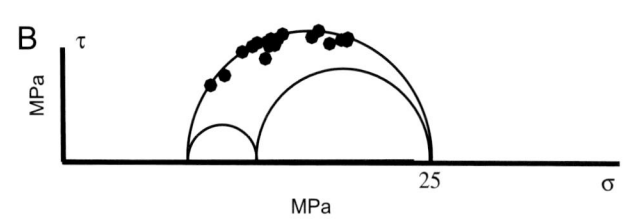

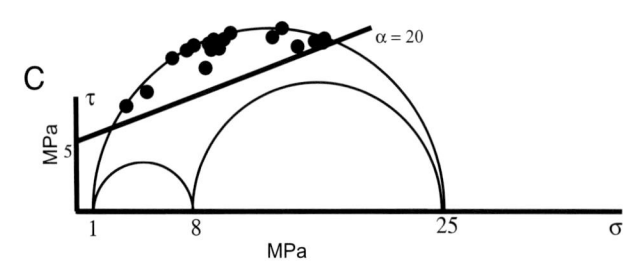

FIGURE 9. Mohr diagram construction used to estimate the magnitude of the principal stresses. (A) Dimensionless Mohr diagram obtained from the inversion. The stress ratio determines the relative size of the Mohr circles. Black dots represent the orientations of activated fault planes. (B) The position of σ_1 on the horizontal axis can be determined by calculating σ_v (i.e., the inversion suggests $\sigma_1 = \sigma_v$). (C) The size of the Mohr diagram is increased until all failure planes lie above the assumed failure envelope. Note that the ratio between the principle stresses remains constant. This allows σ_2 and σ_3 to be estimated. Effective stress values are used and assume a normal pore-pressure gradient.

1998; Castillo et al., 2000). In the vicinity of the Rowan fault, the measurements of the contemporary stress tensor indicate that σ_1 is horizontal (i.e., $\sigma_1 = \sigma_{h_{max}}$) and oriented 67°, σ_2 is vertical, σ_3 is horizontal and oriented 327°, and $R = 0.82$ (transtensional). Numerical plate boundary modeling indicates that the present-day regional east-northeast–west-southwest orientation of the maximum horizontal stress ($\sigma_{h_{max}} = \sigma_1$ at present) in the Timor Sea is primarily related to stress focusing at the continent-continent collision occurring around New Guinea (Coblentz et al., 1998).

In contrast to the present transtensional regime, an extensional stress regime in which σ_1 was vertical and $R = 0.28$ was determined for the late Miocene by the paleostress inversion described above. The orientations of the late Miocene horizontal stress axes differ from those of today by an angle of about 20°, with the paleo-$\sigma_{h_{max}}$ (which is σ_2 for the late Miocene) oriented nearly east–west and paleo-$\sigma_{h_{min}}$ oriented nearly north–south.

It is interesting to note that the orientation of the late Miocene stress field determined for the Skua area is similar to the present-day stress state observed further south on the North West Shelf, where $\sigma_{h_{max}}$ is oriented approximately east–west. This may imply that the stress-focusing effect caused by the collision of Australia with New Guinea was either absent or weak in the late Miocene and has progressively increased to the present day. Plate reconstructions by Hall (1996) indicate predominantly transcurrent displacement along the plate boundary at New Guinea during the late Miocene, which would have probably resulted in relatively low horizontal stresses being directed into the Australian plate.

Comparison between the Mohr circles describing the contemporary stress and paleostress tensors indicates that the differential stress (i.e., $\sigma_1 - \sigma_3$) operating at both times may be similar (Figure 10). However, the Mohr circle for the present-day stress tensor is shifted to the right, away from the failure envelope, because of higher magnitudes for all the principal stresses (i.e., higher mean stress). The indicated decrease in the likelihood of fault reactivation because of an increased mean stress is consistent with the decrease and/or termination of fault activity since the late Pliocene (~2.5 Ma), as observed in seismic data. For example, Meyer et al. (2002) observed increasing late Tertiary fault mortalities from about 5 Ma. These authors suggested that this phenomenon was primarily associated with progressive localization of strain onto larger faults but also identified an associated decrease in extension rate at about 2–3 Ma. One explanation for this change could be that the extensional stress regime suggested by the paleostress work was more favorable for reactivation of the Jurassic rift faults, and that the change to the present-day transtensional stress regime resulted in less reactivation and, consequently, less extension.

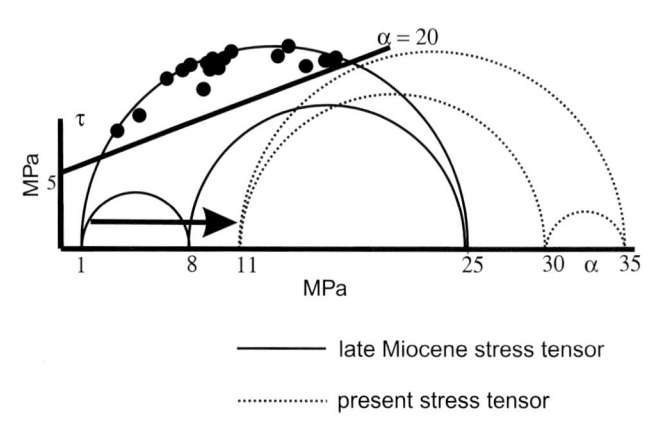

FIGURE 10. Comparison between late Miocene and present-day stress tensors.

Much of the late Tertiary faulting observed in the region is associated with reactivation of buried Jurassic rift faults, commonly resulting in characteristic hourglass structures (Woods, 1988; Nicol et al., 1995). The development of these structures is possible in a stress regime in which σ_1 is horizontal (i.e., present-day configuration) because of the existence of a preexisting weakness plane. However, examples of significant late Tertiary extensional faults that are not apparently related to older faults are also observed. The development of these neoformed normal faults is difficult to explain when σ_1 is horizontal. However, the occurrence of an extensional stress regime (i.e., σ_1 vertical) sometime during the late Tertiary is consistent with both reactivated and neoformed extensional fault systems.

Implications for Trap Integrity and Hydrocarbon Leakage in the Timor Sea

The discussion above suggests a system in which most fault reactivation occurred early on in the late Tertiary deformation history of the region and then progressively decreased with time to the present day. It follows that the leakage of hydrocarbons, because of fault reactivation, would also be likely to decrease to the present. In other words, the trap integrity in the region should have been improving with time.

This notion is consistent with comments by O'Brien et al. (1996, 1998, 1999), who suggested that most traps in the Vulcan Subbasin were breached at about 5.5 Ma in association with the passage of hot brines up reactivated fault zones. These authors present evidence from fission-track and fluid-inclusion data to deduce that this key fluid-flow event was relatively short-lived, lasting between 10,000 and 1,000,000 yr. Hence, most hydrocarbon leakage from traps in the region probably occurred several million years before the present day. Subsequent leakage is suggested to be relatively minor, with traps partially or fully resealing. For example, fluid-inclusion analysis of the Keeling structure indicates that it was filled with oil prior to late Tertiary fault reactivation. Subsequently, the trap has apparently resealed and is now filled with late-stage gas (O'Brien et al., 1996). Present-day fault-related seepage detected in the region probably represents volumetrically insignificant leakage, at least in comparison to probable late Miocene rates of leakage (O'Brien et al., 1999). Instead, the seepage anomalies are probably related to hydrocarbons that are currently being generated and migrating through faults to the surface or show minor leakage from charged traps (O'Brien et al., 1998).

These arguments raise the profile of a model in which most trap breach in the Timor Sea occurred during the late Miocene to late Pliocene, and therefore, stress-based assessments (e.g., Fault Analysis Seal Technology, Coulomb Failure Function) of the likelihood of trap failure caused by fault reactivation should incorporate the paleostress state. However, as the system apparently becomes more sealing toward the present day, traps that have remained on a migration pathway may have received late hydrocarbon recharge. The criticality of trap-bounding faults with respect to the contemporary stress regime may also be a factor in this case.

Other factors that may also impact on hydrocarbon leakage, such as dilation zones (e.g., fault intersections, jogs, and tips), strain localization processes, seal rheology, and diagenesis need to be considered in addition to stress conditions. For example, Gartrell et al. (2002, 2003) suggest that a dilational fault intersection was the critical control on leakage from the Skua field. This structure formed in a downdip location prior to or synchronous with oil charge and continued to act as a passive leak zone that was essentially independent of the stress field and late Tertiary fault reactivation. Preferential strain localization on the Rowan fault during late Tertiary deformation may have acted to protect the Skua field from further deterioration of the trap's integrity (Gartrell et al., 2002). These observations, however, are not inconsistent with higher order controls, such as the active stress regime, that may have an influence on the trap integrity on a more regional level. Indeed, we believe that for the interpreted rheology of the seal rocks, it is reasonable to suggest that if the regional stress state results in a greater tendency for fault reactivation to occur, then the trap integrity in the region is likely to be reduced in general.

Future Work

As stated above, several limitations exist in this work that may compromise the accuracy of the combined techniques and, hence, the validity of the results. A range of strategies are planned to further test the technique and quantify the errors associated with its use. For example, the methodology will be applied to additional seismic data sets from the Timor Sea with greater image quality and a wider range of fault orientations (e.g., Laminaria high). This should give a stronger indication of whether the results are consistent in the regional framework.

More rigorous testing and fine tuning of the methodology may be achieved by comparing the results of the processes involved with known fault geometries that have moved within known applied stress fields. Analog and/or numerical modeling techniques will be employed for this purpose. Errors associated with such detailed fault interpretation from seismic data sets, although somewhat subjective, also need to be considered in greater detail.

CONCLUSION

Our testing of the methodology described in this chapter is at an early stage, and several sources of error related to the quality of the data available, seismic interpretation, restoration techniques, and inversion methodology exist. Notwithstanding this, we believe that without such an approach, paleostress estimations based on simple geometric analysis (such as fault orientations) are prone to greater errors that are less quantifiable, and the opportunities for overlooking important subtleties in the structural history are high, particularly in reactivated settings. Further development and testing of the paleostress estimation method is underway.

The paleostress work described here is based on only a few faults confined to one area, and the faults used to obtain slip information for the paleostress inversion are located in close proximity to each other. A larger number of faults with various fault orientations (e.g., Laminaria High) are to be analyzed in future work to better constrain the late Tertiary to present stress history of the region. However, regional observations suggest that the initial results from the Skua area, with the limitations discussed in this chapter in mind, may be applicable to the Timor Sea region in general, and we tentatively put forward comments on the possible implications of the paleostress results.

Regional observations suggest that stress changes in the Timor Sea region may have been likely during the complex collisional evolution of Australia's northern margin in the late Tertiary. These observations are corroborated by our initial paleostress analysis for late Miocene fault activity in the Skua area of the southern Vulcan Subbasin using fault slip inversion methods. Three-dimensional restoration and fault slip inversion of the Rowan fault and associated postrift faults indicate that an extensional stress regime, with σ_1 vertically oriented, σ_2 oriented almost east–west (84–264°), σ_3 oriented almost north–south (175–355°), and a stress ratio of 0.28, existed in the late Miocene. The estimated late Miocene extensional stress regime differs significantly from the present-day transtensional stress regime in the area, in which σ_1 is horizontal and oriented east-northeast–west-southwest (67–247°), σ_2 is vertical, σ_3 is oriented north-northwest–south-southeast (147–327°), and the stress ratio is about 0.8.

The change from the late Miocene to the present-day stress regime may have been caused by plate boundary forces generated at evolving collisional fronts in New Guinea and Timor. An evolving stress system may help to explain temporal and spatial variations in late Tertiary fault activity observed in the Timor Sea region, particularly the decrease in fault activity over the last 2–3 Ma. The calculated extensional paleostress regime is likely to have been more favorable for reactivating the Jurassic rifting-related faults than the present-day transtensional stress regime. This model suggests that hydrocarbon leakage associated with fault reactivation is less likely to be associated with the present-day stress regime than the paleostress regime.

When combined with charge histories, the variation in stress regimes indicated by this early work appears to reflect a new layer of complexity to fault reactivation and associated hydrocarbon leakage histories in the Timor Sea. Proper consideration of this additional complexity may be necessary to obtain more accurate predictions of trap integrity scenarios and risks in the region.

ACKNOWLEDGMENTS

This work was funded through the Australian Petroleum Collaborative Research Centre Seals Program, and company sponsors to this program are gratefully acknowledged. Structural restoration software (2DMove® and 3DMove) was generously provided by Midland Valley Exploration Ltd. Our gratitude also goes to Schlumberger Oilfield Australia Pty Ltd for the use of GeoFrame™ software, which was used exclusively for interpretation and depth conversion of seismic data in this project. The computer program TENSOR® (Delvaux, 1993) was used to perform the fault slip inversion calculations. This chapter has benefited significantly from reviews and/or comments by John Walsh, Geoff O'Brien, Quentin Fisher, and Peter Boult.

REFERENCES CITED

Anderson, E. M., 1951, The dynamics of faulting: Edinburgh, Oliver and Boyd, 206 p.

Angelier, J., 1990, Inversion of field data in fault tectonics to obtain the regional stress: III. A new rapid direct inversion method by analytical means: Geophysics Journal International, v. 103, p. 363–376.

Angelier, J., 1994, Fault slip analysis and paleostress reconstruction, in P. L. Hancock, ed., Continental deformation: Oxford, Pergamon Press, p. 101–120.

Bott, M. H. P., 1959, The mechanisms of oblique slip faulting: Geological Magazine, v. 96, p. 109–117.

Bradley, D. C., and W. S. F. Kidd, 1991, Flexural extension of the upper continental crust in collisional foredeeps: Geological Society of America Bulletin, v. 59, p. 1416–1438.

Castillo, D. A., D. J. Bishop, I. Donaldson, D. Kuek, M. de Ruig, M. Trupp, and M. W. Shuster, 2000, Trap integrity in the Laminaria high–Nancar trough region, Timor Sea: Prediction of fault seal failure using well-constrained stress tensors and fault surfaces interpreted from 3-D seismic: Australian Petroleum Production and Exploration Association Journal, v. 40, p. 151–173.

Charlton, T. R., A. J. Barber, and S. T. Barkham, 1991, The structural evolution of the Timor collision complex, eastern Indonesia: Journal of Structural Geology, v. 13, p. 489–500.

Coblentz, D. D., S. Zhou, R. R. Hillis, R. M. Richardson, and M. Sandiford, 1998, Topography, boundary forces, and the Indo-Australian intraplate stress field: Journal of Geophysical Research, v. 103, p. 919–931.

Delvaux, D., 1993, The TENSOR® program for reconstruction: Examples from the east African and Baikal rift systems: Terra Nova, Abstract Supplement, v. 5, p. 216.

Delvaux, D., R. Moeys, G. Stapel, A. Melnikov, and V. Ermikov, 1995, Paleostress reconstruction and geodynamics of the Baikal region, Central Asia: Part I. Paleozoic and Mesozoic rift evolution: Tectonophysics, v. 251, p. 61–102.

de Ruig, M. J., M. Trupp, D. J. Bishop, D. Kuek, and D. A. Castillo, 2000, Fault architecture and the mechanics of fault reactivation in the Nancar trough/Laminaria area of the Timor Sea, northern Australia: Australian Petroleum Production and Exploration Association Journal, v. 40, p. 174–193.

Etheridge, M. A., H. Mcqueen, and K. Lambeck, 1991, The role of intraplate stress Tertiary (and Mesozoic) deformation of the Australian continent and its margins: A key factor in petroleum trap formation: Exploration Geophysics, v. 22, p. 123–128.

Fittall, A. M., and R. G. Cowley, 1992, The HV11 3-D seismic survey: Skua-Swift area geology revealed: Australian Petroleum Exploration Association Journal, v. 32, p. 159–170.

Gartrell, A., M. Lisk, and J. Undershultz, 2002, Controls on the trap integrity of the Skua oil field, Timor Sea, in M. Keep and S. J. Moss, eds., The sedimentary basins of Western Australia: 3. Proceedings of Petroleum Exploration Society of Australia, Perth, Australia, p. 390–407.

Gartrell, A. P., Y. Zhang, M. Lisk, and D. Dewhurst, 2003, Enhanced hydrocarbon leakage at fault intersections: An example from the Timor Sea, Northwest Shelf, Australia: Journal of Geochemical Exploration, v. 78–79, p. 361–365.

Hall, R., 1996, Reconstructing Cenozoic southeast Asia, in R. Hall and D. J. Blundell, eds., Tectonic evolution of southeast Asia: Geological Society (London) Special Publication 106, p. 153–184.

Heeremans, M., B. T. Larsen, and H. Stel, 1996, Paleostress reconstruction from kinematic indicators in the Oslo Graben, southern Norway: New constraints on the mode of rifting: Tectonophysics, v. 266, p. 55–79.

Hibsch, C., J. J. Jarrige, E. M. Cushing, and J. Mercier, 1995, Paleostress analysis, a contribution to the understanding of basin tectonics and geodynamic evolution. Example of the Permian/Cenozoic tectonics of Great Britain and geodynamic implications in western Europe: Tectonophysics, v. 251, p. 103–136.

Hillis, R. R., 1998, Mechanisms of dynamic seal failure in the Timor Sea and central North Sea basins, in P. G. Purcell and R. R. Purcell, eds., The sedimentary basins of Western Australia: 2. Proceedings of Petroleum Exploration Society of Australia Symposium, Perth, Australia, p. 313–324.

Jones, R. M., D. N. Dewhurst, R. R. Hillis, and S. D. Mildren, 2002, Geomechanical fault characterisation: Impact on quantitative fault seal risking: Society of Petroleum Engineers/International Society for Rock Mechanics, SPE Paper 78213, 8 p.

Lacombe, O., J. Angelier, P. Laurent, F. Bergerat, and C. Tourneret, 1990, Joint analyses of calcite twins and fault slips as a key for deciphering polyphase tectonics; Burgundy as a case study: Tectonophysics, v. 182, p. 279–300.

Laurent, P., P. Bernard, G. Vasseur, and A. Etchecopar, 1981, Stress tensor determination from the study of e twins in calcite; a linear programming method: Tectonophysics, v. 78, p. 651–660.

Letouzey, J., 1986, Cenozoic paleo-stress pattern in the Alpine foreland and structural interpretation in a platform: Tectonophysics, v. 132, p. 215–231.

Mercier, J. L., E. Carey-Gailhardis, M. Sebrier, 1991, Paleostress determinations from fault kinematics: application to the neotectonics of the Himalayas–Tibet and Central Andes: Philosophical Transactions of the Royal Society of London, Series A, Mathematical and Physical Sciences, v. 337, no. 1645, p. 41–52.

Meyer, V., A. Nicol, C. Childs, J. J. Walsh, and J. Watterson, 2002, Progressive localisation of strain during the evolution of a normal fault population: Journal of Structural Geology, v. 24, p. 1215–1231.

Mildren, S. D., R. R. Hillis, T. Fett, and P. H. Robinson, 1994, Contemporary stresses in the Timor Sea: Implications for fault-trap integrity, in P. G. Purcell and R. R. Purcell, eds., The sedimentary basins of Western Australia: Proceedings of Petroleum Exploration Society of Australia Symposium, Perth, Australia, p. 291–300.

Mildren, S. D., R. R. Hillis, and J. Kaldi, 2002, Calibrating predictions of fault seal reactivation in the Timor Sea: Australian Petroleum Production and Exploration Association Journal, v. 42, p. 187–202.

Nicol, A., J. J. Walsh, J. Watterson, and P. G. Bretan, 1995, Three-dimensional geometry and growth of conjugate normal faults: Journal of Structural Geology, v. 6, p. 847–862.

O'Brien, G. W., and E. P. Woods, 1995, Hydrocarbon related diagenetic zones (HRDZs) in the Vulcan Sub-basin, Timor Sea: Recognition and exploration implications: Australian Petroleum Exploration Association Journal, v. 35, p. 220–252.

O'Brien, G. W., M. Lisk, I. Duddy, P. J. Eadington, S. Cadman, and M. Fellows, 1996, Late Tertiary fluid migration in the Timor Sea: A key control on thermal and diagenetic histories: Australian Petroleum Production and Exploration Association Journal, v. 36, p. 399–427.

O'Brien, G. W., P. Quaife, R. Cowley, M. Morse, D. Wilson, M. Fellows, and M. Lisk, 1998, Evaluating trap integrity in the Vulcan sub-basin, Timor Sea, Australia, using integrated remote-sensing geochemical technologies, in P. G. Purcell and R. R. Purcell, eds., The sedimentary basins of Western Australia: 2. Proceedings of Petroleum Exploration Society of Australia Symposium, Perth, Australia, p. 237–254.

O'Brien, G. W., M. Lisk, I. R. Duddy, J. Hamilton, P. Woods, and R. Cowley, 1999, Plate convergence, foreland development and fault reactivation: Primary controls on brine migration, thermal histories and trap breach in the Timor Sea, Australia: Marine and Petroleum Geology, v. 16, p. 533–560.

Oncken, O., 1988, Aspects of the reconstruction of the stress history of a fold and thrust belt (Rhenish Massif, Federal Republic of Germany): Tectonophysics, v. 152, p. 19–40.

Osborne, M. I., 1990, The exploration and appraisal history of the Skua field, AC/P2–Timor Sea: Australian Petroleum Exploration Association Journal, v. 30, p. 197–211.

Pattillo, J., and P. J. Nicholls, 1990, A tectonostratigraphic framework for the Vulcan Graben, Timor Sea region: Australian Petroleum Exploration Association Journal, v. 30, p. 27–51.

Reches, Z., 1983, Faulting of rocks in three-dimensional strain fields: II. Theoretical analysis: Tectonophysics, v. 95, p. 133–156.

Rouby, D., H. Fossen, and P. R. Cobbold, 1996, Extension, displacement and block rotations in the larger Gulfaks area, northern North Sea, as determined from plan view restoration: AAPG Bulletin, v. 80, p. 875–890.

Rouby, D., X. Hongbin, and J. Suppe, 2000, 3-D restoration of complexly folded and faulted surfaces using multiple unfolding mechanisms: AAPG Bulletin, v. 84, p. 805–829.

Saintot, A., and J. Angelier, 2002, Tectonic paleostress fields and structural evolution of the northwest-Caucasus fold-and-thrust belt from Late Cretaceous to Quaternary: Tectonophysics, v. 357, p. 1–31.

Shuster, M. W., S. Eaton, L. Wakefield, and H. J. Kloosterman, 1998, Neogene tectonics, greater Timor Sea, offshore Australia: Implications for trap risk: Australian Petroleum Production and Exploration Association Journal, v. 38, p. 351–379.

Streit, J. E., 1999, Conditions for earthquake surface rupture along the San Andreas fault system, California: Journal of Geophysical Research, v. 104, no. B8, p. 17,929–17,939.

Wallace, R. E., 1951, Geometry of shearing stress and relation to faulting: Journal of Structural Geology, v. 59, p. 118–130.

Woods, E. P., 1988, Extensional structures of the Jabiru Terrace, Vulcan sub-basin, in P. G. Purcell and R. R. Purcell, eds., The North West Shelf: Perth, Australia Petroleum Exploration Society of Australia, p. 311–330.

Woods, E. P., 1992, Vulcan sub-basin fault styles: Implications for hydrocarbon migration and entrapment: Australian Petroleum Exploration Association Journal, v. 32, p. 138–158.

Woods, E. P., 1994, A salt-related detachment model for the development of the Vulcan sub-basin, in P. G. Purcell and R. R. Purcell, eds., The sedimentary basins of Western Australia: Proceedings of Petroleum Exploration Society of Australia Symposium, Perth, Australia, p. 259–274.

Fault Healing and Fault Sealing in Impure Sandstones

David N. Dewhurst
Commonwealth Scientific and Industrial Research Organization, Division of Petroleum Resources, Australian Petroleum Cooperative Research Center, Kensington, Western Australia, Australia

Peter J. Boult
Australian School of Petroleum, University of Adelaide, Australia and also Department of Primary Industries and Resources South Australia, Adelaide, Australia

Richard M. Jones
Woodside Energy Ltd., Perth, Western Australia, Australia

Stuart A. Barclay
Commonwealth Scientific and Industrial Research Organization, Division of Petroleum Resources, North Ryde, New South Wales, Australia

ABSTRACT

Clay content is a first-order control on the mechanical and fluid-flow properties of fault rocks. The effects of deformation and also diagenesis are modified by the presence of clay in impure sandstones, although our understanding of the results of such changes is not well constrained. Because a lack of data for fault rocks in impure sandstones limits our ability to assess fault seal risk, a study was undertaken to investigate the effects of physical and diagenetic processes on these parameters in the Otway Basin on the southern margin of Australia. Fault rocks formed in impure reservoir sandstones from the eastern and western Otway Basin exhibit distinct geomechanical and capillary properties caused by differing clay content and distribution, overprinted by regional differences in diagenesis and geohistory. In the eastern Otway Basin, grain mixing and shear-induced clay compaction have increased fault capillary threshold pressures relative to host reservoir strata. These processes have led to a greater proportion of rigid framework grain contact, generating increased fault friction coefficients relative to the host reservoir rocks. Fault strands tend to form dense clusters as a result of strain hardening and preferential localization of new faults in weaker reservoir sandstone.

Mechanical and diagenetic processes in fault rocks in impure sandstones from the western Otway Basin have significantly altered physical and geomechanical properties as a result of increased quartz dissolution and precipitation aided by lower clay contents. Here, faults exhibit increased friction coefficient and capillary threshold pressures because of more efficient grain packing, suturing of quartz grains, and fracture healing likely resulting from local diffusive mass-transfer processes.

Phyllosilicate framework fault rocks from both regions appear significantly stronger than their host reservoirs as a direct result of syn- and postdeformational physical and diagenetic processes. These findings have direct implications for understanding the micromechanics of deformation in impure sandstones, for physical property evolution during and postfaulting, and for geomechanical prediction of fault reactivation. In a regional context, the regeneration of fault strength influences stress distribution in regional top seals through localized rotation of stress trajectories and increased differential stress, which has resulted in fracturing and loss of hydrocarbons.

INTRODUCTION

Properties of fault zones are now known to be one of the major influences on hydrocarbon migration and leakage in the subsurface. Although their importance is readily acknowledged, detailed properties of faults in hydrocarbon-bearing systems remain poorly understood. Three critical factors must be considered when assessing fault seal risk (Jones et al., 2002): juxtaposition (both a membrane-sealing and a geometrical issue), fault rock properties (also a membrane-sealing issue), and reactivation (a geomechanical issue). Juxtaposition seals are routinely assessed through the use of Allan diagrams (e.g., Allan, 1989), where hanging-wall and footwall bed cutoffs are compared across a fault surface. Several empirical algorithms are available for assessing fault rock properties and composition, such as the shale gouge ratio (Yielding et al., 1997) and clay smear potential (Lindsay et al., 1993). Fault gouge predictions can also be made using host rock composition combined with fault throw data using triangle plot methods (Knipe, 1997). Knowledge of fault rock properties is vital to understanding the flow behavior and evolution of a faulted reservoir. However, only a few studies have been published on physical properties (e.g., capillary threshold pressures, permeability) and microstructural evolution of fault rocks in relation to hydrocarbon reservoirs (e.g., Antonellini and Aydin, 1994; Gibson, 1994, 1998; Crawford, 1998; Fisher and Knipe, 1998, 2001; Dewhurst and Jones, 2002; Hesthammer et al., 2002; Sperrevik et al., 2002). Furthermore, little work has been undertaken to ground truth in many of the empirical algorithms currently in use. A first-order control on fault rocks formed is clay content (Fisher and Knipe, 1998), and nomenclature of fault types with respect to clay content is shown in Figure 1.

Diagenetic influences compound fault seal prediction uncertainties. Although physical and mechanical processes are likely to influence fault behavior during fault movement, diagenetic gouge alteration can result in fault healing and resealing over geological timescales. The presence of clays has an important influence on fault zone diagenesis, especially with regard to the kinetics of quartz dissolution and precipitation. Hence, impure sandstones are likely to experience different degrees of diagenetic alteration as compared to their cleaner reservoir counterparts. Although large volumes of pore-coating clays retard quartz precipitation (Tada and Siever, 1989), small amounts of clay between grain boundaries can enhance the dissolution process (e.g., Dewers and Ortoleva, 1991; Bjørkum, 1996). The latter author demonstrated that mica and potassium-rich clays (such as illite) significantly enhance grain-contact dissolution for quartz in reservoir sandstones at low intergranular pressures (<1 MPa) and concluded that this was primarily a temperature-driven process. Recent experimental and observational work has focused on the influence clays have in mineral dissolution and healing and sealing of fault zones (e.g., Hickman et al., 1995; Bos and Spiers, 2000). In impure sandstones, quartz dissolution and precipitation have been shown to alter flow properties such as permeability and seal capacity (Fisher and Knipe, 1998, 2001).

Deformation-induced physical and diagenetic changes in rock microstructure can also affect the geomechanical properties of fault rocks. Fault reactivation risk can be assessed using geomechanical methodologies (e.g., Castillo et al., 2000; Jones et al., 2000; Wiprut and Zoback, 2000; Mildren et al., 2002). These techniques integrate the in-situ stress state with fault rock failure envelopes to determine the orientation of fractures likely to be critically stressed and, therefore, more likely to be conductive (e.g., Barton et al., 1995). Given the paucity

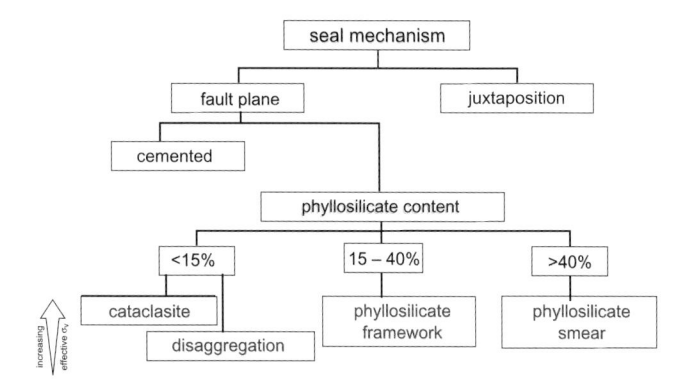

FIGURE 1. Fault rock classification based in part on Fisher and Knipe (1998). Fault rocks can be subdivided into those that have dilated and have been cemented and those whose properties are determined by their clay content.

of fault rock geomechanical property data, some of the above reactivation-risking methodologies assume that all faults are cohesionless, and that fluid flow only occurs along shear fractures. However, recent studies by Dewhurst et al. (2002) and Dewhurst and Jones (2002, 2003) have begun to provide geomechanical data on cataclasites and diagenetically altered fault rocks. These data impact significantly on quantitative fault seal risk through the influence of geomechanical risks (see Jones and Hillis, 2003). Indeed, Jones et al. (2002) demonstrated that knowledge of fault rock geomechanical properties and, specifically, the regeneration of cohesive strength resulting from both physical and diagenetic processes, could completely alter the perception of risk in a particular prospect.

Since the lack of physical property and geomechanical data for naturally occurring intact fault rocks are critical limiting factors in assessing the likelihood of fault sealing, a study was undertaken to investigate the effects of physical and diagenetic processes on these parameters in onshore and offshore oil and gas fields in the Otway Basin on the southern margin of Australia. An understanding of how these properties evolve with continued burial and diagenesis is critical not only to field development in the region but also has implications for a more general application to fault seal risking.

REGIONAL GEOLOGY

The two sets of fault rocks examined in this study are from three wells in the Otway Basin, located onshore and offshore South Australia and Victoria on the southern margin of Australia (Figure 2). The well names are confidential, but their general location is indicated in Figure 2. The basin is one of the best known and most actively explored of the series of Mesozoic rift basins that span the southern coastline of Australia. Regional tectonism is dominated by Late Jurassic to Holocene rifting resulting from the separation of Australia and Antarctica (Geary et al., 2001). A volcanic arc-ridge to the east of the continent had a major influence over sedimentation in the basin (Veevers et al., 1991). Late Jurassic to Early Cretaceous rifting was characterized by the development of a series of northwest-southeast half grabens filled with continental to volcanogenic-influenced sediments. The sediments house proven gas reserves, e.g., in reservoirs in the Pretty Hill Formation (Figure 3) in the Penola trough, located in the onshore part of the western Otway Basin (Figure 2). This early rifting was followed by Albian to Aptian thermal sag and rising heat flow (Mehin and Link, 1997), during which time 2–3 km (1.2–1.8 mi) of highly volcanogenic sediment (the Eumeralla Formation, Figure 3) blanketed the entire basin. The abundance of coals and carbonaceous shales has led to this interval being identified as a key source rock. Within the Pretty Hill Formation, hydrocarbon migration has been modeled (Hill and Boult, 2001), and significant diagenesis has been interpreted (Q. Fisher, Rock Deformation Research Group, 2002, personal communication) to occur toward the end of the deposition of the Eumeralla Formation. A widespread thermal relaxation breakup unconformity separates Early Cretaceous Crayfish Subgroup (Figure 3) sediments from Late Cretaceous sediments.

Rifting shifted to the southwest at the beginning of the Late Cretaceous and was accompanied by significant faulting associated with basement extension and collapse of an oversteepened continental shelf. The main reservoir is the fluvial-to-marine Waarre Sandstone (Figure 3), which occurs along the entire shelf edge. Transgression and renewed rifting ultimately led to the development of true marine conditions across the region and the deposition of the Belfast Mudstone (Figure 3), the top seal to the reservoir sandstone in this area. Significant long-lived overpressures encountered in some wells penetrating the Belfast Mudstone testify to its long-lasting sealing capability despite the occurrence of several Tertiary inversion events across the basin with uplift up to several kilometers (Duddy et al., 2003). Inversion has continued to the present, with Pleistocene uplift documented along several major faults and historical earthquakes in the basin.

All economic petroleum traps in the Otway Basin have some degree of fault dependence. The primary sealing mechanism for both Early (Pretty Hill) and Late (Waarre) Cretaceous plays discussed in this chapter is juxtaposition of nonreservoir hanging-wall rock against footwall reservoir. Typical fault rocks include clay smears (Jones et al., 2000), cemented cataclasites (Jones et al., 2000; Dewhurst and Jones, 2002), phyllosilicate framework fault rocks (PFFRs) and dilated cemented fault rocks (see Fisher and Knipe, 1998, for definitions). However, although the presence of sealing faults is well documented for Otway Basin traps, several paleocolumns have also

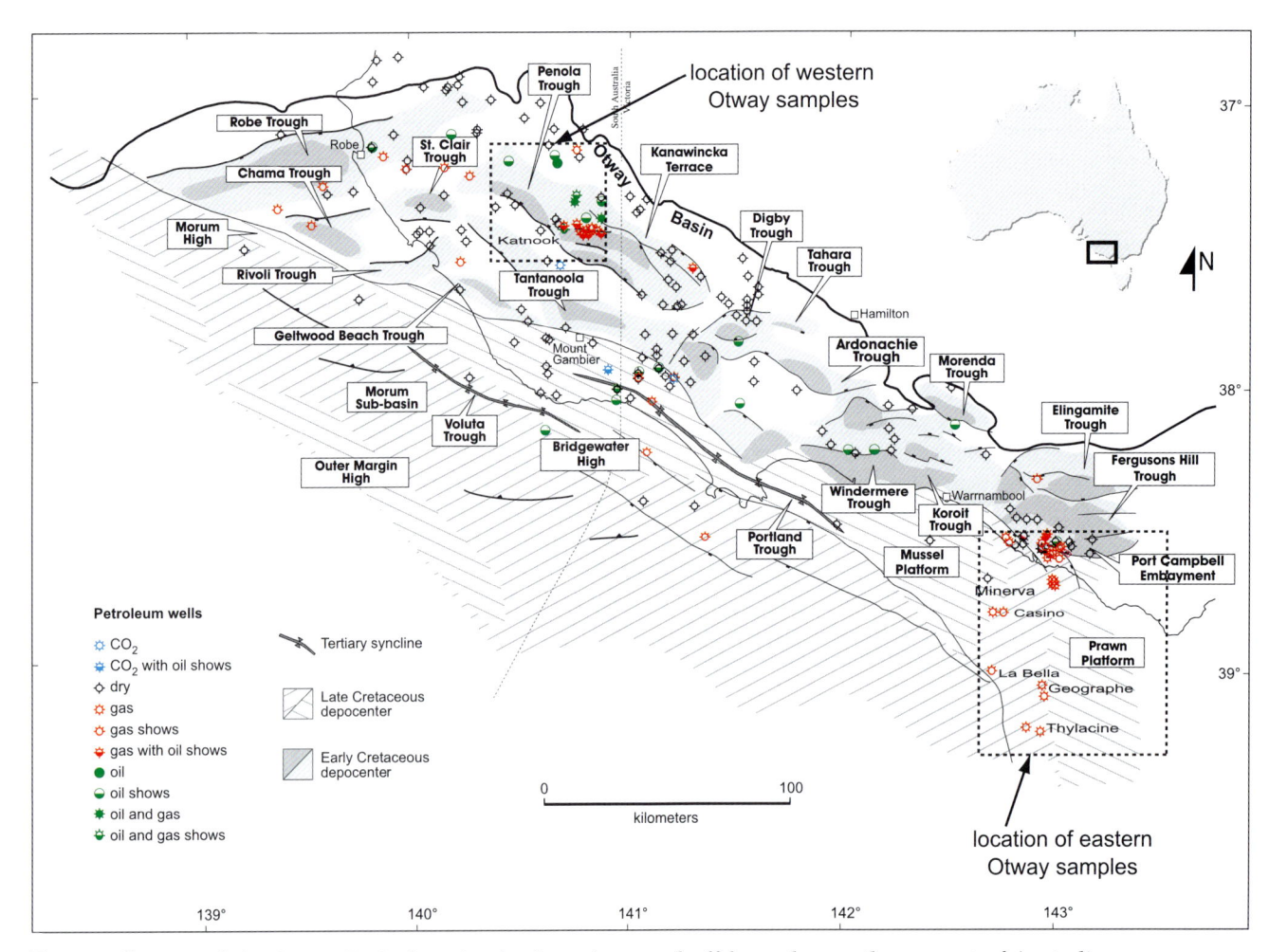

FIGURE 2. Map of the Otway Basin location both onshore and offshore the southern coast of Australia.

been found, which have been ascribed to recent reactivation of faults (Jones et al., 2000) and resultant top seal fracture (Boult et al., 2002). This has implications for exploration, and the risks associated with this can be assessed by geomechanical methods (e.g., Mildren et al., 2002; Mildren et al., 2005). From a production viewpoint, the continuity of fault cementation in the Pretty Hill Formation in the western Otway Basin is highly variable, and production histories tend to show little evidence of fault influence in this area. However, proprietary modeling data of the clay smears and PFFRs in the Waarre Formation in the eastern Otway Basin suggest that such faults are likely to influence fluid flow on a production timescale.

METHODOLOGY

The essential details of the experimental techniques used in this study are presented below. For further details, the reader is referred to Dewhurst and Jones (2002) and Dewhurst et al. (2002).

Experimental Geomechanics

Samples of 50 mm (2 in.) length and 25 mm (1 in.) diameter, saturated with 3.5% NaCl pore fluid, were deformed undrained in a standard triaxial cell with full independent control of cell pressure and axial load. Because of the small amounts of samples available, multistage triaxial tests were run (Fjær et al., 1992). Vertical core plugs were deformed to within a few megapascals of peak strength (5–10%) at a set-confining pressure then unloaded. Confining pressure was then increased, followed by further application of axial load until again close to failure. This cycle was repeated until the desired number of steps (commonly four but as much as seven, ranging between 5 and 35 MPa [725 and 5075 psi]) had been reached. The final cycle was taken to failure and residual strength. Proximity to failure was noted through monitoring of the load-axial displacement curve recorded during deformation, along with displacement on radial gauges and undrained pore-pressure response. Failure envelopes are based on the maximum differential stress applied (except for the last stage where peak strength was achieved) and are not corrected for

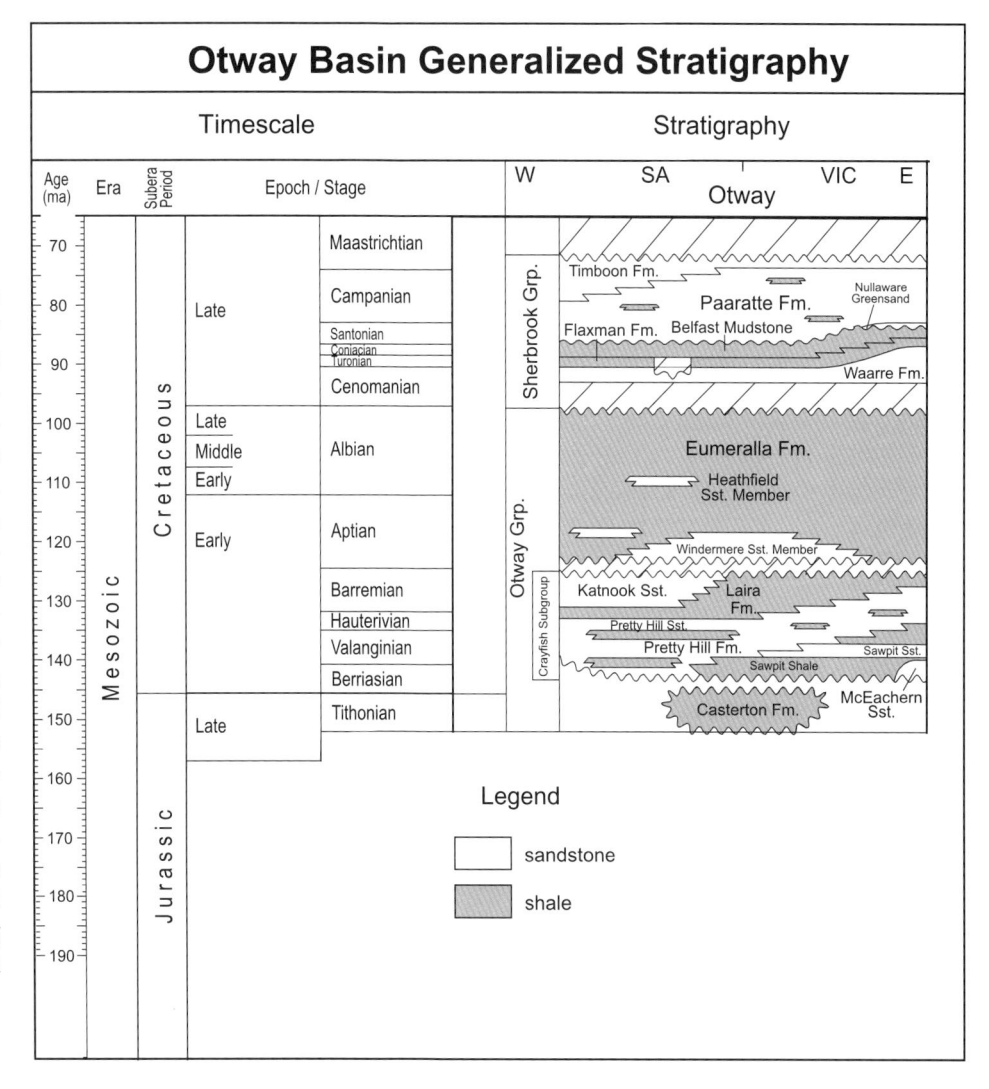

FIGURE 3. Generic stratigraphic column detailing units referred to in the text. The Pretty Hill Formation is of Early Cretaceous age, whereas the Waarre Formation is Late Cretaceous.

being a few megapascals (as much as 5–10 MPa) below the actual peak stress. Errors introduced using the multistage test technique were estimated from tests where peak strength was attained. As such, absolute results from these tests tend to overestimate friction coefficient by about 0.03 and underestimate cohesive strength by about 0.5 MPa. Cores of both fault rocks and associated reservoir sandstones were tested for comparative purposes. Where damage zone faults or deformation features were present as single strands, they were oriented at 30° to σ_1 in the triaxial rig in the optimum orientation for failure.

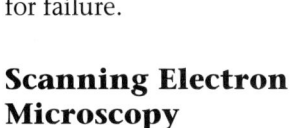

Scanning Electron Microscopy

Scanning electron microscopic (SEM) examination of the microstructure of the faults and reservoir rocks was performed on resin-impregnated, carbon-coated polished thin sections on a Phillips XL30 SEM equipped with a secondary electron detector, a backscattered electron detector, a cathodoluminescence detector, an x-ray detector and digital image capture. From hereon in, backscattered SEM images will be termed BSEM images, and cathodoluminescence images will be CL images. All images are taken with the top of the micrograph aligned with the top of the core.

Mercury Injection Porosimetry

Capillary pressure analysis was conducted using a Micromeritics Autopore III mercury porosimeter. This equipment is capable of injecting mercury in user-defined pressure increments of as much as 60,000 psi (~413 MPa) into a cleaned, dry sample. Directional injection was ensured through resin sealing of cubic samples completely traversed by fault gouge on five sides, which allows footwall injection perpendicular to the fault gouge (see Bolton et al., 2000, for more details). This method has some assumptions associated with it, namely, that both the sandstone and the fault fill are homogeneous (A. Brown, 2003, personal communication). The method of Bolton et al. (2000) results in a bimodal pore-size distribution as mercury first penetrates the reservoir pores and then, at a higher pressure, the fault network (although the smallest pores in the sandstone will also be intruded at this time). This results in two inflection points on cumulative intrusion curves but is commonly better highlighted as two separate peaks on the incremental intrusion curves. The start of the lower peak is considered to be the reservoir threshold pressure, and the start of the second is the fault threshold pressure. Column heights and density of formation water plus hydrocarbon phases under subsurface conditions were calculated via manipulation of industry standard equations and nomographs (e.g., Schowalter, 1979). However, hydrocarbon-brine interfacial tensions at reservoir conditions were calculated from the data published by Firoozabadi and Ramey (1988).

MICROSTRUCTURE AND CAPILLARY PROPERTIES OF FAULTS AND RESERVOIRS

Waarre Formation, Eastern Otway Basin

Fault rocks recovered from core retrieved from wells drilled in this region come from a depth of about 2050 m (6725 ft) below the seafloor and are currently at their maximum experienced in-situ temperature of 90–100°C. In hand specimen, the host reservoir sandstone is light gray, fine grained, and typically massive. Anastomosing linked dark fault strands can be observed in clusters throughout the sequence (Figure 4) with as much as 10 cm (4 in.) of throw. The faults are highlighted in core through the dragging and smearing of clay laminae into the fault planes, forming a highly complex fault system. Faults with larger throws generally exhibit thicker gouge than those with smaller throws.

Backscattered scanning electron microscopic images of the reservoir sandstone (Figure 5A, B) reveal a quartz and K-feldspar matrix, with a significant quantity of kaolinite. The edge of a fault runs through the top left of Figure 5A, showing lower porosity than the surrounding reservoir. Kaolinite occurs as a primary pore fill and, in authigenic form, replacing feldspars, although its distribution is nonuniform, because not all pores are occluded (Figure 5C). Grain size is generally less than 250 μm, pore body apertures range as much as about 200 μm, and sorting is moderate. Minor amounts of heavy minerals are present, including zircon, barite, and native gold. Figure 5B indicates that K-feldspar dissolution is advanced in places, and that where grain fracturing is visible, it is generally confined to feldspar grains. Generally, most quartz grains impinge on one another. However, large areas comprising continuous kaolinite are also present without visible framework grain support. Such a microstructure is likely to influence the geomechanical properties of the reservoir. Higher magnification BSEM images (Figure 5C) show the interaction between quartz grains and the extent of feldspar dissolution. Comparison of these data with CL images (Figure 5D) allows the distinction of detrital and authigenic quartz cements as well as healed fractures. Figure 5C and 5D reveals that authigenic quartz cement occurs in both fractures and as detrital grain-coating overgrowths. Interpenetration of original detrital grains caused by quartz dissolution at the point of contact does not commonly occur. Most quartzgrains have overgrowths, and these are typically greater than 50 μm thick. Small feldspar overgrowths can also be observed in these images (Figure 5C, D).

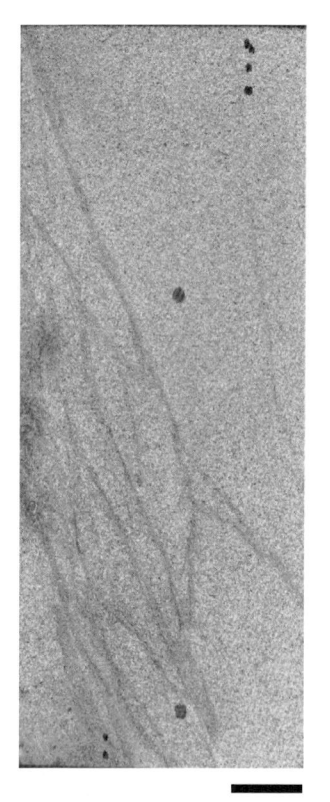

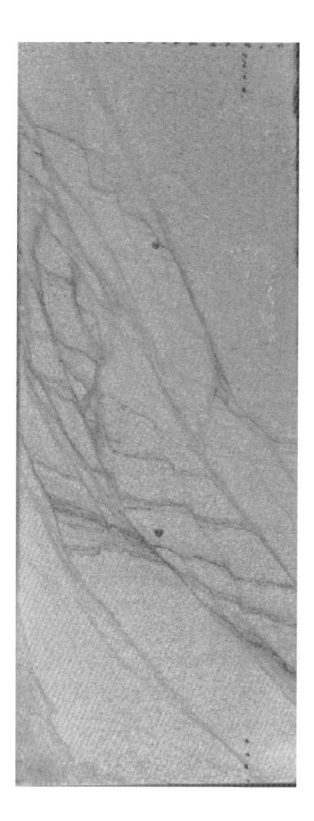

FIGURE 4. Appearance of PFFRs in core in the Waarre Sandstone (sample 3F). On the left, faults appear as anastomosing dark bands in massive sandstone. On the right, clay and organic laminae are dragged into fault zones with throws between a few millimeters and >10 cm (>4 in.). Some smaller, thinner faults tip out downward into the larger, thicker faults. Both images show the complete face of a slabbed 5-in. (12.5-cm) core.

The PFFRs postdate feldspar dissolution and kaolinite formation but predate quartz precipitation. These faults exhibit shear-enhanced compaction of kaolinite in the fault zone relative to the reservoir (Figure 6A), where this clay is pore filling. Quartz and other framework grains are more densely packed in the fault zone as compared to the reservoir (Figure 6A–C), although shear-induced mixing of grains tends to result in most quartz grains being surrounded by clay (Figure 6D). Hence, kaolinite is more evenly distributed in the faults as compared to the reservoir. Minor cataclasis is also noted in these fault rocks (Figure 6D). Postfaulting conversion of kaolinite to illite can also be observed in both reservoir and fault rocks, but this is an infrequent instead of pervasive occurrence. Comparison of fault BSEM images and CL images (Figure 6E–H) reveal that quartz overgrowths are present in limited volumes as compared to the much larger and more extensive overgrowths seen in the reservoir rock. They generally form thin (<10-μm) rims around a volumetrically minor number of detrital quartz grains (Figure 6F, H). Localized

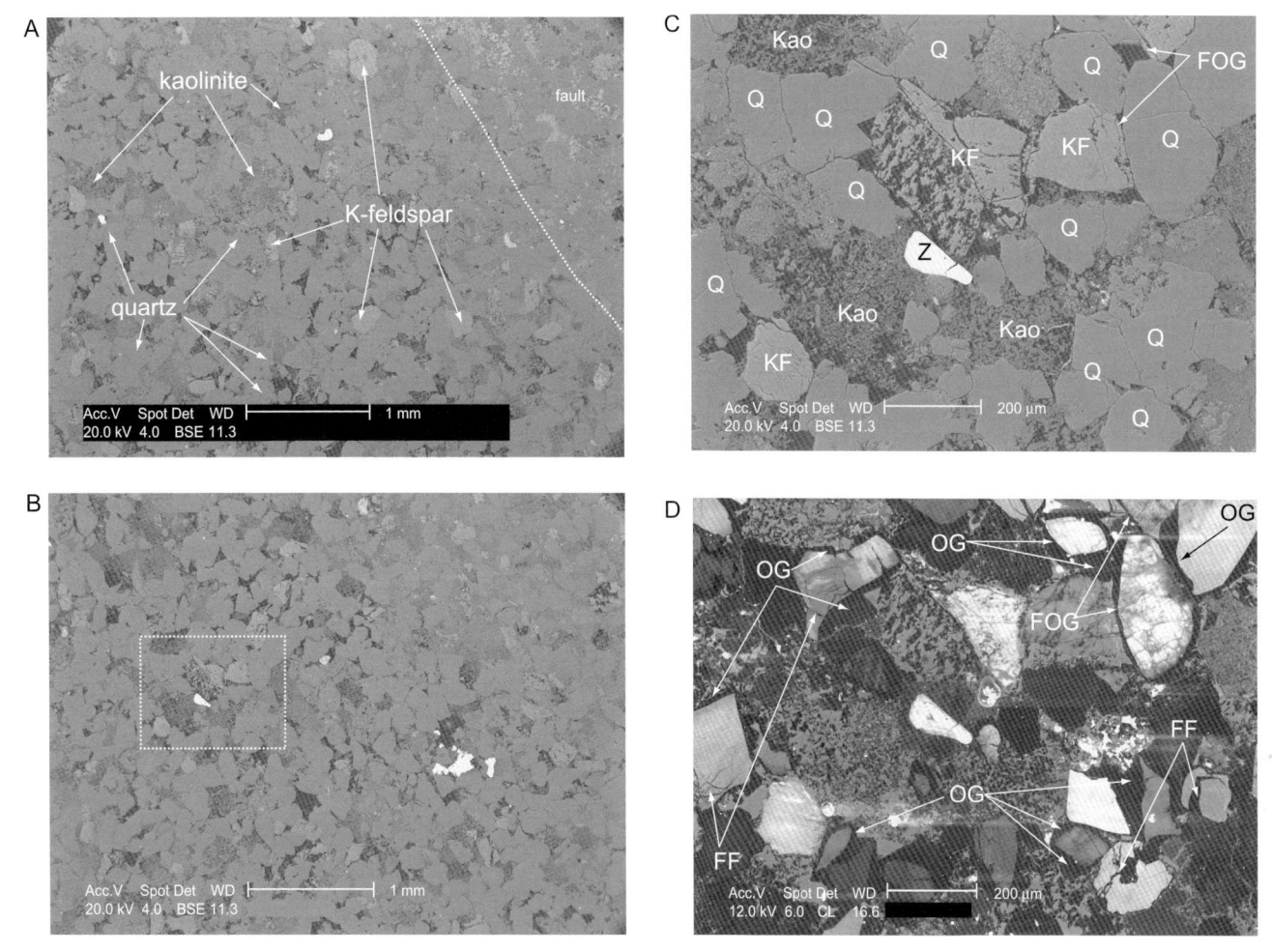

FIGURE 5. (A) Backscattered electron microscopic image of the Waarre Sandstone reservoir (sample 4R), showing for reference the main minerals present. Quartz is dark gray, K-feldspar light gray, and kaolinite is dark and speckled in the pore spaces and replacing feldspar. (B) Backscattered electron microscopic image showing occasional pores unfilled by kaolinite and feldspar dissolution providing patchy connected pathways in this gas reservoir. Box illustrates location of (C) and (D). (C, D) Backscattered electron microscopy and cathodoluminescence images, respectively, of the same area from the Waarre Sandstone reservoir. Quartz overgrowths (OG) are present as are healed fractures (fracture fill, FF) in quartz grains (Q). Minor feldspar overgrowths (FOG) also occur. Some feldspars (KF) have undergone partial dissolution, and kaolinite (Kao) has both replaced feldspar and precipitated in pore space. Z is a zoned zircon.

grain fracturing and quartz healing have also occurred in these faults.

Mercury porosimetry analyses (Figure 7) indicate that PFFR threshold pressures reach about 140–1040 psi (0.96–7.18 MPa) relative to host reservoir threshold pressures of about 60–70 psi (0.42–0.48 MPa). Conversion of threshold pressures noted above suggests gas column heights (in excess of that of the reservoir) of between 8 and 98 m (26 and 321 ft) for these fault rocks (Table 1). Permeability has also been estimated from regression analyses of threshold pressure against permeability in fault rocks performed by Sperrevik et al. (2002). Permeability values are shown in Table 1; these PFFRs are estimated to be one to two orders of magnitude less permeable than the surrounding reservoir. Compositions of these fault rocks are shown in Table 2.

Pretty Hill Formation and Sawpit Sandstone, Western Otway Basin

Fault rocks recovered from this area were from an onshore well at a depth of about 2635 m (8645 ft) and are currently at in-situ temperatures of about 106°C. Core from wells drilled in the region reveals numerous subparallel anastomosing phyllosilicate seams that isolate pods of deformed and undeformed sandstone (Figure 8). Dip slip on most deformation bands occurs on a scale of about 1 mm; however, polished slickenside surfaces (Figure 8) indicate that a component of strike-slip movement is present at least locally.

In hand specimen, the Pretty Hill Formation reservoir sandstone is light gray, fine grained, and relatively homogenous, although its composition at the basin scale

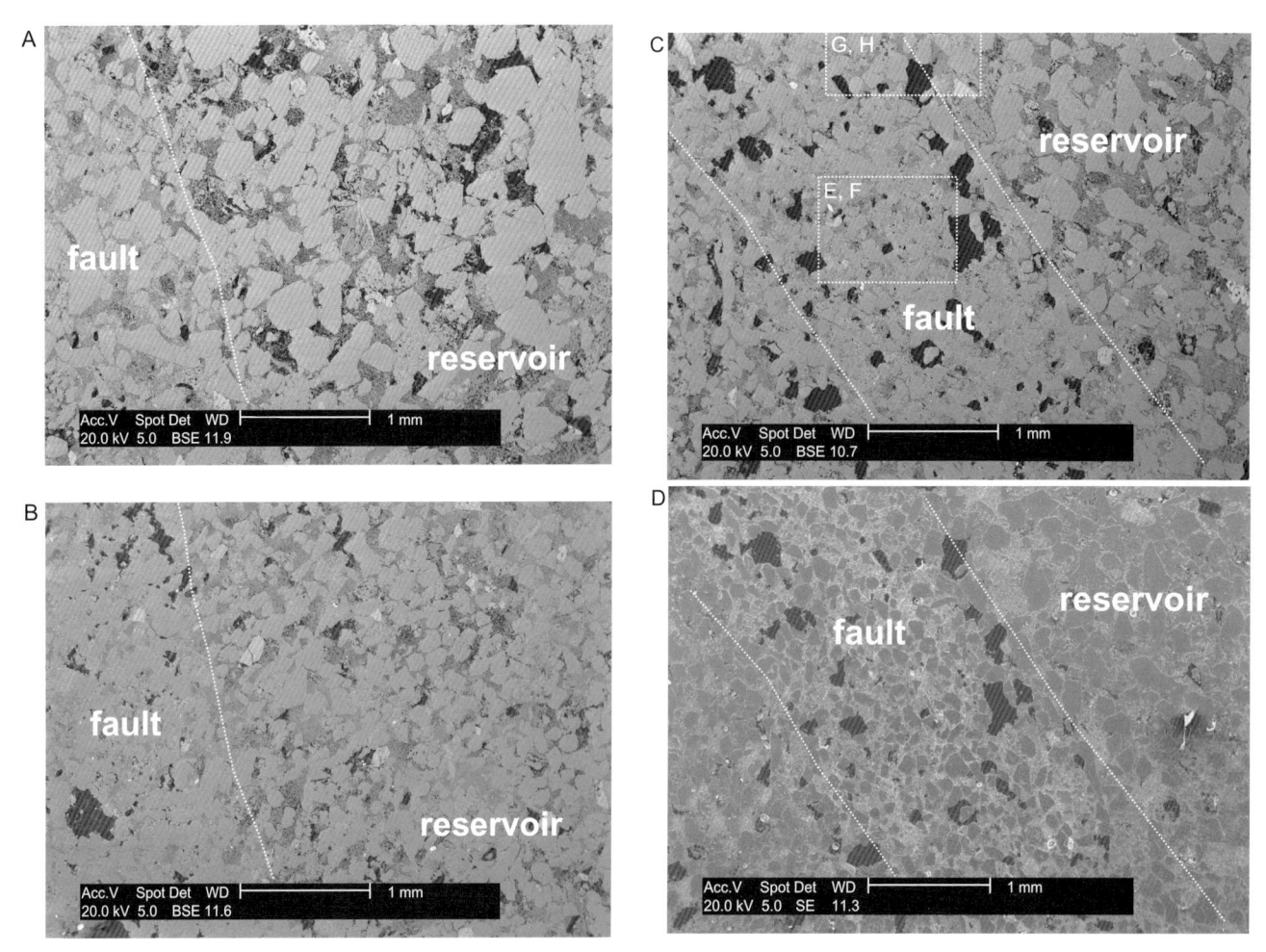

FIGURE 6. (A, B) Backscattered electron microscopic images of examples of PFFRs from the Waarre Sandstone reservoir (sample 3F). Faults run down the left-hand side of both images and can be distinguished from the reservoir by denser packing of grains and shear enhanced compaction in kaolinite. Kaolinite is commonly vermiform and microporous in the reservoir but is tight and has lost its open structure in the faults. (C, D) Backscattered electron microscopic and secondary electron images, respectively, of the same area. The PFFR runs from top left to bottom right of the image. Comparison of the two images shows that whereas grain densification has occurred in the fault, most quartz grains are separated by small amounts of clay. A minor component of cataclasis may also occur in the fault. The boxes outline the areas of the following images. (E, F) Backscattered electron microscopy and cathodoluminescence images (respectively) in a fault in the Waarre Sandstone reservoir. Compacted kaolinite (Kao) and grain densification are evident compared to the reservoir. Few quartz grains (Q) have overgrowths (OG), and where they do occur, they are thin. Quartz also occurs as fracture fill (FF) cement. (G, H) Backscattered electron microscopy and cathodoluminescence images, respectively, in a PFFR again showing quartz occurring as scarce and thin overgrowths and fracture fills.

is not uniform as a result of variable labile rock fragment abundance (Little and Phillips, 1995). Backscattered scanning electron microscopic images of the Pretty Hill Formation reservoir show a grain framework dominantly comprising quartz with lesser amounts of K-feldspar (Figure 9A), the former of which has been partially or completely kaolinitized. K-feldspar has been extensively dissolved, leaving relict smectite rims that originally coated the feldspar grains. Grain contact quartz dissolution has occurred locally, and a few minor overgrowths are present. The Pretty Hill Formation PFFRs have undergone a greater degree of diagenetic alteration than that seen in the Waarre Formation fault rocks (Figure 9B). Compacted clays occlude fault pore networks, and the suturing of quartz grains, grain interpenetration, and stylolites are more abundant. Cataclasis and comminution of both K-feldspars and quartz grains are also noted (Figure 9C). Associated with these cataclastic textures are significant degrees of quartz dissolution and precipitation in fractures, seen in both BSEM and CL images (Figure 9D, E).

Capillary properties for these diagenetically altered PFFRs are shown in Figure 10A. Although a bimodal distribution is evident from the incremental curve, it is difficult to assign an exact threshold pressure, because the wide range of fault pore-throat apertures generates a

FIGURE 6. (cont.)

relatively shallow Hg injection curve without a clear threshold pressure inflection. However, a range of 2000–3000 psi (13.8–20.7 MPa) has been estimated from the injection curve (Figure 10A), although emphasis perhaps should be placed on the difference in pore-size distributions between these fault rocks, the associated reservoir, and the Waarre PFFRs. These threshold pressures equate to a seal capacity to gas of between 170 and 257 m (558 and 843 ft). Fault permeability estimated from threshold pressures are four to five orders of magnitude lower than those calculated for the reservoir sandstone (Table 1). The threshold pressure for the host rock here is also relatively high at about 300 psi (2.1 MPa). Other Pretty Hill reservoir sandstones (Figure 9A) have lower capillary entry pressures (~60 psi [0.4 MPa]; Figure 10B), similar to those, in fact, in the Waarre Formation. Compositions of these fault rocks are also shown in Table 2.

Fault and Reservoir Geomechanics

Multistage triaxial tests were performed on PFFRs and associated reservoir rocks from the Waarre Formation and the Pretty Hill Formation. Data are presented illustrating both failure envelopes defined by the stress conditions at failure, and for the sake of clarity, only the Mohr circles describing the lower failure envelope have been shown. The multistage nature of these tests dictates that they were run so as to approach failure but were halted before macroscopic fracture generation. In the case of the first stage of each test, sample strength was unknown, and with confining pressures being low, rapid brittle failure can occur. Hence, caution was exercised when stopping the first stage so as not to destroy the restricted amount of sample available for testing. Therefore, the first point on the failure envelope is consistently low and, as such, has not been used to define the failure envelope.

Mohr circles and failure envelopes for three Waarre Formation core plugs and the PFFRs they host are shown in Figure 11. Cohesive strengths (the shear stress axis intercept) for the three reservoir samples are relatively consistent at about 10–12 MPa (1451–1742 psi). However, friction coefficients (μ; defined as the failure envelope gradient) are abnormally low for sandstones and range between 0.28 and 0.41. Typically, sandstones will have μ ranging from 0.60 to 0.85 (Byerlee, 1978).

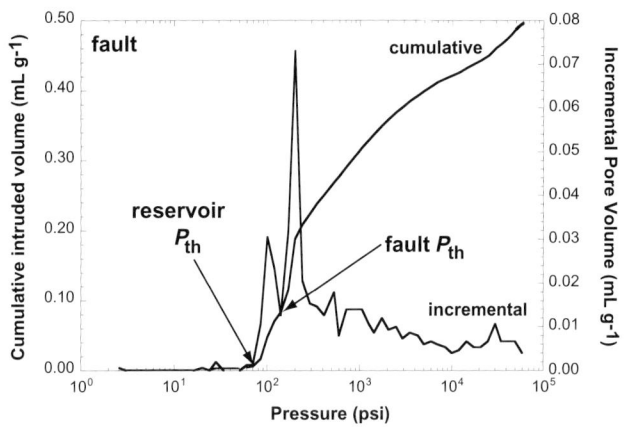

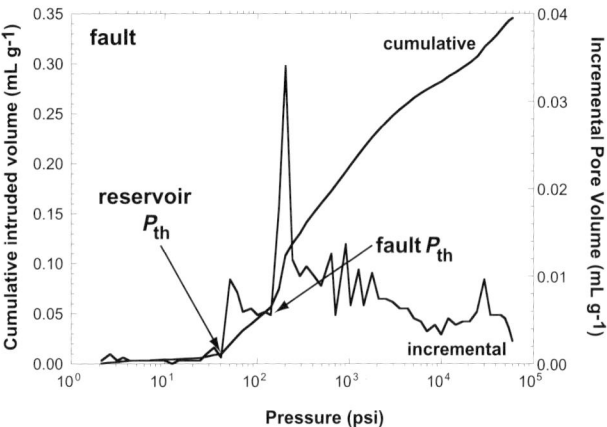

FIGURE 7. Two examples of cumulative and incremental capillary pressure curves for faults and reservoir sandstone in the eastern Otway Basin. Threshold pressure determinations across faults yield both reservoir and fault threshold pressures (see text). In these examples, reservoir threshold pressures are 40–70 psi, whereas fault threshold pressures are about 105–140 psi.

Cohesive strengths for the PFFRs are about 9–15 MPa (1306–2177 psi), similar to that of the reservoir rocks, but in all cases, the friction coefficients are considerably greater than the respective reservoir (0.39–0.59). Hence, in each case, the failure envelope has rotated counterclockwise and shows that under all stress conditions, the faults are stronger than the reservoir rocks in which they are located.

The geomechanical properties of the PFFR in the Pretty Hill Formation from the western Otway Basin and its associated reservoir rocks are significantly different (Figure 12). Here, the PFFRs have a cohesive strength approaching 15 MPa (2177 psi) and a friction coefficient of 0.86. These geomechanical properties are significantly higher than the range seen in the eastern Otway Waarre Formation PFFRs (0.39–0.59). The Pretty Hill Formation reservoir sandstone also has high cohesive strength (~12.5 MPa; ~1814 psi) and a friction coefficient similar to that of the PFFR (μ = 0.85). However, the Pretty Hill Formation reservoir sandstone is known to be variable in composition. Dewhurst and Jones (2002) tested a sample of Pretty Hill Formation reservoir sandstone from a different well and documented a lower cohesive strength (~9 MPa; ~1306 psi) and friction coefficient (0.67). These values are included here (Figure 12) to better constrain geomechanical properties for the Pretty Hill Formation. The failure envelope for the PFFR from the Pretty Hill Formation again lies above those of the surrounding reservoir rocks under all stress conditions. Failure envelopes for the lower (Waarre) and higher temperature (Pretty Hill) fault rocks are compared in Figure 13. Under all stress conditions, the failure envelope for the higher temperature fault rock lies above those for the lower temperature PFFRs from the eastern Otway Basin.

DISCUSSION

Data presented herein point to the influence of both physical and diagenetic processes on the microstructural evolution of faults in impure sandstones, concomitant with changes in both their geomechanical and petrophysical properties. The microstructure and petrophysical properties of PFFRs typically differ significantly from their host reservoir rocks as a result of deformation-induced mixing of grains and shear-enhanced compaction of clays. Such differences may be further enhanced by fracturing and diagenetic alteration. In the context of these fault rocks, temperature-driven diagenetic processes mainly enhance quartz dissolution and precipitation. The results have several implications for predicting petrophysical and geomechanical evolution in faulted impure sandstone reservoirs. Results may also impact regional prospectivity and interpretation of production history.

Micromechanics, Diagenesis, and Petrophysical Property Evolution in PFFRs

The development of clusters of deformation bands in impure sandstones (Figures 4, 8) indicates that complex cycles of strain softening and strain hardening occur, similar to processes seen during cataclasis in clean sandstones (e.g., Main et al., 2001). PFFRs from both formations appear to be clustered, although the offsets on the fault strands in the Waarre Formation are greater. In the Waarre Formation, initial deformation forms the first shear band. Slip is localized by weak clays, until porosity collapse in kaolinite allows the more rigid quartz grains to come into contact (Figure 6A–D). Decimeter-scale slip can accumulate (Figure 4), although slip is likely to vary depending on the local distribution of kaolinite. For example, large areas of continuous

TABLE 1. Geomechanical, capillary, and fluid-flow properties of reservoirs and fault rocks. P_{th} is air-mercury threshold pressure, H_{max} is seal capacity in terms of gas column height, and k is permeability modeled from the regression analyses given by Sperrevik et al. (2002). The reservoirs in both cases are gas reservoirs, accounting for low but producible k. RT = below rotary table; μ = coefficient of friction; C = cohesive strenght; ϕ = porosity.

Temperature	Pretty Hill Formation, Western Otway Basin 106°C			Waarre Formation, Eastern Otway Basin 90–100°C				
	Sample 1R	Sample 1F	Sample 2R	Sample 2F	Sample 3R	Sample 3F	Sample 4R	Sample 4F
Depth (m RT)	2635	2635	2174	2174	2175	2175	2208	2208
P_{th} (psi)	60	2000–3000	60	1041	70	409	70	140
H_{max} (m)	5	170–257	6	98	7	35	7	8
k (md)	0.17	10^{-5}–10^{-6}	0.17	10^{-4}	0.12	10^{-3}	0.12	10^{-2}
μ	0.67–0.85	0.86	0.33	0.58	0.28	0.39	0.41	0.59
C (MPa)	8.8–12.5	14.8	12.3	15.1	9.1	9.1	10.3	10.0
ϕ (%)	11	4	16	13.5	17	12	19	17

TABLE 2. X-ray diffraction data from fault rocks in the Pretty Hill and Waarre Formations.

Mineral	Pretty Hill Formation		Waarre Formation	
	1F	2F	3F	4F
Quartz	67 (2)	76 (2)	74 (2)	63 (1)
Kaolin	17 (2)	19 (2)	20 (2)	20 (1)
Mica (illite)	1 (1)	4 (1)	5 (1)	3 (1)
Anatase	<1	<1	<1	<1
Orthoclase	<1	<1	<1	<1
Pyrite	–	<1	<1	–
Siderite	1 (1)			14 (1)
Smectite	14 (2)			

kaolinite in the grain framework (resulting from feldspar alteration) may allow larger slip to accumulate in areas of the reservoir where rigid grains dominate the framework. Rigid grain interaction results in strain hardening through an increase in friction angle as can be seen in Figures 11 and 12, where fault rock failure envelopes have steeper gradients and thus lie above those of the reservoir sandstones. This effect has also been documented for cataclastic faults in clean sandstones (e.g., Aydin, 1978; Underhill and Woodcock, 1987). As such, the deformation band strengthens, and a new band initiates in the weaker reservoir. This is likely to be a continuous process resulting in the clusters seen in Figures 4 and 8. Grain fracturing is not observed because of the shallow depth (0.5–1.0 km; 0.3–0.6 mi) at which deformation occurred and perhaps also to the large quantity of clay present in these rocks.

A similar initial physical formation mechanism is envisaged for the higher temperature fault rocks from the Pretty Hill Formation in the western Otway Basin. The essential mechanical difference is that the reservoir framework comprises rigid particles, and as such, faults have small throws (millimeter-scale). Cataclastic deformation resulted because of fault reactivation at depth under high stress conditions. Eventually, the formation of major slip surfaces localizes deformation and accumulates strain through repeated slip along the same surface, resulting in a highly polished fault plane (e.g., Figure 8) (cf. Aydin and Johnson, 1978). Essentially, during PFFR formation in impure sandstones, physical processes result in frictional strengthening on short timescales, allowing the formation of microfault clusters.

Diagenetic processes such as authigenic quartz precipitation occur over much longer timeframes than fault slip events (Walderhaug, 1996) and are therefore unlikely to affect the mechanics of faulting during a deformation event. However, given long periods and high temperatures, diagenetic alteration in siliciclastic fault rocks further alters petrophysical properties (cf. Figures 7, 10;

FIGURE 8. Appearance of PFFRs in core in the Pretty Hill Formation (sample 1F). A slip surface here has localized deformation, and slickensides indicate a strong strike-slip component of movement. Damage is most intense close to the slip surface where pods of sandstone are isolated by anastomosing, clay-rich stylolites.

tures (90–100°C) has resulted in the formation of only minor intrafault quartz overgrowths (Figure 6C, D). Thick clay coatings on quartz grains are known to inhibit quartz cementation (Cecil and Heald, 1971) and are likely to have affected the distribution of quartz overgrowths in these faults. However, some grain-contact dissolution has occurred in both reservoir and PFFRs, which, together with the overgrowths, accounts for the cohesive strength displayed by these fault rocks (Figure 11).

The Pretty Hill Formation PFFRs have undergone more intense diagenesis as a result of both higher temperatures (~106°C) and the distribution of clays in the reservoir before deformation (Figure 9A). Grain contact quartz dissolution stylolites, and precipitation of authigenic quartz on newly created cataclastic fracture surfaces are far more extensive in the Pretty Hill Formation PFFRs investigated in this study (Figures 8, 9). The late, deep fault reactivation episode in these rocks resulted in grain fracturing (Figure 9D, E), and the clean surface area of these fractures have allowed them to act as sinks for quartz cement. The cement volume in the fault is greater than that in the reservoir (cf. Figure 9A with Figure 9D, E) as finer grain size caused by cataclasis allows more rapid quartz precipitation (Fisher and Knipe, 1998). Overgrowths in the Pretty Hill Formation fault rocks are conspicuous by their absence, likely because of clay coatings on quartz grains resulting from deformation-induced grain mixing. Lower temperatures

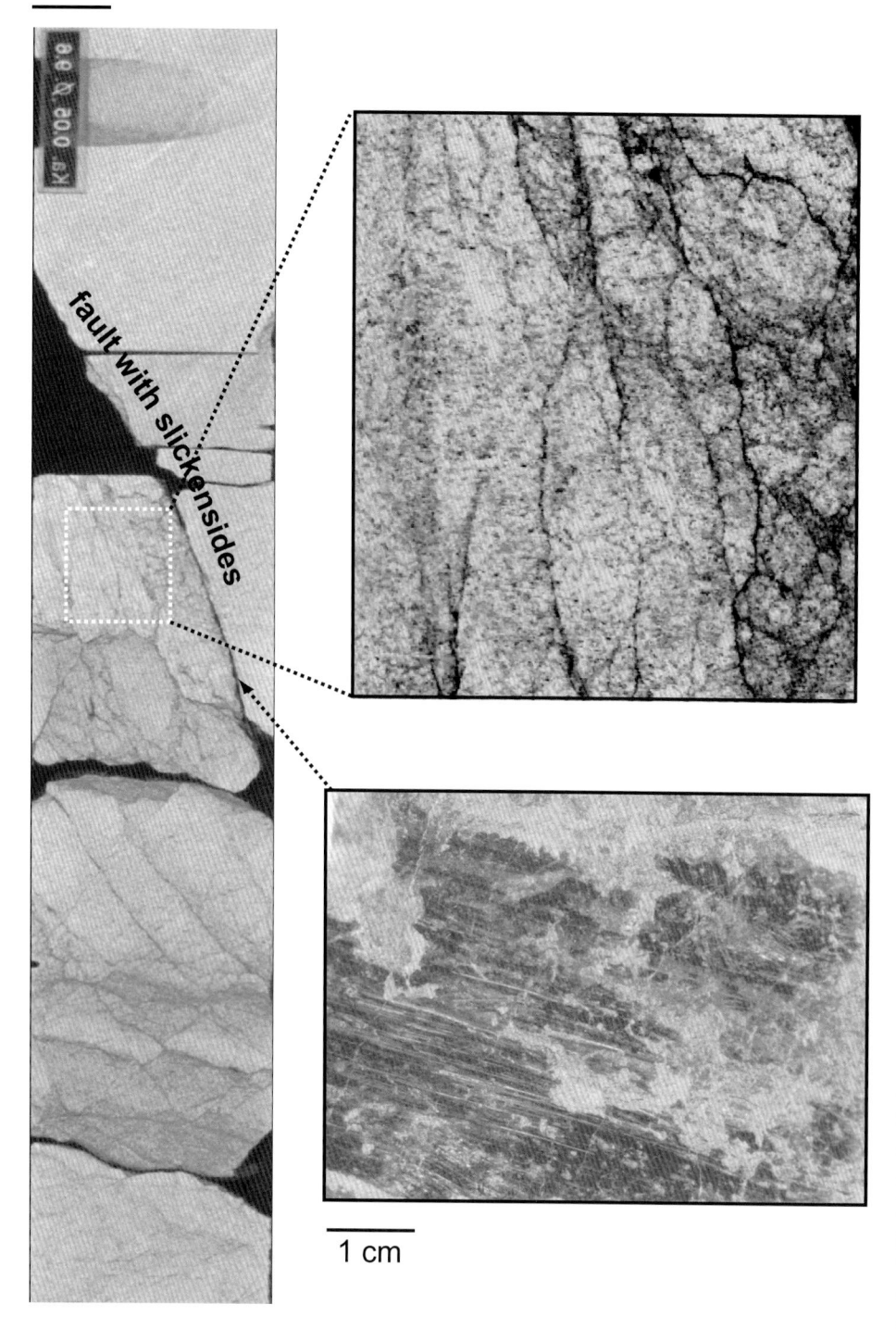

Table 1) and may lead to the regeneration of fault strength (Figures 11, 12). Physical processes such as porosity collapse and incorporation of kaolinite into the deformation bands (Figure 6) change petrophysical properties. Fault seal capacity is increased (Figure 7), and permeability is decreased in the Waarre Formation fault rocks as compared to the reservoir (Table 1). The more uniform distribution of kaolinite in the Waarre Formation PFFRs and the lower maximum in-situ tempera-

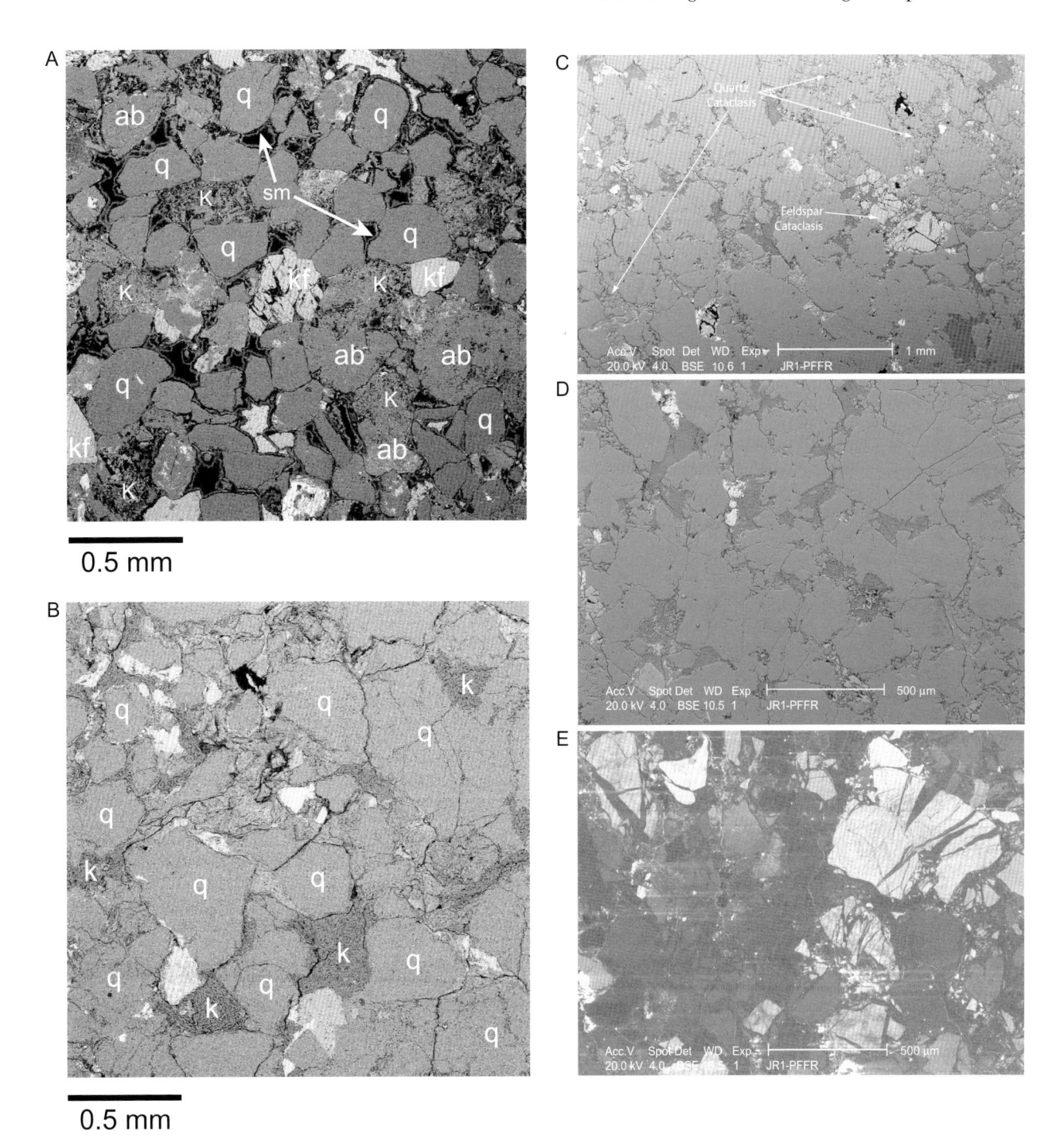

FIGURE 9. (A) Backscattered electron microscopic image of an arkosic reservoir sandstone from the Pretty Hill Formation (sample 1R), showing minor quartz (q) dissolution and some kaolinitization (K) of feldspar (albite [ab] and K-feldspar [kf]). Pore-rimming clays are smectite (sm) (modified from Dewhurst and Jones, 2002). (B) Backscattered electron microscopic image of a PFFR in the Pretty Hill Formation (sample 1F) showing sutured quartz grain boundaries and compacted kaolinite in pores. (C) Backscattered electron microscopic image of a PFFR showing cataclasis of both quartz and feldspar indicating deformation at high effective stress, consistent with late fault reactivation at depth. (D) Backscattered electron microscopy and (E) cathodoluminescence images, respectively, of the same area in a Pretty Hill Formation PFFR. Comparison of the images shows intense compactive cataclasis in quartz and quartz cementation of fractures.

and deformation-induced clay coats inhibited quartz overgrowth formation in the Waarre Formation PFFRs, and grain fracturing is sparse, because deformation occurred at shallow depth. Occasional overgrowths are seen in the Waarre Formation reservoir rocks, because the kaolinite distribution here is patchy (Figure 5A, B),

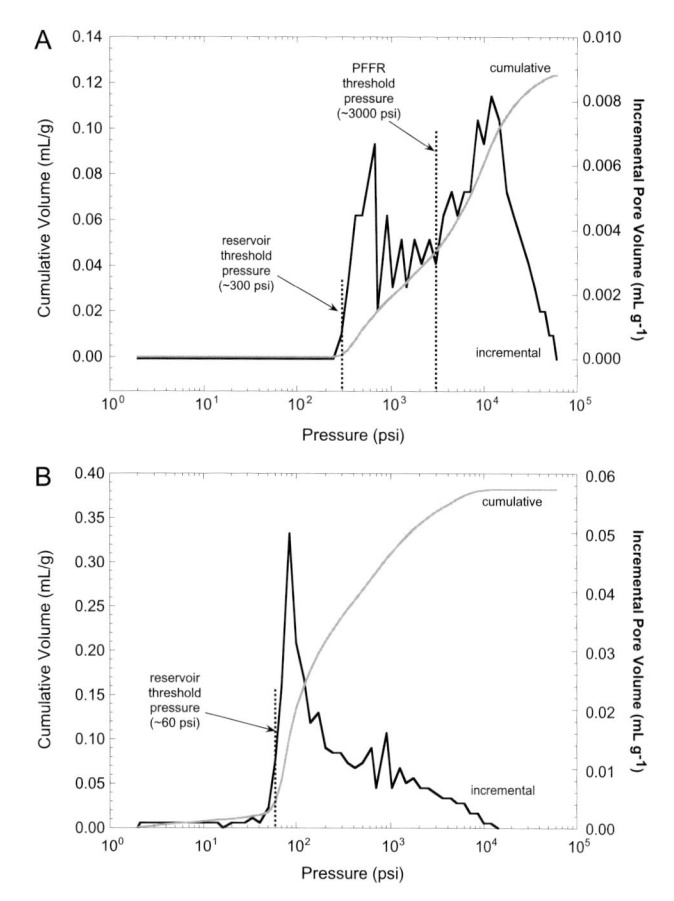

FIGURE 10. Cumulative and incremental capillary pressure curves for PFFRs (sample 1F) and reservoir sandstones (sample 1R) in the Pretty Hill Formation. (A) Reservoir threshold pressures here of 300 psi are high because of alteration in proximity to a fault. PFFR threshold pressure is difficult to determine (see text) but probably lies between 2000 and 3000 psi. (B) Pretty Hill Formation reservoir rock away from faults with a threshold pressure of 60 psi.

but do not occur in the higher temperature Pretty Hill Formation reservoir sandstone caused by the pore-lining nature of the smectitic clays (Figure 9A), which prevents authigenic quartz precipitation.

Clays may have enhanced grain-contact quartz dissolution in the Pretty Hill Formation PFFRs (e.g., Fisher and Knipe, 1998). This process has facilitated stylolitization and grain suturing, generating a gouge comprised of interlocking quartz grains with surrounding pores reduced in size and occluded by clays (Figures 8, 9). The resulting texture has produced a fault with significantly increased seal capacity and cohesive strength (Figures 12, 13) relative to its reservoir (Figure 10) and the lower temperature Waarre Formation fault rocks (Figure 7; Table 1). These results are consistent with data presented by Fisher and Knipe (1998), in that it appears that the PFFRs that formed as a result of mechanical processes have low capillary threshold pressures (although still higher than the initial reservoir rock), whereas those that have undergone diagenetic modification postfaulting have significantly increased seal capacity.

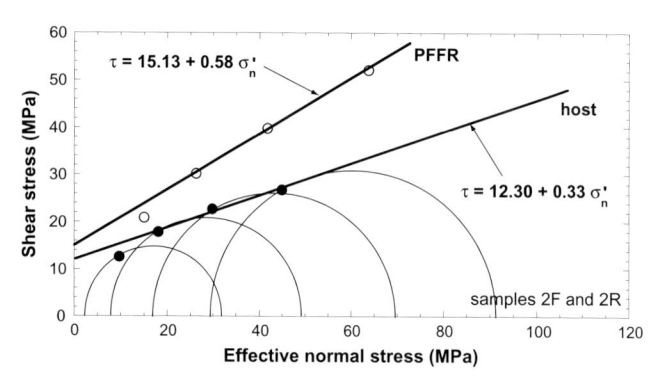

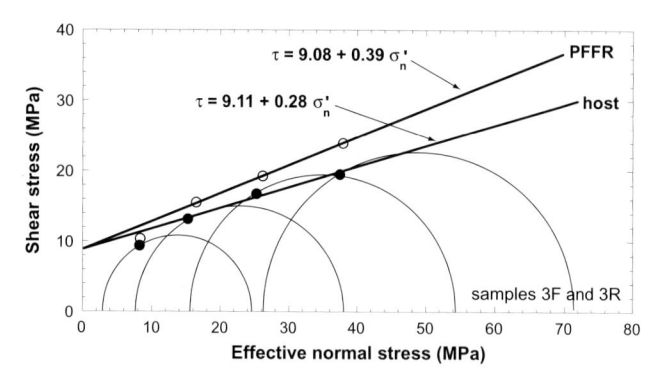

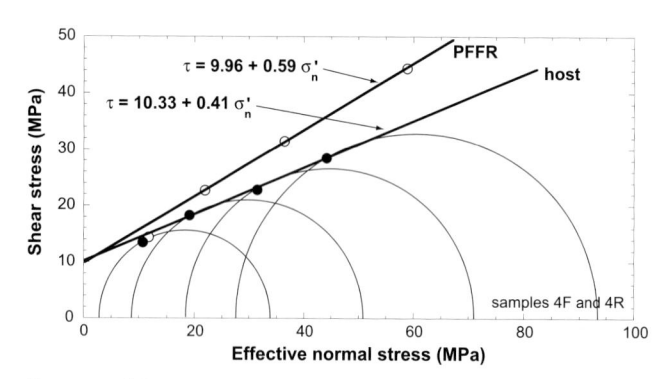

FIGURE 11. Geomechanical properties of three samples of reservoir sandstones and associated PFFRs from the Waarre Formation in the eastern Otway Basin. Mohr circles are shown on the lower failure envelope only for clarity. Symbols represent stress conditions at failure. In each case, the failure envelope for PFFRs lies above that of the reservoir, indicating that the fault rock is stronger than the reservoir. Note also that all fault rocks have cohesive strength at least comparable to those in the reservoir. Friction coefficients are more variable between the three individual reservoir samples, likely the result of patchy distribution of kaolinite.

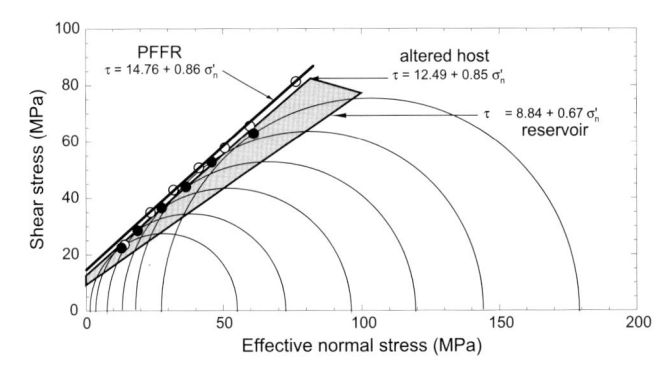

FIGURE 12. Geomechanical properties of reservoir sandstones and associated PFFRs from the Pretty Hill Formation in the western Otway Basin. Again, the PFFRs are stronger under all conditions than their respective reservoir rocks. A range of geomechanical properties is shown for the Pretty Hill Formation reservoir (shaded area) as composition and diagenetic alteration are variable. The lower bound is from Dewhurst and Jones (2002).

IMPLICATIONS FOR GEOMECHANICAL PREDICTION OF FAULT REACTIVATION

Fault Cohesion

Methods derived from theoretical geomechanics are being increasingly employed to assess the likelihood of fault reactivation in tectonically active areas in the contemporary stress field (e.g., Castillo et al., 2000; de Ruig et al., 2000; Jones et al., 2000; Wiprut and Zoback, 2000; Dewhurst et al., 2002; Mildren et al., 2002; Hunt and Boult, 2005). Important differences exist between the various techniques (R. R. Hillis and E. J. Nelson, 2004, personal communication), especially those that consider that all faults are cohesionless, weak, and obey Byerlee (1978) friction laws, i.e., μ = 0.60–0.85 (e.g., Castillo et al., 2000; de Ruig et al., 2000; Wiprut and Zoback, 2000). Recent studies have demonstrated that diagenetically altered geomechanical properties of cataclasites can impact on quantitative fault reactivation risking (Dewhurst and Jones, 2002), and that knowledge and application of fault geomechanical properties can completely alter the perception of risk in a particular prospect (Jones et al., 2002). Data presented as part of this study indicate PFFR strength recovery through healing significantly alters bulk geomechanical properties. Therefore, it is critical to understand how fault geomechanical properties may alter physical and diagenetic processes.

Diagenetic reactions may alter fault zone strength through mineral transformations and cementation, resulting in crack healing and sealing concomitant with changes in hydromechanical properties and grain size. Experimental work has shown that temperature can influence the frictional properties of rocks (Chester, 1994), and that temperature-driven reactions affect fault chemomechanical behavior (Olsen et al., 1998). Fault healing and strengthening in cataclastic quartz-rich rocks, especially those with enhanced grain-contact quartz dissolution caused by clays, are the result of a positive feedback loop between grain-size reduction and dissolution (Bos and Spiers, 2000). Experimental work by Tenthorey et al. (2003) show that both the friction coefficient and cohesive strength increase during short-term circulation of hydrothermal fluids through static quartzofeldspathic faults, attributed to increased lithification and grain-contact area. Increasing grain-contact area lowers fault normal stress and thus leads to increased friction coefficients. Most of the observed strength recovery occurred early in these experiments, consistent with a grain-contact quartz dissolution mechanism and local redistribution and precipitation of quartz on newly created clean fracture surfaces. Grain-contact quartz dissolution results in the replacement of adhesive grain contacts with welded grain contacts (cf. Figures 5, 6 with Figure 9B), thus increasing cohesive strength (Figure 13) (Fredrich and Evans, 1992). Although only some of the experiments quoted above have been conducted at temperatures equivalent to those encountered in sedimentary basins, long geological time spans can be substituted for the effect of temperature in terms of mineral dissolution and precipitation (Karner et al., 1997). It has been suggested

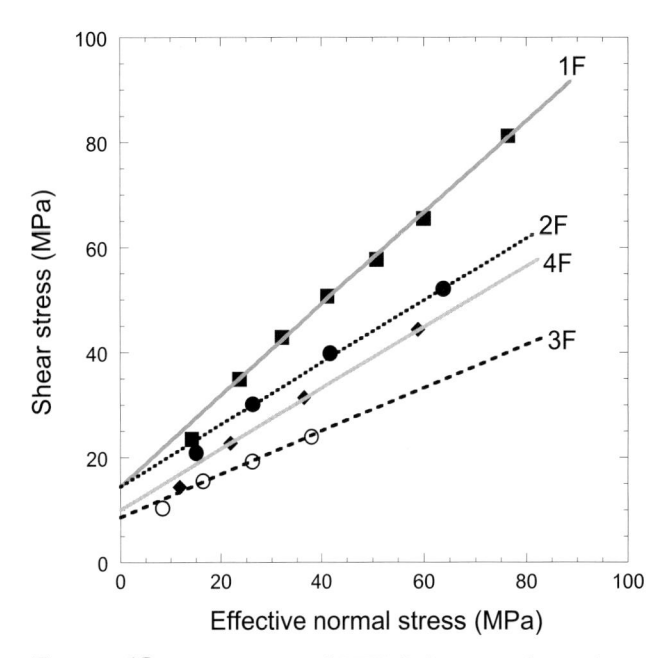

FIGURE 13. Comparison of PFFR failure envelopes from Figures 11 and 12. The PFFR from the Pretty Hill Formation, which has undergone more intense diagenesis, is stronger than the Waarre Formation rocks that experienced lower temperatures. 1F and 2F are from the Pretty Hill Formation; 3F and 4F are from the Waarre Formation.

that strength recovery in cataclastic fault rocks can be geologically rapid at sedimentary basin temperatures (~6000 yr, Q. Fisher, Rock Deformation Research Group, 2002, personal communication).

Although is it likely that clay-rich fault gouges and neoformed faults have close to zero cohesion, strength recovery through mineral dissolution and precipitation is well documented (e.g., Fournier, 1996; Tenthorey et al., 2003), especially on geological timescales. Furthermore, even the seemingly archetypal weak fault, the San Andreas fault (e.g., Rice, 1992; Zoback and Healy, 1992, although see also Scholz, 2000), has been shown to have gouge in the fault core containing both veins and fractured vein fragments, suggesting significant cohesive strength recovery between episodic slip events (Chester et al., 1993). The presence of veins and vein fragments is crucial, because this indicates directly that these fractures formed in tension in the fault core, features that could not form if the fault rocks were cohesionless. Similarly, the laboratory-derived data on PFFRs in this study, coupled with those on cataclastic (Dewhurst and Jones, 2002) and cemented fault rocks (Dewhurst and Jones, 2003), show that strength recovery is an important and widespread process, even at temperature conditions encountered in sedimentary basins. This indicates that geomechanical reactivation-risking methodologies that fail to consider cohesive strength recovery are incomplete at best.

FAULTS STRONGER THAN RESERVOIRS

A little envisaged geological scenario is when faults are actually stronger than their host reservoirs. In this instance, the assumption that all faults are cohesionless and that the fault is the weakest link in the system may be erroneous, and our understanding of trap behavior through time will therefore be incorrect. All the PFFRs examined in this study have failure envelopes that lie above those of their respective reservoir rocks, indicative of higher strength in the fault rock. This suggests that future deformation will be further localized in the reservoir, and this may, in fact, lead to either new faults forming in optimal orientations (e.g., Sibson, 1985; Fournier, 1996) or fault zone broadening (e.g., Faulkner and Rutter, 2001). Whether fault rocks stronger than the surrounding reservoir actually prevent fault reactivation is still open to debate (e.g., Walsh et al., 2001). If faults are strong and misoriented with respect to the stress field in which reactivation occurs, then it is likely that new optimally oriented faults will form. However, although optimally oriented strong faults, at first glance, might seem likely to resist reactivation, the presence of a fault discontinuity and the competency contrast provided by strong fault rocks may act to concentrate stress around the preexisting fault (e.g., Ramsey, 1967), increasing the likelihood of reactivation at the interface between fault and intact rock (Streit, 1999). This hypothesis is supported by studies that demonstrate the existence of a competency contrast at fault-reservoir boundaries and was noted to be a key influence in experimental cataclasite reactivation (Dewhurst and Jones, 2002). Some researchers contend that fault reactivation is primarily controlled by fault geometry (size and connectivity), and that fault rock geomechanical properties are only a secondary influence (Walsh et al., 2001). Numerical modeling by these authors showed that large faults with strong fault rocks were more efficient at localizing fault displacements, possibly along the boundary between the fault rock and host. However, Walsh et al. (2001) also note that for this to be the case, the stress field orientation must remain unchanged through time, because it has long been established that fault reactivation under a rotated stress regime will occur only along optimally oriented faults (Sibson, 1985).

REGIONAL IMPLICATIONS

Knowledge of the properties and distribution of shear bands in a reservoir is important from a production viewpoint. For example, Hesthammer et al. (2002) noted a rapid pressure drop and declining production in a North Sea field, which was attributed to the presence of low-porosity, high-threshold pressure shear bands whose sealing properties had been enhanced by temperature-driven processes. These authors noted three to four orders of magnitude permeability difference between the impure sandstone reservoir and the faults therein, a difference that increased to five orders of magnitude when diagenetic redistribution of quartz had occurred at temperatures of about 120°C.

Data presented herein suggests that faults from both the Pretty Hill and Waarre formations have the capacity to restrict hydrocarbon flow over both geologic and production timescales, as a result of both physical processes and textural changes that have increased capillary seal capacity. This phenomenon would seem more likely to occur in the higher temperature Pretty Hill Formation PFFRs because of diagenetic seal enhancement. However, this assessment must be viewed with caution, because published and proprietary studies note that gouge cement continuity is highly variable, and field evidence from production histories in the western Otway shows no effect of faults on production in the area. However, in the Penola trough (western Otway Basin, Figure 2), strong faults, identified through local stress reorientations, have been discovered in several wells (Boult et al., 2002). For understanding the trapping and leakage mechanisms in this area, the critical observation here is that strong faults in the reservoir appear to be associated with local stress trajectory perturbations in the top seal. This causes variation of differential stress in the top seal that is controlled by the intersection with the prospect-bounding fault (Hunt and Boult, 2005). Local areas of high differential stress may cause the fracturing of the top seal and the initiation of seal-penetrating structural permeability, thus allowing leakage of part or all the entrapped hydrocarbons. In the eastern Otway Basin, modeling of intrareservoir flow baffling in the Waarre Formation is supported by proprietary studies of accumulations that suggest that clay smear and PFFRs are very likely to generate pressure baffles over the production life cycle.

Knowledge of fault strength may yet prove to be extremely significant for understanding and predicting top seal integrity where the regional stress field is perturbed by a discordant body of relatively stronger fault rock in a generally layer cake sedimentary sequence. Stress perturbations cause localized high differential stress zones that can lead to rock failure. A top seal only needs a single, continuous crack of 0.001-in. (0.035-mm) aperture to exist for 1000 yr to leak away all trapped oil in a 150-million-bbl field (Downey, 1984). In the Otway Basin, the top and internal seals to Early Cretaceous Pretty Hill Formation reservoirs deform in a relatively brittle manner under the current stress regime compared to those of the Late Cretaceous Waarre Formation. This may explain the lack of paleocolumns in the latter and their common occurrence in the former.

CONCLUSIONS

Although some similarities exist between the fault rocks in the two investigated reservoir formations in the eastern and western Otway Basins, significant differences also exist in micromechanics and physical and geomechanical properties that can be attributed mainly to clay content and temperature. Initial mechanical processes, such as grain mixing and clay compaction, occur in both sets of faults, and this increases seal capacity and reduces permeability. High clay contents and homogenous distribution resulting from grain mixing during the faulting process, combined with temperatures of about 90–100°C, have resulted in minor quartz cementation in Waarre Formation fault rocks. Lower clay contents and higher temperatures (106°C) in the Pretty Hill Formation PFFRs led to small amounts of clay emplaced at grain contacts, which enhanced quartz dissolution and stylolitization. This process further lowered permeability and significantly increased threshold pressures.

Both sets of fault rocks are stronger geomechanically than their host sediments, and this has direct bearing on methods for predicting fault reactivation and surrounding top seal integrity. The regeneration of cohesive strength of fault rocks should be considered when using geomechanical reactivation risking techniques, because the assumption that all faults in sedimentary basins are cohesionless is flawed. Similarly, care should be taken when applying these techniques, because the prime assumption is that the fault is always the weakest link. Fault healing has been shown to increase fault strength, and if reactivation occurs in a different stress field to the one the faults initiated in, then new, optimally oriented faults may form in the reservoir. In an exploration context, the presence of diagenetically altered strong faults affects stress trajectories in overlying top seals, which can lead to local differential stress concentration, top seal fracturing, and the subsequent loss of hydrocarbon accumulations.

In a regional context, PFFRs in the western Otway Basin seem to exert little effect on production from the Pretty Hill Formation reservoirs. This contrasts with PFFR influence in the eastern Otway Waarre Formation, where proprietary studies suggest that intrareservoir hydrocarbon baffling over a production timescale is highly probable. The difference here is not the result of absolute physical properties (e.g., seal capacity, which, in fact, is greater in the Pretty Hill Formation) but because of the continuity (or lack thereof) of the fault rocks responsible for producing intrareservoir pressure baffles.

ACKNOWLEDGMENTS

This work was performed as part of the Australian Petroleum Cooperative Research Center Research Programme on Hydrocarbon Sealing Potential of Faults and Cap Rocks. Samples were provided by Origin Energy, Primary Industries and Resources of South Australia, and Woodside Energy Limited. Sponsors of this program (Woodside Energy Limited, Origin Energy, OMV, Japan National Oil Corporation, Globex (now Marathon), Santos, Exxon/Mobil, BHP-Billiton, Chevron-Texaco, Conoco-Phillips, Anadarko, and Statoil) are thanked for permission to publish. Helpful discussions with Bronwyn Camac, Dave Cliff, Andrew Davids, Quentin Fisher, Richard Hillis, Suzanne Hunt, and Scott Mildren are duly acknowledged. We also thank the three reviewers, Hege Nordgard Bolas, Quentin Fisher, and Alton Brown, for their clear, insightful, and constructive comments, which significantly enhanced this chapter.

REFERENCES CITED

Allan, U. S., 1989, Model for hydrocarbon migration and entrapment within faulted structures: AAPG Bulletin, v. 73, p. 803–811.

Antonellini, M., and A. Aydin, 1994, Effect of faulting on fluid flow in porous sandstones: Petrophysical properties: AAPG Bulletin, v. 78, p. 355–377.

Aydin, A., 1978, Small faults formed as deformation bands in sandstone: Pure and Applied Geophysics, v. 116, p. 913–930.

Aydin, A., and A. Johnson, 1978, Development of faults as zones of deformation bands and as slip surfaces in sandstone: Pure and Applied Geophysics, v. 116, p. 931–942.

Barton, C. A., M. D. Zoback, and D. Moos, 1995, Fluid flow along potentially active faults in crystalline rock: Geology, v. 23, p. 683–686.

Bjørkum, P. A., 1996, How important is pressure is causing dissolution of quartz in sandstones: Journal of Sedimentary Research, v. 66, p. 147–154.

Bolton, A. J., A. J. Maltman, and Q. J. Fisher, 2000, Anisotropic permeability and bimodal pore size distributions of fine-grained marine sediments: Marine and Petroleum Geology, v. 17, p. 657–672.

Bos, B., and C. J. Spiers, 2000, Effect of phyllosilicates on fluid-assisted healing of gouge-bearing faults: Earth and Planetary Science Letters, v. 184, p. 199–210.

Boult, P. J., B. A. Camac, and A. W. Davids, 2002, 3D fault modelling and assessment of top seal structural permeability—Penola trough, onshore Otway Basin: Australian Petroleum Production and Exploration Association Journal, v. 42, p. 151–166.

Byerlee, J. D., 1978, Friction of rocks: Pure and Applied Geophysics, v. 116, p. 615–626.

Castillo, D. A., D. J. Bishop, I. Donaldson, D. Kuek, M. J. De Ruig, M. Trupp, and M. W. Schuster, 2000, Trap integrity in the Laminaria high–Nancar trough region, Timor Sea: Prediction of fault seal failure: Australian Petroleum Production and Exploration Association Journal, v. 40, p. 151–173.

Cecil, C. B., and M. T. Heald, 1971, Experimental investigations of the effects of grain coatings on quartz overgrowth: Journal of Sedimentary Petrology, v. 41, p. 582–584.

Chester, F. M., 1994, Effects of temperature on friction: Constitutive equations and experiments with quartz gouge: Journal of Geophysical Research, v. 99, p. 7247–7261.

Chester, F. M., J. P. Evans, and R. Biegel, 1993, Internal structure and weakening mechanisms of the San Andreas fault: Journal of Geophysical Research, v. 98, p. 771–786.

Crawford, B. R., 1998, Experimental fault sealing; shear band permeability dependency on cataclastic fault gouge characteristics, *in* M. P. Coward, T. S. Daltaban, and H. D. Johnson, eds., Structural geology in reservoir characterization: Geological Society (London) Special Publication 127, p. 27–47.

De Ruig, M. J., M. Trupp, D. J. Bishop, D. Kuek, and D. A. Castillo, 2000, Fault architecture in the Nancar trough/Laminaria area of the Timor Sea, northern Australia: Australian Petroleum Production and Exploration Association Journal, v. 40, p. 174–193.

Dewers, T., and P. Ortoleva, 1991, Influences of clay minerals on sandstone cementation and pressure solution: Geology, v. 19, p. 1045–1048.

Dewhurst, D. N., and R. M. Jones, 2002, Geomechanical, microstructural and petro-physical evolution in experimentally reactivated cataclasites: Applications to fault seal prediction: AAPG Bulletin, v. 86, p. 1383–1405.

Dewhurst, D. N., and R. M. Jones, 2003, Influence of physical and diagenetic processes on fault geomechanics and reactivation: Journal of Geochemical Exploration, v. 78–79, p. 153–157.

Dewhurst, D. N., R. M. Jones, R. R. Hillis, and S. D. Mildren, 2002, Microstructural and geomechanical characterisation of fault rocks from the Carnarvon and Otway basins: Australian Petroleum Production and Exploration Association Journal, v. 42, p. 167–186.

Downey, M. W., 1984, Evaluating seals for hydrocarbon accumulations: AAPG Bulletin, v. 68, p. 1752–1763.

Duddy, I. R., B. Erout, P. F. Green, P. V. Crowhurst, and P. J. Boult, 2003, Timing constraints on the structural history of the Otway Basin and implications for hydrocarbon prospectivity around the Morum high, South Australia: Australian Petroleum Production and Exploration Association Journal, v. 43, p. 39–59.

Faulkner, D. R., and E. H. Rutter, 2001, Can the maintenance of overpressured fluids in large strike slip fault zones explain their apparent weakness?: Geology, v. 29, p. 503–506.

Firoozabadi, A., and H. J. Ramey, 1988, Surface tension of water-hydrocarbon systems at reservoir conditions: Journal of Canadian Petroleum Technology, v. 27, p. 41–48.

Fisher, Q. J., and R. J. Knipe, 1998, Fault sealing processes in siliciclastic sediments, *in* G. Jones, Q. J. Fisher, and R. J. Knipe, eds., Faulting, fault sealing and fluid flow in hydrocarbon reservoirs: Geological Society (London) Special Publication 147, p. 117–134.

Fisher, Q. J., and R. J. Knipe, 2001, The permeability of faults within siliciclastic petroleum reservoirs of the North Sea and Norwegian continental shelf: Marine and Petroleum Geology, v. 18, p. 1063–1081.

Fjær E., R. M. Holt, P. Horsrud, A. M. Raaen, and R. Risnes, 1992, Petroleum related rock mechanics: Developments in Petroleum Science, v. 33, 338 p.

Fournier, R. O., 1996, Compressive and tensile failure at high fluid pressure where pre-existing fractures have cohesive strength, with application to the San Andreas fault: Journal of Geophysical Research, v. 101, p. 25,499–25,509.

Fredrich, J. T., and B. Evans, 1992, Strength recovery along simulated faults by solution transfer processes, *in* W. Wawersik, ed., Proceedings of the 33rd U.S. National Rock Mechanics Symposium: Rotterdam, Balkema, p. 121–130.

Geary, G. C., A. E. Constantine, and I. S. A., Reid, 2001, New perspectives on structural style and petroleum prospectivity, offshore eastern Otway Basin: Petroleum Exploration Society, Eastern Australian Basins Symposium, v. 1, p. 507–517.

Gibson, R. G., 1994, Fault zone seals in siliciclastic strata of the Columbus basin, offshore Trinidad: AAPG Bulletin, v. 78, p. 1372–1385.

Gibson, R. G., 1998, Physical character and fluid flow properties of sandstone-derived fault zones, in M. P. Coward, T. S. Daltaban, and H. D. Johnson, eds., Structural geology in reservoir characterization: Geological Society (London) Special Publication 127, p. 83–97.

Hesthammer, J., P. A. Bjørkum, and L. Watts, 2002, The effect of temperature on sealing capacity of faults in sandstone reservoirs: Examples from the Gullfaks and Gullfaks Sør fields, North Sea: AAPG Bulletin, v. 86, p. 1733–1751.

Hickman, S., R. Sibson, and R. Bruhn, 1995, Introduction to special section: Mechanical involvement of fluids in faulting: Journal of Geophysical Research, v. 100, p. 12,831–12,840.

Hill, A. J., and P. J. Boult, 2001, Maturity modelling, hydrocarbon occurrence and shows, in P. J. Boult and J. E. Hibburt, eds., The petroleum geology of South Australia, v. 1: Otway Basin: Department of Primary Industries and Resources, Petroleum Geology of South Australia Series, chapter 9, p. 1–30.

Hunt, S. P., and P. J. Boult, 2005, Distinct-element stress modeling in the Penola trough, Otway Basin, South Australia, in P. Boult and J. Kaldi, eds., Evaluating fault and cap rock seals: AAPG Hedberg Series, no. 2, p. 199–213.

Jones, R. M., and R. R. Hillis, 2003, An integrated, quantitative approach to assessing fault seal risk: AAPG Bulletin, v. 87, p. 507–524.

Jones, R. M., P. J. Boult, R. R. Hillis, S. D. Mildren, and J. G. Kaldi, 2000, Integrated hydrocarbon seal evaluation in the Penola trough, Otway Basin: Australian Petroleum Production and Exploration Association Journal, v. 40, p. 194–212.

Jones, R. M., D. N. Dewhurst, R. R. Hillis, and S. D. Mildren, 2002, Geomechanical fault characterization: Impact on quantitative fault seal risking: Society of Petroleum Engineers–International Society for Rock Mechanics, SPE paper 78213, 8 p.

Karner, S. L., C. J. Marone, and B. Evans, 1997, Laboratory study of fault healing and lithification in simulated fault gouge under hydrothermal conditions: Tectonophysics, v. 277, p. 41–55.

Knipe, R. J., 1997, Juxtaposition and seal diagrams to help analyse fault seals in hydrocarbon reservoirs: AAPG Bulletin, v. 81, p. 187–195.

Lindsay, N. G., F. C. Murphy, J. J. Walsh, and J. Watterson, 1993, Outcrop studies of shale smears on fault surfaces: International Association of Sedimentologists Special Publication 15, p. 113–123.

Little, B. M., and S. E. Phillips, 1995, Detrital and authigenic mineralogy of the Pretty Hill Formation in the Penola trough, Otway Basin: Implications for future exploration and production: Australian Petroleum Exploration Association Journal, v. 35, p. 538–557.

Main, I., K. Mair, O. Kwon, S. Elphick, and B. Ngwenya, 2001, Experimental constraints on the mechanical and hydraulic properties of deformation bands in porous sandstones: A review, in R. E. Holdsworth, R. A. Strachan, J. Magloughlin, and R. J. Knipe, eds., The nature and tectonic significance of fault zone weakening: Geological Society (London) Special Publication 186, p. 43–63.

Mehin, K., and A. G. Link, 1997, Kitchens, kettles and cups of hydrocarbons, Victorian Otway Basin: Australian Petroleum Production and Exploration Association Journal, v. 37, p. 285–300.

Mildren, S. D., R. R. Hillis, and J. Kaldi, 2002, Calibrating predictions of fault seal reactivation in the Timor Sea: Australian Petroleum Production and Exploration Association Journal, v. 42, p. 187–202.

Mildren S. D., R. R. Hillis, D. N. Dewhurst, P. J. Lyon, J. J. Meyer, and P. J. Boult, 2005, FAST: A new technique for geomechanical assessment of the risk of reactivation-related breach of fault seals, in P. Boult and J. Kaldi, eds., Evaluating fault and cap rock seals: AAPG Hedberg Series, no. 2, p. 73–85.

Olsen, M. P., C. H. Scholz, and A. Leger, 1998, Healing and sealing of a simulated fault gouge under hydrothermal conditions; implications for fault healing: Journal of Geophysical Research, v. 103, p. 7421–7430.

Ramsey, J. G., 1967, The folding and fracturing of rocks: New York, McGraw-Hill, 568 p.

Rice J. R., 1992, Fault stress states, pore pressure distributions and the weakness of the San Andreas fault, in B. Evans and T.-F. Wong, eds., Fault mechanics and transport properties of rocks: London, Academic Press, p. 475–503.

Scholz, C. H., 2000, Evidence for a strong San Andreas fault: Geology, v. 28, p. 163–166.

Schowalter, T. T., 1979, Mechanisms of secondary hydrocarbon migration and entrapment: AAPG Bulletin, v. 63, p. 723–760.

Sibson, R. H., 1985, A note on fault reactivation: Journal of Structural Geology, v. 7, p. 751–754.

Sperrevik, S., P. A. Gillespie, Q. J. Fisher, T. Halvorsen, and R. J. Knipe, 2002, Empirical estimation of fault rock properties, in A. G. Koestler and R. Hunsdale, eds., Hydrocarbon seal quantification: Norwegian Petroleum Society Special Publication 11, p. 109–125.

Streit, J. E., 1999, Conditions for earthquake surface rupture along the San Andreas fault system, California: Journal of Geophysical Research, v. 104, p. 17,929–17,939.

Tada, R., and R. Siever, 1989, Pressure solution during diagenesis: Annual Reviews of Earth and Planetary Sciences, v. 17, p. 89–118.

Tenthorey, E., S. F. Cox, and H. F. Todd, 2003, Evolution of strength recovery and permeability during fluid-rock reaction in experimental fault zones: Earth and Planetary Science Letters, v. 206, p. 161–172.

Underhill, J. R., and N. H. Woodcock, 1987, Faulting mechanisms in high porosity sandstones; New Red Sandstone, Arran, Scotland, in M. E. Jones and R. M. F. Preston, eds., Deformation of sediments and sedimentary rocks: Geological Society (London) Special Publication 29, p. 91–105.

Veevers, J. J., C. McA. Powell, and S. R. Roots, 1991, Review of sea floor spreading around Australia: I. Synthesis of the patterns of spreading: Australian Journal of Earth Sciences, v. 38, p. 373–389.

Walderhaug, O., 1996, Kinetic modeling of quartz cementation and porosity loss in deeply buried sandstone reservoirs: AAPG Bulletin, v. 80, p. 731–745.

Walsh, J. J., et al., 2001, Geometric controls on the evolution of normal fault systems, in R. E. Holdsworth, R. A. Strachan, J. Magloughlin, and R. J. Knipe, eds., The nature and significance of tectonic fault zone weakening: Geological Society (London) Special Publication 186, p. 157–170.

Wiprut, D., and M. D. Zoback, 2000, Fault reactivation and fluid flow along a previously dormant normal fault in the northern North Sea: Geology, v. 28, p. 595–598.

Yielding, G., B. Freeman, and D. T. Needham, 1997, Quantitative fault seal prediction: AAPG Bulletin, v. 81, p. 897–917.

Zoback, M. D., and J. H. Healy, 1992, In situ stress measurements to 3.5 km depth in the Cajon Pass scientific research borehole: Implications for the mechanics of crustal faulting: Journal of Geophysical Research, v. 97, p. 5039–5057.

A Regional Analysis of Fault Reactivation and Seal Integrity Based on Geomechanical Modeling: An Example from the Bight Basin, Australia

S. D. Reynolds
Australian School of Petroleum, The University of Adelaide, Adelaide, South Australia, Australia

E. Paraschivoiu
Primary Industries and Resources of South Australia, Adelaide, South Australia, Australia

R. R. Hillis
Australian School of Petroleum, The University of Adelaide, Adelaide, South Australia, Australia

G. W. O'Brien
Australian School of Petroleum, The University of Adelaide, Adelaide, South Australia, Australia

ABSTRACT

The Bight Basin is a major frontier basin of Jurassic–Cretaceous age, which is currently undergoing renewed exploration interest. Although only limited data is available for understanding the petroleum systems in the basin, several observations indicate that poor fault seal integrity may represent a key exploration risk. The presence of a paleo-oil column in the Jerboa-1 well, interpreted gas chimneys, oil slicks, and asphaltite strandings indicate that seal failure caused by fault reactivation is potentially a significant issue in the Bight Basin. Thus, in this study, we investigated the likelihood that faults in the Bight Basin will undergo sufficient structural reactivation to induce fault seal failure, under the regional in-situ stress field. Fault reactivation risk was assessed for two sets of faults that represent extensional events of Late Jurassic (Sea Lion faults) and Late Cretaceous age (Tiger faults).

Analysis of in-situ stress data suggests that the region is currently under a strike-slip or normal stress regime. Interpretation of borehole breakouts from six wells indicates the average maximum horizontal stress orientation is 130°N. Although the magnitudes of the three principal stresses could not be unequivocally constrained, plausible ranges of values were determined based on well data. Pore pressure in wells in the region is hydrostatic except in Greenly-1, where moderate overpressure occurs.

This study assesses the risk of fault reactivation using the fault analysis seal technology (FAST) technique. The FAST technique evaluates the increase in pore pressure (ΔP) required to cause reactivation as a measure of fault reactivation risk. In all cases investigated, faults striking 40($\pm$15)°N of any dip are the least likely to be reactivated. Thus, traps requiring such faults to be sealing are the least likely to be breached. Fault reactivation risk for the strike-slip and normal stress regimes have been plotted in map view on a series of fault orientations for the Sea Lion and the Tiger faults using a range of hypothetical dips. The results for these hypothetical dips clearly demonstrate the importance of knowing both the strike and dip of a particular fault when conducting a three-dimensional fault seal analysis, because the risk can range from relatively low risk at 25° dip to relatively high risk at 70° dip, with differences being more significant for certain fault orientations.

INTRODUCTION

Assessing trap integrity is of major importance in hydrocarbon exploration. Entrapment and preservation of oil and gas accumulations is dependent on the sealing potential of the rocks surrounding the reservoir, as a result of their capillary properties, geometries, and mechanical integrity. Fault-bound hydrocarbon traps rely on both the sealing potential of the cap seal and the fault seal. Cap and fault seals can be breached by several mechanisms, including membrane failure, fault juxtaposition, and dynamic reactivation (Watts, 1987; Jones et al., 2000). Although we acknowledge that it is important to have a full understanding of all the mechanisms of seal failure in a basin, in this study, we only focus on the failure of the fault seal caused by dynamic reactivation.

Fault reactivation within the in-situ stress field has been demonstrated to control leakage of hydrocarbons from the subsurface in several regions around the world. For example, trap breaching in the North Sea (Gaarenstroom et al., 1993; Wiprut and Zoback, 2000), the Gulf of Mexico (Finkbeiner et al., 2001), and the Timor Sea (Castillo et al., 1998, 2000; Hillis, 1998; de Ruig et al., 2000; Mildren et al., 2002) has been related to faulting and fracturing associated with the in-situ stress field. This study investigates the use of a regional-scale geomechanical model in assessing the risk of fault reactivation to gain some understanding on seal integrity and fault-related risks in the Bight Basin.

The Bight Basin is a largely unexplored basin located along the southern margin of Australia. Exploration costs in the basin are particularly high because of the significant water depth over much of the basin (as much as 5000 m [16,400 ft]) and the remote location. Hence, a real need exists to understand all of the risks associated with hydrocarbon exploration to minimize the overall risk prior to drilling. Fault seal failure may represent a key exploration risk in the Bight Basin as suggested by the presence of a paleo-oil column in the Jerboa-1 well, interpreted gas chimneys, oil slicks, and asphaltite strandings (Ruble et al., 2001; Struckmeyer et al., 2002). Assessment of the fault reactivation risk uses the contemporary in-situ stress field and, hence, provides an approximation for the present-day leakage of hydrocarbons.

REGIONAL GEOLOGICAL FRAMEWORK OF THE BIGHT BASIN

The Bight Basin is an important frontier petroleum exploration province, located along the southern margin of Australia in the Great Australian Bight (Figure 1). However, only 10 exploration wells have been drilled to date in a basin that spans an area of more than 800,000 km^2 (308,881 mi^2), straddling the border of Western Australia and South Australia and extends onshore and offshore in water depths more than 5000 m (16,400 ft). Broad-scale regional geophysical surveying and limited drilling have revealed that the central eastern part of the basin contains four principal depocenters, namely, the Ceduna, Duntroon, Eyre, and Recherche subbasins. Two thin platforms, the Madura and Couedic shelves, are located along the northern and eastern margins of the basin (Totterdell et al., 2003).

The basin was formed as a result of Jurassic–Cretaceous extension and subsidence following the rifting between the Antarctic and Australian plates. According to Totterdell et al. (2003), a first episode of upper crustal extension, with a northwest–southeast to north–south extension direction occurred during the Middle–Late Jurassic to Early Cretaceous (Figure 2) and resulted in oblique to strongly oblique extension and

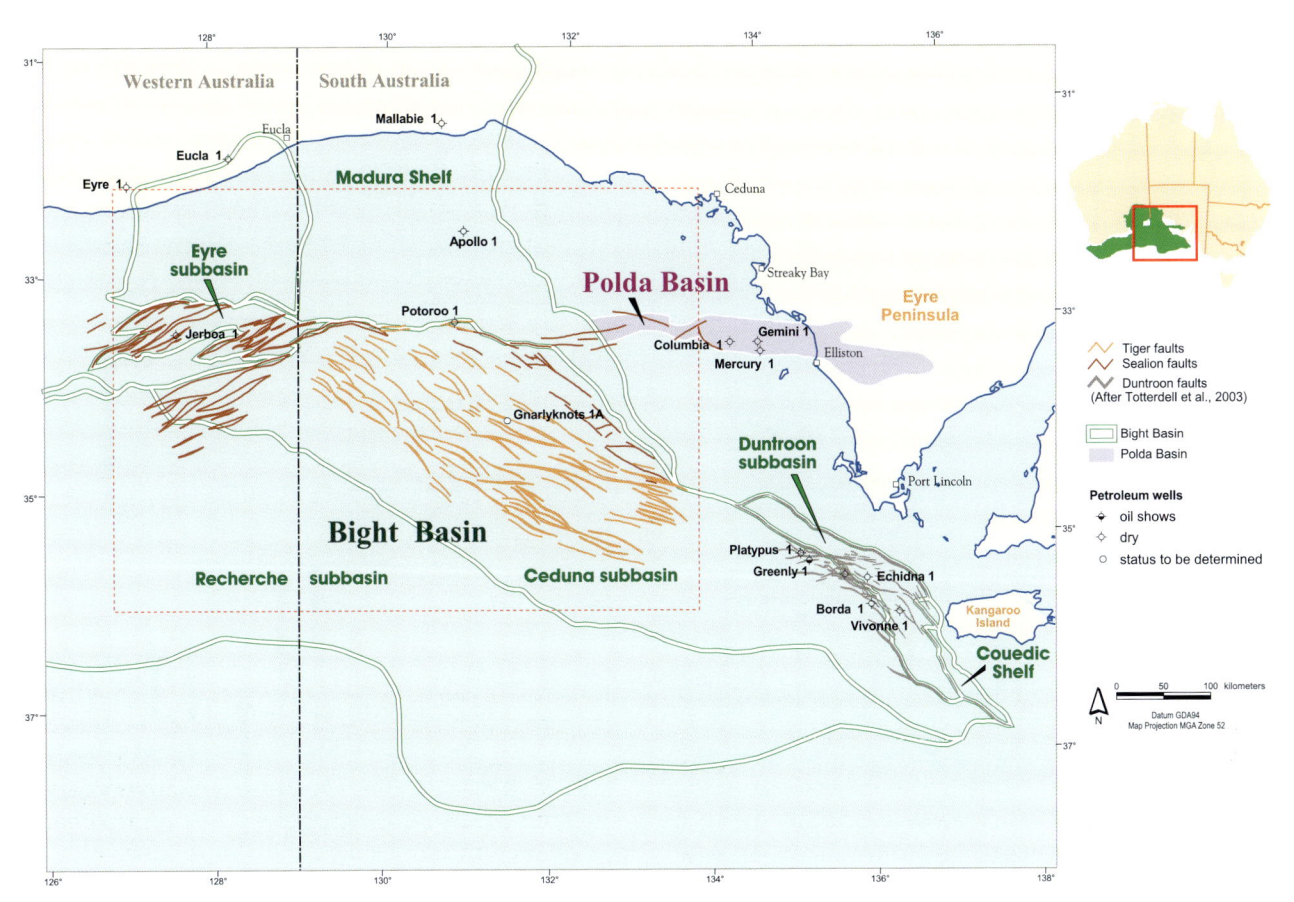

FIGURE 1. Regional map of the South Australian sector of the Great Australian Bight, showing locations of subbasins, key wells, and regional tectonic elements. Sea Lion, Tiger, and Duntroon fault polygons are after Totterdell et al. (2003). Red rectangle denotes area displayed in Figure 9.

the formation of en echelon half graben in the Eyre, inner Recherche, eastern Ceduna, and Duntroon subbasins. The half-graben-bounding faults (Sea Lion faults) appear to strike in an east to east-northeast direction. Early Cretaceous postrift thermal subsidence was followed by a phase of accelerated subsidence, which commenced in the late Albian and continued until continental breakup in the late Santonian–early Campanian. A system of gravity-driven, detached extensional, and contractional structures developed in the Cenomanian as a result of deltaic progradation in the Ceduna subbasin, whereas a system of northwest–southeast extensional faults (Tiger faults) developed during the Turonian–Santonian extension phase. The late Santonian breakup was followed by another period of thermal subsidence and the establishment of a passive margin. The fault trends analyzed in this study are the Late Jurassic Sea Lion faults and the Turonian–Santonian Tiger faults (Figure 1), interpreted by Totterdell et al. (2003).

The platformal portion of the basin appears to have acted as a clastic bypass margin as sediment from the interior was dumped into the rapidly subsiding rift system to the south. The sedimentary section is as much as 15 km (9 mi) thick and is comprised of fluvial to paralic sediments of Late Jurassic–Early Cretaceous age disconformably overlain by nearshore marine to nonmarine Late Cretaceous sediments. A regional unconformity at the top of the Campanian–Maastrichtian Hammerhead supersequence separates the Bight Basin from the thin, transgressive sandstones and massive, open-marine carbonates of the Tertiary Eucla Basin.

The regional sequence-stratigraphic framework has allowed the identification of at least six potential petroleum systems in the Jurassic–Cretaceous sedimentary section (Totterdell et al., 2000). Only the Late Jurassic synrift, lacustrine system (Sea Lion–Minke system) was proven to be effective in the Eyre subbasin, where a paleo-oil column was inferred in the Jerboa-1 well based on the analysis of grains containing oil inclusions. Here, charge is estimated to have occurred during the late Maastrichtian to early Eocene, during a period of fault reactivation and erosion in the subbasin (Blevin et al., 2000).

The Cretaceous systems presented by Totterdell et al. (2000) are conceptual and await validation through further drilling. They include a Berriasian lacustrine system (Southern Right system), an Aptian marine system (upper Bronze Whaler system), a middle Albian marine system (Blue Whale system), a Cenomanian–Santonian marine system (White Pointer–Tiger system), and a Santonian–Campanian marine-deltaic system

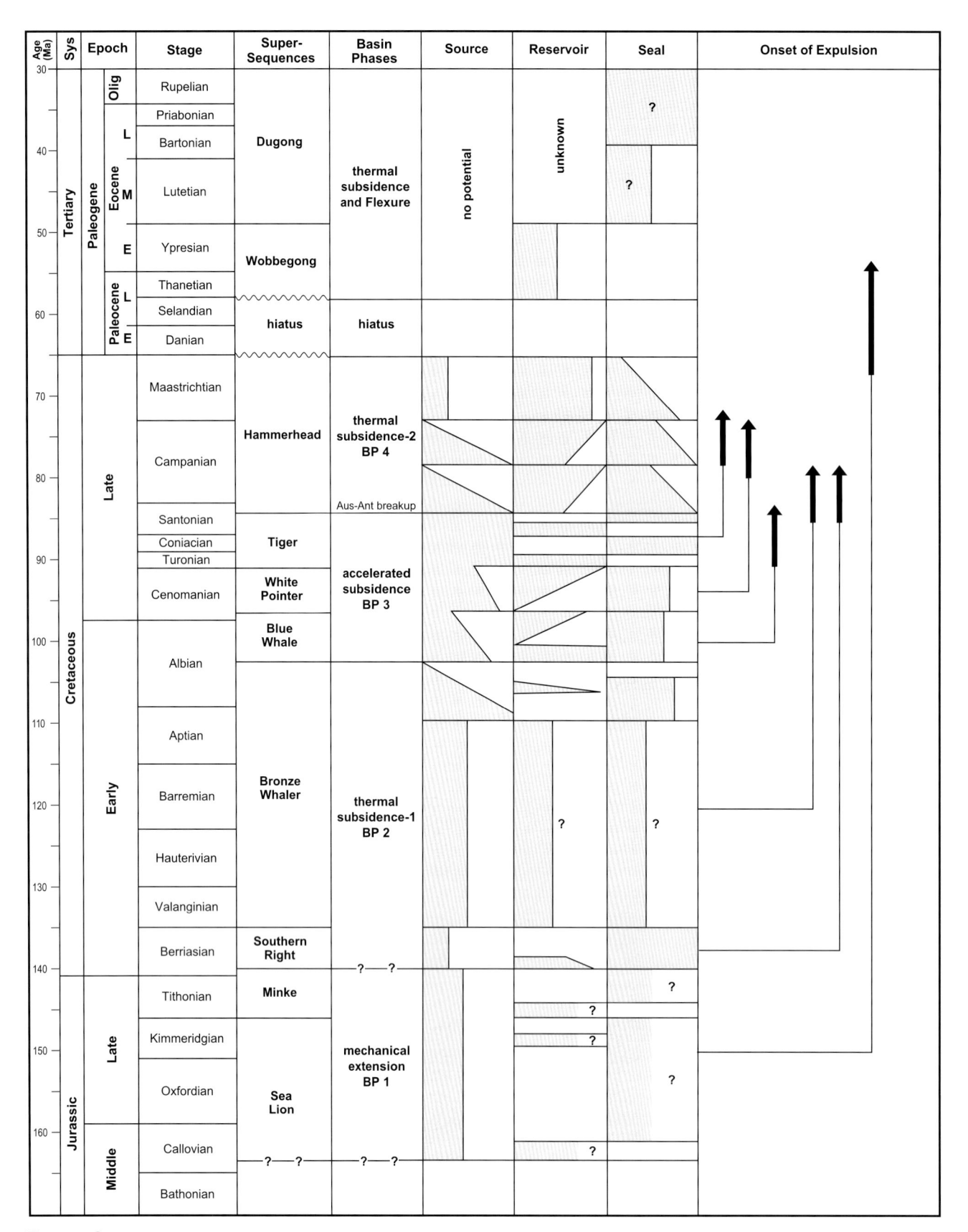

FIGURE 2. Bight Basin correlation chart showing the relationships between sequence stratigraphy and basin phases, modified after Totterdell et al. (2003). Distribution of source, reservoir, and seals in the petroleum systems are modified after Totterdell et al. (2000), and modeled timing of onset of expulsion (arrows) are modified after Struckmeyer et al. (2002).

(Hammerhead system). Oil and gas expulsion from the Albian to Campanian source rocks is modeled to have begun during the Cenomanian to Maastrichtian, although expulsion continued, in phases, into the Tertiary and up to the Holocene in certain parts of the basin (Struckmeyer et al., 2001).

The exploration wells drilled to date encountered only minor oil and gas shows, and the majority of intervals intersected are immature for oil generation. However, the source rock quality in the deeper parts of the basin is inferred to range from good to excellent, and modeling shows that adequate maturities were reached to generate and expel liquid and gaseous hydrocarbons (Struckmeyer et al., 2001).

EMPIRICAL EVIDENCE FOR FAULT SEAL FAILURE IN THE BIGHT BASIN

Identifying the principal exploration uncertainties in the Bight Basin region is difficult. Several observations indicate that poor fault seal integrity may represent a key exploration risk in the Bight Basin. Based on petrophysical analyses of fluid-inclusion samples, Ruble et al. (2001) demonstrated the presence of a 15-m (49-ft)-thick paleo-oil column in Callovian to Kimmeridgian sands in the Jerboa-1 well in the Eyre subbasin (Figure 3). The hypothesis is that Jerboa-1 was charged, probably from a Late Jurassic–Early Cretaceous petroleum system but was later breached during a period of fault reactivation during the Late Cretaceous. Vertical migration of hydrocarbons into the Tertiary sequences may also have occurred subsequent to this Late Cretaceous breaching (Ruble et al., 2001). The fault reactivation and subsequent vertical migration may have been the result of the far-field effects of the collision of the Australian and Asian plates in the late Tertiary.

Other empirical indicators of seal failure are present throughout the Bight Basin. These include the presence of numerous gas chimneys in the Duntroon and Ceduna subbasins, some of which correlate spatially with water column geochemical sniffer anomalies and observations of oil slicks across the Bight Basin using synthetic aperture radar data (Struckmeyer et al., 2002). Other indicators include the well-known presence of asphaltite strandings in the region (Sprigg and Wooley, 1963; Edwards et al., 1998) and colloquial evidence for a relationship between the timing of earthquakes and the occurrence of major strandings in the area. Overall, earthquake activity has been typically focused in the Duntroon and eastern Ceduna subbasins. Most of the earthquakes in the Bight Basin are shallow, with epicenters in the upper 10 km (6 mi) of the sedimentary section, and many are within the top 5 km (3 mi), well within the syn- or postrift sections section. Magnitudes are typically in the range 2–3.5, with some events as high as 4.6 have been recorded on the Couedic Shelf. In addition, a series of earthquakes with magnitudes ranging between 4.2 and 5.2 were recorded along the far southern edge of the Recherche subbasin. The empirical evidence for fault seal failure prompted us to analyze the stress field in the Bight Basin and investigate the likelihood that mapped fault arrays in the subbasins in the region will undergo sufficient structural reactivation to induce fault seal failure.

METHODOLOGY FOR EVALUATING FAULT REACTIVATION IN THE BIGHT BASIN

We have used the results of the in-situ stress field analysis in the Bight Basin (Reynolds et al., 2003) and applied the fault analysis seal technology (FAST) technique

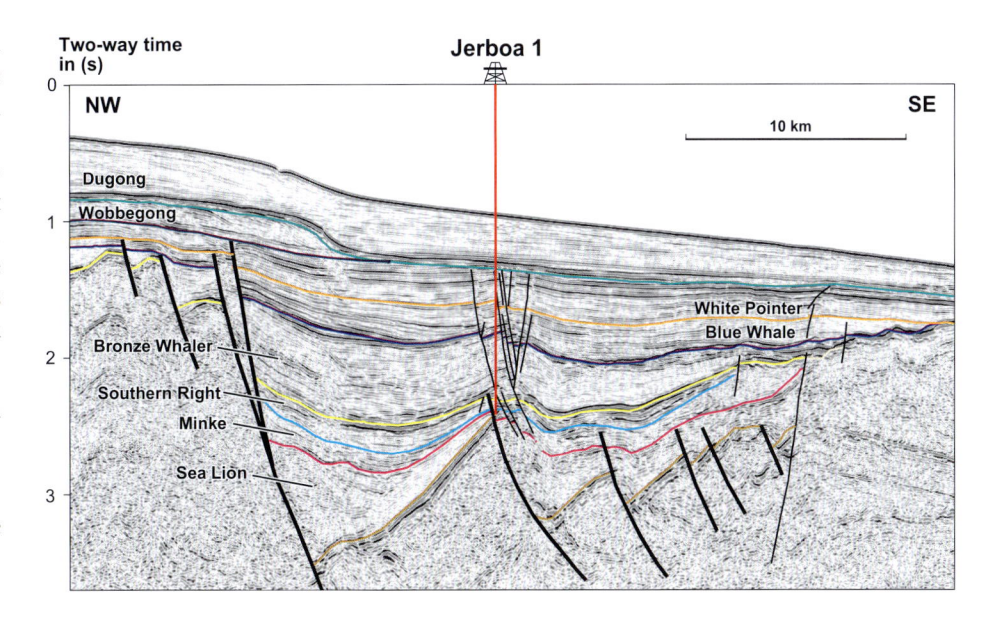

FIGURE 3. Seismic line through the breached trap tested by the Jerboa-1 well in the Eyre subbasin, western Bight Basin. Reactivation faults stop at the base of the Tertiary (modified after Totterdell et al., 2000).

to determine the risk of fault reactivation of the Late Jurassic Sea Lion faults and the Turonian–Santonian Tiger faults. In the FAST technique, the risk of fault reactivation is determined using the stress tensor (Mohr circle) and fault rock strength (failure envelope). Brittle failure is predicted along optimally oriented faults if the ratio of shear to effective normal stress exceeds the coefficient of frictional sliding, which can be illustrated when the Mohr circle intersects the failure envelope. All fault orientations plot within the Mohr circle, and those closest to the failure envelope are at greatest risk of reactivation. The horizontal distance between each fault plane and the failure envelope indicates the increase in pore pressure (ΔP) required to cause reactivation and is used as the measure of the likelihood of fault reactivation in the FAST technique. A small ΔP infers a high likelihood of reactivation, and a large ΔP infers a low likelihood of reactivation. The ΔP value for each plane can be plotted on a stereonet as poles to planes. The risk of reactivation of any pre-existing fault orientation is then read from the stereonet. A composite Griffith–Coulomb failure envelope has been assumed in this study. No fault rock failure envelopes are available for the area, and thus, a cohesive strength of 5 MPa and friction angle of 0.6 have been assumed. For a more detailed discussion on the FAST methodology, see Mildren et al. (2002) and Mildren et al. (2005).

The assessment of fault reactivation risk is commonly applied at prospect scale, where the in-situ stress field is better constrained than at basin scale, and depth-converted fault geometries are available from seismic interpretation. However, assessing fault reactivation at a regional, basin scale can follow similar principles, but more care should be exercised in evaluating the reactivation risk, because more uncertainty is placed on the results when basing the interpretation on just one set of parameters averaged and extrapolated over large areas. Fault reactivation predictions presented herein may not be accurate in areas where the regional stress field has been perturbed by local structures, such as faults and contrasting material properties. The inclusion of local stress perturbations is beyond the scope of this chapter and is not actually possible at the current level of knowledge of the area.

IN-SITU STRESS IN THE BIGHT BASIN

The most important step in assessing the risk of stress-induced fault reactivation that could lead to fault seal failure is to construct a well-constrained geomechanical model. The geomechanical model consists of the in-situ stress field, fault rock properties, and pore pressure. The in-situ stress field is made up of three principal stresses, which are assumed to be the vertical stress, the maximum horizontal stress, and the minimum horizontal stress. This assumption is reasonable given the generally flat-lying seabed surfaces of sedimentary basins. To define the in-situ stress field, the orientation and magnitude of the three stress components must be determined.

The in-situ stress field was determined from assessing the drilling and logging data acquired from the nine open-file exploration wells drilled in the Bight Basin (Reynolds et al., 2003). Of these nine wells, six are clustered in a tight group in the Duntroon and eastern Ceduna subbasins. To gain a better understanding of the regional stress field in the region, additional wells from the adjacent Polda Basin were included in this study (Figure 1). Stress information from the Australian stress map database was also included in the study. The reader is referred to Reynolds et al. (2003) for a more detailed description of the stress field determination in the Bight Basin.

The water depth across the Bight Basin ranges from less than 100 m (3300 ft) on the shelf to more than 5000 m (16,400 ft) in the deepest parts of the Recherche subbasin. This variation poses a significant problem when attempting to analyze the in-situ stress field for the entire basin. At any location and depth in the basin, the stress caused by the weight of the water column contributes to the magnitude of the in-situ stress field. To overcome the problem associated with water depth, effective stress (total stress minus pore pressure) has been used in this study, instead of the total stress, which is more typically used.

Stress Orientations

The maximum horizontal stress (S_{Hmax}) orientation in the Bight Basin was determined by interpreting borehole breakout directions in logs from a four-arm dipmeter (high-resolution dipmeter tool, HDT) from four wells and image log data (Formation Microscanner, FMS) from two wells. The average S_{Hmax} orientation calculated from the six wells is 130°N. The reader is referred to Reynolds et al. (2003) for a more detailed description of the stress field determination in the Bight Basin.

The S_{Hmax} orientations for the Bight Basin and surrounding regions are plotted in Figure 4. Stress orientations could only be determined for the wells that are located on the eastern side of the Bight Basin. The average S_{Hmax} orientation of 130°N for the available wells in the Bight Basin is consistent with S_{Hmax} orientations in the Otway Basin farther to the east (Figure 4). Stress trajectories, which are essentially regionally averaged stress orientations, for the Australian stress field have been calculated by Hillis and Reynolds (2000) and plotted in Figure 4 to obtain a better understanding of

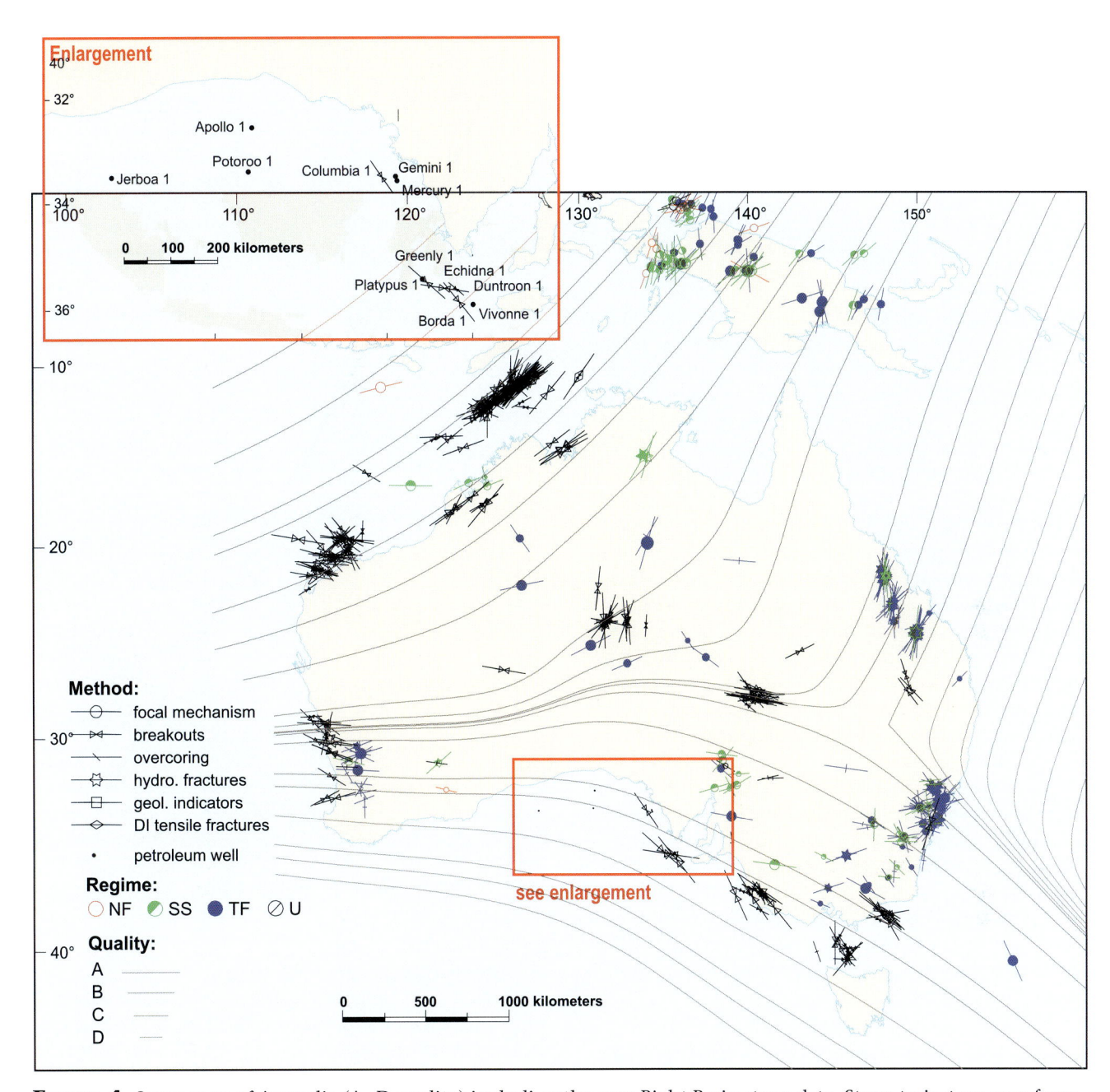

Analysis of Fault Reactivation and Seal Integrity Based on Geomechanical Modeling   63

FIGURE 4. Stress map of Australia (A–D quality) including the new Bight Basin stress data. Stress trajectory map from Hillis and Reynolds (2000) has been plotted to highlight the regional trends across the Australian continent. The S_{Hmax} orientation for the Bight Basin is reasonably consistent with the stress trajectories in the region. Enlargement: Stress map of the Bight Basin showing A–D quality stress indicators. Orientation of vector represents the S_{Hmax} orientation, and length of vector represents the data quality. Wells with no data or E quality data are represented by a dot. NF = normal faulting stress regime; SS = strike-slip faulting stress regime; TF = reverse (thrust) faulting stress regime; U = unknown stress regime.

the regional stress field over the entire Bight Basin. The previously calculated stress trajectories on the eastern side of the Bight Basin are consistent with the new S_{Hmax} orientation determined from the wells in the region. On the western side of the Bight Basin, the stress trajectories indicate a more east–west orientation, which is caused by the influence of the east–west S_{Hmax} orientation in the Perth region to the west (Hillis and Reynolds, 2000; Reynolds and Hillis, 2000). Because of the lack of available data in the western Bight Basin, we are unable to verify if the S_{Hmax} orientation rotates to an east–west orientation in the western part of the Bight Basin. As such, we have used the average S_{Hmax} orientation of 130°N for the entire Bight Basin. Additional drilling in the western Bight Basin would allow the true stress orientations in that area to be determined.

Vertical Stress Magnitude

The vertical or overburden stress (S_v) at a specified depth can be equated with the pressure exerted by the weight of the overlying rocks and expressed as

$$S_v = \int_0^z \rho(z)g\,dz \qquad (1)$$

where $\rho(z)$ is the density of the overlying rock column at depth z, and g is the acceleration caused by gravity. Vertical stress magnitudes were determined using density log data for a total of 10 wells in the Bight and Polda basins. The density logs were initially filtered for bad hole conditions using the bulk density correction density (DRHO) and the caliper data. Vertical stress calculations require that the density log be integrated from the surface (here sea level, assuming the water column has a density of 1.03 g/cm^3). However, the density logs are not commonly run from the surface. The average density from the surface to the top of the density log run can be estimated by converting check-shot velocity data to density using the Nafe–Drake velocity-density transform (Ludwig et al., 1970). To account for the variation in water depth, the vertical stress profiles have been calculated as effective vertical stress (S'_v), assuming normally pressured sediments (Figure 5). The effective vertical stress in the Bight Basin is closely approximated by the power law function

$$S'_v = 10.46 z^{1.179} \qquad (2)$$

where effective vertical stress is in megapascals and z is the depth in kilometers below seabed.

Minimum Horizontal Stress Magnitude

A total of seven leak-off tests and eight formation integrity tests were performed in four wells over the Bight region. Formation integrity tests are not reliable indicators of the minimum horizontal stress, because no fracture is created at the wellbore wall, and hence, they were not used to constrain the minimum horizontal stress magnitude. The reliable leak-off test pressures were plotted along with the formation integrity tests as effective stress magnitudes to compare wells in varying water depths. The lower bound to the effective pressures from the leak-off tests suggests that the effective minimum horizontal stress (S'_{hmin}) gradient is approximately 6 MPa/km. Because of the lack of leak-off data, especially below 2000 m (6600 ft), the S'_{hmin} gradient for the Bight Basin cannot be well constrained. Nevertheless, it is clear from the results obtained that the magnitude of S'_{hmin} is less than that of S'_v (Figure 6). Hence, the Bight Basin is either in a strike-slip faulting ($S'_{hmin} < S'_v < S'_{Hmax}$) or normal faulting ($S'_{hmin} < S'_{Hmax} < S'_v$) stress regime.

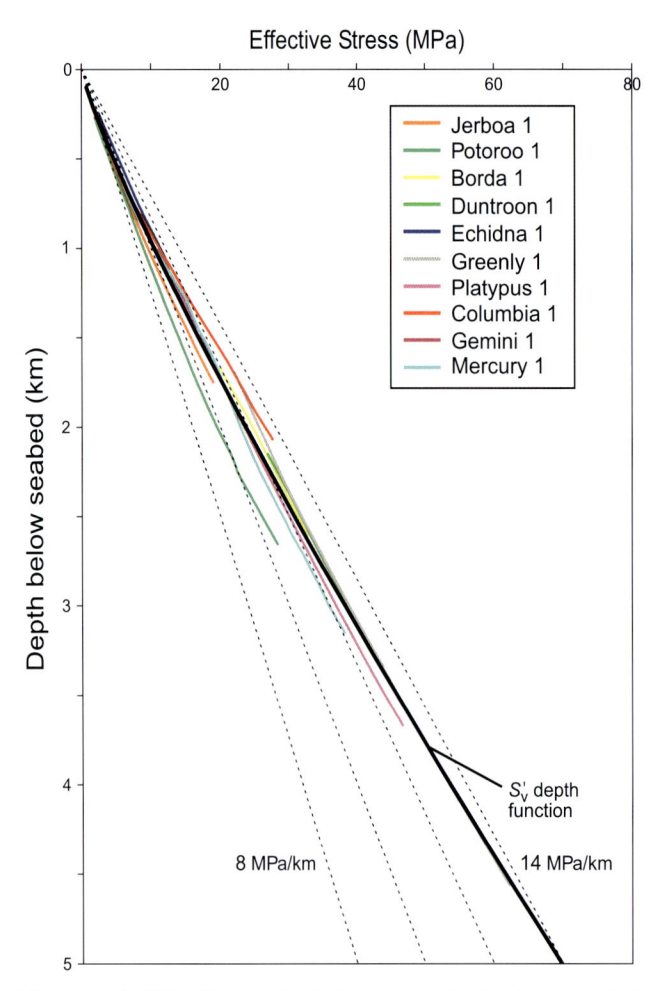

FIGURE 5. Effective vertical stress magnitudes from seabed for the Bight Basin. A power law function has been used to approximate the effective vertical stress.

Maximum Horizontal Stress Magnitude

The magnitude of S_{Hmax} is the most difficult component of the stress tensor to quantify. Many of the methods commonly applied for constraining S_{Hmax} could not be applied in the Bight Basin because of a lack of relevant data. The occurrence of borehole breakouts and drilling-induced tensile fractures could not be used to constrain S_{Hmax} because drilling-induced tensile fracture was not present in the image logs, and rock strength data were not available. Hydraulic fracture test-based techniques could not be applied because no extended leak-off tests or minifracture tests have been undertaken. Nonetheless, broad limits can be placed on S_{Hmax} based on the frictional limits to stress beyond which

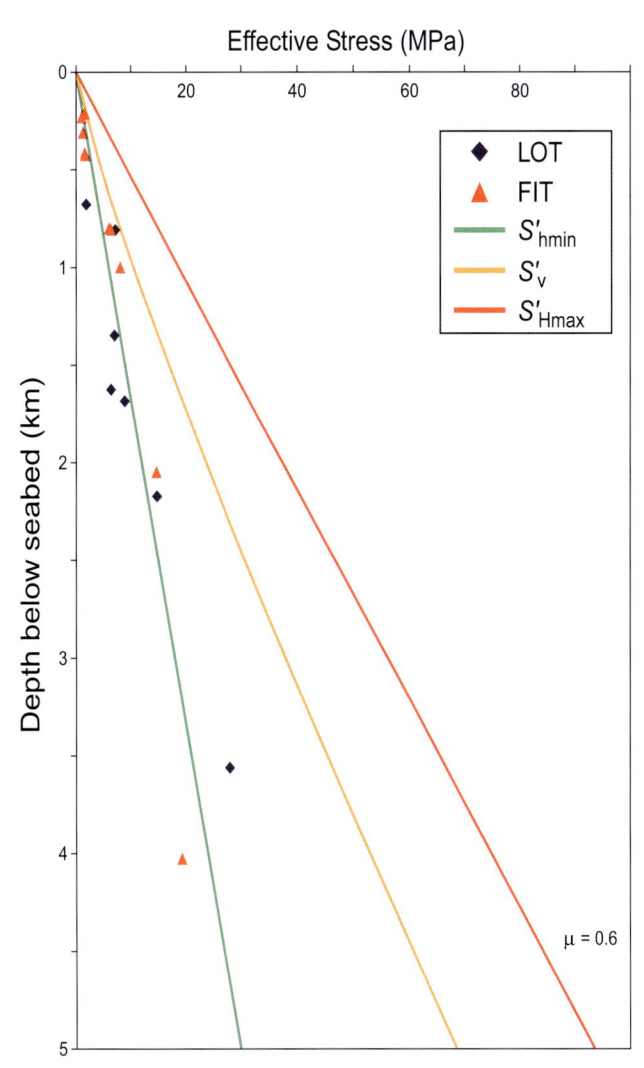

FIGURE 6. Effective stress-depth plots for the Bight Basin. S'_{hmin} is represented by the lower bound to effective pressures from leak-off tests, and S'_{Hmax} has been determined from frictional limits. S'_v has been calculated using the power function described by equation 2. LOT = leak-off test; FIT = formation integrity test.

faulting occurs. The magnitude of the effective maximum horizontal stress (S'_{Hmax}) was calculated to remove the effect of the water depth. The magnitude of S'_{Hmax} can be constrained by assuming that the ratio of the maximum to minimum effective stress cannot exceed that required to cause faulting on an optimally oriented, preexisting fault (Sibson, 1974; Jaeger and Cook, 1979; Zoback and Healy, 1984).

The frictional limit to stress is given by

$$\frac{S'_1}{S'_3} \leq \left\{ \sqrt{(\mu^2 + 1)} + \mu \right\}^2 \quad (3)$$

where μ is the coefficient of friction that the crust can support until an optimally oriented preexisting fault slips to regulate the stress magnitudes; S'_1 is the effective maximum principal stress; and S'_3 is the effective minimum principal stress.

For a typical value of $\mu = 0.6$,

$$\frac{S'_1}{S'_3} \leq 3.12 \quad (4)$$

This relationship can be used to estimate the magnitude of S'_{Hmax} in seismically active regions (Zoback and Healy, 1984) and could provide an upper bound to S'_{Hmax} in the relatively seismically passive regions such as the Bight Basin.

In the Bight Basin, S'_{hmin} is most likely less than the effective vertical stress, implying that $S'_{hmin} = S'_3$. The frictional limits to S'_{Hmax} have been determined following equation 4 and are shown in Figure 6, assuming normally pressured sediments. The maximum S'_{Hmax} gradient, based on frictional limits, is 18.7 MPa/km, implying a strike-slip faulting ($S'_{hmin} < S'_v < S'_{Hmax}$) stress regime. A normal faulting ($S'_{hmin} < S'_{Hmax} < S'_v$) stress regime cannot be ruled out, however, because of the lack of data constraining the magnitude of S'_{Hmax}. Consequently, in our analysis of fault reactivation and seal breach risk, three cases (Table 1) have been considered: a strike-slip faulting ($S'_{hmin} < S'_v < S'_{Hmax}$) stress regime case, a normal faulting ($S'_{hmin} < S'_{Hmax} < S'_v$) stress regime case, and a case on the boundary of strike-slip-normal faulting stress regimes. The magnitude of the in-situ stress field for the three cases was determined at a depth of 1000 m (3300 ft) below seabed.

Pore Pressure

Pore pressure measurements were only conducted in the Jerboa-1 and Greenly-1 wells. Repeat formation tests (RFTs) in the two wells indicate hydrostatic pore

TABLE 1. Parameters used in the three cases to model fault reactivation and seal integrity in the Bight Basin. The cases cover a range of possible values of S'_{Hmax} within the frictional limits. The magnitude values have been calculated for a depth of 1000 m (3300 ft) below seabed.

Case	S'_{Hmax} (MPa)	S'_v (MPa)	S'_{hmin} (MPa)	Fault Regime	S_{Hmax} Orientation
I	18.7	10.5	6.0	strike slip	130°N
II	10.5	10.5	6.0	strike slip-normal	130°N
III	8.5	10.5	6.0	normal	130°N

pressures to a depth of approximately 3600 m (11,800 ft) (10.2 MPa/km). Below 3600 m (11,800 ft), the RFT measurements in Greenly-1 indicate a moderate overpressure of 11.8 MPa/km. To obtain a better understanding of the pore-pressure distribution, the mud weights have been considered as an indicator for pore pressure (Figure 7). In general, most of the region is normally pressured, with only a small indication of overpressure below 3600 m (11,800 ft) in Greenly-1. Hydrostatic pressures are assumed in our analysis of fault reactivation risk.

FAULT REACTIVATION AND SEAL INTEGRITY IN THE BIGHT BASIN

In case I (strike-slip stress regime), vertical faults striking between approximately 100 and 160°N are the most likely to be reactivated (Figure 8b). Hence, traps requiring such faults to be sealing are the most likely to be breached in the in-situ stress field. Vertical faults striking 130°N are located between that conjugate shear pair and are also at high risk of reactivation and breach. Faults striking between 75 and 180°N show little reduction in their risk of reactivation with decreasing dip until shallow dips (<40°) are attained. Faults striking 40°N and with any dip (and horizontal planes) are the least likely to be reactivated. Hence, traps requiring such faults to be sealing are the least likely to be breached in the in-situ stress field. At 1-km (0.6-mi) depth, and assuming the failure envelope in Figure 8a, vertical 160 and 100°N-trending faults require an increase in pore pressure of only slightly in excess of 2 MPa for reactivation and seal breach.

Both case II (strike-slip-normal stress regime) and case III (normal stress regime) show significantly less range in ΔP values than in case I (Figure 8b, c). The ΔP values in cases II and III range between 5.8 and 10 MPa. In case II, faults striking 40°N of any dip are the least likely to be reactivated. In case III, however, horizontal faults with dips as much as 30° are the least likely to be reactivated. In general, most fault orientations and dips in both cases II and III show a similar propensity to be reactivated.

In all three cases, faults striking 40(±15)°N of any dip are the least likely to be reactivated. The magnitude of ΔP required to reactivate faults of this orientation decreases from case I to case III. Thus, traps requiring such faults to be sealing are the least likely to be breached in all three of the stress scenarios investigated.

The fault reactivation risk calculated in this study has been plotted on a series of fault orientations for the Sea Lion (Late Jurassic) rift faults and the Tiger (Late Cretaceous: Turonian to Santonian) postrift faults in the Eyre and Ceduna subbasins (Figure 1). These fault polygons are based on the interpretation of Totterdell et al. (2003). The FAST results for case I (strike slip) and case III (normal) for these faults are summarized in map view in Figure 9. In the absence of a depth-converted interpretation of the fault planes, the fault polygons have been assigned a range of hypothetical dips, specifically 25, 40, 55, and 70°, so that an impression can be gained as to how these variously dipping fault arrays would behave under a range of stress conditions. Figure 9 clearly demonstrates the importance of knowing both the strike and dip of a particular fault when conducting a three-dimensional FAST analysis. The northwest-trending faults in the Ceduna subbasin range from relatively low risk at 25° dip to relatively high risk at 70° dip.

IMPLICATIONS FOR EXPLORATION PROSPECTIVITY

The evaluation of the stress field and the potential for fault reactivation has important implications for petroleum exploration in the Bight Basin. Although an

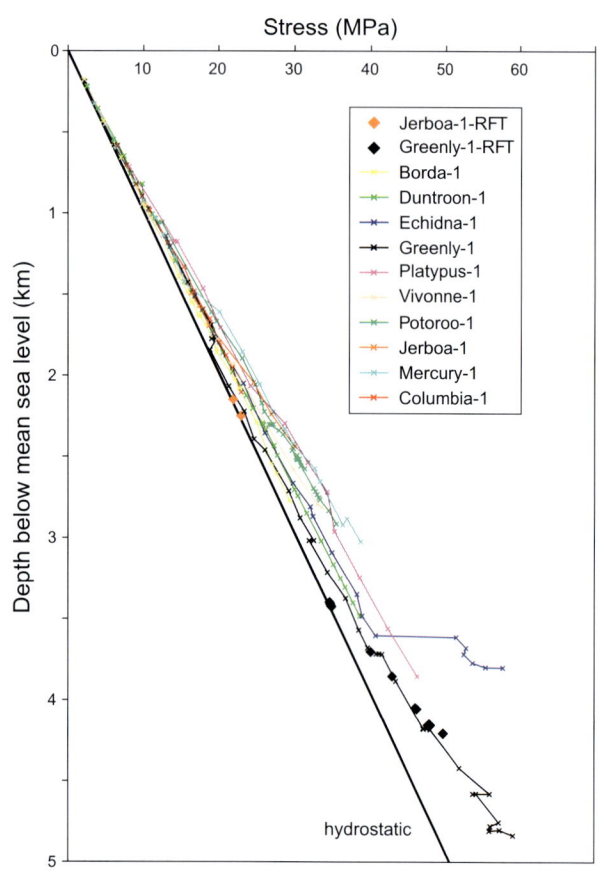

FIGURE 7. Stress-depth plot showing mud weights used in each well and RFTs in Jerboa-1 and Greenly-1. Most of the Bight Basin appears normally pressured. Note that depth is from mean sea level.

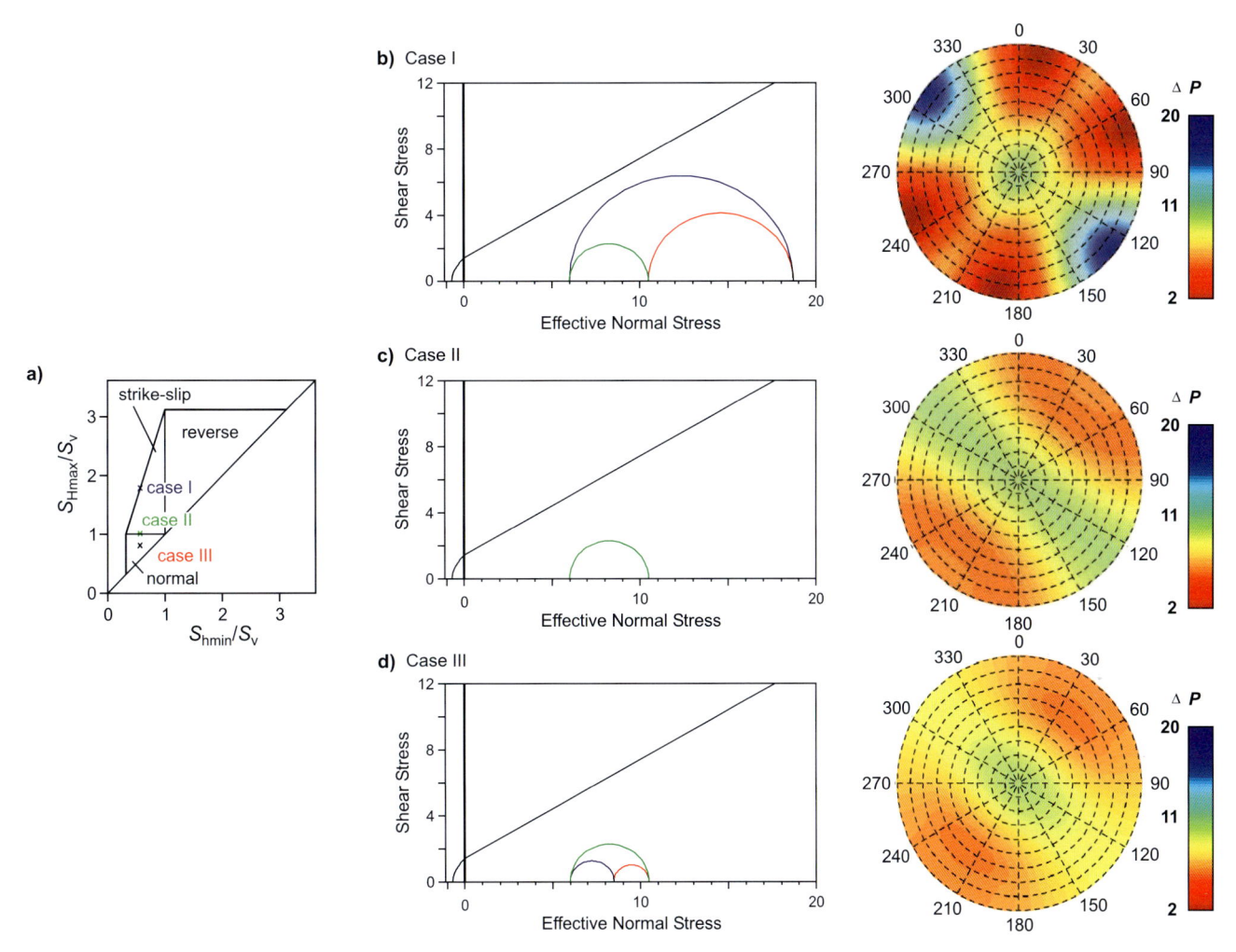

FIGURE 8. Risk of fault reactivation. (a) Location in stress space of the three in-situ stress cases evaluated (Table 1). (b–d) Left-hand side is Mohr's circle of stress and failure envelope (assumed) used to calculate the likelihood of reactivation for each case. Right-hand side is the likelihood of fault and fracture plane reactivation, represented as poles to planes, for the three cases (Table 1). Numerical values on scale refer to increase in fluid pressure required to cause reactivation (ΔP). Equal-angle, lower hemisphere stereographic projection of poles to planes.

accurate assessment of the fault reactivation risk is not possible based on available information, some trends can be established and discussed. Several precautions are required when assessing the results from this study. First, hydrostatic pressures are assumed in our analysis of fault reactivation risk, and hence, in areas where overpressures are anticipated, these predictions would need to be modified. Second, given the complex history of changes in the stress field that controls the structural evolution of sedimentary basins, the in-situ stress field, as constrained herein, cannot be extrapolated back in time and applied to previous structural events. Knowledge of the in-situ stress field can only elucidate contemporary tectonic events.

As shown in the previous section, faults that dip at 25° (Figures 8, 9) have little risk of reactivation, irrespective of the fault orientation and the type of stress field present. However, as the dips increase to 40°, it becomes apparent that the risk of fault seal failure becomes greater in the northwest-trending fault arrays in the Ceduna subbasin. In contrast, the more east- to northeast-trending faults in the Eyre subbasin are at low risk of reactivation. At dips of 55°, the risk of fault seal failure appears to be low in the Eyre subbasin, where the faults have a generally northeast trend. However, in the more northwest-trending fault arrays, the risk of reactivation is much higher, especially in a strike-slip stress regime. An exception is the small, east–west-trending, intrabasinal Sea Lion faults that occur in the overall, northwest-trending Sea Lion faults in the Ceduna subbasin. These faults have a low risk of reactivation compared to the northwest-trending faults that dominate this part of the Bight Basin. At fault dips of 70°, a high risk of reactivation of the northwest-trending fault sets of the Ceduna subbasin is present for all stress regimes. The exception is again the small, more east–west-trending fault arrays. The Eyre subbasin appears to have a low risk of reactivation, particularly

a) Case I (strike-slip regime)

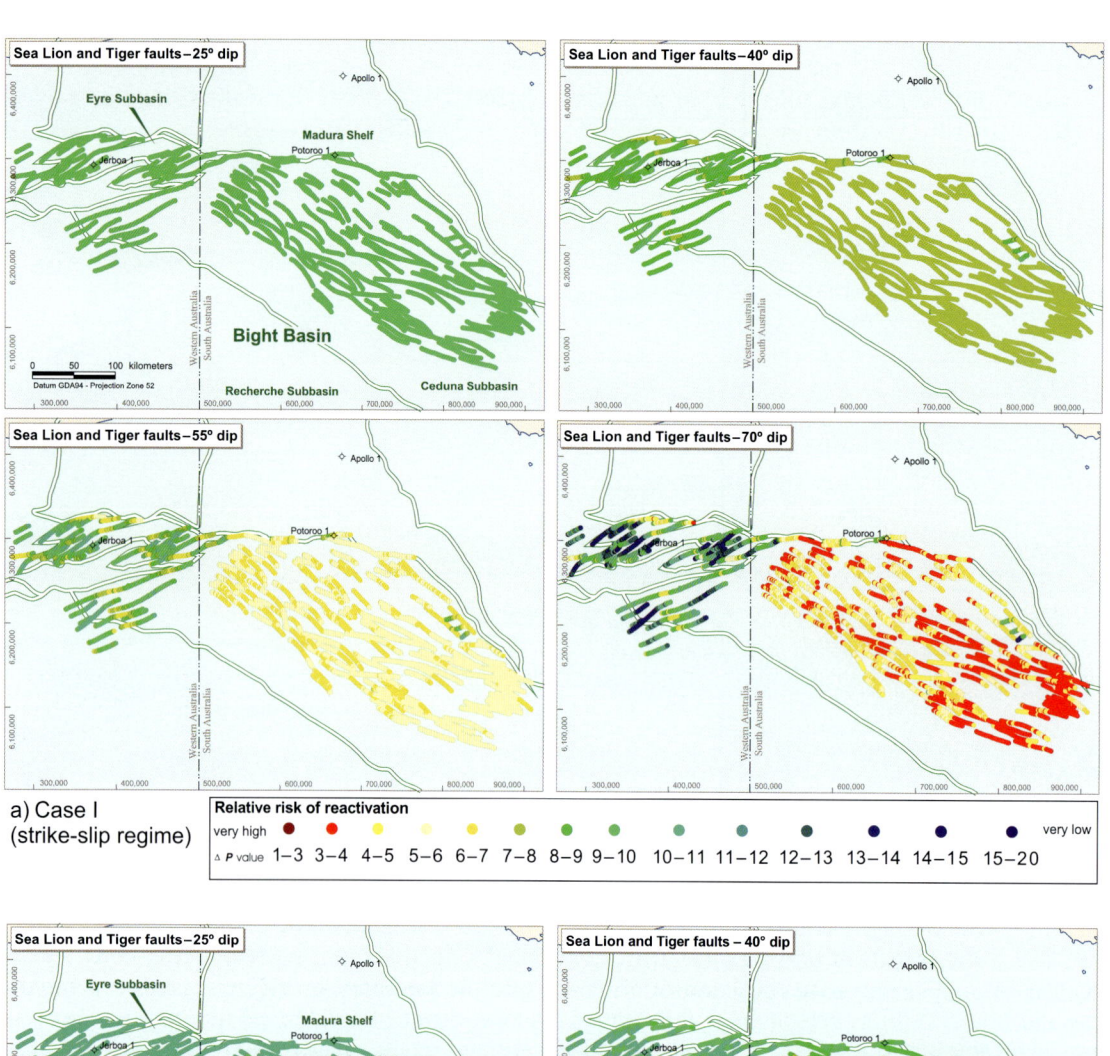

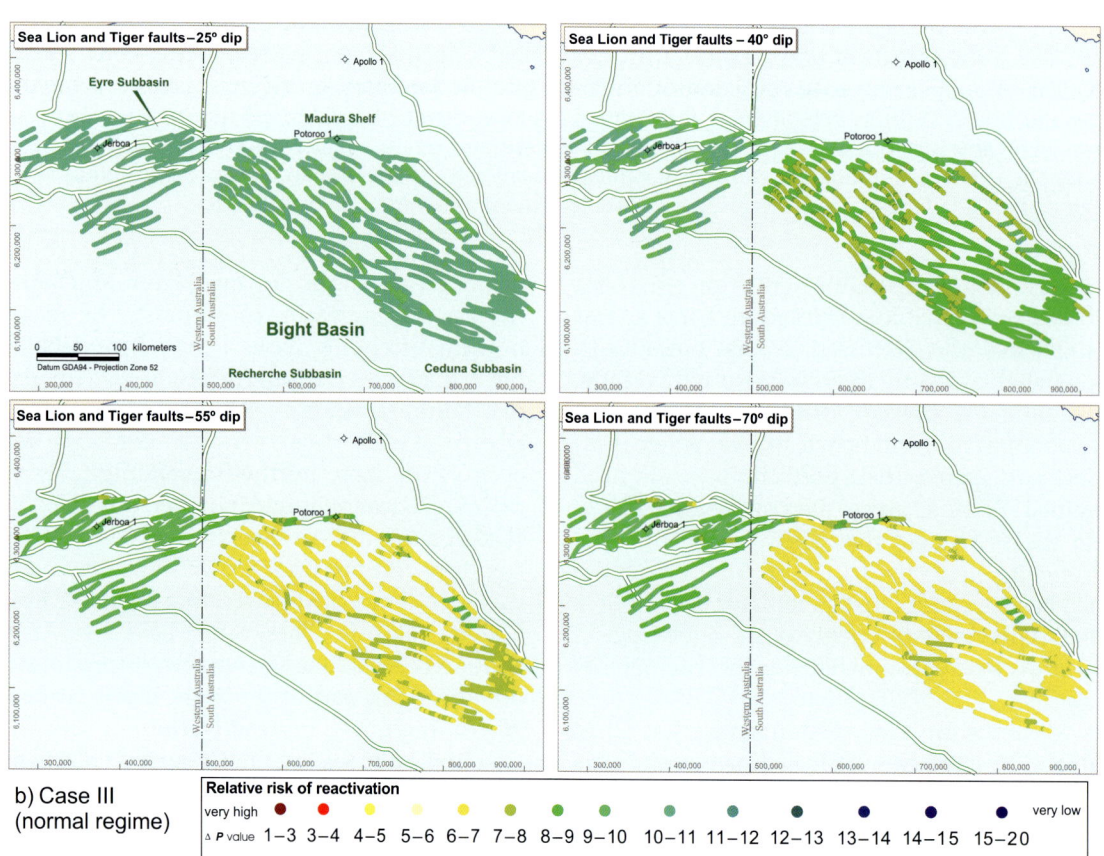

b) Case III (normal regime)

in the strike-slip-normal and strike slip stress regimes. It should be noted, however, that stress trajectories calculated from the Australian stress map database indicate a more east–west S_{Hmax} orientation for the Eyre subbasin than the 130°N S_{Hmax} orientation used for our calculations. Therefore, the results for the Eyre subbasin should be used with care.

Overall, the results suggest that in a strike-slip-normal or normal stress regime, little risk of reactivation exists for fault systems that trend east–west or northeast. For these fault trends, almost no sensitivity to fault dip is present. Clearly, the rift faults of the Eyre subbasin and the intrabasinal, east–west-trending faults of the Ceduna subbasin all have a low risk of fault reactivation. In contrast, the results suggest that the rift and postrift faults of the Ceduna subbasin and, by inference, the Duntroon subbasin have a relatively high risk of reactivation once dips exceed approximately 40° for either a strike-slip-normal or normal stress regime. Traps with the lowest risk in the Ceduna and Duntroon subbasins with respect to reactivation are those with lower ($\lesssim 40°$) dips on the bounding faults or those with a more east–west orientation.

If we consider earthquake data, it appears that the most common earthquakes occur in the eastern Ceduna subbasin and in the Duntroon subbasin, broadly through the area with northwest-trending fault arrays. However, the rest of the Bight appears to be largely aseismic, and this may suggest that the stress regime in the central and western Bight is less conducive to reactivation. Thus, the trends of fault reactivation risk determined herein appear broadly consistent with the earthquake data.

The distribution of oil slicks that have been mapped in the region (Struckmeyer et al., 2002) via the use of satellite-based synthetic aperture radar are generally more common in the deep-water Ceduna subbasin and eastern parts of the Bight Basin, along the northwest fault arrays. Seismic gas chimneys are very common in the eastern Ceduna and Duntroon subbasins and correlate spatially with water column sniffer anomalies. These chimneys typically relate to the northwest-trending fault arrays in the region, so this may support some loss of fault seal integrity in this area.

Interestingly, the only confirmed paleo-oil column in the region was located in the Jerboa-1 well in the Eyre subbasin. This area is predicted, under the present-day stress regime, to be of high fault seal integrity. However, this trap was actually breached in the Late Cretaceous (Ruble et al., 2001). This emphasizes the fact that the stress predictions only relate to the present day and not to paleoreactivation events, when the stresses may have been quite different. It also emphasizes the fact that the timing of hydrocarbon migration (and probably the nature of the hydrocarbon charge) is very important in relation to trap reactivation. Relatively low to moderate fault seal integrity may be beneficial in regions that are now experiencing a high gas charge because this may help to reduce the risk of gas flushing in traps in such areas, which were previously charged with oil (O'Brien and Woods, 1995).

These observations highlight the fact that the results presented here should not be used in isolation. It does appear clear that relatively steeply dipping faults with a generally northwest trend will be prone to reactivation, whereas more east–west- or northeast-trending faults are unlikely to become reactivated, irrespective of dip. To better determine how critical the present-day stress environment is to hydrocarbon prospectivity in the Bight Basin, a complete fault seal assessment that should take into consideration such observations should be integrated with other aspects of the petroleum system, such as the generation history, remote-sensing results, and direct hydrocarbon indicator mapping.

For individual traps, the assessment of fault breaching risk requires detailed prospect studies, so that a geomechanical analysis can be applied to clearly defined, depth-converted fault planes interpreted from seismic data. In addition, in-situ stress field characteristics should be based on fieldwide measurements and the failure envelope constrained by properties specific to the analyzed rocks.

ACKNOWLEDGMENTS

Special thanks to Jennie Totterdell and Barry Bradshaw of Geoscience Australia for permission to use their fault interpretation and for providing the digital fault files, to Peter Boult for valuable comments and suggestions, and to Primary Industries and Resources of South Australia Publishing Services for assistance with graphic files. Fugro Multi Client Services are thanked for permission to publish the seismic image in Figure 3. We thank David Castillo, Isabelle Moretti, and Signe Ottesen for their constructive comments regarding this manuscript.

FIGURE 9. Fault reactivation risks calculated using FAST technique applied to Sea Lion and Tiger fault polygons shown in Figure 1, assuming constant dips of 25, 40, 55, and 70°, respectively, for (a) case I (strike-slip stress regime) and (b) case III (normal stress regime). Results should be used with care in the western part of the Bight Basin and in areas of overpressure.

REFERENCES CITED

Blevin, J. E., J. M. Totterdell, G. A. Logan, J. M. Kennard, H. I. M. Struckmeyer, and J. B. Colwell, 2000, Hydrocarbon prospectivity of the Bight Basin—Petroleum systems analysis in a frontier basin, in 2nd Sprigg Symposium—Frontier Basins, Frontier Ideas, Adelaide, June 29–30, 2000: Geological Society of Australia, Abstracts no. 60, p. 24–29.

Castillo, D. A., R. R. Hillis, K. Asquith, and M. Fisher, 1998, State of stress in the Timor Sea area, based on deep wellbore observations and frictional failure criteria: Application to fault-trap integrity, in P. G. Purcell and R. R. Purcell, eds., The sedimentary basins of Western Australia: 2. Proceedings of Petroleum Exploration Society of Australia Symposium, Perth: p. 326–341.

Castillo, D. A., D. J. Bishop, I. Donaldson, D. Kuek, M. de Ruig, M. Trupp, and M. W. Shuster, 2000, Trap integrity in the Laminaria high–Nancar trough region, Timor Sea: Prediction of fault seal failure using well-constrained stress tensors and fault surfaces interpreted from 3D seismic: Australian Petroleum Production and Exploration Association Journal, v. 40, p. 151–173.

de Ruig, M. J., M. Trupp, D. J. Bishop, D. Kuek, and D. A. Castillo, 2000, Fault architecture and the mechanics of fault reactivation in the Nancar trough/Laminaria area of the Timor Sea, Northern Australia: Australian Petroleum Production and Exploration Association Journal, v. 40, p. 174–193.

Edwards, D., D. M. McKirdy, and R. E. Summons, 1998, Enigmatic asphaltites from the southern Australian margin: Molecular and carbon isotopic composition: Petroleum Exploration Society of Australia Journal, v. 26, p. 106–129.

Finkbeiner, T., M. Zoback, P. Flemings, and B. Stump, 2001, Stress, pore pressure, and dynamically constrained hydrocarbon columns in the South Eugene Island 330 field, northern Gulf of Mexico: AAPG Bulletin, v. 85, p. 1007–1031.

Gaarenstroom, L., R. A. J. Tromp, M. C. D. Jong, and A. M. Brandenburg, 1993, Overpressures in the Central North Sea: Implications for trap integrity and drilling safety, in J. Parker, ed., Petroleum geology of northwest Europe: Proceedings of the 4th Conference, Geological Society (London): p. 1305–1313.

Hillis, R. R., 1998, Mechanisms of dynamic seal failure in the Timor Sea and central North Sea basins, in P. G. Purcell and R. R. Purcell, eds., The sedimentary basins of Western Australia: 2. Proceedings of the Petroleum Exploration Society of Australia Symposium, Perth, Western Australia: p. 313–324.

Hillis, R. R., and S. D. Reynolds, 2000, The Australian stress map: Journal of the Geological Society (London), v. 157, p. 915–921.

Jaeger, J. C., and N. G. W. Cook, 1979, Fundamentals of rock mechanics: London, United Kingdom, Chapman and Hall, 593 p.

Jones, R. M., P. Boult, R. R. Hillis, S. D. Mildren, and J. Kaldi, 2000, Integrated hydrocarbon seal evaluation in the Penola trough, Otway Basin: Australian Petroleum Production and Exploration Association Journal, v. 40, p. 194–211.

Ludwig, W. J., J. E. Nafe, and C. L. Drake, 1970, Seismic refraction, in A. E. Maxwell, ed., The sea: Ideas and observations on progress in the study of the seas, New concepts of sea floor evolution: New York, Wiley-Interscience, v. 4, p. 53–84.

Mildren, S. D., R. R. Hillis, and J. Kaldi, 2002, Calibrating predictions of fault seal reactivation in the Timor Sea: Australian Petroleum Production and Exploration Association Journal, v. 42, p. 187–202.

Mildren S. D., R. R. Hillis, D. N. Dewhurst, P. J. Lyon, J. J. Meyer, and P. J. Boult, 2005, FAST: A new technique for geomechanical assessment of the risk of reactivation-related breach of fault seals, in P. Boult and J. Kaldi, eds., Evaluating fault and cap rock seals: AAPG Hedberg Series, no. 2, p. 73–85.

O'Brien, G. W., and E. P. Woods, 1995, Hydrocarbon-related diagenetic zones (HRDZs) in the Vulcan sub-basin, Timor Sea: Recognition and exploration implications: Australian Petroleum and Exploration Association Journal, v. 35, p. 220–252.

Reynolds, S. D., and R. R. Hillis, 2000, The in situ stress field of the Perth Basin, Australia: Geophysical Research Letters, v. 27, p. 3421–3424.

Reynolds, S., R. Hillis, and E. Paraschivoiu, 2003, In situ stress field, fault reactivation and seal integrity in the Bight Basin, South Australia: Exploration Geophysics, v. 34, p. 174–181.

Ruble, T. E., G. A. Logan, J. E. Blevin, H. I. M. Struckmeyer, K. Liu, M. Ahmed, P. J. Eadington, and R. A. Quezada, 2001, Geochemistry and charge history of a paleo-oil column: Jerboa-1, Eyre sub-basin, Great Australian Bight, in K. C. Hill and T. Bernecker, eds., Eastern Australasian Basins Symposium. A refocused energy perspective for the future: Petroleum Exploration Society of Australia Special Publication, p. 521–529.

Sibson, R. H., 1974, Frictional constraints on thrust, wrench and normal faults: Nature, v. 249, p. 542–544.

Sprigg, R. C., and J. B. Wooley, 1963, Coastal bitumen in South Australia with special reference to observations at Geltwood beach, south-eastern South Australia: Transactions of the Royal Society of South Australia, v. 86, p. 67–103.

Struckmeyer, H. I. M., J. M. Totterdell, J. E. Blevin, G. A. Logan, C. J. Boreham, I. Deighton, A. A. Krassay, and M. T. Bradshaw, 2001, Character, maturity and distribution of potential Cretaceous oil source rocks in the Ceduna sub-basin, Bight Basin, Great Australian Bight, in K. C. Hill and T. Bernecker, eds., Eastern Australasian Basins Symposium: A refocused energy perspective for the future: Petroleum Exploration Society of Australia Special Publication, p. 543–552.

Struckmeyer, H. I. M., A. K. Williams, R. Cowley, J. M. Totterdell, G. Lawrence, and G. W. O'Brien, 2002, Evaluation of hydrocarbon migration and seepage in the Great Australian Bight: Australian Petroleum Production and Exploration Association Journal, v. 42, p. 371–384.

Totterdell, J. M., J. E. Blevin, H. I. M. Struckmeyer, B. E. Bradshaw, J. B. Colwell, and J. M. Kennard, 2000, A new sequence framework for the Great Australian Bight: Starting with a clean slate: Australian Petroleum Production and Exploration Association Journal, v. 40, p. 95–117.

Totterdell, J. M., B. E. Bradshaw, J. B. Willcox, 2003, Structural and tectonic setting, in G. W. O'Brien, E. Paraschivoiu, and J. E. Hibburt, eds., Petroleum geology of South Australia, v. 5: Great Australian Bight, South Australia, Department of Primary Industries and Energy Resources: Petroleum Geology of South Australia Series 5, ch. 4: www.pir.sa.gov.au/dhtml/ss/section.php?sectID=1775&tempID=8 (accessed September 23, 2003).

Watts, N. L., 1987, Theoretical aspects of cap-rock and fault seals for single and two phase hydrocarbon columns: Marine and Petroleum Geology, v. 4, p. 274–307.

Wiprut, D., and M. D. Zoback, 2000, Fault reactivation and fluid flow along a previously dormant normal fault in the northern North Sea: Geology, v. 28, p. 595–598.

Zoback, M. D., and J. H. Healy, 1984, Friction, faulting and in situ stress: Annales Geophysicae, v. 2, p. 689–698.

| 6 |

Mildren, S. D., R. R. Hillis, D. N. Dewhurst, P. J. Lyon, J. J. Meyer, and P. J. Boult, 2005, FAST: A new technique for geomechanical assessment of the risk of reactivation-related breach of fault seals, *in* P. Boult and J. Kaldi, eds., Evaluating fault and cap rock seals: AAPG Hedberg Series, no. 2, p. 73–85.

FAST: A New Technique for Geomechanical Assessment of the Risk of Reactivation-related Breach of Fault Seals

Scott D. Mildren,[1] Richard R. Hillis, Paul J. Lyon, Jeremy J. Meyer,[1]
Australian Petroleum Cooperative Research Center, Australian School of Petroleum, University of Adelaide, Australia

David N. Dewhurst
Australian Petroleum Cooperative Research Center, Commonwealth Scientific and Industrial Research Organization Petroleum, Australian Resources Research Center, Perth, Western Australia

Peter J. Boult
Australian School of Petroleum, University of Adelaide, Australia and also *Department of Primary Industries and Resources South Australia, Adelaide, Australia*

ABSTRACT

Postcharge fault reactivation may cause fault seal breach. We present a new methodology for assessment of the risk of reactivation-related seal breach: fault analysis seal technology (FAST). The methodology is based on the brittle failure theory and, unlike other geomechanical methods, recognizes that faults may show significant cohesive strength. The likelihood of fault reactivation, which is expressed by the increase in pore pressure (ΔP) necessary for fault to reactivate, can be determined given the knowledge of the in-situ stress field, fault rock failure envelope, pore pressure, and fault geometry. The FAST methodology was applied to the fault-bound Zema structure in the Otway Basin, South Australia. Analysis of juxtaposition and fault deformation processes indicated that the fault was likely to be sealing, but the structure was found to contain a residual hydrocarbon column. The FAST analysis indicates that segments of the fault are optimally oriented for reactivation in the in-situ stress field. Microstructural

[1]*Present address:* JRS Petroleum Research, Adelaide, Australia.

Copyright ©2005 by The American Association of Petroleum Geologists.
DOI:10.1306/1060757H23163

evidence of open fractures in a fault zone in the subsurface in an offset well and an SP (self-potential) anomaly associated with a subseismic fault cutting the regional seal in the Zema-1 well support the interpretation that seal breach is related to fracturing.

INTRODUCTION

Fault sealing caused by juxtaposition and deformation processes has received considerable attention, and techniques for the analysis of such, e.g., Allan diagrams, juxtaposition diagrams, and shale smear algorithms, are widely applied (Allan, 1989; Knipe, 1997; Bretan et al., 2003). Well-constrained lithological and juxtaposition data, tied to seismically observable fault zones, can locate potentially leaking sand-on-sand contacts across faults and predict whether such sand-on-sand contacts are likely to be sealed because of deformation processes such as cataclasis or shale smearing (Jev et al., 1993; Hippler, 1997; Fisher and Knipe, 1998). However, whereas such analyses can define the sealing potential of faults that have been inactive since hydrocarbon charge, they do not incorporate the potential for seal breach because of fault reactivation subsequent to charge.

Abundant evidence shows that faults and fractures provide high-permeability conduits for fluid flow during deformation in the brittle crust (e.g., Sibson, 1994; Barton et al., 1995; Dewhurst et al., 1999). Juxtaposition or deformation process seals may be breached if the fault is reactivated subsequent to hydrocarbons charging the trap (Jones and Hillis, 2003). Seal breach caused by fault reactivation has been recognized as a critical risk in the Australian context. For example, in the Timor Sea region, Neogene fault reactivation related to collision between the Australian and Southeast Asian plates has breached many Jurassic or older paleotraps (O'Brien and Woods, 1995; Hillis, 1998; Shuster et al., 1998).

The relative likelihood of fault reactivation can be assessed given the knowledge of the prevailing stress field, fault orientation, pore pressure, and the failure envelope for the fault rocks. Morris et al. (1996) defined slip tendency based on the ratio of shear stress to normal stress acting on a fault surface. Ferrill et al. (1999) defined dilation tendency based on the normal stress acting on a fault plane normalized to the differential stress. Calculations of slip and dilation tendency were used to assess the likelihood of fault reactivation at the proposed high-level radioactive waste repository site at Yucca Mountain, Nevada. Wiprut and Zoback (2000) determined the increase in pore pressure required to induce reactivation of a normal fault in the Visund field, northern North Sea, assuming cohesionless frictional failure. Finkbeiner et al. (2000) assessed the height of hydrocarbon columns that could be sustained without inducing shear or tensile failure on trap-bounding faults in the Gulf of Mexico. These and related studies of the relationship between stresses and fault reactivation and permeability (e.g., Barton et al., 1995; Hickman et al., 1997) assume that the failure envelopes for fault rocks are described by a cohesionless friction law of the Byerlee (1978) type. However, frictional sliding experiments on cohesionless joints or saw-cuts through rocks of the type summarized by Byerlee (1978) do not describe the failure envelopes for cemented fault rocks that may exhibit significant cohesion (Dewhurst and Jones, 2002; Jones et al., 2002).

An alternative geomechanical parameter for assessing the risk of fault reactivation and associated seal breach is presented herein that can incorporate the cohesive strength of faults and, in a single parameter, express the risk of shear or tensile failure. The methodology is applied to the Zema structure in the Otway Basin, South Australia, which contains a fault-bound residual column. The technique can be readily modified to consider areas where shear or tensile failure of intact cap rock presents the key geomechanical risk, either because fault rocks are stronger than intact cap seal rocks or because preexisting faults are misoriented for reactivation. Prior to introducing the new approach, this chapter outlines the background concept of structural permeability and summarizes previously used geomechanical parameters for assessing fracture-related seal breach.

STRUCTURAL PERMEABILITY

Our approach to the geomechanical risking of reactivation-related fault seal breach is based on the concept of structural permeability (Sibson, 1996). Structural permeability is the permeability created by the interaction of various brittle structures (tensile fractures, shear fractures, and hybrid fractures; Figure 1). Such structures are generally created by the pressure of the infiltrating fluids and can be represented by failure criteria expressed in terms of pore pressure (Sibson, 1996) (Table 1). The theory of hybrid fracture generation is relatively contentious. Hybrid fractures can be considered to be multiple jointing events instead of tensile fractures with a shear component (Engelder, 1999). However, the Coulomb–Mohr envelope is still considered to be a valid predictor of failure in the tensile region as shown by Brace's experiments (Brace, 1960; Engelder, 1999). The method presented herein is removed from the theory of hybrid fracture generation and is used to predict brittle failure instead of the form by which it is manifested.

FIGURE 1. The effective normal (σ'_n) and shear (τ) stresses leading to shear and tensile fracturing assuming a cohesionless Coulomb failure envelope ($\tau = \mu_s\sigma'_n$) for shear reactivation of a preexisting fracture and a composite Griffith ($\tau^2 - 4T\sigma'_n - 4T^2 = 0$)– Coulomb ($\tau = C + \mu_i\sigma'_n$) failure envelope for intact rock. The upper diagrams are schematic illustrations of the orientations of tensile and shear fractures in a rock sample. Shear fracturing occurs where differential stress ($\sigma_1 - \sigma_3$, the diameter of Mohr circle) is relatively large compared to the tensile or cohesive strengths and tensile fracturing at relatively lower differential stress. Symbols are as in Table 1; μ_s is the static friction coefficient along an existing plane of weakness; and μ_i is the internal coefficient of rock friction for intact rock.

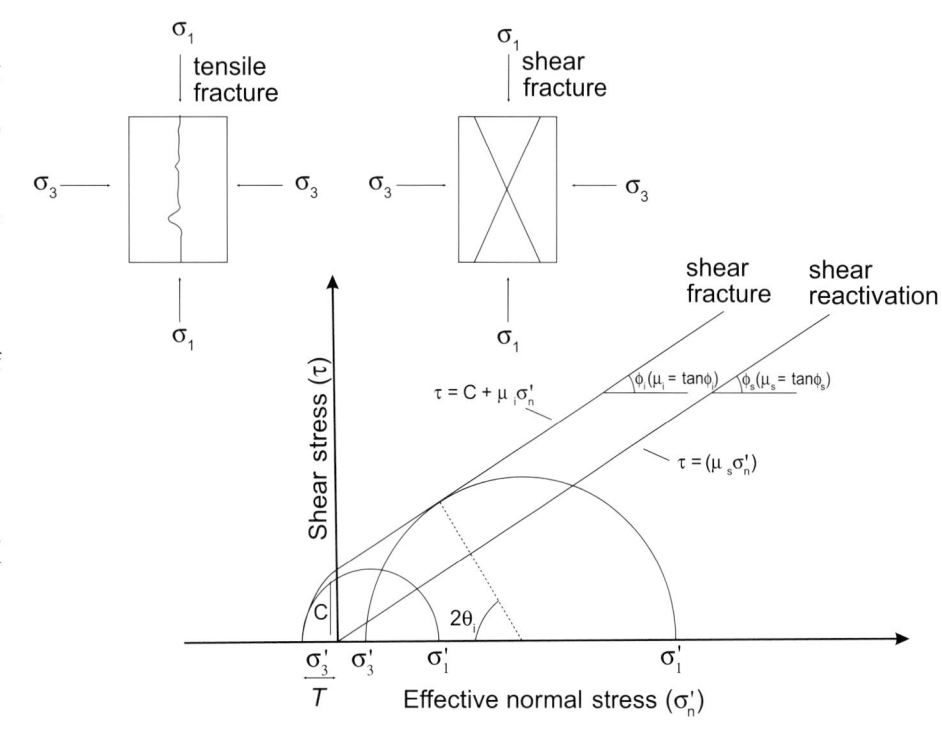

Fracture-related seal failure has widely been considered to occur solely because of tensile fracturing (also termed natural hydraulic fracturing), whereby increasing pore fluid pressure reduces the minimum effective stress to below the tensile strength of the rock (e.g., Palciauskas and Domenico, 1980; Ozkaya, 1984; Bell, 1990; Engelder and Lacazette, 1990; Miller, 1995). Hence, Watts' (1987) explanation of seals that fail by fracturing is hydraulic seals. Such tensile or natural hydraulic fracturing has, for example, been invoked to consider cap rock leakage in the North Sea (Caillet, 1993; Gaarenstroom et al., 1993; Caillet et al., 1997; Grauls, 1997). As discussed below, tensile fracturing can only occur with increasing pore-fluid pressure if differential stress is relatively low (Figure 1; Table 1).

Barton et al. (1995) combined in-situ stress measurements with information on the orientations of hydraulically conductive fractures and faults in three wells in the southeastern United States and demonstrated that fractures and faults optimally oriented for shear reactivation are the most important permeability conduits. Wiprut and Zoback (2002) analyzed four fields in the northern North Sea, concluding that faults that are critically stressed in the current stress field (i.e., capable of slipping) tend to leak, whereas those that are not critically stressed are more likely to be sealing. Given the evidence suggesting that both tensile and shear fractures rupture to cause seal breaching, it is critical that any methodology for assessing the risk of seal breach because of reactivation incorporate the influence of both these elements of structural permeability.

Following Sibson (1996), we assume a composite Griffith–Coulomb failure envelope (Figure 1). Hence, tensile failure is predicted where differential stress is relatively low (Sibson, 1996). If $4T < (\sigma_1 - \sigma_3) < 6T$, hybrid tensile-shear failure is predicted, and if $(\sigma_1 - \sigma_3) > 6T$, shear failure is predicted. If the cohesive strength of a reactivated fault zone is zero, the failure envelope passes through the origin of the normal and shear stress plot, and reactivation in shear is the only possible

TABLE 1. Brittle failure criteria expressed in terms of pore pressure (*P*) and necessary differential stress conditions. Criteria assume a Composite Griffith–Coulomb failure criterion, whereby *C* (cohesion) ~ 2*T* (tensile strength). The criteria apply to intact rock or reactivation of preexisting fractures, provided that the appropriate values of *T*, *C*, and μ (coefficient of friction) are used. After Sibson (1996).

Failure Mode	Criterion	Condition
Tensile (hydraulic)	$P = \sigma_3 + T$	$(\sigma_1 - \sigma_3) < 4T$
Hybrid tensile and shear	$P = \sigma_n + (4T^2 - \tau^2)/4T$	$4T < (\sigma_1 - \sigma_3) < 6T$
Shear	$P = \sigma_n + (C - \tau)/\mu$	$(\sigma_1 - \sigma_3) > 6T$

mode of failure, irrespective of the differential stress. Following experimental data by Handin (1969), Sibson (1996, 1998) recognized the influence of shear reactivation and noted that tensile fractures can only form and provide conduits for fluid flow, where

- rocks are intact and devoid of faults
- existing faults become severely misoriented for shear reactivation
- existing, favorably oriented, faults have regained cohesive strength because of cementation

We add to this last point, "or where the faulting process has resulted in a zone with significant cohesive strength." Faulting may involve processes such as cataclasis with quartz cementation that result in significant cohesion and, therefore, tensile strength (as seen, for example, in the case study presented below from the Otway Basin).

GEOMECHANICAL PARAMETERS FOR ASSESSING THE RISK OF REACTIVATION-RELATED SEAL BREACH

The first introduced geomechanical parameters for assessing the risk of fracture-related seal breach assumed that such was caused by tensile failure of the cap rock (Watts, 1987; Caillet, 1993; Gaarenstroom et al., 1993), and thus, they were based on the criterion for tensile failure (Table 1). For example, Gaarenstroom et al. (1993) introduced the concept of retention capacity, which is given by the difference between the minimum horizontal stress and pore pressure (Figure 2). Retention capacity is thus the effective minimum horizontal stress (σ'_h). A positive retention capacity (or σ'_h) reflects the additional pore pressure (or hydrocarbon column height) that can be developed prior to tensile failure by natural hydraulic fracturing. If retention capacity is zero, then tensile fractures would develop if the rock had no tensile strength. Retention capacity only considers the risk of tensile (and not shear) failure and does not incorporate (tensile) rock strength.

In recognition of the fact that critically stressed shear fractures present conduits for fluid flow, Morris et al. (1996) introduced the concept of slip tendency, which is the ratio of shear stress to effective normal stress acting on a fault and expresses the likelihood of slip on a cohesionless fault (Figure 2). To assess risk caused by both shear and tensile fractures, Ferrill et al. (1999) used slip tendency and dilation tendency. The latter risks the likelihood of dilation (tensile reactivation) of a fault on a linear scale from zero (if σ_1 is normal to the fault) to one (if σ_3 is normal to the fault; Figure 2). Slip and dilation tendency can be used together to geomechanically risk the likelihood of fault seal breach caused by shear and tensile fracturing. However, two separate parameters, neither of which incorporates rock strength, must be assessed.

The Coulomb failure function was used by Castillo et al. (2000) to risk fault seal breach caused by reactivation in the Australian Timor Sea. The Coulomb failure function is the difference between the shear stress acting on a fault and that required to cause failure on a cohesionless fault (Figure 2). A negative Coulomb failure function thus implied a stable fault, whereas a positive Coulomb failure function was associated with low fault seal integrity. Wiprut and Zoback (2002) used the critical pressure perturbation to risk fault seal breach caused by reactivation in the northern North Sea. The critical pressure perturbation is the increase in pore pressure required to reduce the effective normal stress to the value that would cause slip on a cohesionless fault (Figure 2). The Coulomb failure function and the critical pressure perturbation both incorporate the coefficient of sliding friction on a fault in assessing its risk of shear reactivation. However, they do not allow for any cohesive strength on preexisting faults nor for the development of tensile fractures.

FAST: A NEW GEOMECHANICAL PARAMETER FOR ASSESSING THE RISK OF REACTIVATION-RELATED SEAL BEACH

We propose a new geomechanical parameter for assessing the risk of reactivation-related seal breach, the fault analysis seal technology (FAST), which allows for the input of a failure envelope with cohesion. Jones et al. (2002) and Dewhurst and Jones (2002) have demonstrated that fault rocks may show significant postdeformation lithification caused by cementation that results in the regaining of cohesive and, therefore, tensile strength. Thus, knowledge of the fault rock failure envelope should be incorporated into predictions of fault reactivation.

The likelihood of fault reactivation and associated seal breach can be assessed by the FAST method given the knowledge of the stress field, pore pressure, fault orientation, and the failure envelope for the fault rocks. The in-situ stress field can be determined by a variety of wellbore geomechanical techniques. Density and check-shot velocity data yield the vertical stress, borehole breakouts and drilling-induced tensile fractures yield the orientation of the horizontal stresses, leak-off and extended leak-off tests yield the minimum horizontal stress, and the maximum horizontal stress can be determined by the occurrence or nonoccurrence of breakouts and drilling-induced tensile fractures and

FAST: A New Technique for Geomechanical Risk Assessment of Fault Breach   **77**

A Retention Capacity $\boxed{\sigma_3 - P_p}$

B Slip Tendency $\boxed{\dfrac{\tau}{\sigma_n'}}$

C Dilation Tendency $\boxed{\dfrac{\sigma_1' - \sigma_n'}{\sigma_1' - \sigma_3'}}$

D Coulomb Failure Function $\boxed{\tau - \mu(\sigma_n - P_p)}$

E Critical Pressure Perturbation $\boxed{\Delta P}$

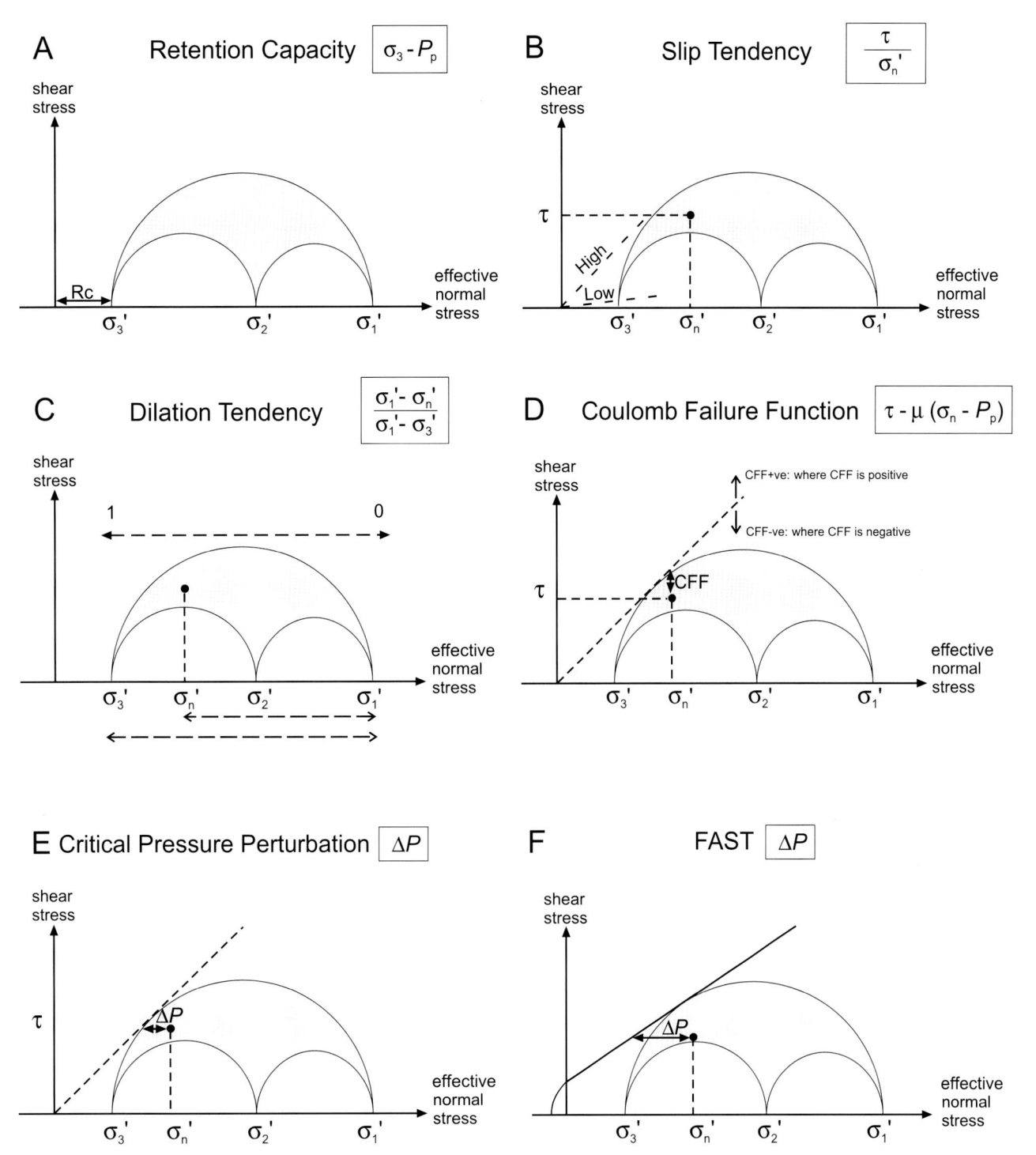

F FAST $\boxed{\Delta P}$

FIGURE 2. Geomechanical risking parameters. See text for full discussion and references. Retention capacity risks the likelihood of tensile failure of intact cap rock. Other techniques risk the reactivation of a preexisting fault, the orientation of which is specified by the dot in the 3-D Mohr circles. The position of the fault in the gray-shaded area in the 3-D Mohr circles is defined by the relative orientation of the fault and the principal stresses. CFF = Coulomb failure function.

knowledge of rock strength (see, e.g., Bell, 1996; Moos and Zoback, 1990, for detailed discussion of these techniques). Knowledge of the fault failure envelope can be determined from laboratory testing of intact fault rocks (Handin and Jaeger, 1957; Handin, 1969; Jaeger and Cook, 1976; Dewhurst and Jones, 2002). To compensate for fault plane heterogeneity or when no strength data are available, sensitivity analysis of fault failure envelopes should also be undertaken. Fault orientation (dip and strike) is determined from depth-converted seismic interpretation. For three-dimensional (3-D) seismic data, the risk of reactivation can be mapped

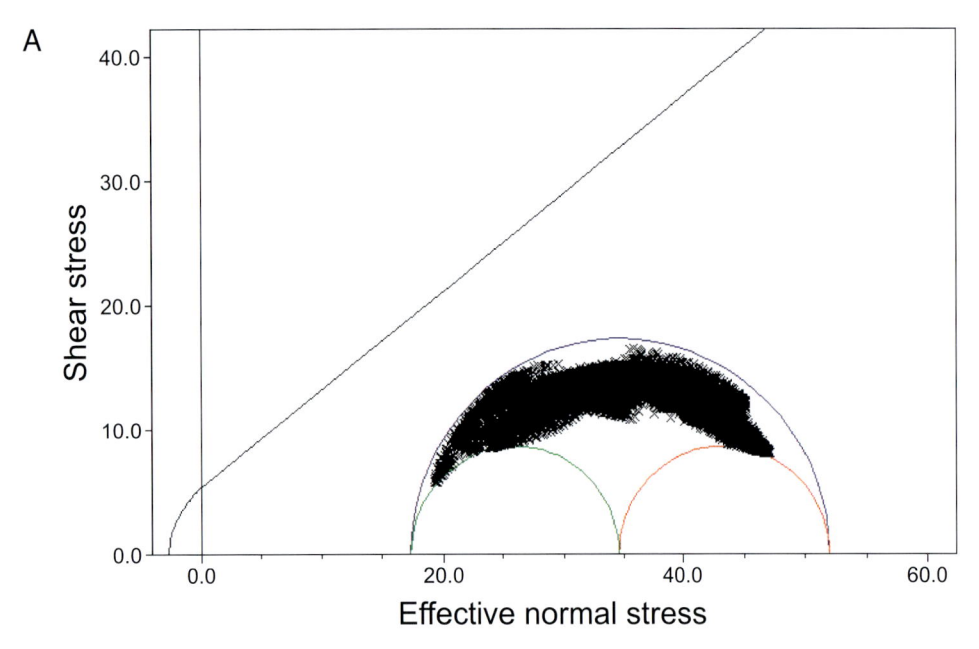

FIGURE 3. The FAST methodology. (A) Three-dimensional Mohr circle for Otway Basin state-of-stress and failure envelope for laboratory-tested cataclasite. (B) Polar diagram of normals to planes colored by increase in pore pressure (ΔP) required for reactivation. The crosses represent the orientations of individual elements of the Zema fault.

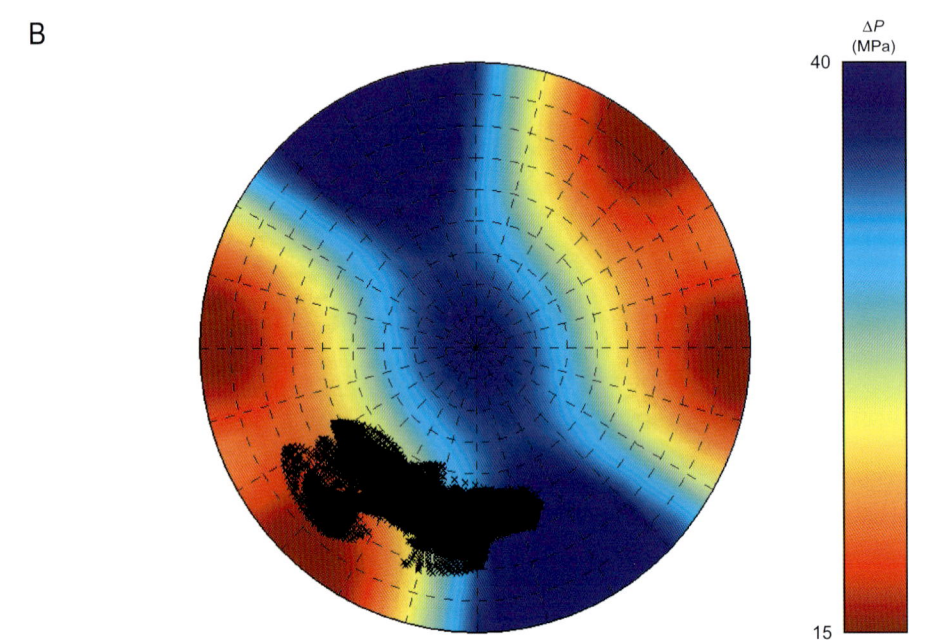

onto the interpreted fault plane geometry in 3-D. For two-dimensional (2-D) seismic data, the dip and dip azimuth of the fault are determined from the offset between reflector terminations. A centerline point in the mapped fault polygon can then be assigned the fault dip and strike and, in turn, the risk of reactivation.

Given the requisite information, three stages to assessing reactivation risk thus exist.

1) A 3-D Mohr diagram representing the state-of-stress and failure envelope for the fault is constructed (Figure 3). The risk of reactivation of a plane of any orientation is expressed by the increase in pore pressure (ΔP) required to cause its reactivation, i.e., horizontal distance on a 3-D Mohr diagram between a fault plane and the failure envelope.
2) The reactivation risk (ΔP) for all planes is plotted on a polar diagram of normals to planes (Figure 3).
3) The appropriate reactivation risk (ΔP) is mapped either onto the fault plane (3-D data; Figure 4) or fault polygon centerline points (2-D data).

APPLICATION TO THE ZEMA STRUCTURE, OTWAY BASIN, SOUTH AUSTRALIA

The Zema-1 well in the Otway Basin, South Australia, intersected a 69-m (226-ft) paleogas leg and a 15-m (49-ft) paleo-oil leg in the Lower Cretaceous Pretty Hill Formation (Lyon et al., 2005). The Zema structure has fault-dependent closure in the footwall of the Zema fault. The strike of the Zema fault varies between east–west and northwest–southeast, and it dips approximately 70° to the north (Lyon et al., 2005). Detailed analysis suggested that juxtaposition and fault deformation processes together were likely to provide an adequate seal for the observed paleocolumn (Jones et al., 2000; Lyon et al., 2005). Hence, the existing methodologies for assessing fault seal for inactive faults did not consider the observed paleocolumn, and the propensity for reactivation-related breach of the fault seal at Zema-1 was investigated.

FIGURE 4. Zema fault in 3-D colored by increase in pore pressure (ΔP) required for reactivation.

The in-situ stress tensor in the area was constrained, as described in more detail by Jones et al. (2000), using density and check-shot log data (for vertical stress), leak-off tests and one extended leak-off test (minimum horizontal stress), and the occurrence of drilling-induced tensile fractures (maximum horizontal stress). Maximum horizontal stress orientation of 156°N was inferred from breakouts in the nearby Katnook-3 well. Pore pressures are hydrostatic. In the depth range of interest of 2500–3000 m (8200–10,000 ft), the following stress gradients apply:

- Minimum horizontal stress (σ_h) = 16.1 MPa/km
- Overburden stress (σ_v) = 22.4 MPa/km
- Maximum horizontal stress (σ_H) = 28.7 MPa/km
- Pore pressure (P) = 9.8 MPa/km
- Maximum horizontal stress (σ_H) orientation = 156°N

Unusually, core is available through a fault zone in the reservoir (Pretty Hill Formation) in the offset Banyula-1 well. The core intersected cataclasites that were tested in a standard triaxial cell with deformation features oriented at 30° to σ_1. The failure envelope derived from testing of the cataclasites, as described in more detail by Dewhurst and Jones (2002) is

$$\tau = 5.40 + 0.78\sigma'_n$$

where τ is the shear stress at failure, σ'_n is the effective normal stress (i.e., $\sigma_n - P$), and pressures are in megapascals. We have used the failure envelope for this core in our analysis, although we recognize that the core is through the reservoir and not the seal, which may have a different rheology (see discussions in Boult et al., 2003; Dewhurst and Jones, 2003). Such potential variation in fault rock failure envelopes is a key driver for the sensitivity analysis (in very weak and strong fault rocks) described in the following section.

The entire region is covered by 2-D seismic data and a 3-D survey covers approximately half of the Zema structure. The geometry of the fault plane was determined by depth conversion and subsequent interpretation of the 2-D and 3-D seismic data over the Zema structure. Applying the FAST methodology to the above data demonstrates that significant elements of the Zema fault are optimally oriented for strike-slip reactivation in the in-situ stress field (Figure 3). The most at-risk segments are those striking approximately northwest–southeast (Figure 4). Given that juxtaposition and fault deformation processes are likely to have created an adequate seal, and that significant sections of the trap-bounding fault are optimally oriented for reactivation, we interpret that fault reactivation is a likely candidate for seal breach and, thus, the presence of the paleocolumn in the Zema-1 well.

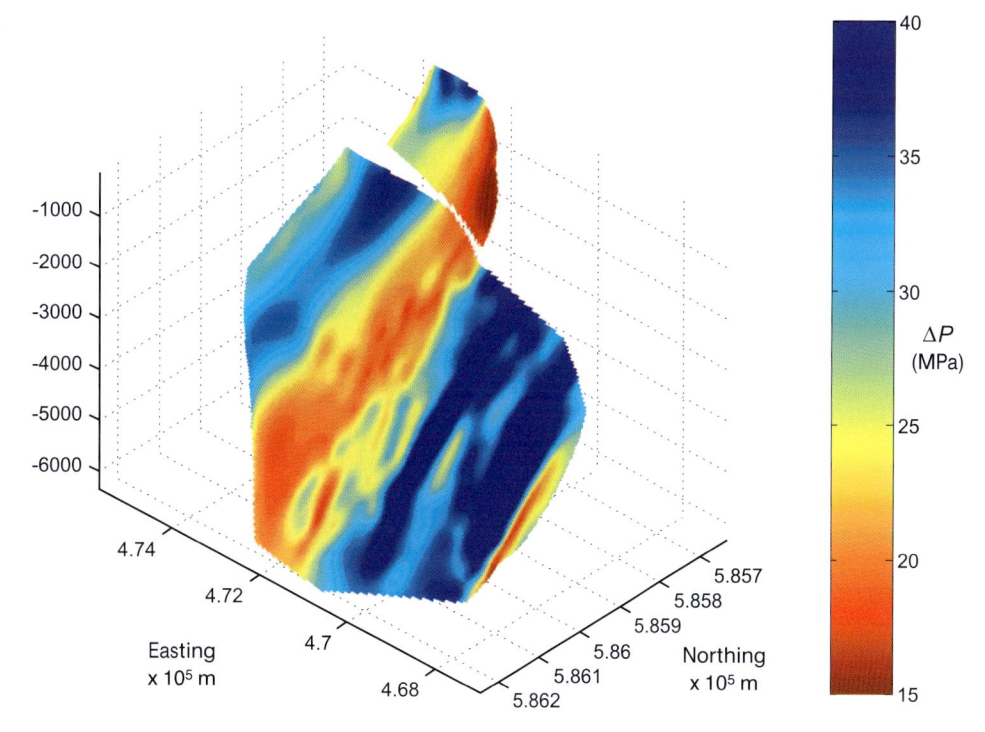

Two pieces of additional evidence are consistent with the interpretation that fault reactivation is responsible for seal breach in the Zema-1 well. First, microstructural analysis of cataclasite samples from the fault zone in the offset Banyula-1 well indicates the presence of open fractures (Jones et al., 2000). Illites span the fractures that must be open in the subsurface.

Second, a minor fault (not apparent on seismic data) is interpreted in the Laira Formation regional seal from dipmeter data in the Zema-1 well. This minor fault is associated with a significant SP anomaly, indicating that fault zones do indeed provide a permeable zone in the regional seal (Figure 5). We are unable to determine whether this subseismic fault intersects the Zema reservoir. The main Zema fault is interpreted to cut the Zema-1 well near the base of the overlying Eumeralla Formation, and its SP effect is harder to ascertain because of the more frequent permeable sands in the Eumeralla Formation.

80 Mildren et al.

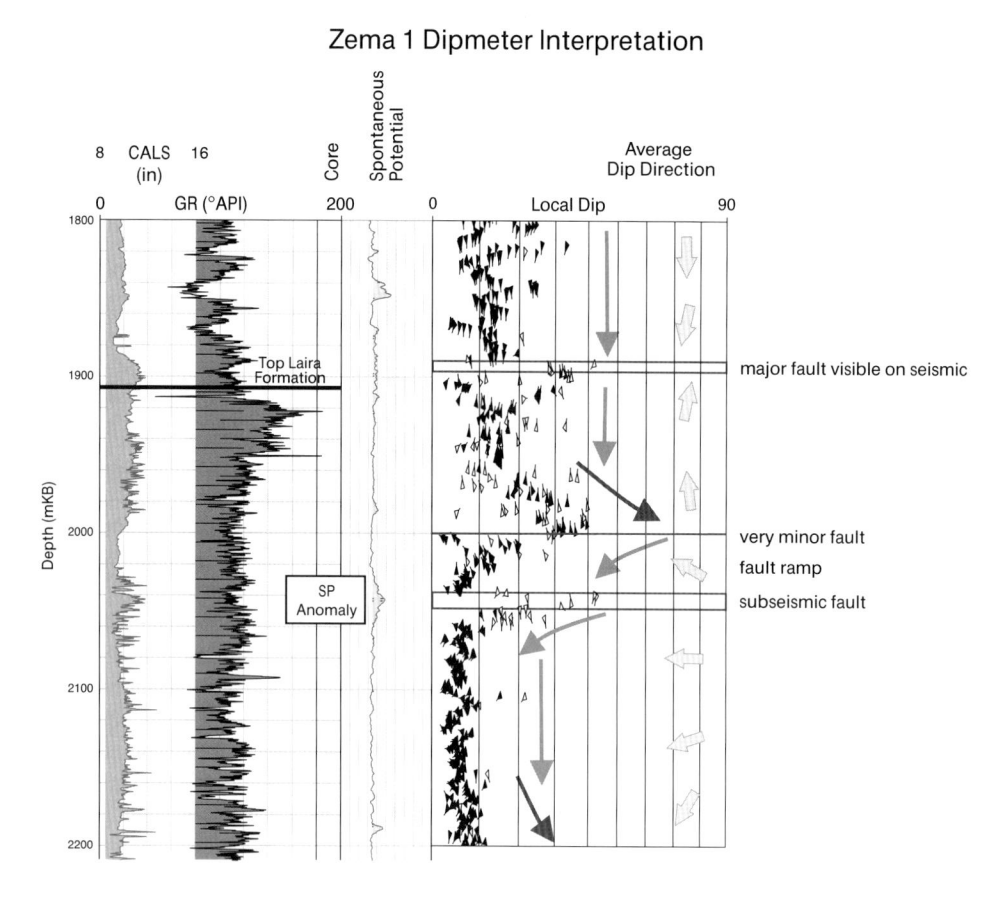

FIGURE 5. SP anomaly associated with fault cut through the Laira Formation regional seal in Zema-1. GR = gamma ray; CALS = caliper logs; mKB = meters below Kelly bushing.

Healy, 1984) in strong rocks (coefficient of friction of 0.8). The following end-member cases were considered to assess the robustness of the assessment of reactivation risk:

- weak fault rocks ($\tau = 0.3\sigma'_n$) and most likely stress regime above
- strong fault rocks ($\tau = 20 + \sigma'_n$) and most likely stress regime above
- lower σ_H (22.4 MPa/km, transitional strike-slip or normal fault stress regime) and failure envelope as determined from lab testing
- higher σ_H (37.1 MPa/km, frictional limit for coefficient of friction of 0.8) and failure envelope as determined from lab testing

SENSITIVITY ANALYSIS

A sensitivity study of the results was also undertaken despite a failure envelope being available for fault rocks in the Otway Basin. This is, in part, to illustrate the application of sensitivity analysis to areas where failure envelopes are not available. Furthermore, the cataclastic zone tested in the Otway Basin is from an offset well and in the reservoir and not the cap rock. The failure envelope may change significantly along the fault plane because of changes in the rock types through which the fault cuts or because of laterally variable fluid flow and diagenetic processes along the fault. Hence, the robustness of the results was considered with respect to very weak and very strong fault rock failure envelopes.

The maximum horizontal stress is generally the least well-constrained component of the in-situ stress field. Hence, the sensitivity of the results to a range in the magnitude of maximum horizontal stress was also assessed. The most likely stress regime of the area (above) is a strike-slip regime, whereby $\sigma_H > \sigma_v > \sigma_h$. The robustness of the results were tested for a much lower maximum horizontal stress value equal to the vertical stress, such that the stress regime would be transitional between strike-slip and normal ($\sigma_H \sim \sigma_v > \sigma_h$) and also for the maximum value that maximum horizontal stress could attain, i.e., the frictional limit (Zoback and

Differences in the risk of reactivation in each of the above four scenarios exist (Figure 6). However, major similarities are present in the most at-risk fault orientations that allow important generalizations to be made. A conservative approach to the sensitivity analysis is to avoid fault planes that are prone to reactivation in any single scenario and to focus on planes not prone to reactivation in any of the scenarios. Planes suitably oriented for reactivation in any single scenario strike between 100 and 210°N and dip greater than 40° (Figure 6). Planes not suitably oriented for reactivation in any of the scenarios strike between 40 and 90°N and dip greater than 60° (Figure 6).

The differences between the reactivation risk plots for each scenario essentially reflect the different mode of brittle reactivation that predominates in each case. Shear failure tends to predominate in the high maximum horizontal stress (hence, high differential stress) case and the weak fault rock case, because ($\sigma_1 - \sigma_3$) > $6T$. Hence, planes oriented approximately 30° to σ_1 and containing the σ_2 direction are prone to reactivation. Tensile failure is more significant in the low maximum horizontal stress (hence, low differential stress) case and the strong fault rock case, because

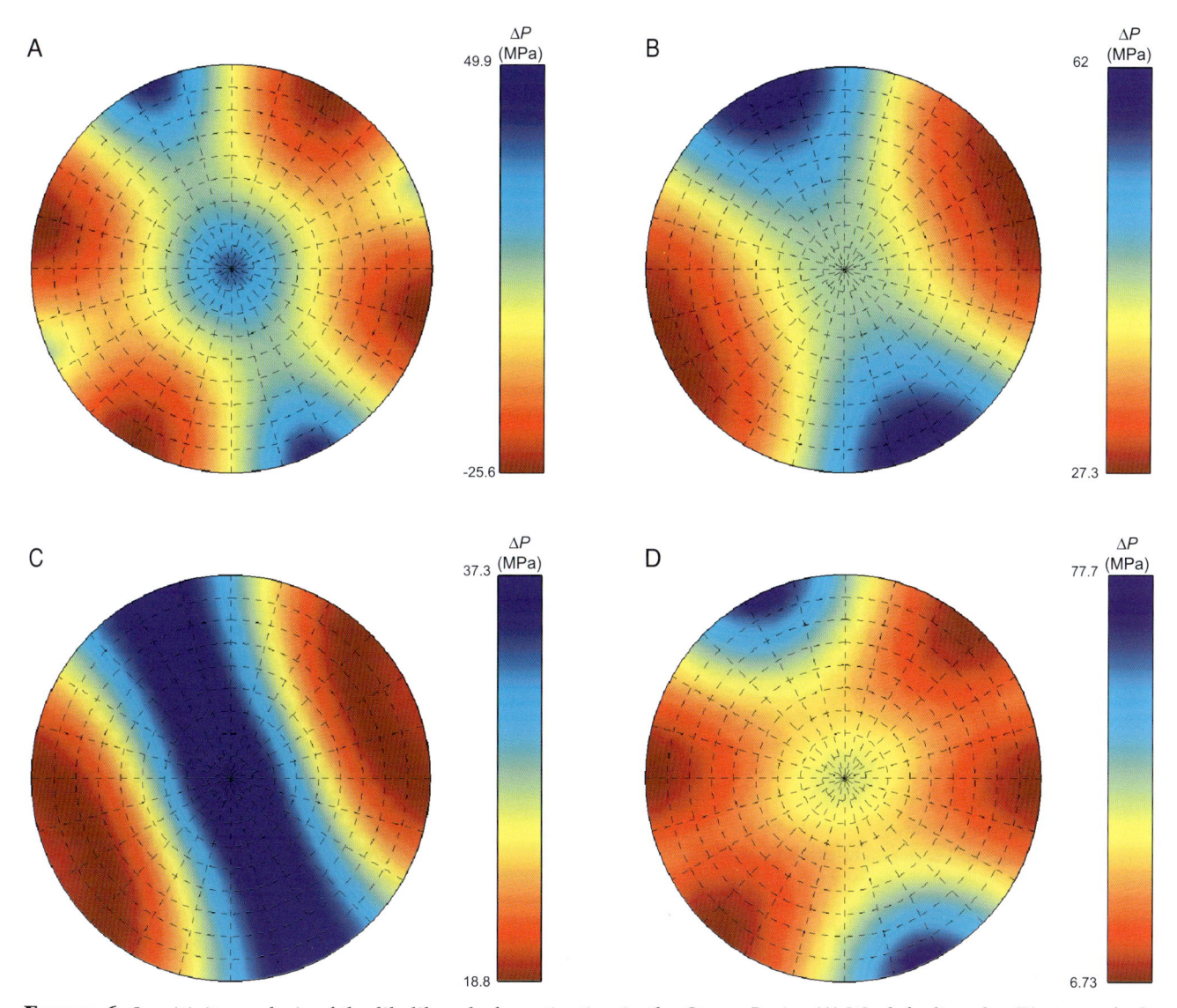

FIGURE 6. Sensitivity analysis of the likelihood of reactivation in the Otway Basin. (A) Weak fault rocks; (B) strong fault rocks; (C) lower limit σ_H; and (D) upper limit σ_H. Plots are polar diagrams of normals to planes colored by ΔP values.

$(\sigma_1 - \sigma_3) < 4T$. Hence, planes orthogonal to σ_3 are prone to reactivation.

These differences also highlight the advantage of the FAST methodology over the Coulomb-based risking algorithms: critical pressure perturbation (CPP) and Coulomb failure function (CFF). Where differential stress is high or fault rock strength is weak, the FAST method will produce almost identical results to the CPP method. A small differential stress or strong failure envelope increases the risk of fault orientations critically oriented for tensile failure relative to those for shear failure. Therefore, CPP and CFF may overestimate the risk associated with shear failure and ignore the risk associated with fault orientations critical for tensile failure.

Significant differences exist between the ΔP values in the four scenarios. Major increases in pore pressure are required to reactivate even optimally oriented faults in the lower bound σ_H and strong fault cases. Significant portions of the fault are at stresses beyond failure in the weak fault case. Although extreme scenarios have been used, these variations illustrate that the errors in this technique preclude it being used for predicting hydrocarbon column heights. A 1-MPa variation in ΔP is equivalent to the buoyancy pressure associated with 1 km (0.6 mi) of oil column, assuming the hydrostatic gradient is 9.8 MPa/km ($\sim$1 g cm^{-3}) and oil gradient 8.8 MPa/km ($\sim$0.9 g cm^{-3}). As shown by Figure 6, the potential errors in the methodology are greater than 1 MPa. This applies to any geomechanical methodology because of the errors inherent in estimating in-situ stresses and rock failure parameters. We believe that the value of the technique lies in its application to areas where reactivation-related breach is suspected to be an issue and in the relative risking of fault-bound prospects in such an area. The ΔP values need to be calibrated with reference to the occurrence of intact and breached columns in specific basins. For example, in the Timor Sea, ΔP values less than 10 MPa represent a significant risk of reactivation-related seal breach (Mildren et al., 2002).

DISCUSSION

The considerable evidence that fault and fracture reactivation leads to fluid redistribution has been summarized by Sibson (1992, 1994), Muir-Wood (1994), and Dewhurst et al. (1999). Sibson's (1992) fault-valve model provides a mechanism whereby the cycling of tectonic shear stress and/or fluid pressure is linked to episodic fault instability and, in turn, episodic fluid redistribution. Cementation or hydrothermal precipitation may lead to faults resealing during periods of stability. The observed episodic discharge of hydrocarbons from overpressured compartments (e.g., Hunt, 1990) is consistent with this model. At the geological timescale, hydrocarbons clearly leak episodically and not continually up trap-bounding faults, because if leakage was continual at rates greater than charge, accumulations would not develop. In the Australian context, the presence of paleocolumns witnesses the fact that trap-bounding faults do seal over significant periods and, thus, that most trap-bounding faults are not permanently open conduits for leakage (O'Brien and Woods, 1995). The model followed herein, like Sibson's (1992) fault-valve model, assumes that the observed episodic breaching of faults and associated fluid redistribution is associated with reactivation.

In some cases, fault rocks may be stronger than the surrounding rocks (Dewhurst and Jones, 2002; Jones et al., 2002), and hence, the risk of failure of intact cap rock must be considered as well as the risk of fault reactivation (Boult et al., 2002). Indeed, the risk of failure of intact cap rock may be greater than the risk of reactivating misoriented faults where the intact cap rock is weaker (Streit, 1999; Hillis and Nelson, 2005). To assess the risk of cap rock failure, it is simply necessary to add an intact rock failure envelope to the analysis. The risk of failure of intact cap rock is the increase in pore pressure that can be sustained prior to failure by the point on the Mohr circle closest to the failure envelope. In the case of cap rock analysis, a single value is provided for the (crest of the) prospect (as is the case with retention capacity; Figure 2), whereas the risk of fault reactivation varies with the geometry of the fault. The strength of fault rock material obtained at Banyula-1 suggests that sand-sand fault contacts are strong in the Otway Basin. The sensitivity analysis reveals the risk of fracturing to be very high when using a weak failure envelope. It is possible that the strength of the cap rock is weaker than the fault, and the generation of cap rock fractures at Zema is also a possibility.

It is necessary to consider the location of at-risk fault segments (low-ΔP zones) with respect to the trap as a whole. If segments of the fault with low ΔP are not coincident with the top of the structure, then reactivation may not lead to breaching of the entire column.

Reactivation must postdate hydrocarbon charge for it to cause seal breaching. If postcharge reactivation occurred in the geological past and within a paleostress regime that differed significantly from that of the present day, it should not be risked with reference to the in-situ stress field. The risk of reactivation may be assessed with reference to the in-situ stress field if leakage is associated with present-day geochemical anomalies such as those witnessed by marine geochemical sniffers and airborne laser fluorescence (Bishop and O'Brien, 1998) or if it is associated with faults that cut a young seabed and land surface. In the North Sea, the significance of the in-situ stress field to the assessment of fault leakage is witnessed by Heffer and Fox's (1996) compilation indicating that nonsealing faults are strongly preferentially oriented in the in-situ maximum horizontal stress direction. Although in many cases, the in-situ stress field is appropriate, if leakage occurred in a paleostress field different to that of the present day, then that paleostress field must be applied to consider the risk of reactivation (Gartrell and Lisk, 2005).

All geomechanical methodologies for risking reactivation-related seal breach assign risk to seismically mapped faults. Hence, it is an implicit assumption that leakage associated with reactivation occurs on the seismically mapped structures or on structures with similar orientations. Structures that are not seismically imaged may be prone to reactivation and contribute to seal breach. Indeed, the subseismic fault cutting the Laira Formation regional seal in the Zema example is permeable. Hence, at the prospect scale, it is an assumption that seal-breaching fractures are parallel or subparallel to the mapped fault.

CONCLUSIONS

Analysis of cross-fault lithological juxtaposition and of deformation processes are routinely used to assess the likelihood of fault seal. However, these techniques cannot incorporate the potential for seal breach caused by fault reactivation subsequent to charge. Faults that are suitably oriented to be reactivated in the prevailing stress field provide conduits for fluid flow. The concepts of brittle failure and structural permeability provide the basis of a technique to assess whether seismically mapped faults are likely to be reactivated in the in-situ stress field and thus associated with seal breaching.

Faults may show significant cohesive strength caused by postdeformation cementation. The existing geomechanical parameters used to assess the likelihood of reactivation-related seal breach assume that preexisting faults have no cohesive strength. Hence, we have introduced a new methodology (FAST) that allows input of a laboratory-derived fault rock failure envelope or a range of likely fault rock failure envelopes. The likelihood of reactivation is expressed by the increase in pore pressure (ΔP) required to cause the fault to reactivate given the orientation in the in-situ stress field. This translates to an advantage over the Coulomb-derived geomechanical methodologies because it incorporates tensile failure in scenarios where either fault rock material is very strong or the differential stress is small. The technique can be readily modified to consider areas where failure of intact cap rock presents the key geomechanical risk.

Sensitivity studies incorporating variable fault strength parameters are required until fault strength can be mapped in detail across fault planes, possibly by relating strength to damage processes using gouge estimates. Once this has been achieved, the identification of across-fault leakage vs. along-fault leakage can be determined.

The FAST methodology was applied to the fault-bound Zema structure in the Otway Basin, South Australia. Juxtaposition and fault deformation processes indicated that the fault was likely to be sealing, but the structure was found to contain a residual hydrocarbon column. The FAST analysis indicates that segments of the fault are optimally oriented for reactivation in the in-situ stress field. Microstructural evidence of open fractures in a fault zone in the subsurface in an offset well and an SP anomaly associated with a subseismic fault cutting the regional seal in the Zema-1 well support the interpretation that seal breach is related to fracturing.

The FAST methodology provides a powerful tool for predrill assessment of the risk of fault reactivation-related seal breach in all stress scenarios and fault strengths. However, limitations to such geomechanical techniques exist. First, they assume that reactivation occurs in the in-situ stress field as can be determined from wellbore data. Second, predrill risk can only be assigned to seismically mapped faults (and not subseismic faults). Finally, the errors associated with the techniques preclude them from being used to assess likely hydrocarbon column heights. Nonetheless, these techniques provide a powerful method for the relative ranking of fault-bound prospects in an area where reactivation-related breach presents an exploration risk.

REFERENCES CITED

Allan, U. S., 1989, Model for hydrocarbon migration and entrapment within faulted structures: AAPG Bulletin, v. 73, p. 803–811.

Barton, C. A., M. D. Zoback, and D. Moos, 1995, Fluid flow along potentially active faults in crystalline rock: Geology, v. 23, p. 683–686.

Bell, J. S., 1990, The stress regime of the Scotian Shelf offshore eastern Canada to 6 kilometers depth and implications for rock mechanics and hydrocarbon migration, in V. Maury and D. Foourmaintraux, eds., Rock at great depth: Proceedings of the International Symposium, Pau, France, August, 1989: Rotterdam, Balkema, v. 3, p. 1243–1265.

Bell, J. S., 1996, In situ stresses in sedimentary rocks: Part 1. Measurement techniques: Geoscience Canada, v. 23, p. 85–100.

Bishop, D. J., and G. W. O'Brien, 1998, A multi-disciplinary approach to definition and characterisation of carbonate shoals, shallow gas accumulations and related complex near-surface sedimentary structures in the Timor Sea: Australian Petroleum Production and Exploration Association Journal, v. 38, p. 93–113.

Boult, P. J., B. A. Camac, and A. W. Davids, 2002, 3D fault modelling and assessment of top seal structural permeability—Penola trough, onshore Otway Basin: Australian Petroleum Production and Exploration Association Journal, v. 42, p. 151–166.

Boult, P. J., Q. Fisher, S. R. J. Clinch, R. Lovibond, and C. D. Cockshell, 2003, Geomechanical, microstructural, and petrophysical evolution in experimentally reactivated cataclasites: Applications to fault seal prediction: Discussion: AAPG Bulletin, v. 87, p. 1681–1683.

Brace, W. F., 1960, An extension of the Griffith theory of fracture to rocks: Journal of Geophysical Research, v. 65, p. 3477–3480.

Bretan, P., Y. Yielding, and H. Jones, 2003, Using calibrated shale gouge ratio to estimate hydrocarbon column heights: AAPG Bulletin, v. 87, p. 397–413.

Byerlee, J. D., 1978, Friction of rocks: Pure and Applied Geophysics, v. 116, p. 615–626.

Caillet, G., 1993, The caprock of the Snorre field, Norway: A possible leakage by hydraulic fracturing: Marine and Petroleum Geology, v. 10, p. 42–50.

Caillet, G., N. C. Judge, N. P. Bramwell, L. Meciani, M. Green, and P. Adam, 1997, Overpressure and hydrocarbon trapping in the Chalk of the Norwegian Central Graben: Petroleum Geoscience, v. 3, p. 33–42.

Castillo, D. A., D. J. Bishop, I. Donaldson, D. Kuek, M. De Ruig, M. Trupp, and M. W. Shuster, 2000, Trap integrity in the Laminaria high–Nancar trough region, Timor Sea: Prediction of fault seal failure using well-constrained stress tensors and fault surfaces interpreted: Australian Petroleum Production and Exploration Association Journal, v. 40, p. 151–173.

Dewhurst, D. N., and R. M. Jones, 2002, Geomechanical, microstructural and petrophysical evolution in experimentally reactivated cataclasites: Application to fault seal prediction: AAPG Bulletin, v. 86, p. 1383–1405.

Dewhurst, D. N., and R. M. Jones, 2003, Geomechanical, microstructural, and petrophysical evolution in experimentally reactivated cataclasites: Applications to fault seal prediction: Reply: AAPG Bulletin, v. 87, p. 1684–1686.

Dewhurst, D. N., Y. Yang, and A. C. Aplin, 1999, Permeability and fluid flow in natural mudstones, *in* A. C. Aplin, A. J. Fleet, and J. H. S. Macquaker, eds., Muds and mudstones: Physical and fluid flow properties: Geological Society (London) Special Publication 158, p. 23–43.

Engelder, T., 1999, The transitional-tensile fracture: A status report: Journal of Structural Geology, v. 21, p. 1049–1055.

Engelder, T., and A. Lacazette, 1990, Natural hydraulic fracturing, *in* N. Barton and O. Stephansson, eds., Rock joints: Rotterdam, A. A. Balkema, p. 35–44.

Ferrill, D. A., J. Winterle, G. Wittmeyer, D. Sims, S. Colton, A. Armstrong, and A. P. Morris, 1999, Stressed rock strains groundwater at Yucca Mountain, Nevada: Geological Society of America Today, v. 9, p. 1–8.

Finkbeiner, T., M. Zoback, P. B. Flemings, and B. B. Stump, 2000, Stress, pore pressure, and dynamically constrained hydrocarbon columns in the South Eugene Island 330 field, northern Gulf of Mexico: AAPG Bulletin, v. 85, p. 1007–1031.

Fisher, Q. J., and R. J. Knipe, 1998, Fault sealing processes in siliciclastic sediments, *in* G. Jones, Q. J. Fisher, and R. J. Knipe, eds., Faulting, fault sealing and fluid flow in hydrocarbon reservoirs: Geological Society (London) Special Publication 147, p. 117–134.

Gaarenstroom, L., R. A. J. Tromp, M. C. De Jong, and A. M., Brandenburg, 1993, Overpressures in the central North Sea: Implications for trap integrity and drilling safety, *in* J. R. Parker, ed., Proceedings of the 4th Conference on the Petroleum Geology of Northwest Europe: Geological Society (London), v. 2, p. 1305–1313.

Gartrell, A. P., and M. Lisk, 2005, Potential new method for paleostress estimation by combining three-dimensional fault restoration and fault slip inversion techniques: First test on the Skua field, Timor Sea, *in* P. Boult and J. Kaldi, eds., Evaluating fault and cap rock seals: AAPG Hedberg Series, no. 2, p. 23–36.

Grauls, D. J., 1997, Minimum principal stress as a control of overpressures in sedimentary basins, *in* J. P. Hendry, P. F. Carey, J., Parnell, A. H. Ruffell, and R. H. Worden, eds., Geofluids II: Extended Abstracts of the Geofluids II Conference, Belfast, United Kingdom, March 1997, p. 219–222.

Handin, J., 1969, On the Coulomb–Mohr failure criterion: Journal of Geophysical Research, v. 74, p. 5343–5348.

Handin, J., and R. V. Jaeger, 1957, Experimental deformation of sedimentary rocks under confining pressure: Tests at room temperature on dry samples: AAPG Bulletin, v. 41, p. 1–50.

Heffer, K., and R. Fox, 1996, Geomechanics, faults and fluid flow, *in* Faulting, fault sealing and fluid flow in hydrocarbon reservoirs: Abstracts of the Conference, Leeds, United Kingdom, September 1996: Leeds, Geological Society Petroleum Group, p. 49–50.

Hickman, S. H., C. A. Barton, M. D. Zoback, R. Morin, J. Sass, and R. Benoit, 1997, In situ stress and fracture permeability along the Stillwater fault zone, Dixie Valley, Nevada (abs.): International Journal of Rock Mechanics and Mining Sciences and Geomechanics, v. 34, p. 414.

Hillis, R. R., 1998, Mechanisms of dynamic seal failure in the Timor Sea and central North Sea, *in* P. G. Purcell and R. R. Purcell, eds., The sedimentary basins of Western Australia: 2. Proceedings of Petroleum Exploration Society of Australia Symposium, Perth, Western Australia, p. 313–324.

Hillis, R. R., and E. J. Nelson, 2005, In situ stresses in the North Sea and their applications: Petroleum geomechanics from exploration to development, *in* A. G. Dore and B. Vining, eds., Petroleum geology: Northwest Europe and global perspectives—Proceedings of the 6th Petroleum Geology Conference: Geological Society, London, p. 551–564.

Hippler, S. J., 1997, Microstructures and diagenesis in North Sea fault zones: Implications for fault-seal potential and fault-migration rates, *in* R. C. Surdam, ed., Seals, traps, and the petroleum system: AAPG Memoir 67, p. 103–113.

Hunt, J. M., 1990, Generation and migration of petroleum from abnormally pressured fluid compartments: AAPG Bulletin, v. 74, p. 1–12.

Jaeger, J., and N. G. W. Cook, 1976, Fundamentals of rock mechanics, 2d ed.: London, Chapman and Hall, 593 p.

Jev, B. L., C. H. Kaars-Sijpesteijn, M. P. A. M. Peters, N. L. Watts, and J. T. Wilkie, 1993, Akaso field, Nigeria: Use of integrated 3-D seismic, fault slicing, clay smearing and RFT pressure data on fault trapping and dynamic leakage: AAPG Bulletin, v. 77, p. 1389–1404.

Jones, R. M., and R. R. Hillis, 2003, An integrated, quantitative approach to assessing fault-seal risk: AAPG Bulletin, v. 87, p. 507–524.

Jones, R. M., P. J. Boult, R. R. Hillis, S. D. Mildren, and J. Kaldi, 2000, Integrated hydrocarbon seal evaluation in the Penola trough, Otway Basin: Australian Petroleum Production and Exploration Association Journal, v. 40, no. 1, p. 194–211.

Jones, R. M., D. N. Dewhurst, R. R. Hillis, and S. D. Mildren, 2002, Geomechanical fault characterization: Impact on quantitative fault seal risking. Society of Petroleum Engineers/International Society for Rock Mechanics Rock Mechanics Conference, Irving, Texas, October 2002: SPE Paper 78213.

Knipe, R. J., 1997, Juxtaposition and seal diagrams to help analyze fault seals in hydrocarbon reservoirs: AAPG Bulletin, v. 81, p. 187–195.

Lyon, P. J., P. J. Boult, R. Hillis, and S. D. Mildren, 2005, Sealing by shale gouge and subsequent seal breach by reactivation: A case study of the Zema prospect, Otway Basin, *in* P. Boult and J. Kaldi, eds., Evaluating fault and cap rock seals: AAPG Hedberg Series, no. 2, p. 179–197.

Mildren, S. D., R. R. Hillis, and J. Kaldi, 2002, Calibrating predictions of fault seal reactivation in the Timor Sea: Australian Petroleum Production and Exploration Association Journal, v. 42, p. 187–202.

Miller, T. W., 1995, New insights on natural hydraulic fractures induced by abnormal high pore pressure: AAPG Bulletin, v. 79, p. 1005–1018.

Moos, D., and M. D. Zoback, 1990, Utilization of observations of well bore failure to constrain the orientation and magnitude of crustal stresses: Application to continental, Deep Sea Drilling Project, and Ocean Drilling Program boreholes: Journal of Geophysical Research, v. 95, p. 9305–9325.

Morris, A., D. A. Ferrill, and D. B. Henderson, 1996, Slip-tendency analysis and fault reactivation: Geology, v. 24, p. 275–278.

Muir-Wood, R., 1994, Earthquakes, strain-cycling and the mobilization of fluids, in J. Parnell, ed., Geofluids: Origin, migration and evolution of fluids in sedimentary basins: Geological Society (London) Special Publication 78, p. 85–98.

O'Brien, G. W., and E. P. Woods, 1995, Hydrocarbon-related diagenetic zones (HRDZs) in the Vulcan sub-basin, Timor Sea: Recognition and exploration implications: Australian Petroleum Production and Exploration Association Journal, v. 35, p. 220–252.

Ozkaya, I., 1984, Computer simulation of hydraulic fracturing in shales— Influences on primary migration: Journal of Petroleum Technology, v. 36, p. 826–828.

Palciauskas, V. V., and P. A. Domenico, 1980, Microfracture development in compacting sediments: Relation to hydrocarbon-maturation kinetics: AAPG Bulletin, v. 64, p. 927–937.

Shuster, M. W., S. Eaton, L. L. Wakefield, and H. J. Kloosterman, 1998, Neogene tectonics, greater Timor Sea area, offshore Australia: Implications for trap risk: Australian Petroleum Production and Exploration Association Journal, v. 38, p. 351–379.

Sibson, R. H., 1992, Implications of fault-valve behaviour for rupture nucleation and occurrence: Tectonophysics, v. 211, p. 283–293.

Sibson, R. H., 1994, Crustal stress, faulting and fluid flow, in J. Parnell, ed., Geofluids: Origin, migration and evolution of fluids in sedimentary basins: Geological Society (London) Special Publication 78, p. 69–84.

Sibson, R. H., 1996, Structural permeability of fluid-driven fault-fracture meshes: Journal of Structural Geology, v. 18, p. 1031–1042.

Sibson, R. H., 1998, Conditions for rapid large-volume flow, in G. B. Arehart and J. R. Hulston, eds., Water-rock interaction: Proceedings of the 9th International Symposium-WRI-9, Taupo, New Zealand, March–April 1998: Rotterdam, Balkema, p. 35–38.

Streit, J. E., 1999, Conditions for earthquake surface rupture along the San Andreas fault system, California: Journal of Geophysical Research, v. 104, p. 17,929–17,939.

Watts, N. L., 1987, Theoretical aspects of cap rock and fault seals for two-phase hydrocarbon columns: Marine and Petroleum Geology, v. 4, p. 274–307.

Wiprut, D., and M. D. Zoback, 2000, Fault reactivation and fluid flow along a previously dormant normal fault in the northern North Sea: Geology, v. 28, p. 595–598.

Wiprut, D., and M. D. Zoback, 2002, Fault reactivation, leakage potential, and hydrocarbon column heights in the northern North Sea, in A. G. Koestler and R. Hunsdale, eds., Hydrocarbon seal quantification: Norwegian Petroleum Society Special Publication 11, p. 17–35.

Zoback, M. D., and J. H. Healy, 1984, Friction, faulting and "in situ" stress: Annales Geophysicae, v. 2, p. 689–698.

Seals: The Role of Geomechanics

Gary D. Couples
Institute of Petroleum Engineering, Heriot-Watt University, Edinburgh, Scotland, United Kingdom

ABSTRACT

Geomechanical analysis is a mechanism for understanding the complex interactions that occur in a deforming system, leading to the ability to predict how that deformation impacts the key physical properties of the rocks, such as those related to fluid flow. Specifically, geomechanics enables us to determine how assemblages of rocks will respond to a loading arrangement, provided that we can also stipulate the complete suite of mechanical behaviors for all of the components of the system. The roles of geomechanical processes, in terms of how they influence the fluid system, are crucial elements in many applied subject areas and especially so in the consideration of seals. In fact, the interaction between geomechanical processes and pore fluids is bidirectional, via effective stress, and through permeability, which is itself primarily controlled by geomechanical processes that alter the pore network during deformation. By acknowledging this bidirectional interaction, we can consider how both rock deformation and fluid flow represent the transfer of energy through complex natural systems. The nonreversable coupling between these two processes leads to highly nonlinear system responses, such as the formation and operation of seals. Poroplasticity, which integrates the data and concepts derived from decades of rock mechanics testing, is a material description that provides the critical link to allow us to make realistic geomechanical predictions about seals. This coupled geomechanical + fluids approach, based on poroplasticity, is applied here to explain the formation and predict the capacity of top seals. The approach is also used to show how our understanding of fault seals can be improved by considering the evolution of deformation-altered materials both during and after faulting. In both situations (top seals and fault seals), the creation (or failure) of seals is primarily related to alterations of the pore system by mechanical processes (which may assist in setting the stage for a chemical or diagenetic overprint). Throughout, geomechanical processes play a first-order role in governing seal behavior.

INTRODUCTION

Geomechanical processes play a dominant role in creating seals and in controlling their functionality. Rocks and their contained pore fluids together transfer mechanical energy through a basin across a range of space- and timescales. These components of natural systems are linked by multiple interaction channels, with bidirectional feedbacks throughout. The transfer of energy through rocks + fluids systems can be delayed or rerouted as a consequence of the formation of seals, which retard fluid movement. Conversely, seal failure

Copyright ©2005 by The American Association of Petroleum Geologists.
DOI:10.1306/1060758H21906

can accelerate (previously delayed) energy transfer. Deformation itself can be a sink for energy or a source. Together, the geomechanical + fluids systems represent a major part of the energy budget of basins but a part that is typically not represented by explicit physics in basin modeling. Within basins, primary lithological heterogeneities and additional heterogeneities introduced by deformation events control where and when seals form and when they fail. The action of seals thus plays a key role in governing the way that basins evolve and how they dissipate their energy. From this point of view, a typical basin system is a type example of complexity and nonlinearity.

In this chapter, I seek to examine geomechanics + fluids systems as examples of nonlinear physical systems that develop important emergent behaviors as a direct consequence of the formation and activity of seals. To accomplish this goal, I need to describe an analysis approach that is focused on geomechanical processes and how these are coupled with the pore-fluid system. Specifically, I wish to consider how geomechanical changes produce seals, or destroy them, as part of the process of transferring and converting energy in a basin. In this context, seals are defined operationally as barriers that inhibit fluid flow and thus influence the transfer of deformational energy because of the way that rock mechanics and fluids are interrelated (see also Holbrook, 2002). The link between rock deformation and seals is caused by the fact that strain alters the pore network, producing changes in the pore volume (porosity), and the way that pores are connected, thus altering the flow characteristics. The pore fluids influence deformation primarily through effective stress. By considering the rate of energy supply, or power input, a rock with a given permeability could result in an operational seal. The context must be considered to identify seals in their role as energy transmission barriers.

My definition of geomechanics emphasizes the need to consider a system of parts, loads, and responses. It is the explicit acknowledgement of the interactions of these system components that characterizes a proper geomechanical analysis. Failure to address the interactions can lead to false understanding that can be applied incorrectly, with potentially serious consequences. In this chapter, I describe the main features of the poroplastic material model and how it represents a synthesis of most of what we have learned about the mechanical responses of rocks from several decades of laboratory testing. I then use the poroplastic model to explain why seals form and why they fail. I also briefly discuss the role of chemical change (vis-a-vis seals) as a process mainly dependent on circumstances created by geomechanics and one that interacts with those deformations. Throughout, I focus on the idea of energy flux in a basin to provide a setting for understanding the role of the seals produced by geomechanics.

DEFINITION OF A SEAL

A seal can be defined as a body of rock that is capable of significantly retarding the movement of fluids. Diffusion is typically not considered to be an important process for moving substantial quantities of fluid, so this definition, in effect, means that a seal is a rock across which a fluid will move only slowly in response to a gradient of potential energy (see Neuzil, 1995, for a lucid discussion concerning dynamic seals). In practice, seals are recognized in one of two ways: in cases where there is a rock mass that bounds an accumulation of a fluid phase, such as oil or gas, or where a rock mass is spatially associated with a change in potential energy of the pore fluids, such as where the pressure-vs.-depth profile has a gradient that differs from hydrostatic (steeper or gentler). In the former case, we would probably say that the seal is capable of preventing (significant) flow of the trapped hydrocarbon phase(s). In the latter case, we would probably say that the seal is effective for all fluids. Interestingly, we recognize seals only in an operational sense: we require evidence of actual fluid retention and retardation. Other rocks with identical properties to those of active (identified) seals are not (normally) classified as sealing if no direct evidence of fluid retention exists. Because of this operational definition, we typically underestimate the frequency of occurrence of rocks with sealing capacity. The definition just given involves adverbs such as "significantly" and "slowly." Thus, the definition (identification) of a seal is qualitative and depends on the context of the physical situation.

That context is provided by a consideration of the rates of the fluid flow (energy dissipation) as compared with the rates associated with (re-)supply of the driving energy. In simple terms, fluid flow is driven by potential energy gradients. If higher potential energy is created in the fluid system at some location in space, the pore fluid(s) will tend to flow away from that site. The rate of flow in any direction and, thus, the rate at which the energy could be dissipated, is dependent on the (commonly directional) permeability of the rock (specifically, the relative permeability applying to the fluid phase under consideration and assuming that additional energy barriers, such as entry pressures, are overcome). This statement is merely a rephrasing of the classical Darcy law, which states that the fluid flux is proportional to the product of the energy gradient multiplied by the (phase-relative) permeability. Thus, the real issue, with regard to identifying operational seals, is to recognize circumstances where the rate of achievable energy dissipation (flow) is less than the rate at which potential energy can be imposed or renewed. This formulation of the definition involves rates of energy supply and dissipation, which means that the notion of seals is linked with power. Specifically, the operational seals are ones that interfere with the dissipation and transmission of power, because they retard the flow of fluids in systems that are not in equilibrium. The converse idea is that seal failure can lead to accelerated power transfer. Geological seals are those that are capable of retarding fluid flows associated with power values typical of basinal processes. Rocks with higher permeabilities can, however, be classified as production seals in a reservoir performance context because of the higher power that can be realized during production activities.

The intrinsic permeability of a rock and the relative permeability and entry pressure (which are functionally related to fluid properties and pore system characteristics; Leverett, 1941; McDougall and Sorbie, 1995, 1997) are the parameters that are important in determining whether a seal is (or could be) operational in the sense of retarding energy transfer (dissipation). It is, of course, the topology and characteristics of the pore network of the rock (Adler et al., 1992; Ferreol and Rothman, 1995; Bakke and Øren, 1997; Hazlett, 1997; Dewhurst et al., 1999; Brown, 2003; Øren and Bakke, 2003) that govern the numerical value of its intrinsic permeability (which is a macroscopic characteristic or emergent property). The intrinsic permeability (in reality, the pore network), along with fluid saturations and the wettability of the solid components of the rock, determines the multiphase properties of the medium (Pereira et al., 1996; Blunt, 1997; Fenwick and Blunt, 1998; Dixit et al., 2000; Manzocchi et al., 2002). The empirical relationships that are commonly used to calculate (estimate) relative permeabilities (e.g., Corey, 1954; Chierici, 1981) are merely macroscopic expressions that, nevertheless, are based on the pore-network controls that underlie all flow properties. Thus, all of the relevant flow characteristics of rocks are intrinsically linked with the nature of the pore network. Therefore, it is important to understand how pore networks change as rocks deform during the accumulation of strain.

The concept that underpins this chapter is that geomechanical events exert a first-order control on the topology and characteristics of the pore network of a rock. Studies of experimentally created and natural damage in porous rocks have shown how, for some deformation modes, especially those of interest to this chapter, the constituent grains and their binding cements can break and then how the individual particles can move during deformation (some representative references include Friedman, 1963; Stearns, 1969; Engelder, 1974; Gallagher et al., 1974; Pittman, 1981; Zhang et al., 1990; Antonellini et al., 1994; Menéndez et al., 1996; Wong et al., 1997; Chester and Chester, 1998; Fisher et al., 2000). The resulting changes to the pore system (e.g., Figure 1a) are the fundamental cause of the reduced permeability associated with such processes. A less-considered type of deformation, normal basin compaction (also called consolidation), also

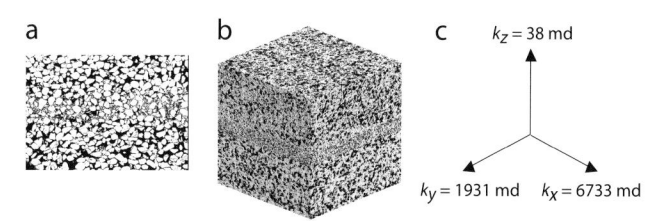

FIGURE 1. (a) A photomicrograph (field of view 9 × 6 mm [0.35 × 0.24 in.]) of porous sandstone containing a shear band (image courtesy of Chris Wibberley; C. A. J. Wibberley, J.-P. Petit, T. Rives, 2005, personal communication). Note major grain-size reduction in the shear zone crossing through the middle of the thin section from left to right. (b) Three-dimensional reconstruction of this matrix + shear band system, created using a multi-component, Markov-chain approach (cube is 10 mm [0.4 in.] on each side). (c) Calculated directional permeabilities of the volume of rock depicted in (b), resulting from the application of a single-phase Lattice–Boltzmann scheme. Unpublished results shown in (b, c) are provided by K. Wu (Heriot-Watt, 2004, personal communication); see also Wu et al. (2004).

rearranges the granular components of the rock (sometimes by processes of recrystallization) and reduces the pore volume, along with modifying the pore shapes and how the pores are connected. Diagenetic changes can also play a role in porosity loss, of course (Bjørlykke and Egeberg, 1993; Walderhaug, 1996; Lander and Walderhaug, 1999; Paxton et al., 2002). However, if chemical changes are contributing to strain (the change of shape and/or volume), then we must consider the relative roles of geomechanics and chemical effects jointly (see below). A joint consideration is also needed if we acknowledge that recrystallization (geochemical processes, generally) can alter a rock's components such that it becomes a different lithology.

The compaction of sediments (which continues after they are classified as rocks) is a permanent deformation, in that the rock deforms in a nonrecoverable fashion (the usual assumed context is vertical, uniaxial strain). Other deformations also involve significant changes to the pore networks of the rocks because of permanent (nonrecoverable) compactant or dilatant volumetric strains (bulk volume loss or gain, respectively) and distortional strains, and these also have associated impacts on the intrinsic permeability (and multiphase characteristics) of the deformed rock (refer to Figure 1). Geochemical processes (such as cementation and dissolution, precipitation of pore-filling clays, etc.) also impact the petrophysical properties, of course; however, because they equally change the mechanical framework of the rock, I suggest that it is useful to consider that any chemical changes that occur do so within a process framework where geomechanical events have both set the stage and may also be continuing. Thus, deformation processes and chemical effects need to be considered together. However, I argue that deformation (a response to imposed gradients of mechanical energy) is typically the dominant process that controls the creation and function of seals. This proposition will be reexamined later in this chapter.

DEFINITION OF GEOMECHANICS

I regard geomechanics as an exercise that is designed to predict rock deformation in a system. Typically, in much of geoscience, geomechanics seeks to make a retrodiction (a "forward," or process-based prediction of past events leading to the "now" observations), which I include in the prediction activity. The emphasis here is on the notion of a system that has three conceptual elements: (1) the component physical parts; (2) the mechanical behavior(s) of the material(s) that comprise the parts and the interfaces between them; and (3) the loading arrangements that cause the parts to respond according to these material laws. Critically, geomechanics requires that we fully consider how the three system elements interact, commonly with strong nonlinearity. Examples of practical application of the geomechanical method (to seals) will be presented later. But first, it is important to examine each of the conceptual elements and then to identify some of the common pitfalls that can be made if the interactions between them are not addressed.

Identifying the component parts of a system is a task that can be a source of major error. A common mistake is to choose a system that is too simple and does not contain enough, or large enough, components to allow the geomechanical analysis to consider how the components interact. In practical terms, it is always necessary to identify and isolate a system that is a finite region; somehow, we must extract (in virtual terms) and separate (in mental terms) this region from the whole of the natural setting. This necessity to consider an unreal case is the primary source of difficulties arising in geomechanics.

In terms of understanding seals, a geomechanical analysis would commonly include several component parts: the rock layers (which represent coherent packages of sediment that were deposited, buried, and compacted and lithified to their current state); possibly the basement; and any faults or other products of deformation that developed earlier in the deformation history or may be deforming at the time of the analysis. Other components would include the pore fluids and, possibly, mass that could be introduced or removed by means of the pore fluids (that is, chemical effects). It is commonly important to include the interfaces between the components as explicit parts, especially if we need to consider possible motions between the solid components (slip).

Identifying the material behavior of each of the parts includes the notion of how each material or interface yields (that is, the conditions [stress and/or strain state] at which each ceases to be able to resist an increase in imposed distortions without suffering permanent change). Although this notion of yielding is conceptually quite clear, laboratory experimentation has identified a range of material responses around the yield point (e.g., Lockner et al., 1992; Menéndez et al., 1996; Wong et al., 1997; Lockner and Beeler, 2002, and references therein; Mair et al., 2002). In simplest terms, many (most?) rocks subjected to laboratory loading exhibit nonlinearities in their response (some departure from linear covariance of two variables, such as stress and strain) before they reach the obvious break in slope on a plot of those variables. In some cases (see Figure 2a), no obvious point is present where the slope abruptly changes; it is merely a progressive change. Do these observations indicate that the notion of yielding is wrong?

I do not think so. They may be suggestive of a transition instead of an instantaneous change (implying, possibly, contributions from multiple physical processes under different conditions). They may also be giving us a hint that there are artifacts associated with laboratory testing. I do not wish to divert into a major debate on this point (although that may subsequently be worthwhile) but instead want to accept as useful the concept that rocks experience a change from not yielding to yielding, and that we can capture that change in a behavioral description that specifies the conditions causing the change. Most of what concerns us (relative to seal formation) involves rock deformation events that progress well beyond initial yield; seal failure may, however, require a more careful consideration of these precursor processes. To limit the scope of this chapter, I will adopt a position in which such precursor damage is assumed to be similar to the damage caused by real yielding (which is appropriate for the dilational mode described below).

The material behavior description also needs to specify how the parts behave before yielding. This is commonly assumed to be well approximated by an elastic material model or a poroelastic one if the possibility of significant fluid pressure changes during the mechanical loading exists, but the precursory yielding noted above is worth mentioning. The key point is that, for realistic geomechanical analysis of systems, the description of material behavior also must specify how the material behaves after it has yielded (i.e., how it continues to yield). A failure analysis (such as is commonly done using a Mohr–Coulomb criterion, commonly based on a peak stress value) is inadequate for many types of geomechanical study (but it seems to be sufficient for some purposes, such as wellbore stability predictions; e.g., Colmenares and Zoback, 2002). It is perhaps worth emphasizing that failure is a subset of the complex suite

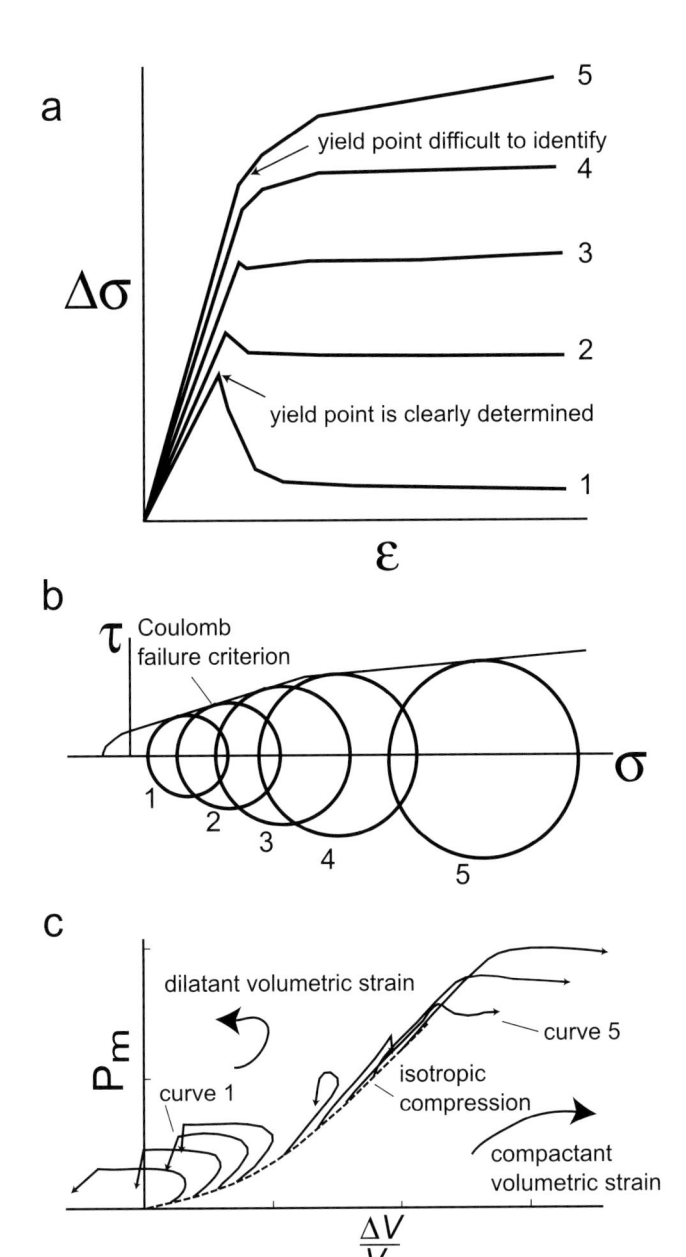

FIGURE 2. Representative (schematic) mechanical data for common porous rocks. (a) Plots of differential stress ($\Delta\sigma$: $\sigma_1 - \sigma_3$) vs. axial strain (ε). Numbers from 1 to 5 represent increasing confining pressure (σ_3). (b) Stress states at yield for the experiments in (a), plotted in typical Mohr–Coulomb fashion. (c) Volumetric changes associated with deformation of such rocks. Note the major change in volumetric behavior (expressed as a fraction: volume change, ΔV, divided by initial volume, V_o) as mean stress (P_m) increases.

of yielding responses. Neither a failure criterion nor a yield analysis, by themselves, are capable of representing how a system of multiple parts will behave after any part of the system has failed or yielded. If pore fluids are included in the system, then the concept of effective stress must be explicitly addressed in the material description. (However, if no seals are present in the system, it is commonly possible to leave fluids out of the analysis, for example, by altering the material density to obtain dry stress components that are equal to what the effective stresses would have been).

The most difficult system element to identify correctly, and particularly, to implement in a model, is the loading arrangement. As noted above, we are forced to undertake geomechanical analyses on extracted, finite regions. One of the most challenging tasks is to discover how to define the region (the model) in such a fashion so as to make it also possible to specify sensible loading conditions on the boundaries of the now-isolated system. It is all too easy to impose boundary conditions that cause the system to respond in a way that is unlike what would happen naturally. The actual form of the system loads (including internal body forces, such as those associated with gravity), in terms of boundary or internal displacements or forces or pressures, is highly dependent on the representation of the system in a mathematical formulation. Unfortunately, it is possible to operate mathematical models in physically implausible (even impossible) ways, but to still obtain an answer (albeit an answer that may have little or no meaning). I will illustrate this problem with a simple example (see below). But before doing that, I wish to reemphasize that the key aspect of performing a useful geomechanical analysis is related to the manner in which we specify boundaries that can be associated with sensible loadings. One of the most important geomechanical skills is to be able to recognize when a model is inappropriately constructed.

The key point underlying my definition of geomechanics is that the elements of the system interact. The example outlined next is intended to show the problems that can arise when the interactions are not addressed, that is, when changes are arbitrarily supposed to occur without reference to the processes that might cause them to occur.

Let us consider a very simple (too simple!) case that has been posed countless times to students learning rock mechanics. Imagine a small volume of rock located somewhere in the subsurface (Figure 3a). The region considered is small, so we only have one material that we shall assume is uniform in its mechanical behavior (i.e., one rule stating how the material responds to loads). Following common practice, we shall assume that the rock's mechanical characteristics are defined by a linear Mohr–Coulomb failure criterion, and that prefailure behavior follows an elastic law. Given our assumption of an Mohr–Coulomb criterion, there is an implication that the initial stresses are reasonable (below the yield point). By adopting Mohr–Coulomb failure as a model, we are implying that failure, when it occurs following additional loading, will be represented by the development of some type of shear zones (fractures or faults); this is the main response for rocks

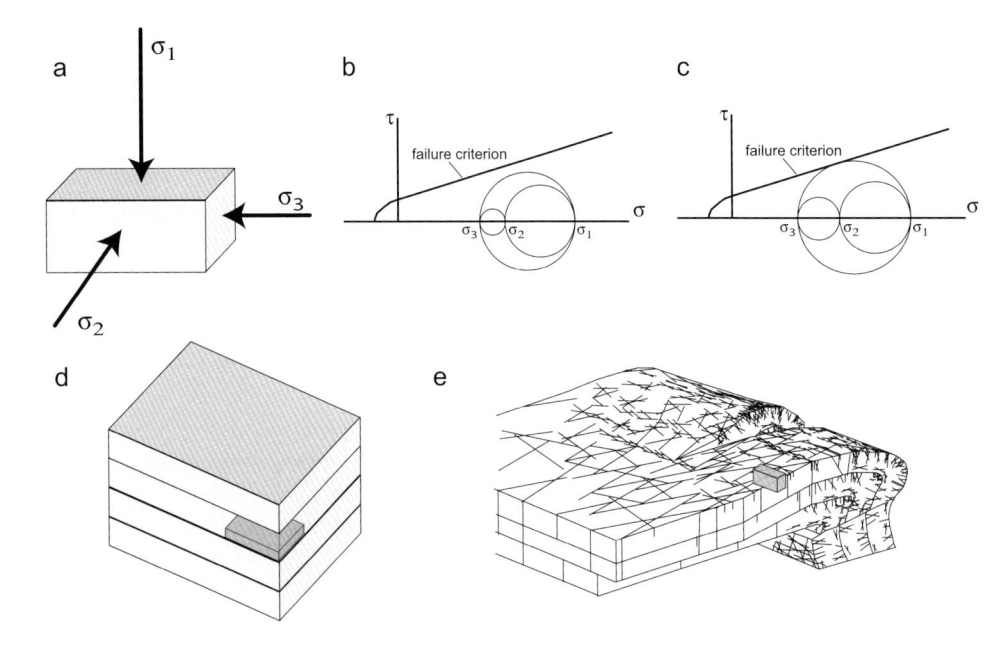

FIGURE 3. Illustration of problems arising in a typical analysis. (a) Small element of rock and assumed initial stress state. (b) Mohr circle representation of stress state in (a). (c) Changes of stress state arbitrarily supposed to occur (see discussion in text). (d) An arbitrary, small rock volume contained within a stack of layers. (e) A small rock volume contained in a complex structure. Fold and fracture model in (e) is modified from Lewis and Couples (1993).

under conditions where Mohr–Coulomb failure occurs. We will also assume a starting point for our analysis in which the material has developed the simple state of stress indicated. (Note that I specifically have not written: "...the material is subjected to a simple state of stress" for reasons that will be explained shortly.) For clarity (although it might be unusual), I choose to assume a simple initial state where the principal stresses are vertical or horizontal, and I identify them with the coordinate directions of a Cartesian system (adopting usual sign conventions and notations):

$$\sigma_v = \sigma_1 = \sigma_z = -\rho g z$$
$$\sigma_1 > \sigma_H = \sigma_2 = \sigma_y > \sigma_3$$
$$\sigma_3 = \sigma_h = \sigma_x (? = \sim 0.6\sigma_1)$$

The initial state of stress can be depicted by means of Mohr circles (Figure 3b). They lie in the stable domain (i.e., underneath the failure criterion already known for this material), so we are satisfied that the model is suitably initialized. Now, the standard analysis goes something like the following: If the σ_x (σ_3) component of stress is reduced, the outer Mohr circle increases in diameter, until it contacts the failure criterion (Figure 3c), at which point the rock fails, and we predict that shear zones will develop (at angles that can be determined graphically or through formulae: the methods are not illustrated here; but all good students learn this skill, although they commonly learn to hate it). This hypothetical situation could occur, and the statement about the rock behavior is correct as far as it goes, but the analysis leaves some important questions unanswered. For example, why did the σ_x component get smaller? What happens after the rock initially yields (fails)? Do the newly formed shears continue to move? How far? Do they propagate into adjacent rocks? Do the adjacent rocks experience a change in stress state? These questions (and others) arise whenever we contemplate deformation in a wider context.

I argue that the simple analysis presented above is not geomechanics. It did not address the interactions of system elements (except in the trivial way that it restates the Mohr–Coulomb criterion, "as the σ_3 stress component is reduced, the material fails"). In fact, the model does not represent a system at all. We learn nothing about what would happen in a real system comprising multiple component parts (even one as simple as a stacked sequence of rock layers), assuming that our original rock region was contained in that system (Figure 3d). It certainly tells us nothing new about how our small block of rock would actually deform or influence surrounding rocks, if it were part of a complex structure (Figure 3e). Geomechanics requires us to consider materials deforming in a realistic context.

Perhaps a more problematic version of this example is when the statement, "If the σ_x component is reduced," is replaced by the statement, "When the σ_x component is reduced." The latter phrase suggests that the hypothetical situation is, in fact, reality. In this instance, the issue of why the stress component is reduced becomes more than a possibility and instead requires a firm answer. That answer can only be provided by considering how the system operates, meaning to determine how the elements of the system interact, with a particular focus on the role of the loading. Considerations about system interactions are missing from the simplistic approach examined above and, instead, are replaced by arbitrary statements about certain aspects ("the stress changes in such-and-such fashion..."). In fact, the stress state of a natural system is the dependent variable, and we make a gross error anytime that we suppose that the stress components change in an arbitrary way. Even the "correct" statement, "If the stresses change in such-and-such a way, the rock behaves like so," is merely a restatement of the material-behavior element of the system.

Unfortunately, we have grown accustomed to using that statement ("the stresses change...") in connection with laboratory testing, where the piece of rock is not part of a natural system. In the laboratory, we operate on a small sample of rock, and we impose loads onto it that we believe can be equated with the resulting state of stress in the sample. In fact, it is essential that we do this to enable us to derive material properties from the experimental results. Leaving aside any concerns about whether our loading actually induces the stress state that we think we create (impose?), laboratory methods are commonly believed to provide us with the ability to create almost any desired state in the sample, allowing us to grow comfortable with the casual word usage criticized above. In contrast, in a natural setting, no strong basis exists for assuming an arbitrary state of stress or how stress might change. Instead, we have to determine how the natural stress state is created or changed as a consequence of any loading and deformation of a system of parts and the responses of the material behaviors, requiring that we acknowledge the interaction issues emphasized above. I encourage the use of proper geomechanical analysis as a way of avoiding the tautology of trivial restatements or the posing of a physically impossible sequence of events.

PRACTICAL MATERIAL DESCRIPTIONS

Real geomechanical analysis, as described above, is becoming more practical as a consequence of improvements in the way that we can represent complex rock behavior through constitutive relationships (i.e., functional interaction between stress and strain). Current best practice methods are based on a poroplastic material model (Couples, 1999; N. G. Higgs, D. R. Parrish, G. Workman, G. D. Couples, 2005, personal communication), which is conceptually related to the Cam Clay and critical state models developed in soil mechanics (Scofield and Wroth, 1968; Atkinson and Bransby, 1978; Wood, 1990). The poroplastic model combines yielding and postyield behavior into a common description. Below, I first show how our long-standing knowledge of rock mechanics, as determined from laboratory testing, is integrated by the poroplastic model, and then I will show how that model is useful for understanding issues such as seal formation and seal failure.

As has been known from laboratory testing for quite a long time (Handin and Hager, 1957; Griggs and Handin, 1960; Handin et al., 1963), the common sedimentary rock types exhibit a progressive change in their macroscopic deformation behavior as a function of the effective confining pressure (Figure 4). At low (relative) effective confining pressure, most common rock types deform by means of discrete shear zones. When the confining pressure is greater, the shear zones become more closely spaced and commonly are slightly less sharp. Macroscopically, the cylindrical rock specimens develop a barrel shape during deformation at the higher confining pressures. The sharp shear zones and planes developed at lower confining pressures are commonly interpreted to represent localization behavior (Rudnicki and Rice, 1975; Olsson, 2000; Issen, 2002), whereas the overall barrel shape that develops at higher pressures is commonly interpreted to be the consequence of a distributed or delocalized material response. However, for many rocks, the deformation mechanism is constant across the localized to delocalized transition (in the case illustrated, all deformation is cataclastic, involving grain breakage and particle motions). Thus, a scale dependency to the identification of localized vs. delocalized deformation is present. This is basically the same issue that arises in the way that many people use the terms brittle and ductile when referring to outcrop observations of deformation.

The experimentally determined stress-strain behaviors of typical sedimentary rocks are also dependent on the confining pressure. At low confining pressure, the yield point is typically well defined (Figure 2a),

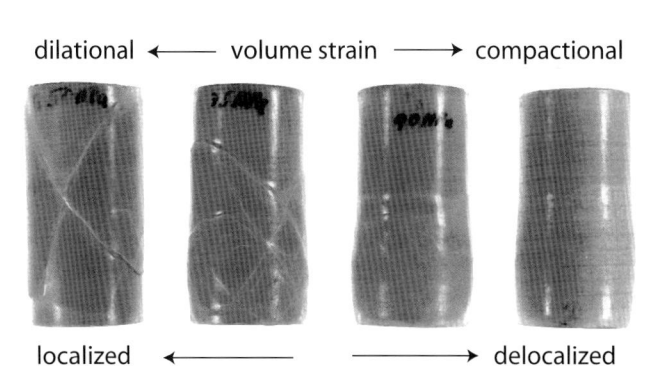

FIGURE 4. Four cylinders of Berea Sandstone (fine-grained, dominantly quartz, porosity approximately 20%) deformed under increasing confining pressure (σ_3), ranging from 50 MPa on the left to 110 MPa on the right. Cylinder diameters are 50 mm (2 in.) (nominal). Shiny surfaces are nearly transparent plastic jackets. Note the change in macroscopic shape of deformed cylinder as a function of confining pressure. Also note the change in (apparent) degree of localization of deformation (shear zones). The leftmost cylinder will produce a stress-strain plot like curve 1 in Figure 2a, whereas the rightmost cylinder will produce a plot like curve 5. Samples provided courtesy of Helen Lewis. See also Wong et al. (1997) and Mair et al. (2002).

whereas at higher confining pressures, it may become less obvious how to identify the yield stress. Postyield characteristics also differ; low confining pressure cases are more likely to be typified by a strain- (or work-) softening response, whereas the higher confining pressure conditions commonly reveal a work-hardening response. If the yield conditions are plotted onto a Mohr–Coulomb diagram, the stability (or yield or failure) criterion line increases at an ever decreasing rate as confining pressure (or mean stress) increases (Figure 2b). If the slope of the criterion is attributed to an internal friction, then the coefficient of internal friction decreases at higher confining pressure. It is worth emphasizing that the Mohr–Coulomb criterion (Handin, 1969) is only a model developed to explain observations (it was later supported theoretically, to some extent, as a possible macroscopic representation of microcrack coalescence; e.g., Brace, 1960). The success or failure of using this criterion to predict any behavior neither validates nor invalidates the experimental observations themselves. My argument is not that the Mohr–Coulomb criterion is invalid, only that it is insufficient for many practical purposes now arising in geomechanical analysis.

An important characteristic of laboratory-induced deformations (and presumably, their natural counterparts) concerns the volumetric strains that occur as the rocks yield. It is commonly observed that the strongly localized type of deformation (e.g., the lower confining pressure cases) produces a volume increase or dilation (Figure 2c), whereas the delocalized type may be more likely to result in a volume loss or compaction, as yielding continues (e.g., Wong and Baud, 1999). These volumetric strains are, of course, extremely important in a complete material description. For example, localization theory (Rudnicki and Rice, 1975; Issen, 2002) is strongly dependent on volumetric strains, and key physical properties (especially flow properties) are also a function of the pore system that is itself altered during strain (Zhu and Wong, 1997; A. A. DiGiovanni, J. T. Fredrich, D. J. Holcomb, W. A. Olsson, 2005, personal communication).

Here, it is worth noting that a wide range of macroscopic physical properties is similarly dependent on the topology of the pore network, which is a function of the arrangement of the grains and cements (solids), which is itself altered by deformation. These microscale processes control the macroscale effects that interest us in applied geoscience and geoengineering. There has been a long-standing interest in the subject of microscale effects in the seismic community, because porosity changes (especially the shapes of the pores, including induced microfractures) also impact rock physics parameters (Gueguen and Palciauskas, 1994; Hudson et al., 1996; S. A. Hall, H. Lewis, X. Macle, 2005, personal communication). It is likely that additional macroscale (emergent) properties will be addressed in this way in the near future. By linking the deformation predictions of geomechanics to predictions of other important properties (fluid flow, acoustic, dielectric, etc.), the value of geomechanical analysis can be considerably increased. In nearly all cases, the focus is related to situations with finite, permanent deformation that has continued after initial yield of the material in question. Thus, real-world requirements demand that the conditions of yielding and postyield behaviors both be understood within a common paradigm of material response.

The plasticity approach (sensu largo), in which the concept of yielding represents a transition between recoverable strain and nonrecoverable strain (damage), largely meets this need (Hill, 1950; Drucker and Prager, 1952; Jenike and Shield, 1959; Dormieux and Maghous, 1999). The typical failure mode observed in many laboratory rock mechanics experiments (e.g., the localization behavior noted above) represents one (important) type of yielding, but other behavior modes are also important and equally must be included, hence, the need to progress beyond Mohr–Coulomb. [Note that Mike Fahy posed this argument to his colleagues, including me, in the late 1970s but never published a standard paper detailing the rationale. He did give a talk on this subject at a Geological Society of America section meeting (Fahy, 1982) and deserves credit for highlighting key issues.] The following discussion of plasticity concepts is intended to provide a context for a description of the poroplastic material type that underpins this chapter. The presentation is quite simplified; further details and additional views can be found in several publications (e.g., Coussy, 1995; van der Veen et al., 1999; Wong and Baud, 1999; de Borst, 2001; Fredrich and Fossum, 2002, and references therein).

Yielding in plasticity is represented by a parameter (commonly labeled F) that is a function of mean effective stress, labeled as P or P_{eff}:

$$P_{\text{eff}} = \frac{1}{3}(\sigma_1 + \sigma_2 + \sigma_3) - P_p$$

where P_p is the pore pressure, and a scalar representation of differential stress (labeled as Q), which is commonly associated with the second invariant of the stress tensor (J_2):

$$J_2 = \sigma_1\sigma_2 + \sigma_2\sigma_3 + \sigma_3\sigma_1$$

The plot of F in P-Q space (Figure 5) is a curve defining the stress conditions at which yielding occurs. Stresses plot as a single point in the P-Q space instead of a circle in the σ-τ space used in Mohr–Coulomb analysis. For stress conditions that lie inside the yield curve in P-Q space, the material is stable (not yielding). The theory

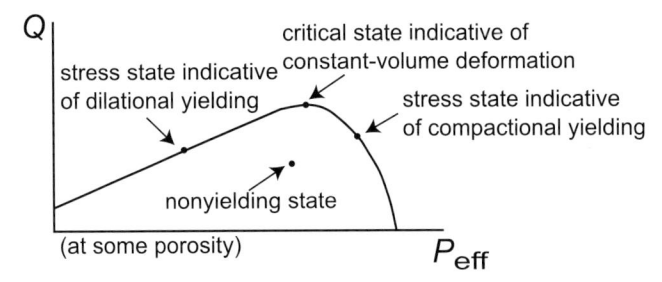

FIGURE 5. Yielding modes as a function of the state, as depicted in P-Q space. The plot of F defines the conditions of yielding.

holds that stress states cannot lie outside the yield curve. These characteristics of the plasticity approach are somewhat analogous to the familiar Mohr–Coulomb representation, and it should be straightforward to relate many ideas from one model to the other.

Postyield strain prediction in plasticity is addressed by inventing another set of axes that are conceptually related to (associated with) the P-Q axes (and which are taken to be parallel to them). The volumetric strain of deformation is represented in this space (Figure 6) by movements of a state point along an axis that is parallel with P, whereas the distortional (shear) strains are represented by motions of the strain-state point parallel to the Q axis. (A higher order scheme is needed to explicitly predict strain components, instead of the scalar shear that can be directly related to the Q scalar representation of stress state.) If a tangent line is drawn to the point of yielding on the P-Q plot, the normal vector at this point is transferred to the strain space, and the incremental strain components are predicted. Unfortunately, this approach fails to accurately reproduce the actual strains observed experimentally; the predicted volumetric strains are too large. The (inaccurate) parallel-space scheme is called the associated flow rule, which gives rise to the name "nonassociated flow rule" (e.g., Zienkiewicz et al., 1975) that is applied to schemes when correction factors are introduced (to change the orientation of the increment vector in strain space). Because the approach deals with the permanent (i.e., plastic) volumetric strains (and hence, porosity changes), it is not surprising that this concept is generally called poroplasticity.

Why do the volumetric strains occur? In a constant-mass system (here, we are referring to a porous rock, so the system is composed of solids [grains plus cement] and the commonly fluid-filled pore space itself), a volumetric strain is related to a change in porosity or void ratio. A volume decrease of the bulk system (porosity loss) represents the material becoming more solid, and the converse also holds. We can appreciate that a more solid material typically has greater strength than it did before the compaction occurred, and it can thereafter carry a larger load (commonly represented by higher values for the deviatoric components of stress, Q, for any value of mean stress, P). Conceptually, we can imagine the material needing to become stronger to allow a continued increase in the intensity of the distortional and/or compactional mechanical energy. Similarly, we can interpret dilational deformations as representing material that is too solid to be able to undergo the imposed distortions, so extra porosity must be created to weaken the material enough to allow cataclastic flow. The initial, porosity-enhancing step to accomplish the dilation (Wong et al., 1997) is represented by the sharp peak that is commonly observed on stress-strain curves for such overly strong materials (Figure 2).

The end point of such volumetric and strength changes is (conceptually) a state where further volume or strength changes are not needed; that is, the material can continue to deform in a steady state. Technically, the term critical state identifies the porosity and stress conditions at which such volume-constant distortional strain increments are possible for a material. These ideas have been widely adopted in soil mechanics research (Scofield and Wroth, 1968; Atkinson and Bransby, 1978; Wood, 1990), from which practical analysis and numerical simulation tools have been developed. These concepts have also been adopted in the geoscience community, particularly for muddy sediments (e.g., Jones and Addis, 1986). Much of the current research effort in theoretical poroplasticity is concerned with developing better predictions of the localization of strain (considered to be a bifurcation phenomenon) that occurs during many laboratory and natural deformations involving both dilation and compaction (Vardoulakis and Sulem, 1995; Desrues et al., 1996; Olsson, 1999; Issen and Rudnicki, 2001; Issen, 2002; Desrues and Viggiani, 2004). A comprehensive review of numerical approaches in rock mechanics, including plasticity and other topics, has recently been published by Jing (2003). Numerical approaches to simulating

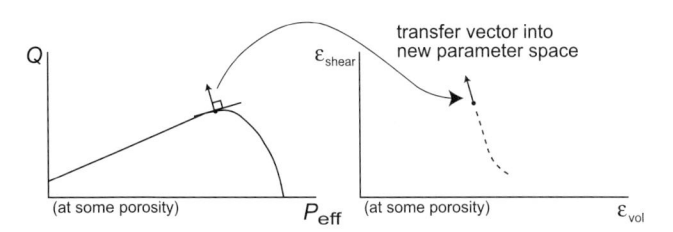

FIGURE 6. Conceptual relationship between stress space (left) and strain space (right). Normal vector in stress space is transferred to strain space to predict the associated flow of material in terms of volumetric and distortional strain increments. Dashed line indicates possible strain path leading to current strain state.

poroplastic deformations are fairly diverse, and no comparable overview has been published (but refer to the various papers, cited throughout this discussion, for some insights). Considerations of micromechanical processes are also being pursued to better understand what macroscopic behaviors mean (e.g., de Buhan and Dormieux, 1999; Chateau and Dormieux, 2002).

Although the poroplastic conceptual model is a convenient way to integrate the main deformation characteristics of porous rocks, our real data on geomechanical behavior is only derived from laboratory tests, where practical choices have historically focused most of our investigation effort into only some aspects of the poroplastic approach. Normally, in laboratory testing, rock samples (typically cylinders) are subjected to a loading arrangement, such that a biaxial (radially symmetric, commonly called triaxial) stress state is thought to be generated in the material. Stresses, and especially strains, are inferred from measurements made externally to the sample: stress (particularly axial stress) is typically calculated by determining the force applied and dividing by the area across which it acts or by directly assuming that the (radial) stress equals the pressure of a fluid surrounding the sample; strain is calculated by displacements on the surface of the sample (strain gauges) or by making measurements of the volume change of the confining fluid or even by determining the motion of the sample ends, sometimes outside of the pressure vessel.

Practical considerations have led to the development of standard experimental protocols that use one of a few simple stress paths during the measurement process (Jaeger and Cook, 1969). These paths may differ considerably from what is natural, and some work has been done to examine other paths (such as strain control). Nevertheless, the general behavior that is observed in these tests is very consistent, allowing significant generalizations to be made concerning the way that any given rock will respond to loads as a function of increasing effective least stress (confining pressure in experimental jargon), strain rate, temperature, etc. In addition, the range of rocks that have been studied also allows us to generalize about the role of lithology (dolostone is quite different than limestone, for example) and how grain size, sorting, packing, cement, and porosity affect the mechanical response of otherwise similar rocks. Old but still valid summaries of these generalizations are found in Handin and Hager (1957), Griggs and Handin (1960), and Handin et al. (1963).

The laboratory conditions that cause yielding or failure can be represented by a criterion. The Mohr–Coulomb depiction and equation is widely assumed to be broadly applicable in rock deformation, and Byerlee's Law is also widely cited as a general portrayal of yielding conditions for crustal rocks (e.g., Brace and Kohlstedt, 1980; Mandl, 1988; Scholtz, 1990). However, neither of these simple relationships is very useful for understanding the conditions that lead to compactional yielding. To address that issue, the Mohr–Coulomb approach was modified to include a cap (e.g., Resende and Martin, 1985). However, even after this change, the criterion is still that— a criterion stating that yielding will occur at such-and-such conditions. What is needed is also a prediction of what happens after yielding. This is because we wish to understand finite deformations that have progressed beyond the initial point of yield, and any predictions about real-world distributions of deformation, and their effects on properties, need to be based on the finite deformation state.

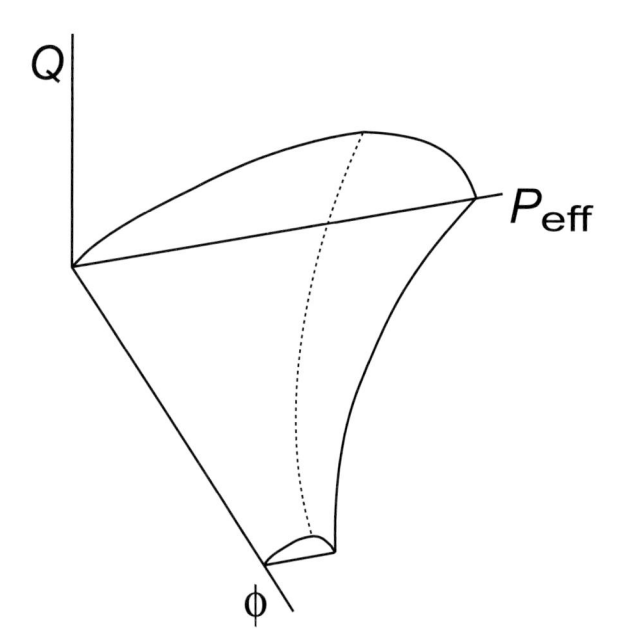

FIGURE 7. Poroplastic yield surface with explicit dependence on porosity (or void ratio) as a state variable.

I have proposed (Jones and Addis, 1986; Couples and Lewis, 1998, 2002; Couples, 1999; Cuss et al., 2003) that a useful way to integrate the wealth of laboratory data is to use a parameter space in which porosity is a state variable (i.e., it controls the mechanical response, instead of being a derived consequence of it). This approach was pioneered in soil mechanics (as noted above) to enable a robust calculation of the substantial volumetric strains that occur as soil compacts under self-weight and other loads. In geomechanics, we also have a need to consider large volumetric strains, especially during basin compaction. I have adapted the soil mechanics concepts to develop a general rock mechanics description for finite deformations of typical geomaterials (Figure 7).

Although the poroplastic graphical representation (stress space) and its related volumetric and distortional strain predictions (in strain space) is reasonably simple to comprehend, the implementation of the theory (as, for example, in a numerical simulator) relies on mathematical expressions involving partial differential equations and tensor notations. These mathematical representations can seem daunting to some, making it unduly difficult to communicate what is a fairly simple concept. In the following section, I work through an example where no mathematics is present; this allows me to demonstrate the primary conceptual basis for and value of the poroplastic description of material behavior. I then introduce examples where poroplastic ideas have been implemented into a numerical simulator, allowing a demonstration of how this approach allows a realization of real geomechanical analysis in a complex system of deformation.

APPLICATIONS OF GEOMECHANICS TO SEALS

Geomechanical analysis is important in two ways relative to the subject of sealing. One is to provide an understanding of the processes that occur during deformation. This use of geomechanics is exemplified in the following discussion related to the formation and failure of top seals. I also show how geomechanical analysis can be applied to a consideration of fault seals and especially toward understanding the distributions of deformation damage created during faulting. Potentially, this understanding could be used to identify classes of faults to which particular empirical fault seal prediction algorithms can be applied. For example, the scheme described by Sperrevik et al. (2002) relies on information about the clay content of the faulted sequence of rocks to estimate the resulting fault rock properties. Geomechanical models of faulting could lead to improvements to that scheme, by considering a greater dependence on the lithology-controlled deformation behavior of the system components, and by considering how the mechanical state evolves in the system during faulting. This approach may lead to the identification of new classes for which further empirical data could be gathered and applied more effectively. The second role for geomechanics, and one that is only on the verge of being practically realized, is to be fully integrated with petrophysical predictions and then coupled to fluid-flow simulations to enable us to address all aspects of the seal problem. An initial example of this use of geomechanics included here indicates the potential soon to be fully realized.

Creation and Failure of Cap Rock Seals

The poroplastic concept, as detailed above, represents a general, macroscopic material description that is well suited to understanding the mechanical basis for many common basin events (Couples and Lewis, 2002), including the formation of pressure seals (and similar arguments that apply to cap rock seals). Let us consider a hypothetical basin in which the sediment accumulation rate is high, so there is a continuing input of mechanical energy driving deformation that, because of consequent seal formation, leads to anomalous regions of high fluid potential energy (Swarbrick et al., 2002). To start, we need to acknowledge the depositional heterogeneity inherent to typical stratigraphic successions, including muddy sequences of rocks, which are commonly associated with overpressure (Aplin et al., 1999; Swarbrick et al., 2000; Almon et al., 2001, 2002; Dawson and Almon, 2002). For demonstration purposes, I further assume that the primary heterogeneity is expressed by vertical variations, hence allowing me to undertake a simple one-dimensional analysis.

The geomechanical analysis (the method) requires us to identify the component parts of the system. These are embedded in a hypothetical sequence of sediments, which, when they were deposited, inherited their primary variations of mineralogy, grain size, packing, etc. from the deposition processes. In our hypothetical system, subsidence continues, and the sequence of layers comprising the system is buried by additional deposition. The entrained pore waters escape up through the sediments as they compact (i.e., as they plastically deform). Each layer of sediment represents a material type that is characterized by its individual material behavior law. What we can say with confidence is that the layers all have a state of stress that is at yield, and on the compactional yield part of the surface, for each material that is compacting (e.g., Figure 5). Because the layers are different materials, their yield surfaces are also different.

At a point in the basin history, the muddy sediments around some layer S have been buried to sufficient depth and become compacted such that the associated permeabilities are also reduced to low values (Figure 8). We assume that layer S has the lowest permeability in this succession at this particular point in the basin history. If loading (e.g., from continued deposition above) is rapid enough (Waples and Couples, 1998; Holbrook, 2002; Miller et al., 2002; Yassir and Addis, 2002), the pore waters continue to be out of equilibrium (the sediments continue to compact at some rate), and they continue to be expelled upward. If the flow retardation caused by the low permeability of layer S is sufficient, the sediments below layer S begin to experience more than a virtual overpressure condition. Because of this real overpressure, the mechanical states in these deeper sediments move away from their previous locations on the yield surface of each material type (Figure 9a). Hence, yielding of these overpressured sediments may cease, keeping a constant porosity, or it will at least slow down (if we acknowledge a rate

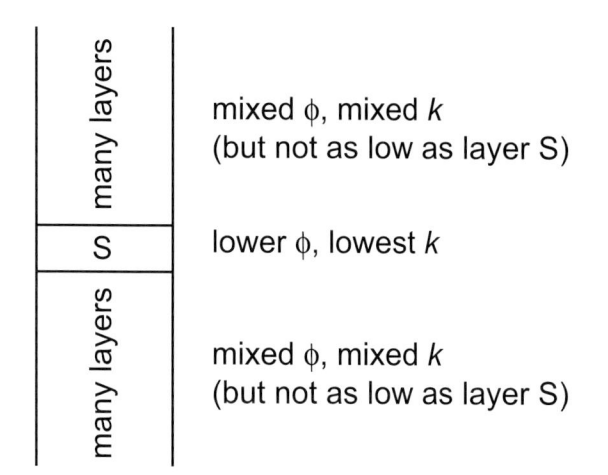

FIGURE 8. Hypothetical stratigraphic column showing layer S (protoseal) in a succession of muddy rocks. Layer S has the lowest permeability (k) and porosity (ϕ) at this stage of basin evolution.

dependence in the poroplastic model; see Couples and Lewis, 2002).

Meanwhile, the muddy rock layers that lie above layer S can continue to compact as before, since none of them is (yet) retarding fluid escape to such an extent that it acts as a seal. However, as these materials above the protoseal keep compacting, their permeabilities are, in fact, reduced. This leads to a situation where the escaping fluid flow (crossing layer S) must also cross numerous subsequent layers of low(er) permeability, which, because of the finite path length and low permeabilities, creates a total pressure drop that is larger than we could attribute to layer S alone. Thus, the overall effectiveness of the sealing interval is increased (i.e., greater pressure differentials are possible), even if the overlying layers are not seals. However, it is also possible that through time, the overlying layers do compact sufficiently that we would classify them as seals, leading to a thicker true seal. By several effects, the seal interval grows in thickness and develops a greater capacity to maintain a pressure differential in a dynamic fluid system.

Once a top seal has formed, probably via the sequence of events as described above, how can its ultimate pressure capacity be predicted? Let us consider the simplest(?) case, where continued overpressure increase (by local continuation of compactional causes somewhere in the overpressured cell or perhaps by lateral transfer; see Yardley and Swarbrick, 1998) is the only causative mechanism for seal failure. For ease of illustration, I depict the yield surface of the seal layer in a two-dimensional P-Q space, assuming that the porosity is some nominal value (Figure 9). As overpressure builds, the mechanical state moves to the left. Pore pressure affects the normal stress components, and hence, the P parameter, but not the deviatoric stress components (if we do not consider here pore-pressure and stress coupling), so the Q parameter is constant. The new state lies inside the yield surface, and compaction ceases (as noted above, this would happen for the rocks lying beneath a seal layer within a pressure cell).

Continued increases in overpressure underneath the seal (and affecting it also via effective stress) continue to move the state to the left. (I illustrate a horizontal trajectory on the schematic plot, but poroelastic considerations, or other types of stress and pore-pressure coupling, would suggest the possibility of a nonhorizontal trajectory.) As the state point moves far to the left, it will contact the yield surface once again. In this location, the probable mode of yielding and failure is dilatant (Couples and Lewis, 1998), with a likely increase in permeability. This behavior represents seal breakdown, and the overpressure magnitude that causes the failure represents the nominal pressure capacity of the seal. Predicting that capacity requires that the yield surface be known for that particular rock at its porosity state, and that the actual state of stress be known (to specify the starting point for the above analysis). We also have to consider the larger context, e.g., the extra pressure capacity associated with overlying low-permeability units, as noted above. Other possible causes of seal failure include tectonic deformations, whose paths in this space may be complex (Couples, 1999). Space limitations do not allow me to address this aspect of seal failure here (see Lewis et al., 2002, for examples of geomechanical simulations that examine this process).

I have described how large pore-pressure increases can cause top seal failure. Do large pore-pressure increases also cause failure of the rocks that lie within an overpressure cell? If so, what happens? The preceding analysis supplies much of the answer. Poroplasticity predicts that large fluid pressures (low effective mean

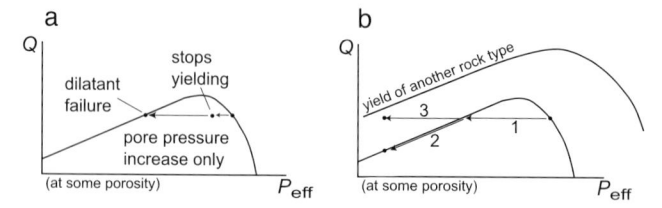

FIGURE 9. Illustration of failure types caused by pore-pressure increases (ignoring poroelastic effects or other forms of P_p and stress coupling). (a) Cessation of compaction and then failure of the seal itself caused by rising pore pressure and, hence, reduction of effective mean stress. (b) Failure of rocks in an overpressured cell (underneath the seal) and changes in load (stress) as pore pressure continues to increase in a closed (undrained) system. Note the possibility that high pore pressures may cause additional rocks to yield and fail.

stress) can cause these rocks to yield. However, in contrast to the situation where we considered top seal failure, rocks that fail inside an overpressure cell cannot bleed off the excess pressure, regardless of the mode of yielding: whether it is compactional or dilational. Only if the bounding seal itself fails can the pore pressure bleed off. Thus, rock failure in this situation (inside the cell) does not lead to a reduction in pore pressure caused by fluid flow. Instead, the failed rocks and the pore fluids still continue to share the overall loading. This situation is analogous to the undrained test conditions that represent one experimental protocol (e.g., Mokni and Desrues, 1999).

The key factor in predicting subsequent events is that the mode of failure that is expected in this situation (at low effective mean stress) is what is described as strain (work) softening. This means that the rocks become weaker after they yield, so they are unable to carry as much load as before. Because the total load has not changed, a greater portion of the total load is transferred to the fluid phase, causing the pore pressure to rise further, moving the state point further to the left (Figure 9b). This situation represents a positive feedback loop, with the possibility for a run-away process. If the top seal capacity is sufficient, the pore pressure inside the cell can continue to rise, while more and more of the rocks underneath the top seal reach their strength limit and fail. This analysis suggests that an overpressured cell could consist of very unstable rocks that might undergo significant deformation brought about by a relatively minor external event (such as even partial overpressure bleed-off, perhaps caused by a small movement on a key fault, etc). Is this the explanation for the occurrence of polygonal fault systems and related sand-injection features (Cartwright, 1994; Cartwright and Lonergan, 1996; Cartwright and Dewhurst, 1998; Lonergan and Cartwright, 1999; Davis et al., 2002; Duranti et al., 2002; Goulty, 2002; Nicol et al., 2003; Huuse et al., 2004)?

The deformation processes that occur as the rocks are weakened lead to a significant change in the state of stress. For simplicity, let us assume that the vertical (total) stress is the largest stress component, and that it remains constant. (Yes, this is the sort of thing I cautioned against doing without good reason. I am adopting the arbitrary approach to avoid complicating the scenario.) The path illustrated in Figure 9 (for yielding rocks in the overpressured cell) produces a change in both mean stress and differential stress (Figure 10a). If the vertical (total) stress is unchanged, these conditions are met if both the minimum and intermediate, horizontal (total) stresses become larger and approach the magnitude of the vertical stress (Figure 10b). Because of the high pore pressures, the effective stresses are all small, but the least (total) stress, as may be estimated from a leakoff test, will be much higher than otherwise expected. We know from compilations of leakoff test data that such apparent stress changes (to higher σ_{min}) are common in overpressured basins (White et al., 2002). The poroplastic model explains that these stress changes are simply a consequence of how the total system is responding. The geomechanical approach, even without reliance on complex mathematics, explains much of what we know about the operation of seals and overpressured rocks and the main stress-state effects related to overpressure.

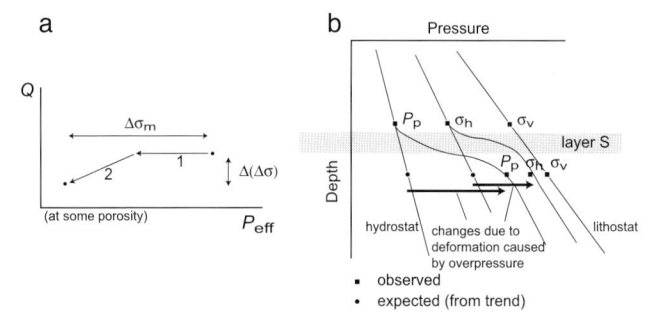

FIGURE 10. Changes in stress state produced during deformation in a sealed overpressure cell. (a) Changes in mean stress (P_{eff}) and stress difference ($\Delta(\Delta\sigma)$) as state point moves along paths 1 and 2 of Figure 9. (b) These stress changes and pore-pressure changes are plotted in standard pressure-depth space.

The preceding discussion has focused on pressure seals. I have used the porosity state as a proxy descriptor for the topology and characteristics of the pore network of the rocks under consideration, which actually determine the pressure-retention capacity of the seal. It is now important to explicitly connect that argument with the ability of rocks to retain hydrocarbons, i.e., to allow them be classified as a top seal (i.e., a phase seal, with or without significant pressure differentials). As noted earlier, the multiphase petrophysical properties of a rock are simply macroscopic parameters governed by the microscale nature of the pore system. Thus, the reasoning applied to predictions about the formation and function of pressure seals is transferrable to the case of hydrocarbon (phase) seals if we exchange the word "porosity" through most of the text in this section with the term "pore network characteristics."

Evolution of Fault Zone Seals

Fault zone sealing issues under active consideration by the research community can be grouped into three categories: juxtaposition, gouge, and critically stressed (Knipe, 1997; Wirput and Zoback, 2002; Yielding, 2002). Geomechanical analysis can be a significant factor in all of these areas. Even the qualitative method outlined in the preceding section on pressure seals and

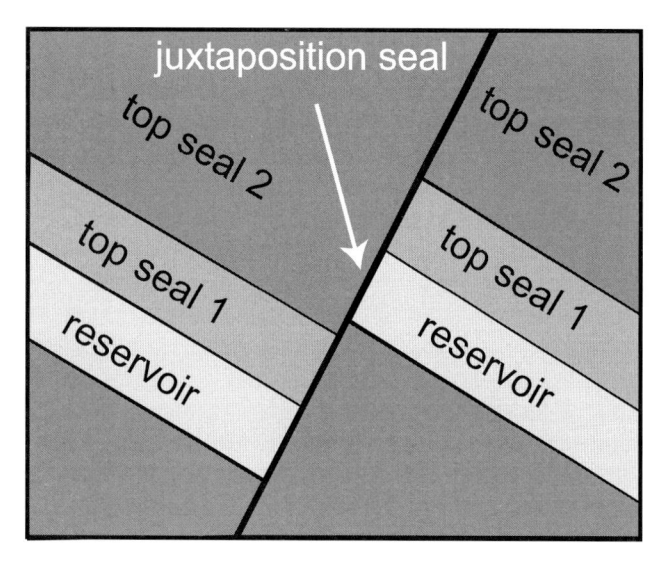

FIGURE 11. Cartoon illustrating spatial relationships in fault-juxtaposition seal. Note that this seal arrangement requires one or more effective top seals.

top seals can be a useful contributor to the prediction task for fault seals. For example, consider the juxtaposition type of fault seal. In the simplest version of the juxtaposition concept, hydrocarbons get trapped in a reservoir layer by both its top seal and by the fact that faulting has displaced another top seal layer (here, top seal 2) into adjacency contact with the reservoir unit (Figure 11). Both of these (muddy) layers need to have achieved a compaction state appropriate for inhibiting hydrocarbon migration if the idea of a juxtaposition seal is to work. The prediction of sealing capacity for these layers can follow a geomechanical analysis almost identical with that presented above.

Up to this point, I have described geomechanical analysis in fairly general terms, with a strong emphasis on the conceptual aspect of the topic. Now, it is appropriate to illustrate how geomechanical understanding can be applied to complex systems. This requires that we adopt a numerical representation that can accommodate the spatial and temporal heterogeneities of materials and the nonlinearities of poroplastic material behaviors. These requirements are met (to a greater or lesser extent) by a range of commercial software simulation products that are based on the finite-element method. The tool used for the results included here is called SAVFEM (Structural Analysis via the Finite-element Method, licensed from Applied Mechanics, Inc.). The scope of this chapter does not warrant an extensive discussion relative to how poroplasticity behavior has been implemented in SAVFEM (N. G. Higgs, D. R. Parrish, G. Workman, G. D. Couples, 2005, personal communication), so here, I proceed by accepting that such tools can adequately represent the mechanical behavior of rocks, as summarized above.

I present two examples of the use of geomechanics to achieve better understanding of fault seal issues. The first example is focused on the subject of fault gouge; specifically, the results are part of a research effort aimed at understanding the distribution of fault rock materials in fault zones. The second example is focused on the topic of faults and stress states. My purpose in introducing this latter example is to show that faulting needs to be considered as part of an interacting system, and that the resulting stress state around faults in an array is quite different from that which is typically assumed to apply in situations where faulting occurs.

As faults move, the wall rocks can be progressively damaged, such that highly strained domains (damage zones) are created around the nominal fault plane (Knipe et al., 1997; Shipton and Cowie, 2003). Depending on the details of the geohistory of the fault, several deformation processes can operate, including injection of sand or clay, cataclasis, smearing, etc. Breaching of lateral ramps (Walsh et al., 2003) and spatial or temporal variations in the other processes all lead to the potential for considerable heterogeneity in terms of the distribution of fault rock types. At present, we do not have a robust process model that enables a good prediction of these fault rock distributions; hence, it is difficult to quantify the flow effects of fault gouge from a first-principles approach.

The classical way to resolve such issues is to undertake outcrop studies and/or look for correlations in data sets. That work has certainly been important in improving our understanding of fault sealing (Childs et al., 1997; Foxford et al., 1998; Yielding, 2002; Bretan et al., 2003). Another complementary method is to develop geomechanical simulations of the faulting process to help rationalize the complexity of the outcrop information. Here, I show one model to represent this new approach. The model is designed to simulate a very specific situation, one where an initially intact layer is sheared in response to fault propagation, idealized as distinct displacements, both from above and below an initially intact layer (Figure 12a). This model, although simple, has the necessary characteristics to meet the criteria for geomechanics. Component parts exist representing the intact layer and also the prefaulted material above and below it. The rocks are assigned poroplastic material properties, and the interfaces are assigned frictional properties. Finally, the system is subjected to boundary loads that are sensible (in this case, the prefaulted materials above and below are translated parallel to the initial fault surfaces, and a confining pressure is imposed to elicit realistic responses in the middle, dark-colored layer).

The simulation procedure (e.g., as implemented in SAVFEM or other tools) calculates the stresses and strains throughout the model at each load step (i.e., small increments of imposed displacement). The initially

FIGURE 12. Shear zone models. (a) Experimental design involving precut cylinders of steel and originally intact central layer (disk). (b) Photomicrograph of resulting deformation in a well-lithified mudstone (collected from South Queensferry, near Edinburgh). The white area is missing material lost during thin-section process. (c) Simulation of deformation in this system, created by a numerical tool with localization behavior. The white lines represent a finite-element mesh. Within the induced shear zone, the darker gray colors represent magnitudes of shear strain. Note the en echelon pattern of damage that is similar to the damage observed in the experiment. Unpublished experimental results in (b) courtesy of S. Uehara (Heriot-Watt, 2004, personal communication). Unpublished numerical result in (c) courtesy of P. Olden (Heriot-Watt, 2004, personal communication).

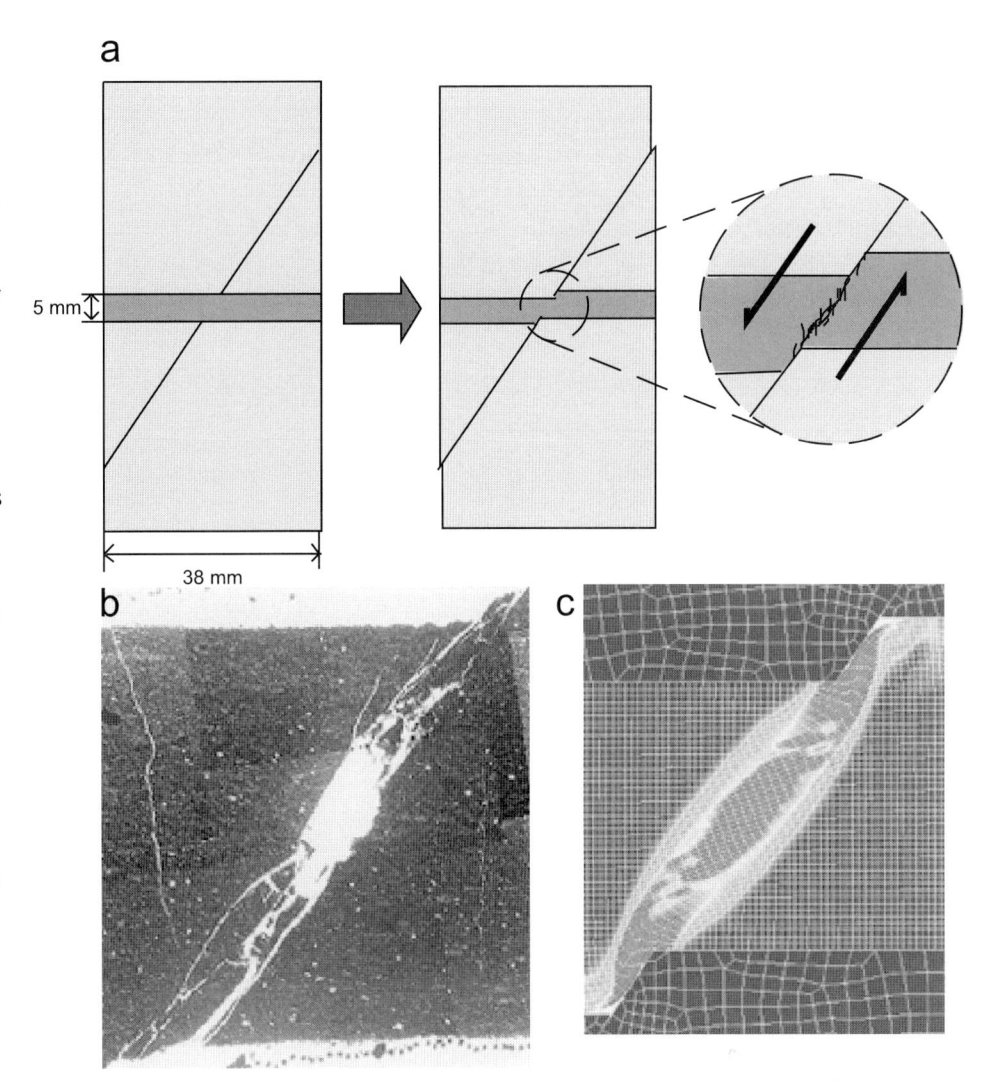

undeformed layer yields during the simulation. In fact, it yields in some places, although it is a homogeneous material at the start. The distribution of strain (the actual is in Figure 12b, and the calculated is in Figure 12c) reveals a well-ordered pattern of subparallel, en echelon shears that delimit alternations of high and low strains (and volumetric strain differences; see below). The calculated deformation is very similar to that observed in similar experimental models of shale smear described by Takahashi (2003). Additional model configurations (not shown here) could address the consequences of several aspects; for example, the preexisting fault surfaces might not be coplanar; they might not have the same orientation; the intact layer could be replaced by multiple layers with contrasting properties; or the boundary displacements could have a component perpendicular to the fault surface. The point that I wish to make is that such geomechanical simulations, although they are based on simple systems, can supply us with a way to investigate the complex parameter space that arises when we wish to understand the geomechanical processes operative around faults.

The second example also uses SAVFEM to simulate a system of deformation, in this case, caused by a flexural loading imposed onto the base of a sequence of layers. In the crestal region of the antiform produced during the simulation, a system of faults develops in the continuum (Figure 13a). In this model, the faults are domains of high plastic strain that I associate with a faultlike process, although the size of the finite elements used in this model means that the shear zones do not develop the same apparent degree of fault zone complexity as that illustrated in the previous example. However, it is instructive to consider the stress state that develops throughout and around the fault array (Figure 13b). The principal stress magnitudes and orientations differ considerably from the Andersonian model that underlies current methods for assessing leakage from the reactivation of so-called critically stressed faults (Jones et al., 2000; Wirput and Zoback, 2002; Jones and Hillis, 2003). This simulation (which has been created in a research program investigating mechanical heterogeneities around faults) shows how geomechanical

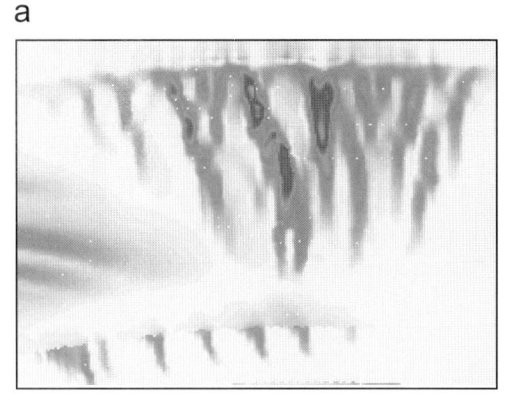

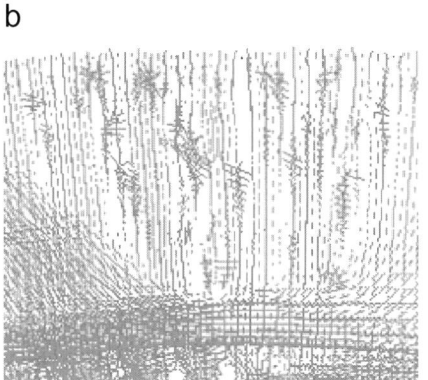

FIGURE 13. A portion of the crestal region of a numerical model deformed in flexure. This model has several layers separated by frictional interfaces. (a) Intensity of total plastic strain (a scalar number), including distortion and volumetric strain. (b) Orientations of stress trajectories for the result depicted in (a). Line lengths are proportional to stress magnitude. Note significant rotations of stress direction and major changes in magnitude (including low stress zones) as a consequence of fault development.

analysis can lead to important insights that will improve the way that we predict fault sealing. Such geomechanical studies are also going to have a key influence in questioning the notion of stress regimes (e.g., Engelder, 1993) and whether that simple idea (dating at least from Anderson, 1951) is helping or hindering our understanding of natural deformations.

To illustrate the emerging potential to develop links between the geomechanical simulations and the prediction of petrophysical properties, I show a single example where the link depends on an arbitrary relationship between the calculated strain state and the resulting permeability. Here, I use the simulation result shown in Figure 12c as input, from which I extract the volumetric strain as a scalar quantity (Figure 14a). Laboratory studies suggest some generalizations concerning the way that permeability is altered as deformation occurs, with dilational modes favoring permeability increase and compactional modes associated with permeability decrease. To represent this generalization, I invent a simple generic relationship (Figure 14b) to illustrate the possible evolution of permeability (here considered to be a scalar parameter) as a function of the deformation. I have chosen to base the strain-to-permeability transform on volumetric strain only, but it could also be important to recognize a dependence

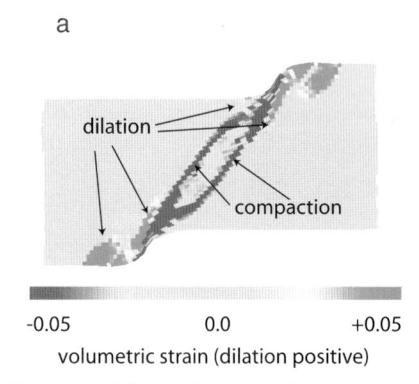

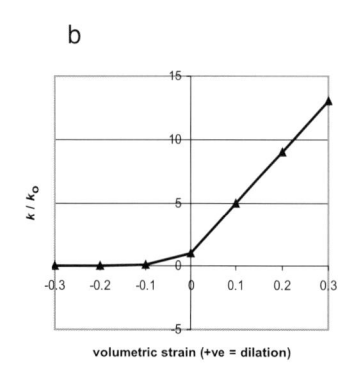

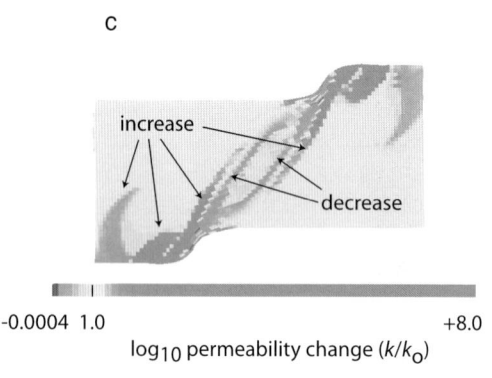

FIGURE 14. Prediction of permeability as a function of simulated deformation. (a) Volumetric strain from the model result shown in Figure 12c. Note the presence of a central lozenge of slightly deformed material, surrounded by thin bands of compacted and dilated material. The compacted zones surrounding the lozenge grow during the simulation sequence (not shown), with the band on the left side growing upward from the base of the lozenge and becoming more strained. The right-side band is almost a symmetrical equivalent. Surrounding the compacted material are bands of dilated material, and strained regions (both compactional and dilational) are also present, where the thin layer is bent at the ends of the shear zone. (b) A possible generic relationship between permeability changes (k/k_o) and volume strains. Compactional strain is arbitrarily assigned an exponential decline curve (permeability ratio is 0.002 at 0.3 compactional strain), whereas the dilational effects are a linear function. (c) Plot (logarithmic) of permeability predicted from the strain distribution. This result illustrates zone-parallel bands of increased permeability, along with zone-parallel bands of decreased permeability. Across-zone permeability (e.g., as can be calculated by a harmonic average) varies with position along the shear zone because of changes in thickness of the strained materials and changes in their permeabilities. Unpublished results provided courtesy of J. Ma (Heriot-Watt, 2004, personal communication).

on the specific strain components. That improvement could result in the ability to predict a full permeability tensor. The calculated permeability distribution for this model is shown in Figure 14c, where we can see that the shear zone consists of a central "lozenge" that has both (slightly) enhanced and degraded permeabilities resulting from the en echelon shears (refer to Figure 12). The lozenge is bounded by thin bands of compactional strain, with degraded permeabilities, and these are themselves bounded by thin bands of dilated material with enhanced permeabilities. The architecture illustrated is not constant but evolves with increased amounts of shear imposed across the whole zone (these changing patterns are not shown because of space limitations). The ability to predict petrophysical (and other) properties, as a consequence of the deformation state calculated by geomechanical simulation, opens a new avenue for research into the prediction of seal behavior.

CHEMICAL EFFECTS

Chemical effects, diagenesis in the wider sense, occur throughout basin evolution. Most of the chemical processes of interest alter the pore structure through dissolution or precipitation events, and petrophysical changes must occur because of these pore system modifications. Thus, chemical changes certainly have an effect on the sealing issue.

However, it is also true that chemical effects are related to the geomechanical processes operating in basins. We know that chemical changes affect the mechanical characteristics of the rocks by either strengthening or weakening them (cement can create a more robust framework of grains, and partly dissolved grains or cements or grains that recrystallize to a new, softer mineral lead to a weaker rock framework) and thus altering the deformation response from what might otherwise be expected. As an example, Lahann (2004) has defined the term "framework weakening" to refer to a range of processes that allow a muddy rock to compact to a greater degree than would be expected for its initial porosity and composition and for a given loading history. In terms of the poroplastic constitutive relationship, such chemical changes move the yield surface F in the P-Q porosity space of Figure 7. A chemically related change of yield surface could cause a material that was deforming to stop or one that was previously stable to start yielding.

Thus, chemical effects influence the geomechanical processes that occur by altering the constitutive relationships of the rocks. However, deformation also influences the chemical processes, by making new, perfect-sized pore throats, by exposing fresh mineral surfaces (Fisher et al., 2000), by changing the energy level of certain grains, etc. Overall, an intimate interaction clearly exists between the chemical changes in a basin and the mechanical deformation that occurs there.

Are chemical and mechanical processes equally important relative to seals? I argue that the answer is no. Sealing is a concept that is relevant only in situations that are dynamic with respect to the concentration of mechanical energy (i.e., where gradients occur). By this, I mean that seals are intrinsic to the retention of excess mechanical energy as represented by elevated fluid pressures (or differences in the potential energy of adjacent domains of a fluid phase). Seal failure is also, as we understand it, a primarily mechanical process being driven by gradients in mechanical energy (of both the rocks and pore fluids). Even compaction is a dominantly mechanical event, but one that chemical processes can assist (e.g., Lahann, 2002). Perhaps one reason for the ongoing debate about the relative roles of mechanics and chemistry is that we may not be careful enough to clearly distinguish between the possible meanings of the term "compaction:" it can be used to refer to a vertical shortening or strain, and it can be used to refer to porosity loss. Commonly, both occur together, and the term can be appropriately used to refer to this joint process. However, when we are debating causative mechanisms, sloppy usage can be a source of miscommunication.

If no gradient of mechanical energy is present, chemical processes can continue and can significantly alter the rocks (porosity loss or gain, strengthening or weakening, etc), but there is also no issue with regard to seals. If, at a later time, overpressure is generated, such that the chemically altered rock must be assessed for its role in terms of seals, the analysis proceeds as before, albeit with a yield surface changed as a consequence of the chemical action. Thus, geomechanical processes are fundamental to the creation and operation of seals, and chemistry is a secondary factor.

COMPLEX SYSTEMS

When fluid flow and deformation are fully coupled, as in many natural situations, we can see how heterogeneities, such as the seals that are produced by rock deformation, impact the transfer of mechanical energy through the system, altering the power transmission through the basin. As in many physical systems that involve complex coupling of processes, the geomechanics + fluids systems investigated here have both positive and negative feedback channels. These are characteristic of physical systems that typically operate in a near-critical state or a state of virtual instability.

Changes in input power can be imagined as the cause of subtle or major shifts in when and how seals operate in a basin or reservoir, depending on whether the system is near a bifurcation point. Bak and Tang (1989), Harper and Szymanski (1991), Crampin (1994),

Main et al. (1994), Heffer (2002), and Van Wagoner et al. (2003) have developed similar ideas (from related perspectives) about the Earth's crust being a complex system. My contribution to this topic has been to emphasize the role of seals in terms of how they alter the transmission of power. I have described some of the feedback processes that tend to keep the system behavior in a normal band. It is perhaps unsurprising, then, that we can successfully develop good rules of thumb about seals that can be applied to many situations around the world. The concepts developed in this chapter should be useful in understanding the physics that underlie these normal cases. They will also help to identify the reasons that lead to anomalies in behavior, where current rules of thumb may not be appropriate.

CONCLUSIONS

Considering the interactions between the components of a system, which include the rock bodies and their contacts, the loads imposed onto the system, and how the rocks react, geomechanical analysis allows us to predict how, where, when, and why deformations occur. The examples I have included here illustrate how geomechanical analysis can be applied to understand the formation of seals (top seals and fault seals) and how those seals subsequently interact with the mechanical and fluid system. Modern numerical implementations of geomechanical concepts allow us to make predictions about the rock properties resulting from the deformations and thus enable a link to fluid-flow simulations to realize a fully coupled approach. Perhaps the most important reason why the geomechanical approach, as defined, is valuable is that it reduces the chance of making errors and drawing false conclusions about deformation processes related to inappropriate assumptions.

ACKNOWLEDGMENTS

My ideas on geomechanics have been influenced by many people, some of whom were enthusiastic supporters and others who served to question my (sometimes) incomplete thoughts. Particular mention in this regard (as supporters) is due to Dave Stearns, Mike Fahy, John Handin, and Helen Lewis. Particular thanks is due to Nigel Higgs for helping me and my team to restart our work in numerical simulation, along with George Workman (Applied Mechanics Inc), who has patiently addressed our technical questions over the operation of SAVFEM. Dave Dewhurst and Suzanne Hunt are thanked for comprehensive and helpful reviews of the manuscript that have helped me to make significant improvements.

REFERENCES CITED

Adler, P. M., C. G. Jacquin, and J. F. Thovert, 1992, The formation factor of reconstructed porous media: Water Resources Research, v. 28, p. 1571–1576.

Almon, W. R., W. C. Dawson, S. J. Sutton, F. G. Ethridge, and B. Castelblanco, 2001, Sequence stratigraphy, petrophysical variation, and sealing capacity in deepwater shales, Upper Cretaceous Lewis Shale, south-central Wyoming, in D. P. Stillwell, ed., Wyoming gas resources and technology, 52nd Annual Field Conference guidebook: Wyoming Geological Association, p. 163–182.

Almon, W. R., W. C. Dawson, S. J. Sutton, F. G. Ethridge, and B. Castelblanco, 2002, Sequence stratigraphy, facies variations and petrophysical properties in deepwater shales, Upper Cretaceous Lewis Shale, south-central Wyoming: Gulf Coast Association of Geological Societies Transactions, v. 52, p. 1041–1053.

Anderson, E. M., 1951, The dynamics of faulting and dyke formation with application to Britain (2nd ed.): London, Oliver and Boyd, 206 p.

Antonellini, M. A. A., A. Aydin, and D. D. Pollard, 1994, Microstructure of deformation bands in porous sandstones at Arches National Park, Utah: Journal of Structural Geology, v. 16, p. 941–959.

Aplin, A. C., A. J. Fleet, and J. H. S. Macquaker, 1999, Muds and mudstones: Physical and fluid flow properties, in A. C. Aplin, A. J. Fleet, and J. H. S. Macquaker, eds., Muds and mudstones: Physical and fluid flow properties: Geological Society (London) Special Publication 158, p. 1–8.

Atkinson, J. H., and P. L. Bransby, 1978, The mechanics of soils: An introduction to critical state soil mechanics: London, McGraw-Hill, 375 p.

Bak, P., and C. Tang, 1989, Earthquakes as self-organised critical phenomena: Geophysical Journal International, v. 94, p. 15,635–15,638.

Bakke, S., and P. E. Øren, 1997, 3-D pore scale modelling of sandstones and flow simulations in the pore networks: Society of Petroleum Engineers Journal, v. 2, p. 136–149.

Bjorlykke, K. O., and P. K. Egeberg, 1993, Quartz cementation in sedimentary basins: AAPG Bulletin, v. 77, p. 1538–1548.

Blunt, M. J., 1997, Pore level modelling of the effects of wettability: Society of Petroleum Engineers Journal, v. 2, p. 494–510.

Brace, W. F., 1960, An extension of the Griffith theory of fracture to rocks: Journal of Geophysical Research, v. 65, p. 3477–3480.

Brace, W. F., and D. L. Kohlstedt, 1980, Limits on lithospheric stress imposed by laboratory experiments: Journal of Geophysical Research, v. 85, p. 6248–6252.

Bretan, P., G. Yielding, and H. Jones, 2003, Using calibrated shale gouge ratio to estimate hydrocarbon column heights: AAPG Bulletin, v. 87, p. 397–413.

Brown, A., 2003, Capillary effects on fault-fill sealing: AAPG Bulletin, v. 87, p. 381–395.

de Borst, R., 2001, Some recent issues in computational

failure mechanics: International Journal for Numerical Methods in Engineering, v. 52, p. 63–95.

de Buhan, P., and L. Dormieux, 1999, A micromechanics-based approach to the failure of porous media: Transport in Porous Media, v. 34, p. 47–62.

Cartwright, J. A., 1994, Episodic basin-wide hydrofracturing of overpressure early Cenozoic mudrock sequences in the North Sea basin: Marine and Petroleum Geology, v. 11, p. 587–607.

Cartwright, J. A., and D. N. Dewhurst, 1998, Layer-bound compaction faults in fine-grained sediments: Geological Society of America Bulletin, v. 110, p. 1242–1257.

Cartwright, J. A., and L. Lonergan, 1996, Volumetric contraction during the compaction of mudrocks: A mechanism for the development of regional-scale polygonal fault systems: Basin Research, v. 8, p. 183–193.

Chateau, X., and L. Dormieux, 2002, Micromechanics of saturated and unsaturated porous media: International Journal of Numerical and Analytical Methods in Geomechanics, v. 26, p. 831–844.

Chester, F. M., and J. S. Chester, 1998, Ultracataclastic structure and friction processes of the Punchbowl fault, San Andreas system, California: Tectonophysics, v. 295, p. 199–221.

Chierici, G. L., 1981, Novel relations for drainage and imbibition relative permeabilities, in Proceedings of the 56th Annual Fall Technical Conference and Exhibition, Society of Petroleum Engineers of the American Institute of Mining Engineers, p. 1–10.

Childs, C., J. J. Walsh, and J. Watterson, 1997, Complexity in fault zones and implications for fault sealing investigations, in P. Moller-Pedersen and A. G. Kostler, eds., Hydrocarbon seals: Importance for exploration and production: Norwegian Petroleum Society Special Publication 7, p. 51–59.

Colmenares, L. B., and M. D. Zoback, 2002, A statistical evaluation of intact rock failure criteria constrained by polyaxial test data for five different rocks: International Journal of Rock Mechanics and Mining Sciences, v. 39, p. 695–729.

Corey, A. T., 1954, The interrelation between gas and oil permeabilities: Producers Monthly, v. 19, p. 38–41.

Couples, G. D., 1999, A hydro-geomechanical view of seal formation and failure in overpressured basins: Oil and Gas Science and Technology, v. 54, p. 785–795.

Couples, G. D., and H. Lewis, 1998, A poro-plastic material description, in A. Mitchell and D. Grauls, eds., Overpressures in petroleum exploration, Workshop Proceedings, Pau, France, April 1998: Bulletin Centre Recherche, Elf Exploration and Production Memoir 22, p. 187–192.

Couples, G. D., and H. Lewis, 2002, A poro-hydromechanical perspective on basin events: Evolving stress states, overpressure, and seal formation and failure, in R. E. Swarbrick, ed., OverPressure 2000 Research Workshop, GeoPressure Technology Ltd., Durham, England (CD-ROM).

Coussy, O., 1995, Mechanics of porous continua: Chichester, Wiley, 472 p.

Crampin, S., 1994, The fracture criticality of crustal rocks: Geophysical Journal International, v. 118, p. 428–438.

Cuss, R. J., E. H., Rutter, and R. F. Holloway, 2003, The application of critical state soil mechanics to the mechanical behaviour of sandstone: International Journal of Rock Mechanics and Mining Sciences, v. 40, p. 847–862.

Davies, R. J., B. R. Bell, J. A. Cartwright, and S. Shoulders, 2002, Three-dimensional seismic imaging of Paleogene dike-fed submarine volcanoes from the northeast Atlantic margin: Geology, v. 30, p. 223–226.

Dawson, W. C., and W. R. Almon, 2002, Top seal potential of Tertiary deepwater Gulf of Mexico shales: Gulf Coast Association of Geological Societies Transactions, v. 52, p. 167–176.

Desrues, J., and C. Viggiani, 2004, Strain localization in sand: An overview of the experimental results obtained in Grenoble using stereophotogrammetry: International Journal for Numerical and Analytical Methods in Geomechanics, v. 28, p. 279–321.

Desrues, J., R. Chambon, M. Mokni, and F. Mazerolle, 1996, Void ratio evolution inside shear bands in triaxial sand specimens studied by computed tomography: Geotechnique, v. 46, p. 527–546.

Dewhurst, D. N., A. C. Aplin, and J. P. Sarda, 1999, Influence of clay fraction on the pore scale properties and hydraulic conductivity of experimentally compacted mudstones: Journal of Geophysical Research, v. 104, p. 29,261–29,274.

Dixit, A. B., J. S. Buckley, S. R. McDougall, and K. S. Sorbie, 2000, Empirical measures of wettability in porous media and the relationship between them derived from pore-scale modelling: Transport in Porous Media, v. 40, p. 27–54.

Dormieux, L., and S. Maghous, 1999, Poroelasticity and poroplasticity at large strains: Oil and Gas Science and Technology, Revue de l'Institut Français du Pétrole, v. 54, p. 773–784.

Drucker, D. C., and W. Prager, 1952, Soil mechanics and plastic analysis or limit design: Quarterly of Applied Mathematics, v. 10, p. 157–165.

Duranti, D., A. Hurst, C. Bell, S. Groves, and R. Hanson, 2002, Injected and remobilised Eocene sandstones from the Alba field, UKCS: Core and wireline log characteristics: Petroleum Geoscience, v. 8, p. 99–107.

Engelder, J. T., 1974, Cataclasis and the generation of fault gouge: Geological Society of America Bulletin, v. 85, p. 1515–1522.

Engelder, J. T., 1993, Stress regimes in the lithosphere: Princeton, New Jersey, Princeton Press, 457 p.

Fahy, M., 1982, Beyond Mohr–Coulomb (abs.): Geological Society of America Section Meeting, College Station, Texas.

Fenwick, D., and M. Blunt, 1998, Three-dimensional modelling of three-phase imbibition and drainage: Advances in Water Resources, v. 25, p. 121–143.

Ferreol, B., and D. H. Rothman, 1995, Lattice–Boltzmann simulations of flow through Fontainebleau sandstone: Transport in Porous Media, v. 20, p. 3–20.

Fisher, Q. J., R. J. Knipe, and R. H. Worden, 2000, Microstructures of deformed and non-deformed sandstones from the North Sea: Implications for the origins of quartz cement in sandstones, in R. H. Worden and S. Morad, eds., Quartz cementation in sandstones: International Association of Sedimentologists Special Publication 29, p. 129–146.

Foxford, K. A., J. J. Walsh, J. Watterson, I. R. Garden, S. C. Guscott, and S. D. Burley, 1998, Structure and content of the Moab fault zone, Utah, U.S.A., and its implications for fault seal prediction, in G. Jones, Q. J. Fisher, and R. J. Knipe, eds., Faulting, fault sealing and fluid flow in hydrocarbon reservoirs: Geological Society (London) Special Publication 147, p. 87–103.

Fredrich, J. T., and A. F. Fossum, 2002, Large-scale three-dimensional geomechanical modelling of reservoirs: Examples from California and the deepwater Gulf of Mexico: Oil and Gas Science and Technology, v. 57, p. 423–441.

Friedman, M., 1963, Petrofabrix analysis of experimentally deformed calcite-cemented sandstones: Journal of Geology, v. 71, p. 12–37.

Gallagher, J. J., M. Friedman, J. W. Handin, and G. M. Sowers, 1974, Experimental studies relating to microfracture in sandstone: Tectonophysics, v. 21, p. 203–247.

Goulty, N. R., 2002, Mechanics of layer-bound polygonal faulting in fine-grained sediments: Journal of the Geological Society (London), v. 159, p. 239–246.

Griggs, D. T., and J. W. Handin, 1960, Observations on fracture and a hypothesis of earthquakes, in D. T. Griggs and J. W. Handin, eds., Rock deformation: Geological Society of America Memoir 79, p. 347–364.

Gueguen, Y., and V. Palciauskas, 1994, Introduction to the physics of rocks: Princeton, New Jersey, Princeton University Press, 294 p.

Handin, J. W., 1969, On the Coulomb–Mohr failure criterion: Journal of Geophysical Research, v. 74, p. 5343–5348.

Handin, J. W., and R. V. Hager, 1957, Experimental deformation of sedimentary rocks: tests at room temperature on dry samples: AAPG Bulletin, v. 41, p. 1–50.

Handin, J. W., R. V. Hager, M. Friedman, and J. N. Feather, 1963, Experimental deformation of sedimentary rocks under confining pressure: Pore pressure tests: AAPG Bulletin, v. 47, p. 717–755.

Harper, T. R., and J. S. Szymanski, 1991, The nature and determination of stress in the accessible lithosphere: Philosophical Transactions of the Royal Society of London, Series A, v. 337, p. 5–24.

Hazlett, R. D., 1997, Statistical characterization and stochastic modelling of pore networks in relation to fluid flow: Mathematical Geology, v. 29. p. 801–822.

Heffer, K., 2002, Geomechanical influences in water injection projects: An overview: Oil and Gas Science and Technology, v. 57, p. 415–422.

Hill, R., 1950, The mathematical theory of plasticity: Oxford, Clarendon Press, 356 p.

Holbrook, P., 2002, The primary controls over sediment compaction, in A. R. Huffman and G. L. Bowers, eds., Pressure regimes in sedimentary basins and their prediction: AAPG Memoir 76, p. 21–32.

Hudson, J. A., E. Liu, and S. Crampin, 1996, Mechanical properties of materials with interconnected cracks and pores: Geophysical Journal International, v. 124, p. 105–112.

Huuse, M., D. Duranti, N. Steinsland, C. G. Guargena, P. Prat, K. Holm, J. A. Cartwright, and A. Hurst, 2004, Seismic characteristics of large-scale sandstone intrusions in the Paleogene of the South Viking Graben, U.K. and Norwegian North Sea, in R. J. Davies, J. Cartwright, S. A. Stewart, M. Lappin, and J. R. Underhill, eds., 3-D seismic technology: Application to the exploration of sedimentary basins: Geological Society (London) Memoir 29, p. 263–277.

Issen, K. A., 2002, The influence of constitutive models on localization conditions for porous rocks: Engineering Fracture Mechanics, v. 69, p. 1891–1906.

Issen, K. A., and J. W. Rudnicki, 2001, Theory of compaction bands in porous rock: Physics and Chemistry of the Earth, Part A, v. 26, p. 95–100.

Jaeger, J. C., and N. G. W. Cook, 1969, Fundamentals of rock mechanics: London, Chapman and Hall, 513 p.

Jenike, A. W., and R. T. Shield, 1959, On the plastic flow of Coulomb solids beyond original failure: Journal of Applied Mechanics, v. 26, p. 599–602.

Jing, L., 2003, A review of techniques, advances and outstanding issues in numerical modelling for rock mechanics and rock engineering: International Journal of Rock Mechanics and Mining Sciences, v. 40, p. 283–353.

Jones, M. E., and M. A. Addis, 1986, The application of stress path and critical state analysis to sediment deformation: Journal of Structural Geology, v. 8, p. 575–580.

Jones, R. M., and R. R. Hillis, 2003, An integrated, quantitative approach to assessing fault-seal risk: AAPG Bulletin, v. 87, p. 507–524.

Jones, R. M., P. J. Boult, R. R. Hillis, S. D. Mildren, and J. Kaldi, 2000, Integrated hydrocarbon seal evaluation in the Penola trough, Otway Basin: Australian Petroleum Production and Exploration Association Journal, v. 49, p. 194–212.

Knipe, R. J., 1997, Juxtaposition and seal diagrams to help analyze fault seals in hydrocarbon reservoirs: AAPG Bulletin, v. 81, p. 187–195.

Knipe, R. J., Q. J. Fisher, G. Jones, M. B. Clennell, A. B. Farmer, A. Harrison, B. Kidd, E. McAllister, J. R. Porter, and E. A. White, 1997, Fault seal analysis: Successful methodologies, application, and future directions, in P. Moller-Pedersen and A. G. Kostler, eds., Hydrocarbon seals: Importance for exploration and production: Norwegian Petroleum Society Special Publication 7, p. 15–38.

Lahann, R., 2002, Impact of smectite diagenesis on compaction modeling and compaction equilibrium, in A. R. Huffman and G. L. Bowers, eds., Pressure regimes in sedimentary basins and their prediction: AAPG Memoir 76, p. 61–72.

Lahann, R. W., 2004, A broader view of framework weakening (abs): AAPG Annual Meeting Program, v. A80.

Lander, R. H., and O. Walderhaug, 1999, Predicting porosity through simulating sandstone compaction and quartz cementation: AAPG Bulletin, v. 83, p. 433–449.

Leverett, M. C., 1941, Capillary behaviour in porous solids: Transactions of the American Institute of Mining Engineers, v. 142, p. 152–169.

Lewis, H., and G. D. Couples, 1993, Influence of structure on reservoir performance of aeolian strata; the Anschutz Ranch East field, western U.S.A., in C. P. North and D. J. Prosser, eds., Characterisation of fluvial and aeolian reservoirs: Geological Society (London) Special Publication 73, p. 117–139.

Lewis, H., P. Olden, and G. D. Couples, 2002, Geomechanical simulations of top seal integrity, in A. G. Koestler and R. Hunsdale, eds., Hydrocarbon seal quantification: Norwegian Petroleum Society Special Publication 11, p. 75–87.

Lockner, D. A., and N. M. Beeler, 2002, Rock failure and earthquakes, in W. H. K. Lee, H. Kanamori, P. C. Jennings, and C. Kisslinger, eds., International handbook of earthquake and engineering seismology: Amsterdam, Academic Press Ltd., chapter 32, p. 505–537.

Lockner, D. A., J. D. Byerlee, V. Kuksenko, A. Ponomarev, and A. Sidorin, 1992, Observations of quasistatic fault growth from acoustic emissions, in B. Evans and T.-F. Wong, eds., Fault mechanics and transport properties of rocks: London, Academic Press Ltd., chapter 1, p. 1—31.

Lonergan, L., and J. A. Cartwright, 1999, Polygonal faults and their influence on deep-water sandstone reservoir geometries, Alba field, United Kingdom central North Sea: AAPG Bulletin, v. 83, p. 410–432.

Main, I. G., P. G. Meredith, J. R. Henderson, and P. R. Sammonds, 1994, Positive and negative feedback in the earthquake cycle: The role of pore fluids on states of criticality in the crust: Annali di Geofisica, v. 37, p. 207–225.

Mair, K., S. Elphick, and I. Main, 2002, Influence of confining pressure on the mechanical and structural evolution of laboratory deformation bands: Geophysical Research Letters, v. 29, no. 10, p. 49-1–49-4 (doi: 10.1029/2001GL013964).

Mandl, G., 1988, Mechanics of tectonic faulting: Models and basic concepts: Amsterdam, Elsevier, 407 p.

Manzocchi, T., A. E. Heath, J. J. Walsh, and C. Childs, 2002, The representation of two phase fault rock properties in flow simulation models: Petroleum Geoscience, v. 8, p. 119–132.

McDougall, S. R., and K. S. Sorbie, 1995, The impact of wettability on waterflooding: Pore-scale simulation: Society of Petroleum Engineers Reservoir Engineering, v. 8, p. 208–213.

McDougall, S. R., and K. S. Sorbie, 1997, The application of network modelling techniques to multiphase flow in porous media: Petroleum Geoscience, v. 3, p. 11–19.

Menéndez, B., W. Zhu, and T.-F. Wong, 1996, Micromechanics of brittle faulting and cataclastic flow in Berea Sandstone: Journal of Structural Geology, v. 18, p. 1–16.

Miller, T. W., C. H. Luk, and D. L. Olgaard, 2002, The interrelationships between overpressure mechanisms and in-situ stress, in A. R. Huffman and G. L. Bowers, eds., Pressure regimes in sedimentary basins and their prediction: AAPG Memoir 76, p. 13–20.

Mokni, M., and J. Desrues, 1999, Strain localisation measurements in undrained plane-strain biaxial tests on Hostun RF sand: Mechanics of Cohesive-Frictional Materials, v. 4, p. 419–441.

Neuzil, C. E., 1995, Abnormal pressures as hydrodynamics phenomena: American Journal of Science, v. 295, p. 742–786.

Nicol, A., J. J. Walsh, J. Watterson, P. A. R. Nell, and P. Bretan, 2003, The geometry, growth and linkage of faults within a polygonal fault system from South Australia, in P. Van Rensbergen, R. Hillis, A. Maltman, and C. Morley, eds., Mobilisation, intrusion and faulting of subsurface sediments: Geological Society (London) Special Publication 216, p. 245–261.

Olsson, W. A., 1999, Theoretical and experimental investigation of compaction bands: Journal of Geophysical Research, v. 104, p. 7219–7228.

Olsson, W. A., 2000, Origin of Luders bands in deformed rock: Journal of Geophysical Research, v. 105, no. B3, p. 3537–3540.

Øren, P. E., and S. Bakke, 2003, Reconstruction of Berea Sandstone and pore-scale modelling of wettability effects: Journal of Petroleum Science and Engineering, v. 39, p. 177–199.

Paxton, S. T., J. O. Szabo, J. M. Ajdukiewicz, and R. E. Klimentidis, 2002, Construction of an inter-granular volume compaction curve for evaluating and predicting porosity loss in rigid-grain sandstone reservoirs: AAPG Bulletin, v. 86, p. 2047–2067.

Pereira, G. G., W. V. Pinczewski, D. Y. C. Chan, L. Paterson, and P. E. Øren, 1996, Pore-scale network model for drainage dominated three-phase flow in porous media: Transport in Porous Media, v. 24, p. 167–201.

Pittman, E. D., 1981, Effect of fault-related granulation on porosity and permeability of quartz sandstones, Simpson Group (Ordovician), Oklahoma: AAPG Bulletin, v. 65, p. 2381–2387.

Resende, L., and J. B. Martin, 1985, Formulation of Drucker-Prager cap model: Journal of Engineering Mechanics, v. 111, p. 855–881.

Rudnicki, J. W., and J. R. Rice, 1975, Conditions for the localization of deformation in pressure-sensitive dilatant materials: Journal of the Mechanics and Physics of Solids, v. 23, p. 371–394.

Scofield, A. N., and C. P. Wroth, 1968, Critical state soil mechanics: New York, McGraw-Hill, 310 p.

Scholtz, C. H., 1990, The mechanics of earthquakes and faulting: London, Cambridge Press, 439 p.

Shipton, Z. K., and P. A. Cowie, 2003, A conceptual model for the origin of fault damage zone structures in high-porosity sandstone: Journal of Structural Geology, v. 25, p. 333–345.

Sperrevik, S., P. A. Gillespie, Q. J. Fisher, T. Halvorsen, and R. J. Knipe, 2002, Empirical estimation of fault rock properties, in A. G. Koestler and R. Hunsdale, eds.,

Hydrocarbon seal quantification: Norwegian Petroleum Society Special Publication 11, p. 109–125.

Stearns, D. W., 1969, Fracture as a mechanism of flow in naturally deformed layered rocks, in A. J. Baer and D. K. Norris, eds., Proceedings of the Conference on Research in Tectonics: Geological Survey of Canada, p. 79–95.

Swarbrick, R. E., M. J. Osborne, G. S. Yardley, G. Macleod, A. Bigge, D. Grunberger, A. C. Aplin, S. R. Larter, I. R. Knight, and H. Auld, 2000, Integrated study of an overpressured central North Sea oil/gas field: Marine and Petroleum Geology, v. 17, p. 993–1010.

Swarbrick, R. E., M. J. Osborne, and G. S. Yardley, 2002, Comparison of overpressure magnitudes resulting from the main generating mechanisms, in A. R. Huffman and G. L. Bowers, eds., Pressure regimes in sedimentary basins and their prediction: AAPG Memoir 76, p. 1–12.

Takahashi, M., 2003, Permeability change during experimental fault smearing: Journal of Geophysical Research, v. 128, no. B5, p. ECV 1-1–ECV 1-15.

van der Veen, H. I., C. Vuik, and R. de Borst, 1999, A mathematical analysis of non-associated plasticity: Computers and Mathematics with Applications, v. 38, p. 107–115.

Van Wagoner, J. C., D. C. J. D. Hoyal, N. L. Adair, T. Sun, R. T. Beauboeuf, M. Deffenbaugh, P. A. Dunn, C. Huh, and D. Li, 2003, Energy dissipation and the fundamental shape of siliciclastic sedimentary bodies: AAPG Search and Discovery Article 40081, www.searchanddiscovery.com/documents/abstracts/annual2003/extend/80607.pdf (accessed April 26, 2004).

Vardoulakis, I., and J. Sulem, 1995, Bifurcation analysis in geomechanics: London, Blackie, 462 p.

Walderhaug, O., 1996, Kinetic modelling of quartz cementation and porosity loss in deeply buried sandstone reservoirs: AAPG Bulletin, v. 80, p. 731–745.

Walsh, J. J., W. R. Bailey, C. Childs, A. Nicol, and C. G. Bonson, 2003, Formation of segmented normal faults: A 3-D perspective: Journal of Structural Geology, v. 25, p. 1251–1262.

Waples, D. W., and G. D. Couples, 1998, Some thoughts on porosity reduction—Rock mechanics, overpressure, and fluid flow, in S. P. Duppenbecker and J. E. Iliffe, eds., Basin modelling: Practice and progress: Geological Society (London) Special Publication 141, p. 73–81.

White, A. J., M. O. Traugott, and R. E. Swarbrick, 2002, The use of leak-off tests as a means of predicting minimum in-situ stress: Petroleum Geoscience, v. 8, p. 189–193.

Wirput, D., and M. D. Zoback, 2002, Fault reactivation, leakage potential, and hydrocarbon column heights in the northern North Sea, in A. G. Koestler and R. Hunsdale, eds., Hydrocarbon seal quantification, Norwegian Petroleum Society Special Publication 11, p. 203–219.

Wong, T.-F., and P. Baud, 1999, Mechanical compaction of porous sandstone: Oil and Gas Science and Technology, v. 54, p. 715–727.

Wong, T.-F., C. David, and W. Zhu, 1997, The transition from brittle faulting to cataclastic flow in porous sandstones: Mechanical deformation: Journal of Geophysical Research, v. 102, no. B2, p. 3009–3025.

Wood, D. M., 1990, Soil behaviour and critical state soil mechanics: Cambridge, Cambridge University Press, 462 p.

Wu, K., N. Nunan, K. Ritz, I. Young, and J. Crawford, 2004, An efficient Markov chain model for the simulation of heterogeneous soil structure: Soil Science Society of America Journal, v. 69, p. 346–351.

Yardley, G. S., and R. E. Swarbrick, 1998, Lateral transfer: A source of additional overpressure?: Marine and Petroleum Geology, v. 17, p. 523–537.

Yassir, N., and M. A. Addis, 2002, Relationships between pore pressure and stress in different tectonic settings, in A. R. Huffman and G. L. Bowers, eds., Pressure regimes in sedimentary basins and their prediction: AAPG Memoir 76, p. 79–88.

Yielding, G., 2002, Shale gouge ratio—Calibration by geohistory, in A. G. Koestler and R. Hunsdale, eds., Hydrocarbon seal quantification, Norwegian Petroleum Society Special Publication 11, p. 1–15.

Zhang, J., T.-F. Wong, and D. M. Davis, 1990, Micromechanics of pressure-induced grain crushing in porous rocks: Journal of Geophysical Research, v. 95, p. 341–352.

Zhu, W., and T.-F. Wong, 1997, The transition from brittle faulting to cataclastic flow: Permeability evolution: Journal of Geophysical Research, v. 102, no. B2, p. 3027–3041.

Zienkiewicz, O. C., C. Humpheston, and R. W. Lewis, 1975, Associated and non-associated viscoplasticity and plasticity in soil mechanics: Géotechnique, v. 25, p. 671–689.

The Influence of Stress Regimes on Hydrocarbon Leakage

Hege M. Nordgård Bolås
Statoil ASA, Rotvoll, Trondheim, Norway

Christian Hermanrud
Statoil ASA, Rotvoll, Trondheim, Norway

Gunn M. G. Teige
Statoil ASA, Rotvoll, Trondheim, Norway

ABSTRACT

Hydrocarbon leakage through faults and fractures commonly limits in-place hydrocarbon reserves. Faulting and fracturing are controlled by effective stress changes, and such changes may therefore alter hydrocarbon column heights. The predictive power of stress history analyses in seal evaluation depends on how accurately the stress history and relationships between effective stress changes and hydrocarbon leakage can be determined.

Stress history and hydrocarbon occurrence were examined in four different overpressured provinces of offshore Norway in the search for such relationships. These provinces have experienced different geological histories and variable amounts of hydrocarbon leakage. Because all these areas received fairly recent hydrocarbon charge, the work focused on the identification of recent geological events that may subsequently have influenced recent stress history, including the present-day stresses. Areas of recent structuring were found to be characterized by more extensive hydrocarbon leakage than areas with less such structuring. This increased frequency of hydrocarbon leakage was interpreted to be the result of shear failure at the trap crests, induced by the combined effects of elevated pore pressures, stress anisotropy, and recent stress changes.

These results suggest that identification of recent stress changes based on the geological history of the study area could aid the prediction of hydrocarbon occurrence. It is inferred that stress history analyses can also reduce the uncertainty involved in seal analyses elsewhere.

Copyright ©2005 by The American Association of Petroleum Geologists.
DOI:10.1306/1060759H23164

INTRODUCTION

Pore pressure and stress in sedimentary basins are closely related. Overpressures in sedimentary basins all over the world are commonly explained by disequilibrium compaction, resulting from rapid vertical loading, a subsequent related rapid increase in the vertical stress, and restricted fluid escape (Swarbrick and Osborne, 1998). Overpressures have also been suggested to result from compressive tectonics (Higgins and Saunders, 1974; Hottman et al., 1979; Unruh et al., 1992; Henning et al., 2002). Yassir and Addis (2002) demonstrated that stress and pore pressure are intrinsically linked, and that a change of one affects the other; tectonic events (which influence the stress) may contribute to overpressure generation, and increased pore pressures result in a reduction of the effective stress.

High overpressures are also commonly associated with high exploration risks, as relationships between high overpressures and seal failure are reported from several sedimentary basins around the world, including the United States Gulf Coast, North Sea, Baram delta, Nile Delta, Sinai basin, and basins in Trinidad, the former Soviet Union, China, and Southeast Asia (Law and Spencer, 1998). In all these basins, hydrocarbon leakage through pressure-induced fracturing has been suggested when the overpressure gradient exceeds an area-specific maximum value. As a general observation, very little hydrocarbon production is reported from reservoirs with fluid pressure gradients in excess of 1.96 g/cm^3 (Law and Spencer, 1998).

A direct link between stress and hydrocarbon occurrence is expected because of the reported relationships between stress and fluid overpressures on the one hand and between fluid overpressures and hydrocarbon leakage on the other. Detailed relationships between stress and hydrocarbon habitat are, however, poorly known, especially because tectonically active regions are also prolific hydrocarbon provinces.

The focus of the investigations reported here was correlations and causal relationships between recent tectonic activity and hydrocarbon preservation in overpressured structural traps. Several regions offshore Norway contain structures with significant hydrocarbon volumes, as well as exploration failures that have been attributed to hydrocarbon leakage. Pliocene–Pleistocene subsidence has resulted in recent hydrocarbon generation in adjacent kitchen areas (Hermanrud and Nordgård Bolås, 2002), suggesting that seal failure in the area was related to recent tectonic activity. This and the fact that the stress state in the various regions has been influenced by different geological processes make the area well suited for analyses of relationships between tectonic activity and hydrocarbon occurrence.

BRIEF GEOLOGICAL HISTORY

Four different regions off the Norwegian Coast have been addressed in this study: the Halten Terrace, the western and eastern parts of the North Viking Graben, and the Central Graben areas (Figure 1). The Viking and Central Graben areas are located in the intracratonic sedimentary basin of the North Sea. The Halten Terrace is situated along the passive margin that separates mainland Norway from the Atlantic Ocean.

The main geological features of these mature hydrocarbon provinces have been known for some time (Ziegler et al., 1986; Spencer et al., 1987; Koch and Heum, 1995; Fraser et al., 2002). Geological profiles through the study areas (Figure 2) illustrate their main geological characteristics. Deltaic and shallow-marine sands of Middle Jurassic age constitute the main reservoirs in the Halten Terrace and Viking Graben areas. Tertiary deep-marine sands also host significant hydrocarbon volumes in the South Viking Graben. In the Central Graben, clastic reservoirs ranging from Early Permian to Eocene, as well as Late Cretaceous chalks, have been proven to be reservoir rocks. Organic-rich shales of Upper Jurassic age are the main source rock in all these areas. Upper Jurassic and Cretaceous shales and mudstones cap the hydrocarbon accumulations in the Viking Graben and Halten Terrace areas, whereas the Cretaceous and Tertiary discoveries are capped by Paleogene mudstones.

All the studied areas experienced widespread faulting that was related to the Cimmerian rifting phase in the Jurassic. This faulting created the structural traps that host most of the Jurassic hydrocarbon discoveries. In the Central Graben, halokinetic movements of the Upper Permian salt started in the Triassic, and salt diapirs formed structural traps in the Cretaceous chalks. This diapirism, which has continued until the present, also led to widespread fracturing and thereby high permeability in the otherwise tight chalks. Except for this salt-induced doming, all the Cretaceous, Tertiary, and Pliocene–Pleistocene strata offshore Norway are mainly characterized by passive subsidence. The maximum horizontal stress is oriented west–east to west-northwest–east-southeast at the Halten Terrace and in the North Viking Graben areas (Nordgård Bolås and Hermanrud, 2002). The horizontal stress orientation in the Central Graben area displays significant scatter (Mueller et al., 2003), testifying to the influence of salt diapirism on the stress state in individual wells in this area.

Overpressures are common in central parts of the North Sea and at the Halten Terrace. The timing of overpressure development in the various regions cannot be accurately constrained. However, because several of the pressure compartments in the study area are bound by sealing faults, it is inferred that overpressuring was linked to the cementation of these faults,

FIGURE 1. Locations of the Halten Terrace, Viking Graben, and Central Graben study areas offshore Norway (red rectangles), displayed on a simplified Late Jurassic regional geological setting. The map is adapted from Nordgård Bolås and Hermanrud (2003) and is based on Brekke et al. (1992) north of 62°N and Fraser et al. (2002) south of 62°N. Lines AA′, BB′, and CC′ are shown in Figure 2.

which, to a large extent, resulted from the late Tertiary subsidence and subsequent heating.

The generation, migration, and entrapment of hydrocarbons commenced in the early Tertiary and increased in the Pliocene–Pleistocene because of rapid subsidence. Hydrocarbon leakage at earlier times therefore did not significantly influence the present hydrocarbon habitat in the area. Recent stress changes, which could have influenced hydrocarbon preservation, have, however, occurred because of local events in the different areas.

RECENT STRESS-GENERATING PROCESSES IN THE STUDY AREAS

The Pliocene–Pleistocene subsidence of the study areas coincided with the uplift of the Baltic shield and the British Isles. As a result, all areas received sediment supply from the east, and the western part of the Viking Graben also received sediment supply from the west (Figure 2).

The westward progradation of the deltaic Pliocene–Pleistocene sedimentary wedges resulted in increased vertical stress in the underlying sediments. This

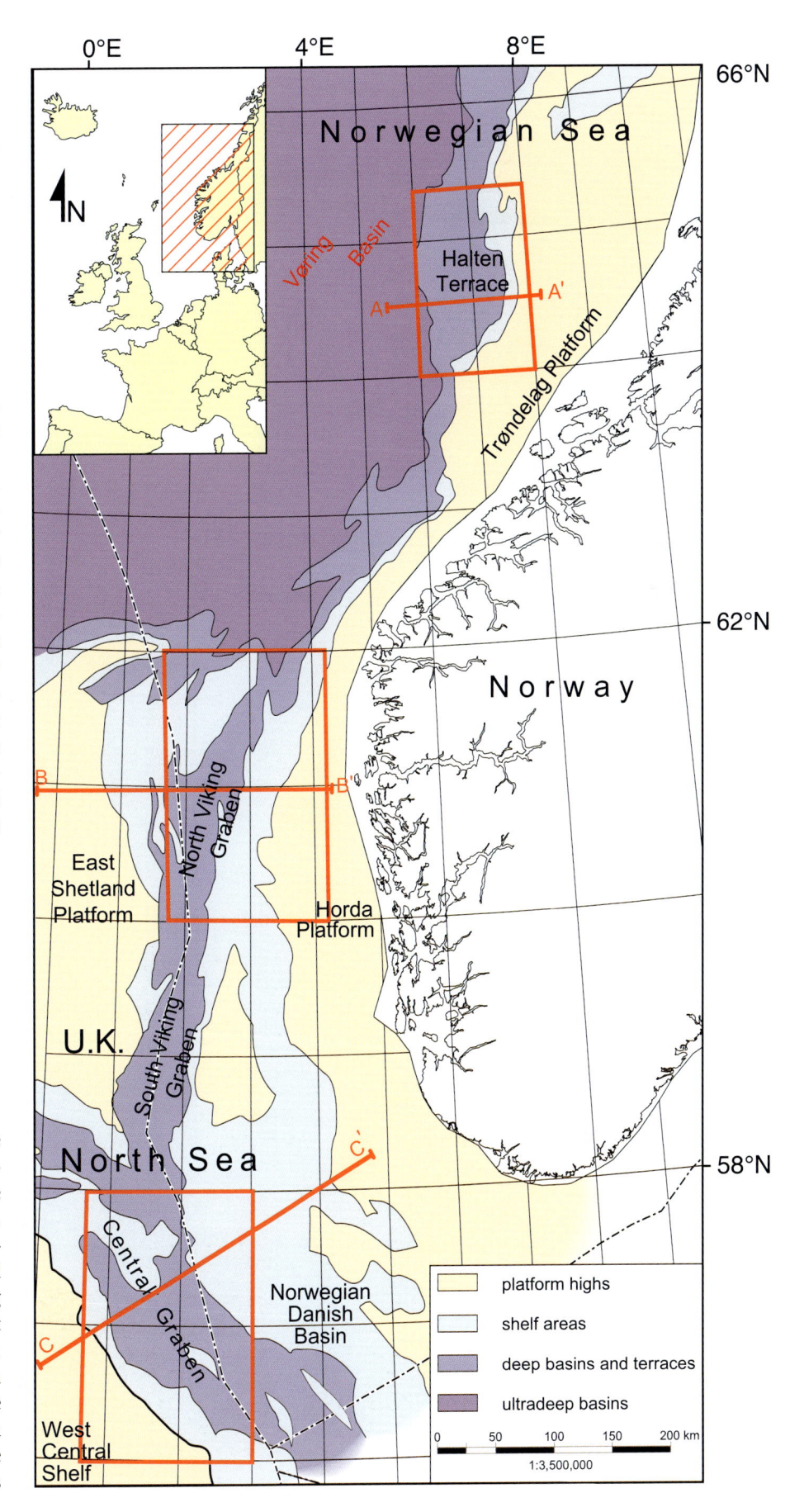

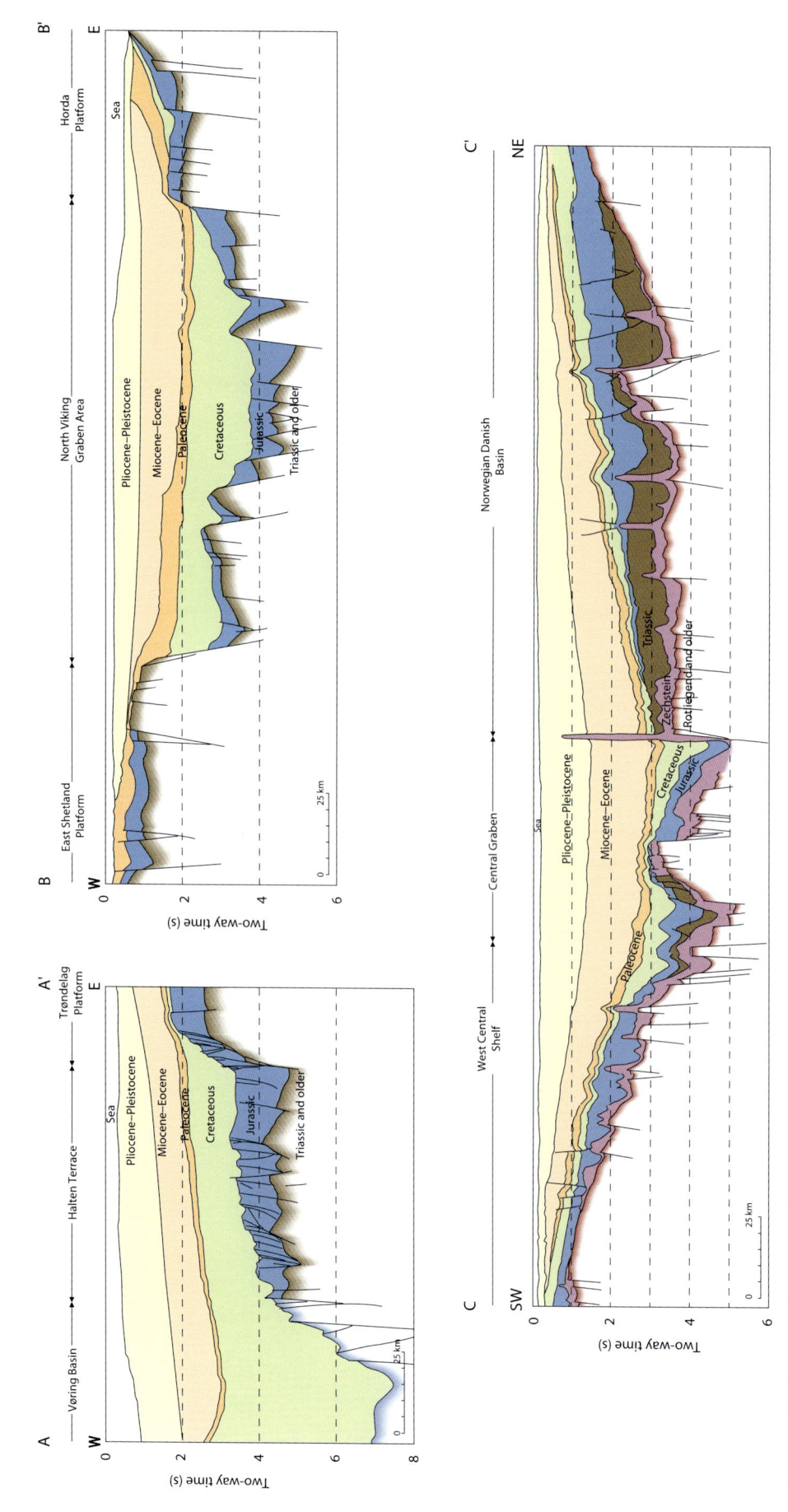

FIGURE 2. Geological profiles through the study areas, showing the general stratigraphy and tectonic setting. The locations of the profiles are shown in Figure 1. The profile from Halten Terrace was adapted from Blystad et al. (1995), and the profiles from the North Viking Graben and Central Graben areas were based on Fraser et al. (2002).

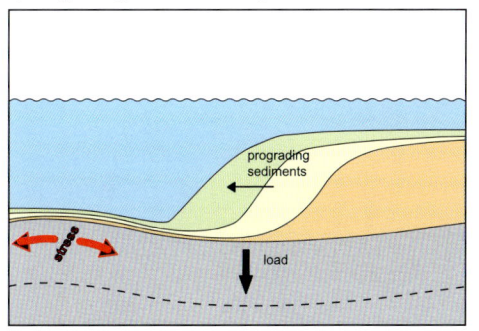

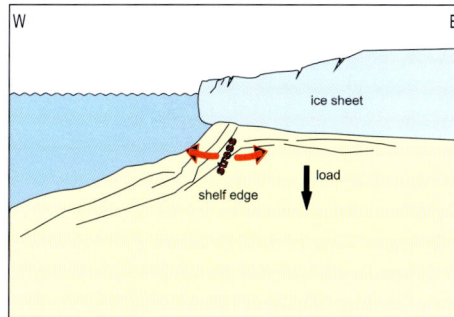

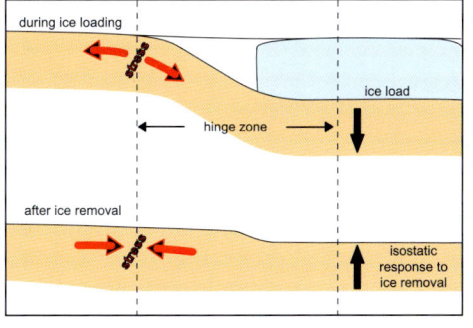

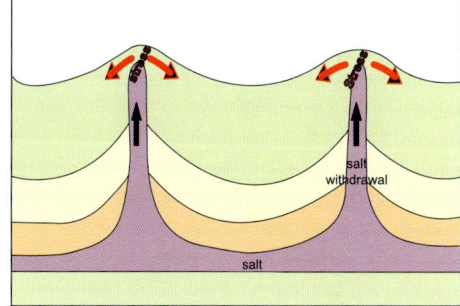

FIGURE 3. Recent stress-generating processes in the study areas: (a) prograding sediment wedges, (b) isostatic uplift, (c) glaciers at the shelf edge, and (d) salt diapirism. Red arrows indicate stress changes, and the black arrows indicate changes in load.

increased vertical stress migrated westward as the shelf moved. The increased sediment load also resulted in an increase in the horizontal stress because of elastic deformation of the underlying sediments. Thus, a stress anisotropy propagated westward in the underlying sediments as a response to the advance of the Pliocene–Pleistocene sedimentary wedges. The advance of these wedges also resulted in crustal flexuring beneath the wedge and in the formation of a forebulge in front of it. Such flexuring further influenced the stress state of the sediments (Figure 3).

Figure 4a shows an isopach map of the Pliocene–Pleistocene deposits (negative values) west of the hinge line (zero line) and estimated postglacial uplift (positive values) to the east of the hinge line. The Pliocene–Pleistocene sedimentary wedges in the Halten Terrace and in the Central Graben areas are more than 1500 m (5000 ft) thick in places. In contrast, the sedimentary wedges of the same age in the Viking Graben only reach a maximum thickness of about 600 m (2000 ft). The sediment wedge-induced stress field was thus less pronounced here than in the Halten Terrace and Central Graben areas.

Ice of significant thickness (as much as 3.5 km [2.1 mi]) covered the Baltic shield several times during the Pleistocene and caused isostatic subsidence (Fjeldskaar et al., 2000). Subsequent ice removals set off isostatic rebound, the last of which is still in progress. These movements thus resulted in crustal flexuring (bending), which influenced the horizontal stress, and thereby initiated the development of stress anisotropy in the sediments (Figure 3b). This process mainly influenced the sediments in the hinge zone, where the curvature of the bend was the tightest. The detailed position and extent of this zone is not well known, although it was probably quite close to the western border of the uplifted area. The position of this hinge zone was given by Doré and Jensen (1996) and is shown in Figure 4a as the area around the zero contour. Note that the hinge zone intersects the eastern part of the North Viking Graben area, whereas the major parts of the Central Graben and the Halten Terrace study areas do not interfere with this zone.

The shelfal areas offshore Norway also experienced repeated Pleistocene glacial advance and withdrawal. Maximum ice thicknesses may have reached 1500 m (5000 ft) (Fjeldskaar et al., 2000). The ice rested on the sediments of the continental shelf periodically, resulting in increased vertical stress underneath and related crustal flexuring where the vertical stress changes caused by the ice load changed abruptly (Figure 3c). Such changes were most pronounced at the shelf edge, because the ice was floating to the west of this edge (Ottesen et al., 2001) and therefore did not influence the stresses in the sediments below the seafloor. The position of the present-day shelf edge offshore Norway is shown in Figure 4b. As is seen from this map, only areas along the passive margin have developed a shelf edge in the study areas. Stress changes related to the advance and withdrawal of glaciers thus mainly affected the Halten Terrace area and were less significant in the Viking Graben and Central Graben areas, where the load from glaciers was more uniformly distributed.

The timing of halokinetic movements of the Permian salt varies in the Central Graben, and some salt diapirs have moved in the Pliocene–Pleistocene (Fraser et al., 2002). Such diapirism flexed the overlying strata at the time of active halokinesis (Figure 3d). This flexuring reduced the horizontal stress. Recent stress

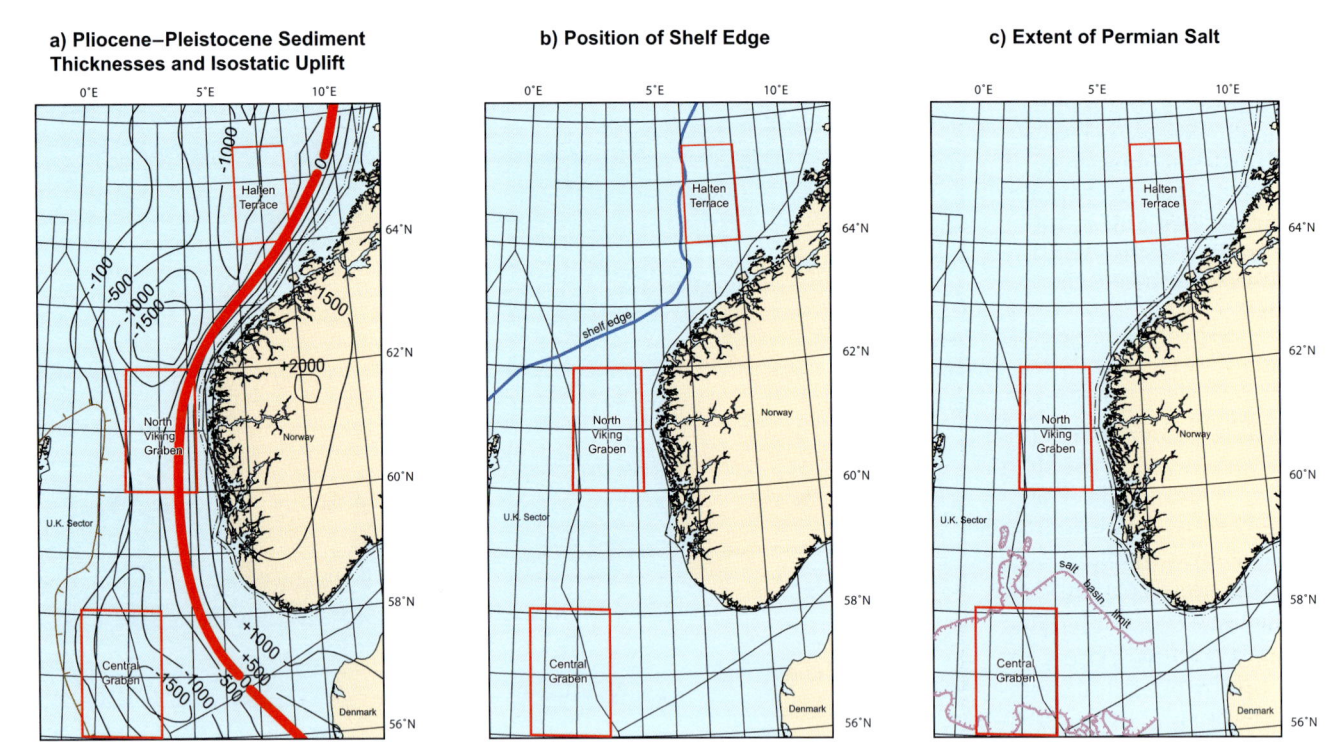

FIGURE 4. Positions of geological features that have led to recent flexuring offshore Norway: (a) isopach of the Pliocene–Pleistocene prograding sediment wedges (negative contours) and the amount of isostatic uplift (positive contours). The hinge zone is the area close to the zero contour (thick red line) and adapted from Doré and Jensen (1996); (b) location of the shelf edge offshore Norway; (c) areal extent of Permian (Zechstein) salt deposits (and thus, diapirism); adapted from Fraser et al. (2002).

anisotropy may thus have developed in the parts of the Central Graben area where salt diapirs have been generated (Figure 4c).

OBSERVATIONS: FLUID PRESSURES AND HYDROCARBON HABITAT

Overpressures have been observed in sediments of a variety of ages, and the pore pressure generally increases with depth toward the Jurassic reservoirs as exemplified by well 34/10-35 in the Gullfaks South field (Figure 5). The locations of hydrocarbon discoveries relative to the distribution of overpressures are shown in Figure 6. It is clear from this figure that overpressures in excess of 20 MPa are quite common in the deepest parts of the basins.

Figure 6 also shows that hydrocarbon discoveries are quite common in the overpressured areas. Exploration wells that are believed to have failed because of seal breaching and subsequent hydrocarbon leakage have been included to demonstrate that the frequency of hydrocarbon discoveries varies between the regions. The exploration failures were attributed to hydrocarbon leakage based on the observations that are listed in Table 1 and on the suggestions made by Gaarenstroom et al. (1993) for the Central Graben.

As rock failure is controlled by the effective stress (the difference between stress and pore pressure), retention capacities were inspected. Retention capacity is defined as the least principal stress minus the pore pressure, where leak-off pressures are commonly used as a measure of the least principal stress (Gaarenstroom et al., 1993; Nordgård Bolås and Hermanrud, 2003). The retention capacity, therefore, equals the minimum horizontal effective stress in extensional and strike-slip stress regimes. The retention capacity is, as such, a measure of the pore-pressure increase a trap can tolerate before the seal fails, provided that (1) tensile (and not shear) failure occurs, and (2) the seal has no tensile strength. The retention capacity is thus not a satisfactory indicator of the risk for seal breaching in general. Such an indicator should reflect the pore-pressure increase that a trap can tolerate before rupturing, based on realistic rock strength assessments, irrespective of the fracturing mode. The fault analysis seal technology (Mildren et al., 2005), which is essentially based on calculating the distance from Mohr's circle to the failure envelope of the rock, provides such a tool.

Despite the obvious shortcomings of retention capacity as a direct indicator of seal integrity risk, we,

FIGURE 5. Typical pore-pressure profile from the Norwegian North Sea (well 34/10-35 in the Gullfaks South field). This pore-pressure estimation is mainly derived from increased mud gas content including trip gas. The maximum pore-pressure gradient of 1.97 g/cm³ (77 MPa) at the top of the reservoir is measured by formation multitester. MSL = mean sea level; MWE = mud weight equivalent.

nevertheless, decided to analyze retention capacities in the study area. This choice was made because the investigations of retention capacities can aid the interpretation of the subsurface stress and recent stress history (see below). In addition, the retention capacity term is commonly used among explorationists, and such data were readily available in our study area.

Pore pressures and leak-off test data from 13 wells from the Halten Terrace and 16 wells from the North Viking Graben area, previously reported by Nordgård Bolås and Hermanrud (2003), form the main basis for this study. These authors calculated retention capacities in each individual well based on leak-off pressures and reservoir pore pressures reported in well completion reports. Retention capacities in individual wells may be too low by 4 MPa or more (Nordgård Bolås and Hermanrud, 2002) because of the inaccuracy of the leak-off pressure as a measure of the least principal stress. Such errors should, however, be equally distributed among the investigated areas and do not preclude the identification of smaller differences in retention capacity characteristics between areas. All the investigated wells represent different pressure compartments.

Gaarenstroom et al. (1993) presented retention capacity data for 29 wells in the United Kingdom Central Graben. However, these authors did not calculate the difference between the leak-off pressure and the pore pressure in individual wells. Instead, they mapped the lower envelope of a large amount of leak-off pressure data in the area and used this envelope as a measure for the least principal stress. We consider this approach unfortunate for two different reasons. First, some of the scatter in the leak-off pressure data of Gaarenstroom et al. (1993) reflect the inaccuracy of leak-off pressure as a measure of the least principal stress. Some values are erroneously low, with the result that the lower envelope will underestimate the least principal stress. Second, overpressured wells will, on average, have higher leak-off pressure values than normally pressured wells because of pore pressure and stress coupling (Hillis, 2001). Analyses of retention capacities in overpressured wells should thus not be based on leak-off pressure data from normally pressured wells.

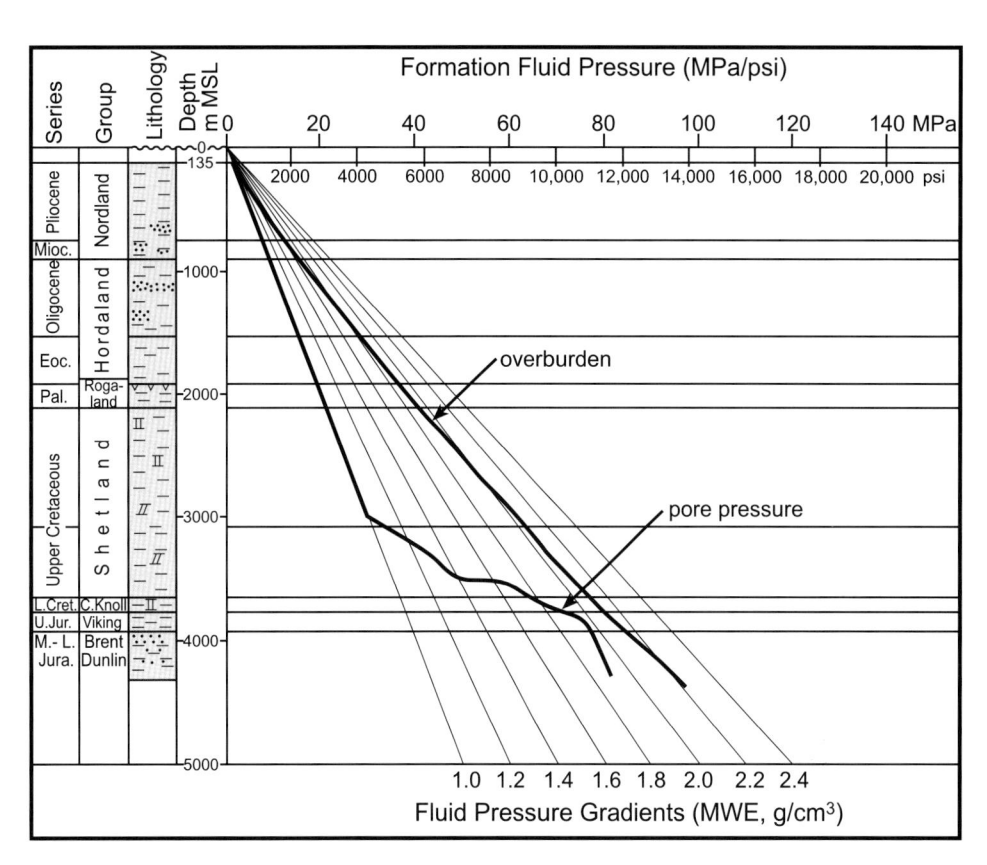

Nordgård Bolås and Hermanrud (2003) investigated the differences between leak-off pressure data from overpressured wells at 3–4.5-km (1.8–2.8-mi) burial depth and the lower envelope of leak-off pressure data (including normally pressured wells) at the Halten Terrace. Their results demonstrated that the former were 5–15 MPa higher than the lower envelope values. Based on these results, we added 10 MPa to the retention capacity data of Gaarenstroom et al. (1993). This addition permits a regional comparison between these data and those of Nordgård Bolås and Hermanrud (2003). This addition, which is appropriate for the highly overpressured wells of Gaarenstroom et al. (1993) at 3–4.5 km (1.8–2.8 mi) of burial depth, may have resulted in too high calculated retention capacities in the less overpressured or normally pressured wells and may also be less accurate at other burial depths. However, such errors do not influence the conclusions of this chapter. The addition of 10 MPa to all retention capacity data in the Central Graben necessarily implies that no individual well is listed with a retention capacity less than

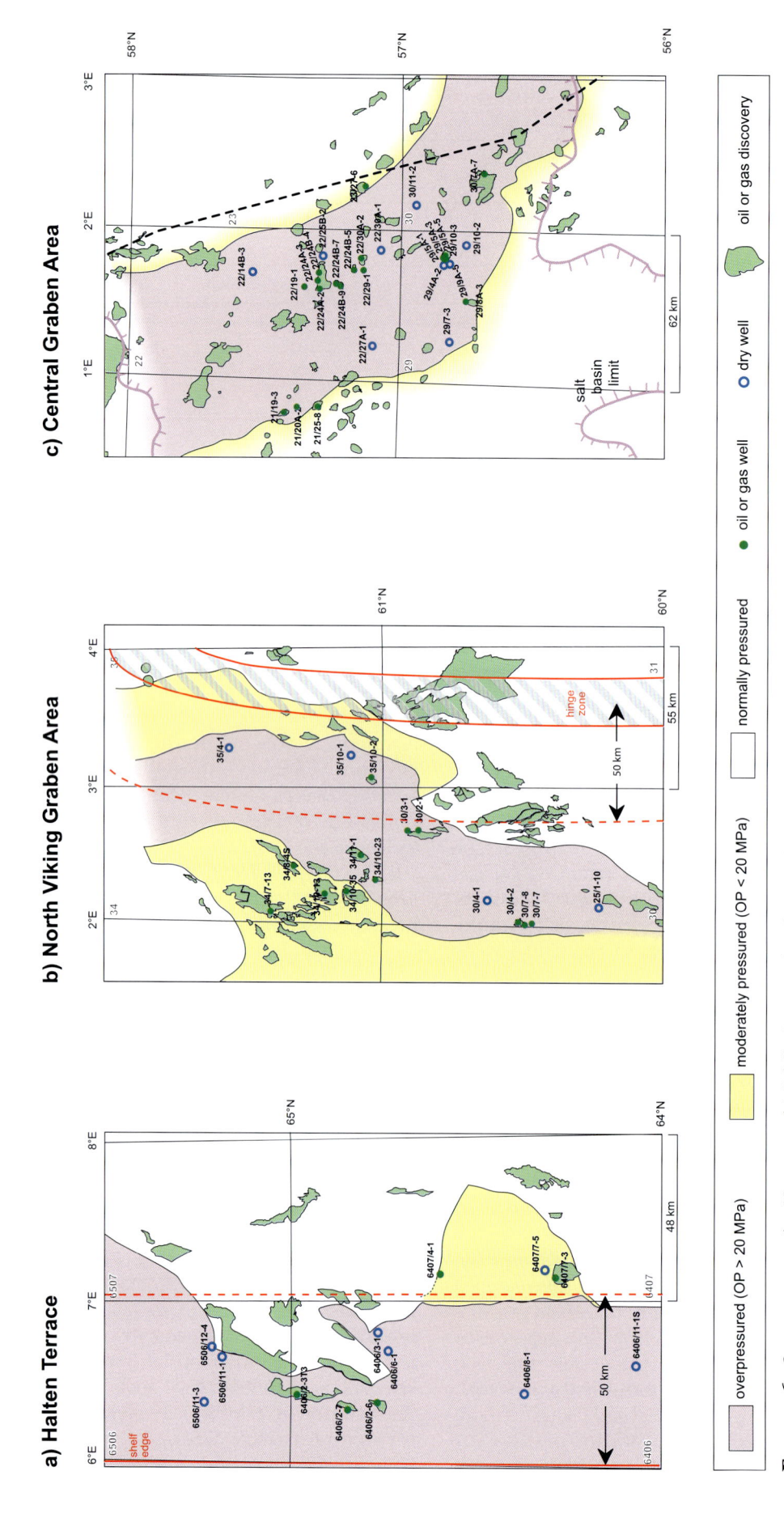

FIGURE 6. Overpressures (at Upper and Middle Jurassic levels) and hydrocarbon occurrences at (a) Halten Terrace, (b) North Viking Graben, and (c) Central Graben. The locations of the shelf edge (Figure 4a), the hinge zone (Figure 4b), and the extension of salt diapirism (Figure 4c) are superimposed to the maps in (a–c), respectively. Dotted lines show limits of areas that are believed to have experienced the most pronounced recent flexuring.

TABLE 1. Well numbers, operators, wellbore content, discovery names, and interpreted reasons for exploration well failures in the Halten Terrace and North Viking Graben areas. Discoveries are defined as technical discoveries (presence of hydrocarbons confirmed by testing), not necessarily commercial volumes of hydrocarbons present in the drilled structure. Hydrocarbon leakage is interpreted as a likely failure reason whenever significant hydrocarbon shows are present.

Well Number	Operator	Wellbore Content	Discovery Name	Interpreted Reason for Exploration Failure
North Viking Graben Area				
25/1-10	Elf	dry, minor shows		hydrocarbon leakage and/or restricted charge
30/2-1	Statoil	gas and condensate	Huldra South	
30/3-1	Statoil	gas and condensate	Huldra North	
30/4-1	BP	dry, minor shows		hydrocarbon leakage and/or restricted charge
30/4-2	BP	gas and condensate	Hild North	
30/7-7	Norsk Hydro	gas and condensate	Hild South	
30/7-8	Norsk Hydro	gas and condensate	Hild Central	
34/7-13	Saga	oil	Vigdis	
34/8-4s	Norsk Hydro	oil and gas	Visund	
34/10-13	Statoil	oil	Gullfaks	
34/10-23	Statoil	gas	Gullfaks Gamma	
34/10-35	Statoil	oil and gas	Gullfaks South	
34/11-1	Statoil	gas and condensate	Kvitebjørn	
35/4-1	Norsk Hydro	dry, shows		hydrocarbon leakage
35/10-1	Statoil	dry, shows		hydrocarbon leakage
35/10-2	Statoil	gas	B-S-375	
Halten Terrace				
6406/2-3T3	Saga	gas and condensate	Kristin	
6406/2-6	Saga	gas and condensate	Ragnfrid S	
6406/2-7	Saga	gas and condensate	Erlend	
6406/3-1	Statoil	dry, shows		hydrocarbon leakage
6406/6-1	Statoil	dry		lack of hydrocarbon charge
6406/8-1	Elf	dry, shows		hydrocarbon leakage
6406/11-1S	Saga	dry, shows		hydrocarbon leakage
6407/4-1	Statoil	gas and condensate	Nameless	
6407/7-3	Norsk Hydro	oil	Njord	
6407/7-5	Norsk Hydro	dry, minor shows		hydrocarbon leakage and/or missing trap elements
6506/11-1	Statoil	dry, shows		hydrocarbon leakage
6506/11-3	Statoil	dry, shows		hydrocarbon leakage
6506/12-4	Statoil	dry, shows		hydrocarbon leakage

10 MPa in Figure 7. This result is not necessarily correct in all wells, but once again, it does not influence the conclusions of this chapter, which deals with basin-wide stress characteristics.

As seen from Figure 7, the retention capacities of the overpressured wells at the Halten Terrace and Central Graben areas are, in general, higher than the corresponding values in the North Viking Graben area. The retention capacities of the overpressured Jurassic structures at the Halten Terrace average around 13 MPa and are never lower than 7 MPa. The retention capacities in the Central Graben (when adjusted as described above) are even higher (18 MPa on average). On the contrary, retention capacities as low as 2 MPa are found in the North Viking Graben area, with an average of 7 MPa for the whole area. When the retention capacity of these wells was calculated on the basis of a lower envelope for the least principal stress, as pioneered by Gaarenstroom et al. (1993), negative retention capacities resulted.

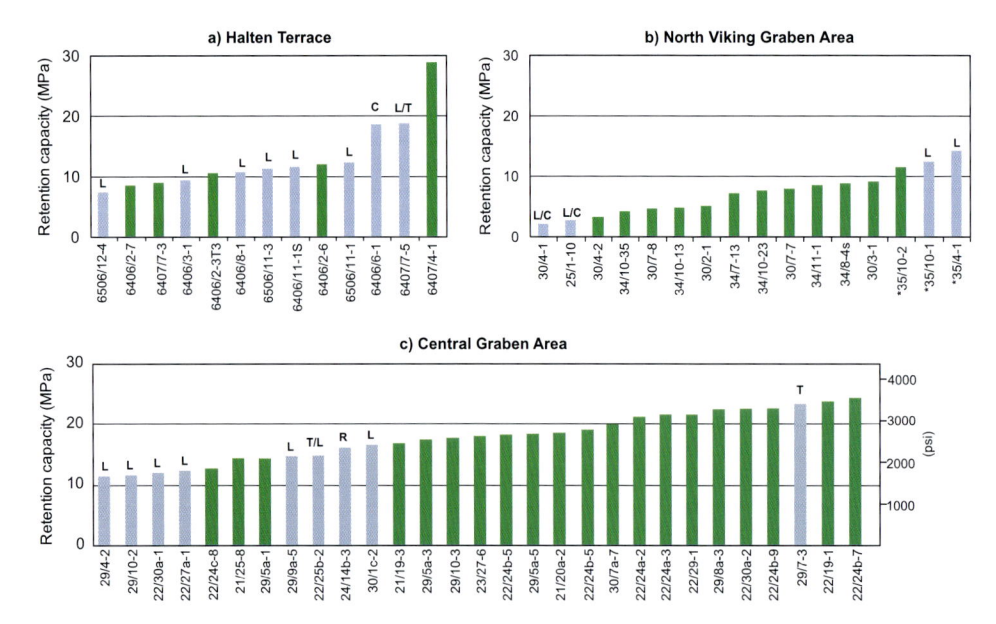

FIGURE 7. Retention capacities from the (a) Halten Terrace, (b) North Viking Graben area, and (c) Central Graben area. Blue bars indicate water-bearing structures, and green bars represent hydrocarbon-bearing structures. The letters L, C, R, and T illustrate whether the water-bearing structures are void of hydrocarbons because of hydrocarbon leakage, lack of charge, lack of reservoir, or lack of trap, respectively. The asterisks indicate wells located on the eastern flank of the North Viking Graben.

It is noted that the lowest retention capacities in the North Viking Graben are found in the western area, and that the structures that have leaked their hydrocarbons in the eastern area have retention capacities similar to those of the structures emptied for hydrocarbons in the Halten Terrace area (the average retention capacity of the wells 35/10-2, 35/10-1, and 35/4-1 is 13 MPa). The average retention capacity in the western part of the North Viking Graben is 6 MPa.

Table 2 summarizes the frequency of traps that are emptied of hydrocarbons because of leakage in the investigated areas. As is seen from this table, the area that is characterized by the lowest retention capacities (the western part of the North Viking Graben) also has the lowest frequency of water-bearing traps that have leaked their hydrocarbons. This result may seem counterintuitive, because it demonstrates that the area with the highest pore pressures relative to the overburden

TABLE 2. Observed relationships between hydrocarbon occurrences and geological processes that promote crustal flexuring offshore Norway.

Observations	*Provinces*			
	Halten Terrace	**Western Part, North Viking Graben**	**Eastern Part, North Viking Graben**	**Central Graben**
Frequency of traps emptied of hydrocarbons due to leakage	high: 54% (7 of 13)	low: 15% (2 of 13)	high: 67% (2 of 3)	medium: 24% (7 of 29)
Average level of retention capacities	high: 13 MPa (minimum 7 MPa)	low: 6 MPa (minimum 2 MPa)	high: 13 MPa (minimum 12 MPa)	high: 18 MPa (minimum 12 MPa)
Flexuring by prograding sediment wedges	+	−	−	+
Flexuring along isostatic hinge line	−	−	+	−
Glacial flexuring	+	−	−	−
Flexuring by salt diapirism	−	−	−	+

Frequency of traps emptied of hydrocarbons due to leakage: high = 50% or more of these structures are emptied of hydrocarbons because of leakage (residual hydrocarbon columns are detected in the traps); medium = 20–50% of these structures are dry because of hydrocarbon leakage; low = less than 20% of these structures are dry because of probable hydrocarbon leakage.

Average levels of retention capacities: high = average retention capacity is equal to or higher than 10 MPa; low = average retention capacity is less than 10 MPa.

The retention capacity values for the Central Graben presented in this table are based on the data from Gaarenstroom et al. (1993) but increased by 10 MPa as justified in the text.

The plus signs indicate the presence of recent flexuring processes, whereas the minus signs indicate that such processes are absent or less pronounced.

stress (i.e., the lowest retention capacities) has preserved hydrocarbons more efficiently than the other areas.

The assumed exposures to recent stress-generating processes are also listed in Table 2. Apparently, the Halten Terrace, the eastern part of the North Viking Graben area, and the Central Graben have all been influenced by at least one geological process that has the potential of generating stress anisotropy by crustal flexuring in the sediments. These provinces also show the highest frequency of water-bearing structures resulting from hydrocarbon leakage. Thus, a correlation exists between the formation of stress anisotropy and the frequency of traps that only contain residual hydrocarbons caused by leakage.

It is also observed that the provinces that have undergone recent crustal flexuring have the highest retention capacities. The Halten Terrace, the eastern part of the North Viking Graben, and the Central Graben areas are all interpreted to have developed recent stress anisotropy because of such flexuring and are associated with average retention capacities of 13 MPa or more. The corresponding average retention capacity level in the western part of the North Viking Graben area is around 6 MPa. Thus, a correlation exists between retention capacity and the (interpreted) recent generation of stress anisotropy. Given these observations, a correlation between hydrocarbon preservation and small retention capacities follows as a necessity.

ARE THERE CAUSAL RELATIONSHIPS BETWEEN GEOLOGICAL PROCESSES, STRESS ANISOTROPY, AND HYDROCARBON PRESERVATION?

Whereas the empirical relationships between hydrocarbon occurrence, retention capacity data, and geological processes listed in Table 2 are interesting on their own, their relevance to hydrocarbon exploration outside the study area hinges on the elucidation of their underlying causal relationships. All four geological processes listed in Table 2 may have caused crustal flexuring and thereby reductions in the horizontal stress at present and/or in the recent past. Such reductions in the horizontal stress lead to increased stress anisotropy, providing that the basin was in an extensional stress regime (where the vertical stress is the largest) prior to the stress perturbations. Reduced horizontal stress and increased stress anisotropy promote shear failure, as visualized by Mohr's circle (Figure 8). Here, a decrease in the least principal stress decreases the minimum effective stress from σ_3 to σ_{3new}, and the diameter of Mohr's circle expands. When the geological conditions cause Mohr's circle to touch the failure envelope, shear failure is initiated (circle 1 expands to circle 2). Shear failure involves both the generation and reopening of faults and fractures, which may or may not have resulted in hydrocarbon leakage, depending on the position of the fracture relative to the hydrocarbon column.

Grollimund (2000) and Grollimund and Zoback (2000, 2001) investigated the magnitudes of the stress perturbations resulting from recent ice deposition and withdrawal. Their numerical modeling suggested that stress perturbations of as much as 40 MPa could have resulted at 3 km (1.8 mi) burial depth. Although these calculations were based on quite coarse descriptions of the involved lithologies and, thus, may have suffered from inaccurate knowledge of the elastic properties of the various rocks, they, nevertheless, demonstrate that the stress perturbations were quite significant and of sufficient magnitude to promote shear failure. Modeling of stress perturbations that may have resulted from the other processes listed in Table 2 have apparently not been performed in the study area.

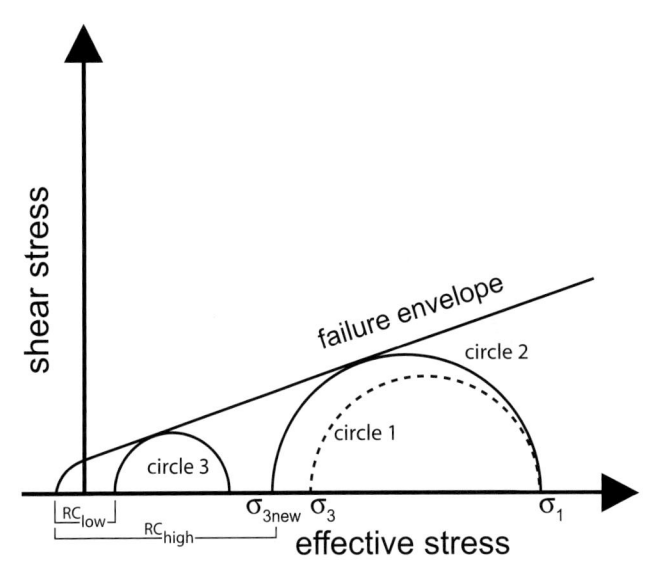

FIGURE 8. Mohr's circle, shear failure envelope, and relationship between shear failure and retention capacity. Circle 1 illustrates an anisotropic stress state of a rock that is not critically stressed with regard to shear failure. σ_1 and σ_3 represent maximum and minimum effective stress in this case. Circle 2 shows the stress state of a rock with increased stress anisotropy and which, therefore, meets the failure criterion. σ_{3new} represents the minimum effective stress for this situation. Circle 3 represents a rock with a more isotropic stress state, where shear failure is initiated. Rocks that experience shear failure in isotropic stress regimes will, in general, be characterized by lower retention capacities (RC_{low}) than rocks that suffer from similar failure in more anisotropic stress regimes (RC_{high}). This general relationship is valid regardless of the assumed failure envelope.

Although their magnitudes and practical significance thus remain speculative, we suggest that these processes may also have influenced the stress fields all the way down to the depth of the Mesozoic reservoirs.

The comparatively high retention capacities (RC_{high}) in the Halten Terrace, the eastern part of the North Viking Graben, and the Central Graben areas may have resulted from shear failure triggered by the recent development of stress anisotropies related to late flexuring. If this increase forced Mohr's circle to intersect the failure envelope (circle 2 in Figure 8), then fracturing would lead to pore-fluid discharge. Loss of fluids would reduce the pore pressures and increase the retention capacity. This increase will proceed until the circle no longer intersects the failure envelope. It follows from this line of argument that rocks that experience shear failure in highly anisotropic stress regimes (circle 2) generally will be characterized by higher retention capacities (RC_{high}) than rocks that are critically stressed in more isotropic stress regimes. A close to isotropic rock stress is illustrated by circle 3 in Figure 8, which has a smaller diameter than circle 2 and a related lower retention capacity (RC_{low}) when the failure criterion is met. This general rule is valid, although highly overpressured rocks tend to have higher leak-off pressures than normally pressured rocks.

Hence, it is suggested that geological processes leading to crustal flexuring and increased stress anisotropy promote shear failure in overpressured traps. Such shear failure poses a risk to hydrocarbon preservation if it happens close to the crest of the traps, and it also leads to reduced pore pressures and thereby precludes low retention capacities from ever developing. Low retention capacities are diagnostic for close to isotropic stress states and, thus, for the absence of recent shear failure and related hydrocarbon leakage. However, traps with an isotropic stress state may still be at risk of leaking hydrocarbons by tensile failure, but this does not seem to be the case for the majority of the traps in the western part of the North Viking Graben area (Hermanrud and Nordgård Bolås, 2002). Based on these observations, we suggest that the level of the minimum retention capacities in a basin may be an indicator of the general leakage mode in that basin.

PREDICTION OF SEAL INTEGRITY FROM STRESS HISTORY ANALYSIS

If the proposed causal relationships between trap failure, on the one hand, and recent stress changes caused by crustal flexuring, on the other, are correct in the study areas, then such relationships should also be expected elsewhere and could thus be applied in seal evaluations regardless of geographical location. To test this hypothesis, wells positioned in areas that were thought to have experienced recent flexuring were separated from wells in areas that have been less exposed to such events.

The structures located closest to the shelf edge in the western part of the Halten Terrace area are interpreted to have experienced most crustal flexuring. Most of the emptied hydrocarbon traps in the overpressured region at the Halten Terrace are located approximately within a 50-km (31-mi) range from the shelf edge. It is therefore possible that the expected flexuring in this area mainly influenced overpressured traps within this 50-km (31-mi) distance (Figure 6a). We have no results, however, from stress modeling to confirm this impact range for crustal flexuring. The 50-km (31-mi) range indicates that wells 6407/7-3, 6407/7-5, and 6407/4-1 are not heavily influenced by crustal flexuring.

In the North Viking Graben area, structures located closest to the hinge zone are expected to have been most exposed to recent flexuring. The three studied wells located to the east in the highly overpressured part of the North Viking Graben area (wells 35/4-1, 35/10-1, and 35/10-2) are located within a 50-km (31-mi) distance to the center of the hinge zone (Figure 6b). These structures may therefore have experienced more recent crustal flexuring than the other structures further to the west in this area (Figure 6b), and the high overpressures seem to have triggered hydrocarbon leakage from two of these eastern structures (wells 35/4-1 and 35/10-1), because these two wells only contain residual hydrocarbons. Some hydrocarbon discoveries are also located even closer to the hinge zone, but these structures have obviously not suffered from seal failure, probably because the pore pressures have not been high enough to trigger extensive leakage of hydrocarbons. Wells from the Central Graben were omitted from this analysis, because information on the stress history of individual wells in this area was not available.

The retention capacities of the wells in areas that we suggest have been influenced by crustal flexuring are shown in Figure 9a. The wells in this figure are sorted with respect to reservoir depth, which increases to the right. The observation that all wells that encountered movable hydrocarbons plot to the right is consistent with the proposal that flexuring-triggered shear failure resulted in hydrocarbon leakage, because the effects of such flexuring diminish with depth (Grollimund, 2000). However, flexuring-induced shear failure may result in reactivations of faults that intersect the pressure compartments in downflank positions, as suggested for the North Sea 35/10-2 structure by Teige and Hermanrud (2004). Thus, a perfect correlation between increasing reservoir depth and hydrocarbon preservation would not be expected. Still, the relationship is

FIGURE 9. Retention capacities for (a) wells penetrating recently flexed reservoirs and (b) wells penetrating reservoirs that have seen little or no recent flexuring. The depths of the reservoirs are increasing toward the right in the plots. The letters L, C, R, and T illustrate whether the water bearing structures are void of hydrocarbons because of hydrocarbon leakage, lack of charge, lack of reservoir, or lack of trap, respectively.

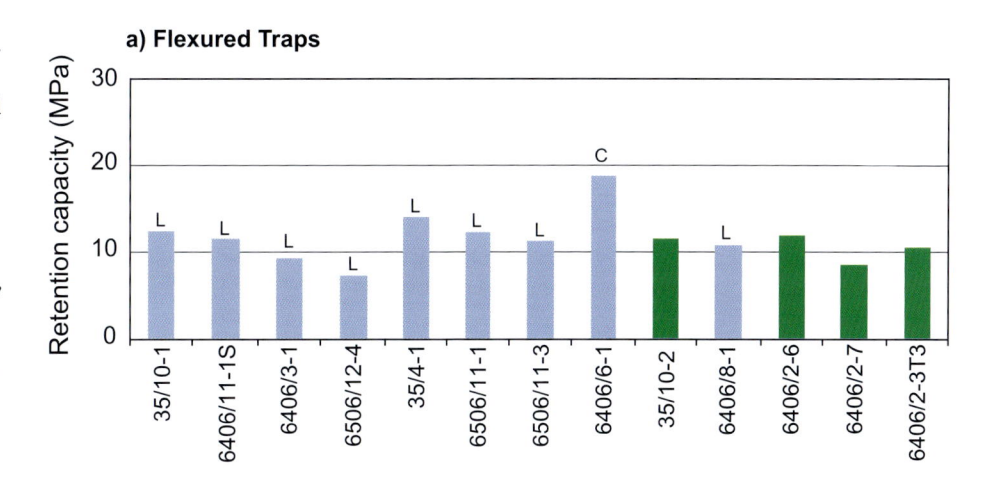

present between reservoir depth and hydrocarbon preservation seen in Figure 9a. With one exception, all reservoirs deeper than that of well 35/10-2 (4240 mKB [13,910 ftKB]) (KB = Kelly Bushing) contain movable hydrocarbons, whereas none of the shallower reservoirs do. Only one of these shallow wells (well 6406/6-1) does not contain residual hydrocarbons indicative of previous filling. This observation suggests that flexuring, to a large extent, resulted in hydrocarbon leakage from the apex of the structure in the shallowest traps, whereas the deeper traps either were shielded from such leakage or leaked hydrocarbons from downflank positions. Apparently, such flexure-induced leakage removed all hydrocarbons below the position of leakage, consistent with the suggestions for the 35/10-2 structure by Teige and Hermanrud (2004).

The retention capacities for the traps that are believed to have experienced little or no recent flexuring are shown in Figure 9b, also with increasing depths toward the right. Two wells in this figure have significantly higher retention capacities than the others: wells 6407/4-1 (with intermediate fluid pressures; Hermanrud and Nordgård Bolås, 2002) and well 6407/7-5 (possibly in pressure communication with sandy intervals in the shallower well 6407/7-3) (Lilleng and Gundesø, 1997). With these exceptions, all retention capacities are, on average, about half of the average value for the flexed reservoirs.

The high frequency of movable hydrocarbons in Figure 9b differs dramatically from that of the flexed reservoirs (Figure 9a). This observation is also consistent with the suggestion that stress changes, resulting from flexuring, triggered hydrocarbon leakage in the investigated areas. The two deepest reservoirs (represented by wells 25/1-10 and 30/4-1, at 4503- and 5185-m (14,773- and 17,011-ft) burial depths, respectively) are void of hydrocarbons. These wells also have the lowest retention capacities among the investigated wells, and both wells encountered well-cemented, low-permeability reservoirs. Hydrocarbon leakage through tensile failure may explain residual hydrocarbons in these wells, where brittle cap rocks may be expected. These two wells (out of a total of 29 traps investigated in the Halten Terrace and in the North Viking Graben in this study) represent the only two candidates for overpressured traps that have leaked all their hydrocarbons through tensile failure in these areas.

The significant difference in hydrocarbon habitat vs. depth and the overall lower retention capacities and higher frequency of hydrocarbon preservation in nonflexured areas all support the proposed causal relationships between recent stress changes, leakage mode, and hydrocarbon preservation. These analyses were not based on data confined to certain geographical areas but, instead, according to the perceived influence of recent crustal flexuring. The results, therefore, support the suggestion that analyses of recent stress changes can effectively aid seal evaluation in new areas.

CONCLUSIONS

Investigations of overpressured Jurassic reservoirs offshore Norway demonstrate a correlation between the frequency of structural traps emptied of hydrocarbons by leakage and recent geological processes that promote crustal flexuring and generate stress anisotropy. Hydrocarbon preservation appears to be favored in provinces less influenced by such processes. Here, the stress state is close to isotropic, as evidenced by very low retention capacities (down to 2 MPa).

It is suggested that the investigated water-bearing traps lost their hydrocarbons because of shear failure and subsequent hydrocarbon leakage, which was triggered by the development of recent stress anisotropies. Recent geological processes that promote stress anisotropy are thus considered to be negative for hydrocarbon preservation in the study area. Regions that have not experienced such processes are less prone to seal failure.

The close relationships between the intensity of recent crustal flexuring and trap integrity, which is strengthened when data from different geographical locations are analyzed collectively, support the suggestion that analyses of recent stress changes can effectively aid seal evaluation in new areas.

ACKNOWLEDGMENTS

We thank Statoil for the permission to publish this chapter. We also thank P. A. Bjørkum, R. Hillis, P. Boult, and D. Dewhurst for constructive comments to earlier versions of this manuscript. Elin Storsten is thanked for graphical support, and S. Clark is thanked for his improvements to the English text.

REFERENCES CITED

Blystad, P., H. Brekke, R. B. Færseth, B. T. Larsen, J. Skogseid, and B. Tørudbakken, 1995, Structural elements of the Norwegian continental shelf: Part II. The Norwegian Sea region: Norwegian Petroleum Directorate Bulletin, v. 8, 45 p.

Brekke, H., J. E. Kalheim, F. Riis, B. Egeland, P. Blystad, S. Johnsen, and J. Ragnhildstveit, 1992, Two-way time map of the unconformity at the base of the Upper Jurassic (north of 69°N) and the unconformity at the base of the Cretaceous (south of 69°N), offshore Norway, including the geological trends onshore: Norwegian Petroleum Directorate Continental Shelf Map No. 1: Norwegian Petroleum Directorate/The Geological Survey of Norway, scale 1:2,000,000, one poster-size sheet.

Doré, A. G., and L. N. Jensen, 1996, The impact of late Cenozoic uplift and erosion on hydrocarbon exploration: Offshore Norway and some other uplifted basins: Global and Planetary Change, v. 12, p. 415–436.

Fjeldskaar, W., C. Lindholm, J. F. Dehls, and I. Fjeldskaar, 2000, Postglacial uplift, neotectonics and seismicity in Fennoscandia: Quaternary Science Reviews, v. 19, p. 1413–1422.

Fraser, S. I., A. M. Robinson, H. D. Johnson, J. R. Underhill, D. G. A. Kadolsky, R. Conell, P. Johannesen, and R. Ravnås, 2002, Upper Jurassic, in D. Evans, C. Graham, A. Armour, and P. Bathurst, eds., The millenium atlas: Petroleum geology of the central and northern North Sea: Geological Society (London), p. 157–189.

Gaarenstroom, L., R. A. J. Tromp, M. C. de Jong, and A. M. Brandenburg, 1993, Overpressures in the central North Sea; implications for trap integrity and drilling safety, in J. R. Parker, ed., Petroleum geology of northwest Europe: Proceedings of the 4th Conference: Geological Society (London), p. 1305–1313.

Grollimund, B., 2000, Impact of deglatiation on stress and implications for seismicity and hydrocarbon exploration: Ph.D. thesis, Stanford University, California, 204 p.

Grollimund, B., and M. Zoback, 2000, Post glacial lithospheric flexure and induced stresses and pore pressure changes in the northern North Sea: Tectonophysics, v. 327, no. 1–2, p. 61–81.

Grollimund, B., and M. Zoback, 2001, Impact of glacially induced stress changes on fault-seal integrity offshore Norway: AAPG Bulletin, v. 87, no. 3, p. 493–506.

Henning, A., M. A. Addis, N. Yassir, and A. H. Warrington, 2002, Pore-pressure estimation in an active thrust region and its impact on exploration and drilling, in A. Huffman and G. Bowers, eds., Pressure regimes in sedimentary basins and their prediction: AAPG Memoir 76, p. 89–105.

Hermanrud, C., and H. M. Nordgård Bolås, 2002, Leakage from overpressured hydrocarbon reservoirs at Haltenbanken and in the northern North Sea, in A. G. Koestler and R. Hunsdale, eds., Hydrocarbon seal quantification: Norwegian Petroleum Society Special Publication 11, p. 221–231.

Higgins, G. E., and J. B. Saunders, 1974, Mud volcanoes; their nature and origin. Contributions to the geology and paleobiology of the Caribbean and adjacent areas: Verhandlungen der Naturforschenden Gesellschaft in Basel, v. 84, no. 1, p. 101–152.

Hillis, R. H., 2001, Coupled changes in pore pressure and stress in oil fields and sedimentary basins: Petroleum Geoscience, v. 7, p. 419–425.

Hottman, C. E., J. H. Smith, and W. R. Purcell, 1979, Relationship among Earth stresses, pore pressure, and drilling problems offshore Gulf of Alaska: Journal of Petroleum Technology, v. 31, no. 11, p. 1477–1484.

Koch, J. O., and O. R. Heum, 1995, Exploration trends of the Halten Terrace, in S. Hanslien, ed., Petroleum exploration and exploitation in Norway: Norwegian Petroleum Society Special Publication 4, p. 235–251.

Law, B. E., and C. W. Spencer, 1998, Abnormal pressures

in hydrocarbon environments, *in* B. Law, G. F. Ulmishek, and V. I. Slavin, eds., Abnormal pressures in hydrocarbon environments: AAPG Memoir 70, p. 1–11.

Lilleng, T., and R. Gundesø, 1997, The Njord field: A dynamic hydrocarbon trap, *in* P. Møller-Pedersen and A. G. Koestler, eds., Hydrocarbon seals: Importance for exploration and production: Norwegian Petroleum Society Special Publication 11, p. 217–229.

Mildren, S. D., R. R. Hillis, D. N. Dewhurst, P. J. Lyon, J. J. Meyer, and P. J. Boult, 2005, FAST: A new technique for geomechanical assessment of the risk of reactivation-related breach of fault seals, *in* P. Boult and J. Kaldi, eds., Evaluating fault and cap rock seals: AAPG Hedberg Series, no. 2, p. 73–85.

Mueller, B., J. Reinecher and K. Fuchs, 2003, The 2003 release of the world stress map: http://www-wsm.physik.uni-karlsruhe.de/pub/introduction/introduction_frame.html (accessed December 21, 2004).

Nordgård Bolås, H. M., and C. Hermanrud, 2002, Rock stress in sedimentary basins— Implications for trap integrity, *in* A. G. Koestler and R. Hunsdale, eds., Hydrocarbon seal quantification: Norwegian Petroleum Society Special Publication 11, p. 17–35.

Nordgård Bolås, H. M., and C. Hermanrud, 2003, Hydrocarbon leakage processes and trap retention capacities offshore Norway: Petroleum Geoscience, v. 9, p. 321–332.

Ottesen, D., L. Rise, K. Rokoengen, and J. Sættem, 2001, Glacial processes and large-scale morphology on the mid-Norwegian continental shelf, *in* O. J. Martinsen and T. Dreyer, eds., Sedimentary environments offshore Norway— Palaeozoic to Recent: Norwegian Petroleum Society Special Publication 10, p. 439–447.

Spencer, A. M., E. Holter, C. J. Campbell, S. H. Hanslien, P. H. H. Nelson, E. Nysæther, and E. G. Ormaasen, 1987, Geology of the Norwegian oil and gas fields: London, Graham and Trotman, 493 p.

Swarbrick, R. E., and M. J. Osborne, 1998, Mechanisms that generate abnormal pressures: An overview, *in* B. E. Law, G. F. Ulmishek, and V. I. Slavin, eds., Abnormal pressures in hydrocarbon environments: AAPG Memoir 70, p. 13–34.

Teige, G. M. G., and C. Hermanrud, 2004, Seismic characteristics of fluid leakage from an under-filled and over-pressured Jurassic fault trap in the Norwegian North Sea: Petroleum Geoscience, v. 10, p. 35–42.

Unruh, J., M. Davidsson, K. Criss, and E. Moores, 1992, Implications of perennial saline springs for abnormally high fluid pressures and active thrusting in western California: Geology, v. 20, p. 431–434.

Yassir, N., and M. A. Addis, 2002, Relationships between pore pressure and stress in different tectonic settings, *in* A. Huffman and G. Bowers, eds., Pressure regimes in sedimentary basins and their prediction: AAPG Memoir 76, p. 79–88.

Ziegler, W. H., R. Doery, and J. Scott, 1986, Tectonic habitat of Norwegian oil and gas, *in* A. M. Spencer, E. Holter, C. Campbell, S. H. Hanslien, P. H. H. Nelson, E. Nysæther, and E. G. Ormåsen, eds., Habitat of hydrocarbons on the Norwegian continental shelf: Norwegian Petroleum Society, p. 3–20.

Investigating the Effect of Varying Fault Geometry and Transmissibility on Recovery: Using a New Workflow for Structural Uncertainty Modeling in a Clastic Reservoir

Signe Ottesen
Statoil, Stavanger, Norway

Chris Townsend
Shell EP Europe Nederlanse Aardolie Maatschappij, Assen, The Netherlands

Kjersti Marie Øverland
Statoil, Stavanger, Norway

ABSTRACT

Assiduous readers may have noticed that the statistical treatment of tectonic heterogeneities in reservoir simulation is not as rigorous as that of their sedimentary counterparts. Indeed, it is common operational practice to carry only one structural model forward to dynamic reservoir simulation. This means that although the uncertainties attached to variations in sedimentary parameters are commonly addressed, those that are caused by structural heterogeneities are neglected, oversimplified, or underrepresented. The main reasons for this are (1) a lack of appreciation of how structural parameter uncertainties may impact predicted reservoir performance; (2) the need for an efficient methodology; and (3) a lack of an easy-to-use, fully integrated software. A new methodology for investigating the effect of structural uncertainty on reservoir production was tested during this study. The method efficiently assesses structural uncertainties by varying description parameters that define structural horizons and faults and then building multiple realizations of the structural model (i.e., a three-dimensional corner-point grid). Fault properties are modeled by combining fault seal algorithms, and representative transmissibility multipliers are calculated for each

Copyright ©2005 by The American Association of Petroleum Geologists.
DOI:10.1306/1060760H23165

realization of the reservoir-simulation grid. The parameters that can be varied include alternative horizon interpretation, velocity models for depth conversion, fault density, fault pattern, fault throw, fault length, fault position, fault thickness, and methodology for fault permeability estimation. This chapter describes the methodology and presents the results from a pilot study of fault uncertainty in a Brent Group reservoir gas field from the Southern Viking Graben in the Norwegian sector of the North Sea. The recovery factor for gas shows an absolute difference of 17.5% (i.e., a variation of 23%) between the very best and very worst cases. The study shows that increasing fault density in the model was the single most important factor in field recovery, and that fault density has an effect on recovery even when the faults did not completely seal. This indicates that subseismic faults, if present, could reduce the recovery factor even further. The choice of fault permeability had the second largest effect on recovery. It must be emphasized that the study focused on investigating the effect of intrareservoir faults, and investigation of the uncertainty in the major block-bounding faults was not included. This study is not a full uncertainty study because no probabilities are attached to the different realizations, but it provides a good basis for such a study.

INTRODUCTION

Traditional uncertainty studies in static volumes address structural uncertainty only through seismic interpretation and depth conversion. Uncertainty in seismic interpretation is commonly limited to uncertainty in seismic picking, and only occasionally is the possibility of an alternative structural interpretation included. Uncertainty in pore volume and recoverable reserves are routinely performed, but in many cases, the actual production lies outside the predicted range in recoverable reserves. Commonly, the factors contributing to the large range in static volumes come from the uncertainty in the structural model and of fault seal behavior; however, because of the time-consuming task of complex structural modeling, commonly, only one structural model is carried forward to dynamic reservoir simulation.

One of the aims of reservoir simulation is to perform a history match and thereby validate an individual model; however, because structural uncertainty is not commonly honored in dynamic simulation, the history match is normally obtained by modifying petrophysical parameters that represent the sediments and by giving the faults homogenous fault transmissibility multipliers (e.g., 1, 0.5, and 0). These modifications are not often based on geological input or knowledge and therefore tend to be structurally unrealistic. As a consequence, these models are able to match the current-day history but then, 3 months later, may have very poor predictive value. This study represents a continuation of the previous work by England and Townsend (1998), and the major advances from their work are the sophistication of the methodology, the inclusion of eight as opposed to four structural parameters, and the use of a real instead of a synthetic reservoir: a Brent Group gas field in the Southern Viking Graben (Norwegian North Sea).

Modern technology used for interpreting three-dimensional (3-D) seismic data, advanced software for modeling reservoir geometries in 3-D, performing fault seal calculations, and visualization tools are impressive and form the basic tools for the modern-day oil company geoscientist. They improve the possibilities for integrated workflows and allow increased understanding of the nature of the subsurface geology, but the 3-D model can never be more accurate than the accuracy of the input data. The outcome of modeling may actually mislead us to have too much faith in our models and, hence, divert the necessary focus of understanding the weaknesses of the model that is necessary to evaluate total uncertainty in reservoir behavior and the economic uncertainty.

We believe that structural uncertainty modeling is a necessity, because input data, data quality, and methodology commonly give nonunique answers for geological model building and reservoir fluid-flow description. The proposed workflow for the structural uncertainty modeling in this chapter aims to bring the structural uncertainty forward into the reservoir simulation model to better quantify the economical consequences of the structural uncertainty and to improve history matching by increasing the geological input used when performing fault seal calculations.

STRUCTURAL UNCERTAINTIES

The three basic components of the structural uncertainty modeling are uncertainty in (1) horizon geometry, (2) fault geometry, and (3) fault seal parameters. In

TABLE 1. "Total Structural Uncertainty" is defined here and is affected by a multitude of geological parameters. VSH = rock volume of shale; SGR = shale gouge ratio.

	Causes	Uncertain Parameters	Input to Structural Uncertainty Modeling
Horizon geometry	seismic data collection and processing	horizon time interpretation	different realizations of horizon picking
	choice of structural model	velocity model	time-shifted horizons because of picking problems
	picking of reflector		different realizations because of different velocity models
	depth conversion		
Fault geometry	seismic data collection and processing	fault density	different realizations varying the fault density and fault pattern
	choice of structural model	fault pattern	input on how to vary dip, position, and throw
	picking of faults	fault dip	
	seismic resolution	fault position	
	associated drag	fault throw	
	throw distribution on one or several faults		
Fault seal parameters	knowledge of fault behavior	VSH parameter modeling	different VSH realizations
	sedimetological model	fault seal methodology	different fault throw/thickness curves
		fault throw/thickness relationship	different fault permeability-SGR curves
		fault permeability-SGR relationship	different shale smear factors
		shale smear factor	

addition to these structural components, uncertainty associated with the type, geometry, and distribution of sediments and their petrophysical properties also affect the calculation of fluid flow across faults in the reservoir. Therefore, it needs to be stressed that a fully integrated approach is required for 3-D uncertainty modeling. Fully integrated modeling can be complex because several of the input parameters are intrinsically linked. Numerous realizations are required to capture the full range of possible outcomes, which can only be achieved using a stochastic workflow. However, because this commonly requires hundreds or even thousands of realizations, experimental design can be used to help perform an uncertainty study in a time-efficient way. The main parameters that need to be evaluated for a total structural uncertainty are outlined along with their sources in Table 1.

The number of variables in the structural uncertainty modeling increases with the increased ability to model fault-sealing properties. Present-day fault description and fault seal calculation is a severe simplification of the observed complex nature of faults in outcrop (i.e., Foxford et al., 1998; Hesthammer and Fossen, 2000). Fault description and seal estimation is, however, necessary to be able to realistically simulate fluid flow in faulted hydrocarbon-bearing reservoirs. Simple algorithms like shale gouge ratio (SGR) (Yielding et al., 1997) and shale smear factor (SSF) (Lindsay et al., 1993), combined with microtectonic core analysis of fault processes and property measurements (Fisher and Knipe, 1998, 2001), at least allow some differentiation of relative sealing capacities between and along different faults (Knipe, 1997; Knai and Knipe, 1998). The use of standard algorithms also allows a standard methodology to be implemented, so that the sealing behavior of different fields can be compared (Ottesen Ellevset et al., 1998). The uncertainty in the sealing algorithms and their different input parameters (Table 1) are, however, so large that investigation of these uncertainties is required.

Experience gained from fault seal modeling has shown that in reservoirs with horizontal shale barriers, the best results are achieved by combining both the SGR and SSF algorithms, and we use the combined method in this study on uncertainties. The method is such that if clay smear is expected to be present, this will override the SGR calculated fault permeability, and a transmissibility multiplier of zero will be used for that part of the fault zone. For the other parts of the fault zone, a fault permeability linked to the SGR value is used to calculate transmissibility multipliers. This factor can be very low but never zero.

THE STRUCTURAL UNCERTAINTY MODELING WORKFLOW

A command line-based, advanced fault modeling software was used in this study, and some of its functionalities has previously been described by Hollund et al. (2001). The software allows for manipulation of fault surfaces, gridded horizons, and reservoir-simulation grids. The manipulation of horizons includes some of the functionality described by Abrahamsen (1992). This study created and tested a workflow to use the horizon and fault manipulation functionality, which efficiently created multiple realizations of the reservoir-simulation grid by varying the structural parameters within the specified range of uncertainty. The workflow that was developed during this study has since been coded into software as an automated workflow and is described by Holden et al. (2003). Figure 1 highlights the proposed structural uncertainty modeling workflow, and step-by-step guidelines are outlined below.

1) Build a base case geological model, which includes seismic horizons, isochored geological horizons, and faults. In this case, the model was built in the depth domain, but a time domain model could also be used.
2) Create a parameterized fault set, meaning that the faults are no longer objects but a coordinate description with a fault displacement operator that can be applied to either surfaces or grids.
3) Unfault the reservoir horizons and build an unfaulted horizon model using the displacement functionalities.
4) Modify the unfaulted horizons according to alternative velocity models or within an uncertainty range that is related to a particular depth-conversion method or seismic interpretation. If the base case model is built in the time domain, the different velocity models are applied using an appropriate tool. If the model is built in the depth domain, then different maps representing the different velocity models can be applied, and stochastic realizations of the horizons can be generated.
5) Build an unfaulted grid for each realization of the unfaulted horizon.
6) Populate the unfaulted grids with sedimentary facies realizations.
7) Refault the unfaulted grids, allowing the parameters fault density, fault pattern, fault throw, fault length, fault position, and fault thickness to vary. The range of uncertainties in these parameters is predefined and controlled using an experimental design scheme (this is further explained in a separate section below).
8) For each structural and sedimentological realization, calculate fault transmissibility multipliers by varying the uncertainties related to the fault seal algorithms, which are rock volume of shale (VSH) parameter modeling, fault throw-thickness relationship, fault permeability-SGR relationship, and SSF cutoff value.

It must be added that several other workflows are possible using the structural uncertainty modeling software, including fully stochastic modeling of uncertainties.

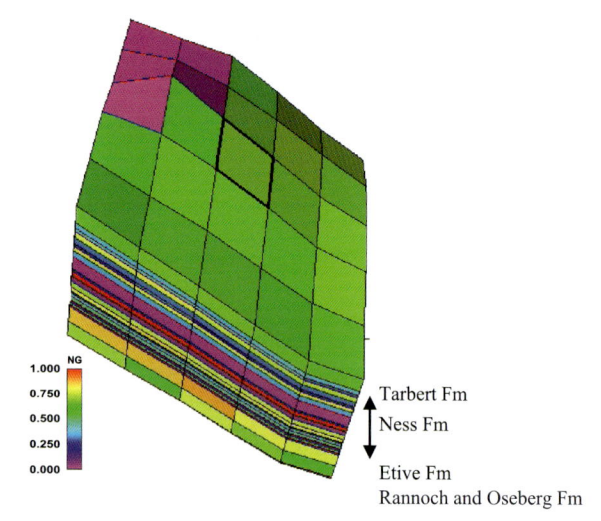

FIGURE 2. Distribution of net to gross (NG) in a few selected cells from the reservoir simulation model.

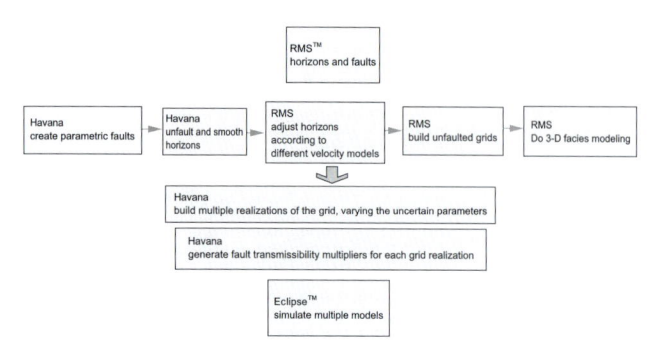

FIGURE 1. A flow diagram of the SUM workflow for investigation of structural uncertainty. The workflow integrates the RMS software for geological modeling, the Havana software for fault and uncertainty modeling, and the Eclipse software for the reservoir simulation.

STRUCTURAL UNCERTAINTY MODELING, CASE STUDY

The case study uses a Brent Group gas–condensate-producing reservoir from the North Sea, and an example of the net-to-gross distribution for this field is

TABLE 2. Modeled parameters and their variables used are listed here and described further in the text. (1), (0), and (−1) refer to the experimental design in Table 3.

Cases 1, 2, and 4	Open (1)	Medium (0)	Low (−1)
Fault density	low	medium	high
Fault throw	base	minimum 20	minimum 50
Fault length		medium	long
Fault position	+100 m (330 ft) updip	as is	−100 m (−330 ft) downdip
Fault throw/thickness relationship	1:200	1:100	1:50
Shale smear factor	3	5	7
Permeability/VSH relationship	high	medium	low
VSH grid	low −5%	medium	high +5%

shown in Figure 2. The stratigraphy consists of a thin eroded Tarbert Formation and uneroded Ness Formation, Etive Formation, and Rannoch Formation. The fault resolution of the seismic data varies but is especially poor at the crest of the structure because of a combination of depth-to-reservoir gas effects in the overburden and multiple energy in the seismic response. The resolution, at best, is estimated to be about 20 m (66 ft) and possibly decreasing to 50 m (160 ft), where the seismic quality is at its poorest. The uncertainty in fault interpretation in the field is therefore considered to be large, and in the early well planning and development phase, it was necessary to estimate the possible production behavior of different fault model scenarios.

Several fault variables were selected and modeled with 20 realizations using the structural uncertainty modeling software. The structural uncertainties addressed in this case study can be divided into two groups: fault and horizon geometry and fault properties. The modeled parameters and their associated uncertainty range are listed in Table 2 and are described in more detail below.

Fault Density

The seismically detectable faults of the field were divided into three categories based on the certainty with which they could be picked. This resulted in three different fault densities, low, medium, and high, that represent a certain, uncertain, and speculative fault pattern, respectively (Figure 3). Examples of the fault categories are shown in the seismic cross sections in Figure 4. The low-density fault model contains only faults that can be picked out on the seismic cross sections with certainty. Because of the rotated nature of the main fault blocks (about 20°) and the large fault throw on the western-bounding fault, it is expected that more faulting should be present. The medium-density fault pattern includes more uncertain near-seismic-resolution faults. The high-density fault pattern contains most interpreted faults, including the presence of near-seismically resolvable faults. However, the poor quality of the seismic data leads to the expectation that some of these features may well be seismic artifacts.

Fault Throw

A result of the poor vertical seismic resolution in the case study data set is that the fault throws from seismic interpretation are very uncertain. At the crest of the structure, the resolution may be as low as 50 m (160 ft), implying that faults barely identifiable on the seismic cross sections may have fault throws of as much as 50 m (160 ft). To test the effect of this uncertainty hypothesis, three scenarios were made using the following fault-throw distribution:

- keeping the interpreted throw values with some unrealistically low values (down to 0.5 m [1.6 ft])

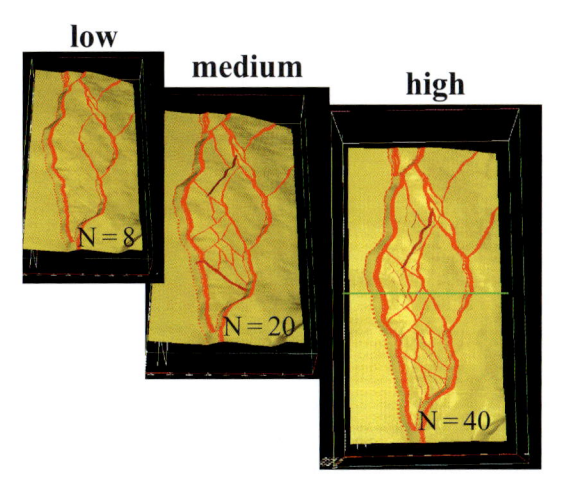

FIGURE 3. Uncertainty in fault pattern and fault density. Green line shows location of Figure 4, and N defines the number of faults.

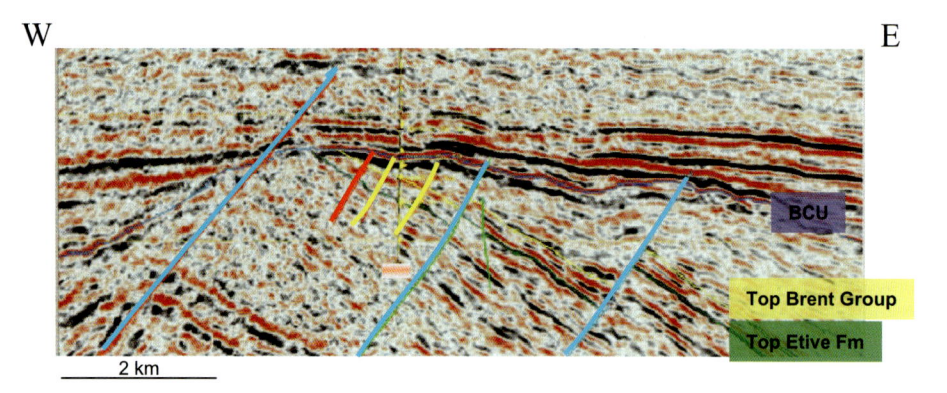

FIGURE 4. Seismic cross section through the structure showing examples of the ranking of faults: certain faults are blue, uncertain faults are yellow, and speculative faults are red. (The thin green fault was not included in the modeling.) Location of the line is shown in Figure 3. The structure is eroded at the crest by the Base Cretaceous Unconformity (BCU).

- all faults having a minimum fault throw of 20 m (66 ft)
- all faults having a minimum fault throw of 50 m (160 ft)

Fault Position

Generally, the location of faults from seismic interpretation is imprecise, and calibration with drilled faults has proven this. The three main sources of error are seismic migration, seismic processing, and interpreted fault position. To evaluate some of the effects of mispositioning, three alternative scenarios were given for fault position:

- Keep the position as originally interpreted.
- Move the faults 100 m (330 ft) updip.
- Move the faults 100 m (330 ft) downdip.

Fault Length

Fault tips mapped from seismic data should indicate the location where the throw is equal to the seismic resolution. As a consequence of this, there should be a portion of each fault that is unmapped, because the throw at this portion of the fault is below seismic resolution. The uncertainty related to this phenomenon has been considered by allowing the fault lengths to be either modeled as originally mapped or modeled with a 200-m (660-ft) extension to each fault tip.

Fault Thickness/Throw

The fault thickness/throw ratio is part of the fault seal transmissibility calculation, and a large uncertainty is related to this factor. In general, a linear relationship is present in fault thickness/throw, but a large degree of variation exists (e.g., Evans, 1990), and in some cases, this relationship is regarded as totally unreliable (Foxford et al., 1998). For this study, the uncertainty is represented by the following three relationships, although we openly acknowledge that this does not cover the full range of uncertainty: 1:200, 1:100, and 1:50.

Shale Smear Factor

The SSF is used according to Lindsay et al. (1993) to describe over what length a shale layer forms a continuous tongue when being displaced along a fault surface. The length of this tongue is assumed to be related to the thickness of the shale layer by the formula

$$L = bT^a$$

where L is the length of the tongue; T is the thickness of the shale layer; and a and b are facies-specific parameters.

If such a tongue covers an area along the fault plane, the fault transmissibility becomes zero; otherwise, it remains unchanged. In this study, $a = 1$ and b is set to 3, 5, or 7. The higher the b value is set and the thicker the shale layer, the larger the area of zero transmissibility becomes.

Fault Permeability-SGR Relationship

Fault permeability has been measured on small faults and deformation bands using core plug measurements, and in addition, the mineral content of

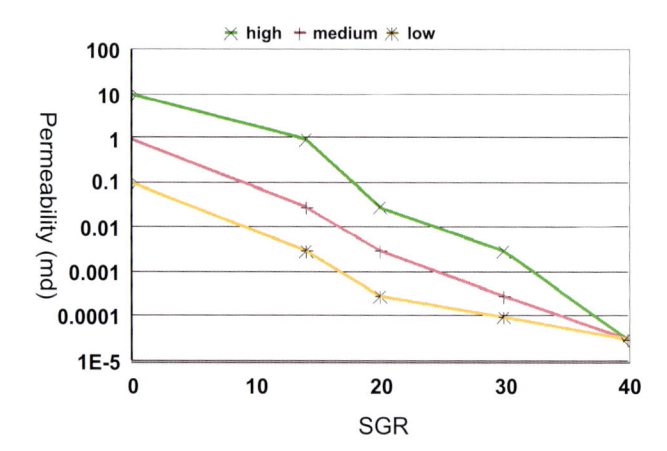

FIGURE 5. The three curves define different permeability-SGR relationships that are used as input for the transmissibility calculations.

TABLE 3. Experimental design is used to reduce the number of experiments from 512 to 20. 0 is medium value, 1 is high value, and −1 is low value as listed in Table 2.

	Experiments	Fault Density	Throw	Length	Position	Throw/ Thickness	Shale Smear Factor	Permeability/ VSH	VSH	Recovery Factor (%) all wells
1	open case	1	0	0	0	1	1	1	0	50
2	medium	0	0	0	0	0	0	0	0	44.7
3	complex	−1	0	0	0	0	0	0	0	42.6
4	low	−1	−1	0	0	−1	−1	−1	0	34.3
5	run 5	−1	−1	1	−1	1	1	1	−1	39.4
6	run 6	−1	1	1	−1	−1	−1	1	1	51.1
7	run 7	1	1	1	1	1	1	1	1	49.2
8	run 8	−1	1	0	−1	1	1	−1	−1	47.0
9	run 9	1	−1	1	1	−1	−1	1	−1	50.8
10	run 10	−1	1	0	1	1	−1	1	−1	49.9
11	run 11	1	1	0	1	−1	−1	−1	−1	51.0
12	run 12	1	1	1	−1	1	−1	−1	−1	50.0
13	run 13	1	1	0	−1	−1	1	1	1	51.6
14	run 14	1	−1	1	−1	−1	1	−1	1	51.8
15	run 15	−1	1	1	1	−1	1	−1	−1	41.4
16	run 16	1	−1	0	1	1	1	−1	1	51.1
17	run 17	1	−1	0	−1	1	−1	1	−1	51.7
18	run 18	−1	−1	0	−1	−1	−1	−1	1	37.9
19	run 19	−1	−1	0	1	−1	1	1	−1	36.2
20	run 20	−1	−1	1	1	1	−1	−1	1	34.4

the fault rocks were also analyzed. A relationship was identified between the permeability and the clay content of these small faults, which was then used to predict fault permeability in the models. Although this is the best available method, it should be noted that it does not include the effect of fault slip plane development or the complexity commonly observed in fault zones with seismic-scale displacements. Three different curves for permeability vs. SGR have been used in this case study and are shown in Figure 5.

Experimental Design and Sensitivities

Factorial experimental designs allow a high number of geological scenarios to be analyzed using a specifically sampled subset. This can save a significant amount of time in both the modeling process and in the simulation of the different scenarios. The implications of using experimental design is based on the modeling of a combination of several input parameters that all have a selected number of confidence levels (e.g., high and low) and a set of scenarios that are a combination of parameters and confidence levels selected from the full-factorial design (i.e., all possible scenarios). The analysis of the results of the different outcomes should determine the relative importance of each input parameter and a representative estimate of the spread in the uncertainty. Clearly, this method will not replace running a full-factorial design, but the results will lie within specific confidence limits.

In this study, eight parameters were analyzed, with most of the parameters having three confidence levels as defined in Table 2. Given these input parameters, a full-factorial design would require 512 possible outcomes to be modeled, whereas the planned experimental design (in Table 3) resulted in only 20 outcomes being modeled and run. This is expected to be sufficient to produce meaningful results. The experimental design used focuses on the open and the low level, whereas only a few experiments use the medium-level parameters. This allows us the opportunity to minimize the number of experiments, but we are still able to define the extreme effects of the variables and also compare the outcomes with a close to base case scenario.

Experiments 1–4 in Table 3 were deterministic models. In experiment 1, all the variables in the open column in Table 2 are combined in one model, whereas in experiments 2 and 4, all variables in the medium and low columns are used, respectively. The complex model (experiment 3) is nearly identical to the medium model, except that the dense fault pattern has been used in this experiment. Although the 20 runs from the factorial experimental design could have also been analyzed to determine the relative effect of each

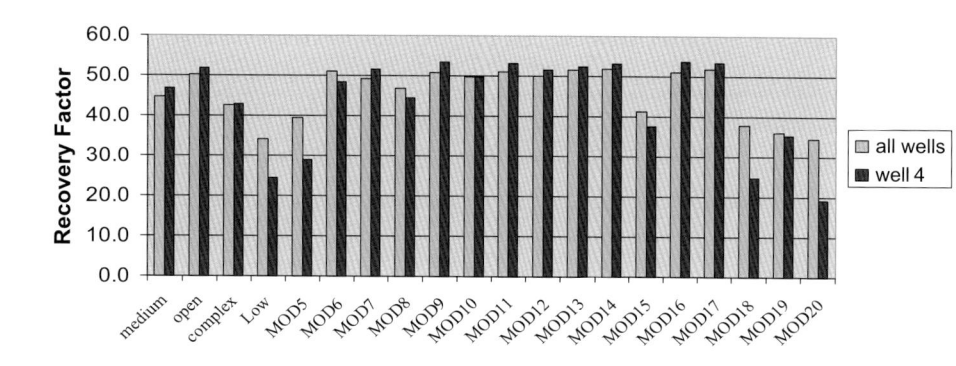

FIGURE 6. Recovery factors for the different realizations.

parameter, they were not used because of time constraints and the lack of access to skills in experimental design. Instead, an additional 12 realizations were built for parameter-sensitivity analysis. In these realizations, only one fault parameter were changed per realization, and the results were compared with the medium case realization from the experimental design.

The Effect on Gas Recovery

The gas recovery factor is a dimensionless parameter used to evaluate gas production, and in this study, it is used for an evaluation of the effect of fault parameters on production. Recovery factors are compared in Figure 6. The recovery factor varies by 17.7, with a spread from 51 to 34.3.

For all experiments, a single, 3-D sedimentological model was used, and therefore, the only effects that can be assessed are the fault parameters. However, these new realizations could be compared with the original model built during the field development stage, where layer-cake stratigraphy was used and not more advanced 3-D facies modeling. The difference between the layer-cake model and the closest faulted realization of the 3-D facies model was significant, indicating that the facies model, by itself, has a major influence on predicting reservoir performance.

Water production was also identified as a major factor in the reservoir performance of the field. The recovery from each run was significantly affected if water-breakthrough occurred. Therefore, to avoid the effect of water production, the recovery from the well least influenced by water production (well 4) was investigated in detail.

The results from well 4 (Figure 6) showed similar trends to the total production, but the effect of fault seal is greater. Well 4 shows a general increase in production with either fewer and/or more open faults, and this effect can also be seen in the first four runs: medium, open, complex, and low. Runs 1–4 almost cover the total range of the uncertainty, with a recovery factor between 34.3 and 50.3%. As these runs cover the range of uncertainty, they alone were used for further analysis. Compared with the medium run, the open-run recovery factor is 6% greater, whereas in the low run, the recover factor is reduced by 11%.

Detailed analysis of the runs revealed that the faults effect the production differently in the different stratigraphic zones. This is interpreted to be caused probably by a combined effect of different fault seal properties in the different reservoir layers and the proximity to the free water level. A high fault density results in a general increase in recovery factor in the Etive Formation (Figure 7). The interpretation of this is that the faults do not completely seal, slowing down the rate at which the water level advances toward the production wells. For the Ness Formation, increasing the number of faults has a negative effect on recovery. Here, a significant component of clay- and coal-rich material is present in the formation, causing very tight fault seals. Therefore, isolated compartments are likely to form that leave behind pockets of unswept hydrocarbons.

Sensitivities

The result of the sensitivity modeling is shown in Figure 8. Explicit modeling of the fault displacement in the grid and the fault seal potential implemented as heterogeneous fault transmissibility multipliers reduces the recovery factor by 7 (Figure 7) when compared with an identical model but without fault transmissibility multipliers. The fault density is the single most influential factor and may account for a variation in recovery factor of as much as 9, although this is a

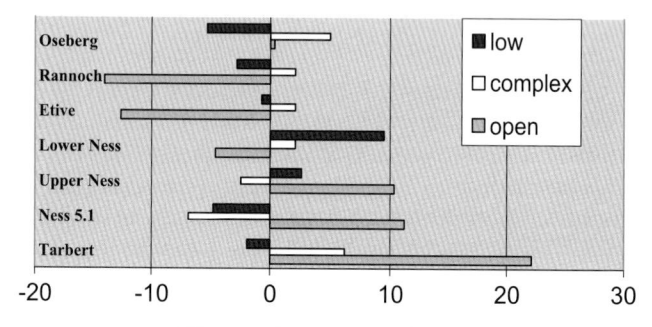

FIGURE 7. The plot shows the relative influence of different fault models on the recovery factor for different intervals of the reservoir. All models are compared with the medium model.

FIGURE 8. Results of the sensitivity analysis, showing absolute variation in the recovery factor as a function of varying one single parameter compared with a medium run.

gas field and the faults are not completely sealing. The second most influential factor is the choice of fault permeability, which, in this case, potentially affects production by as much as 3%. One may be surprised to find that the effect from varying the different SSF cutoff values is minimal, but this may be case specific and should be tested further on other sedimentary sequences.

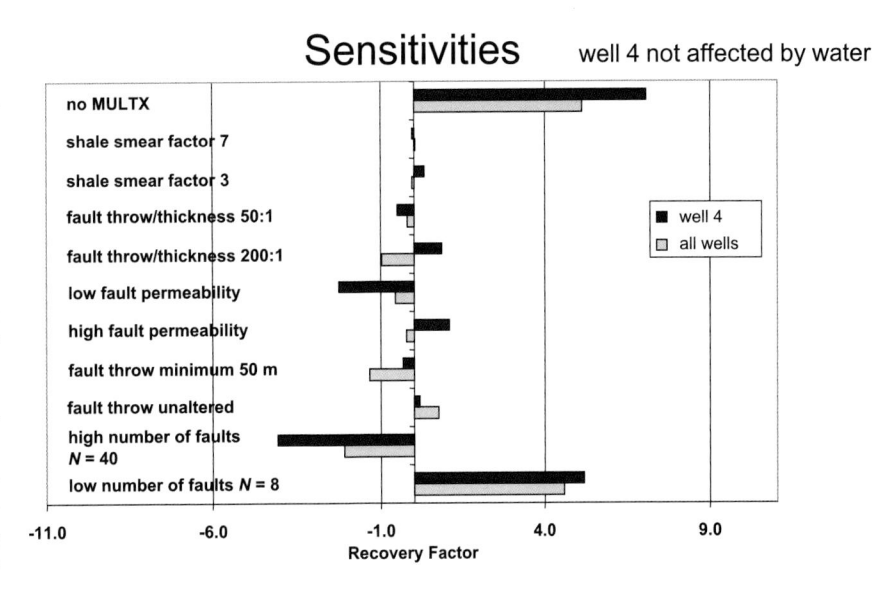

The Effect on Pressure Development

Pressure development is the most important factor in history matching, because it is commonly the most difficult to reproduce. If a good history match on pressures can be found, then the characteristics of that particular realization may give us some important information about the true properties of the reservoir. When this study was carried out, no historical information was available. However, the field came on stream a short while after, and the initial production immediately enabled an improved understanding of the reservoir, because it fell within the predicted ranges determined in this study.

Pressure development curves are plotted for four different fault segments and for runs 1–4 (Figure 9). Based on these pressure plots, it is possible to observe the following:

- Some degree of pressure segmentation exists in all realizations.
- Only the faults in the low-run realization create fault segments with no pressure communication.
- Adding more faults results in poorer pressure communication in the complex run when compared with the medium run, even if the faults do not strongly seal.
- Pressure differences between segments in the range of several hundred bars should be expected from these realizations.

DISCUSSION

As mentioned, this study was carried out before the field was put into production, and no historical production information was available. When the field came on stream a short while after, the pressure dropped rapidly in the wells, clearly indicating that the field is compartmentalized. Even if uncertainty associated with the production behavior, sedimentological models, and erosion still existed, we were able to use the known pressure development from this uncertainty study, and by comparison with the production pressures, we could conclude that the observed pressures indicate

- a large sealing capacity of the segmenting faults
- less compartmentalizing faults than in the realization with high fault density
- that the observed pressure behavior indicates better production conditions than our low case

Because of the knowledge provided by the structural uncertainty study, the initial production immediately allowed for an improved understanding of the reservoir.

The described workflow is only one possible workflow for the structural uncertainty modeling. Other workflows are possible by the use of the structural uncertainty modeling method (Holden et al., 2003), and it is also encouraging to see other publications on the use of other methods and tools for the structural uncertainty modeling (i.e., Thore et al., 2002). However, we do not believe that the method presented in this article or those of Thore et al. (2002) have solved all of the issues. These should be considered as initial studies highlighting the problems and possibilities in modeling structural uncertainties in hydrocarbon reservoirs. Ideally, the structural uncertainty modeling should be integrated with total field uncertainty, and especially, the uncertainty in sedimentary and tectonic heterogeneities should be modeled and investigated at the same level of detail. Bailey et al. (2003) is starting on this important work by investigating the effects on

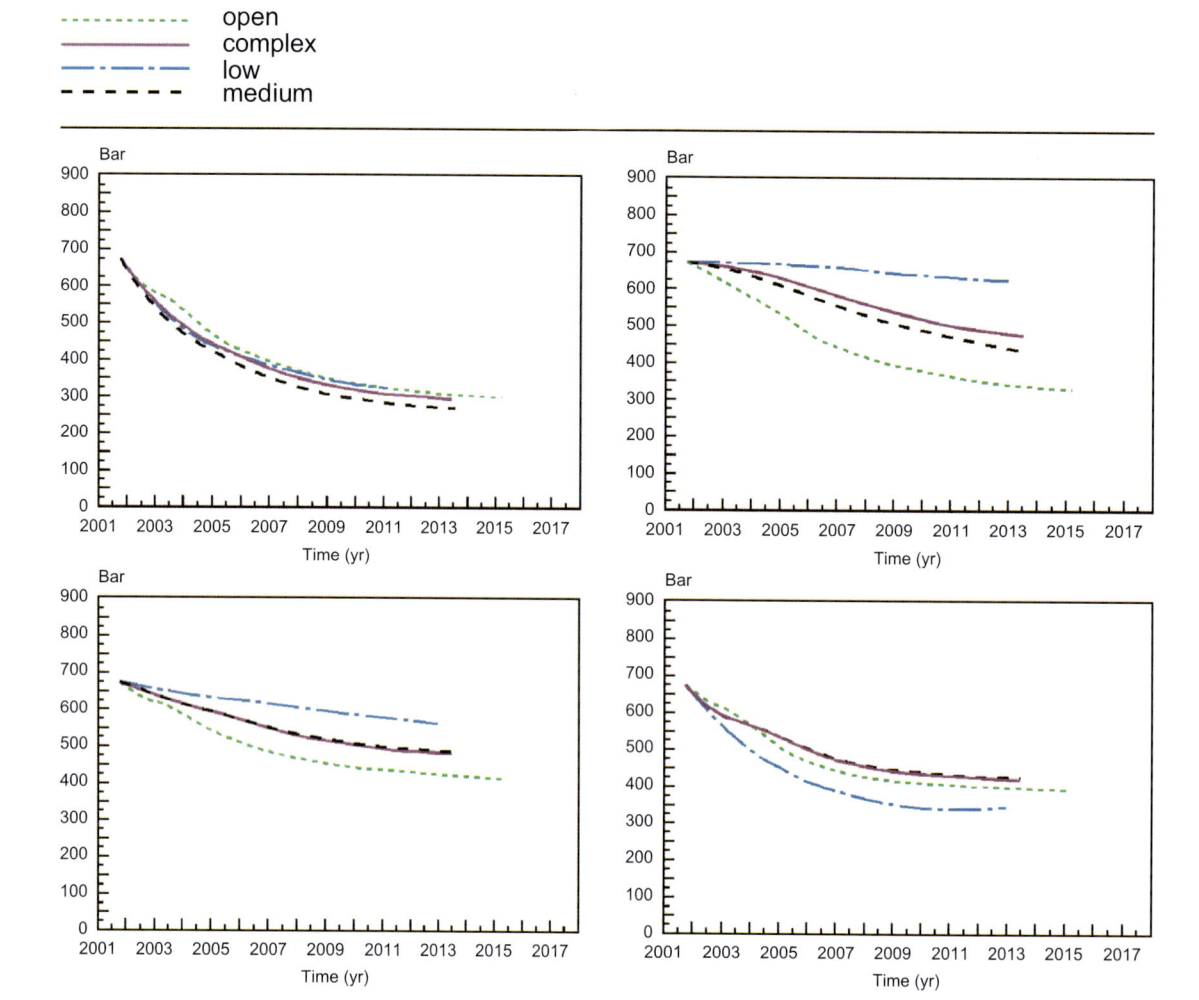

FIGURE 9. Pressure plots for four different fault segments and for four different realizations of the structural model.

field connectivity by varying both fault and channel parameters, but unfortunately, their study did not include fault property calculation and fluid-flow modeling.

Probably, the most important result of this study was the recognition that fault density has the most significant effect on both hydrocarbon productivity and water flow in the reservoir (Figure 7). It must be stressed that this applies for faults with a low permeability, but which do not completely seal. This is in general agreement with several other published studies, but it is not always fully appreciated by geoscientists and reservoir engineers. England and Townsend (1998) concluded that fairly open, noncontinuous subseismic faults created an improved sweep in the reservoir and delayed water production. In a detailed 3-D modeling study on the Gullfaks field in the Norwegian sector of the northern North Sea, the interpreted faults were assigned transmissibility multipliers calculated by the use of SGR algorithms. Initial history matches based on only the largest 150 faults were generally poor, and only with the inclusion of 300 near-seismic resolution faults, an acceptable history match was reached (Jacobsen et al., 2000). However, this generally perceived conclusion that the realistic inclusion of all faults above seismic resolution will lead to improved reservoir forecasting seems to contradict one of the main conclusions in Damsleth et al. (1998, p. 295), where they stated that "subseismic faults have negligible effect on recovery or on the shape of the production profile, unless they are almost completely sealing." The term "almost completely sealing" may be misleading here. More precisely, it was shown that faults with transmissibility multipliers of 0.01 did not give any effect, but transmissibility multipliers of 0.0005 did. We can compare this with the multipliers used for the Gullfaks field and find that apart from where there is self-juxtapostition of the Tarbert Formation, the major part of the fault gouge is estimated to be phyllosilicate framework fault rocks with multipliers of 0.001 and below. This shows that these studies are not contradictory, and we believe that transmissibility multipliers of 0.001 to below 0.0005, although low, may actually be a realistic representation of the sealing effects of phyllosilicate framework fault rocks.

It must be emphasized that fault seal and fluid flow in and across fault zones are difficult to predict because of the large and rapid variation in fault rock properties and fault rock thickness seen in analog field data (Foxford et al., 1998). Core material of large faults in reservoirs is very sparse and does not exist in the case studied. In most reservoirs, assumptions are made about larger faults by studying core-scale faults. However, the larger seismic faults could have physical properties different from the small faults measured. Moreover, physical constraints are lacking in the current method for calculating transmissibility multipliers, because it is not yet possible to model the relative permeability in the fault zones (Manzocchi et al., 2002). This current study has, however, used the best available technology for the modeling of fault transmissibility multipliers and provides a range of reservoir behavior dependent on structural uncertainty. In Lia et al. (1997), the sealing capacity of faults is, by far, the most influential reservoir characteristic. Three different scenarios were presented in their paper, with (1) completely sealing, (2) perfect flow across faults, and (3) partial flow across the fault zones. The difference between the favorable and the unfavorable scenarios was approximately 15% of the total oil production. The results of the case study presented in this article also support this conclusion, because it has been shown that when comparing faults that have been modeled as completely open with those assigned a realistic sealing effect, a significant effect is present on the simulation results, i.e., the recovery factor changes by 5.3 (Figure 7). Another interesting observation is the effect on the reservoir production of the individual parameters that control the way the fault transmissibility is calculated. These all had very little effect on the total recovery of the reservoir, with the possible exception of fault permeability. These results emphasize the importance of fault density and including faults with associated realistic permeability values.

We know from outcrop studies (i.e., figure 4 in Bailey et al., 2003) that faults of many different scales exist, and we expect subseismic faults to exist both as swarms in the fault damage zone of the seismically resolvable faults and as more isolated subseismic faults distributed in the reservoir. The question should therefore be posed: If the realistic modeling of all seismic-scale faults is important, then what about the inclusion of subseismic-scale faults, especially if these faults are modeled with a sealing component? One must remember that in deep fields, fault throw on subseismic faults may be as large as seismically resolvable faults in fields of shallower depth. Much attention was previously paid to document the existence, numbers, and fractal dimensions of these faults (i.e., Yielding et al. 1992), and tools are available with algorithms for modeling subseismic faults with petrophysical properties. This topic was not investigated during our case study but should be addressed in further studies. The subseismic faults in the damage zone may be modeled using fault statistics from well log data, whereas 3-D stress analysis is necessary to predict likely locations for more isolated subseismic faults or fault zones.

CONCLUSIONS

The structural uncertainty modeling workflow presented here allows for the structural uncertainty modeling to be completed in an acceptable timeframe of oil-company model building.

The case study clearly demonstrates that the fault parameters are important and should be modeled, and that the effect of their uncertainty on recovery should be investigated. The production in this field may vary by as much as 17% because of the estimated uncertainty in the structural input to the model. To improve reservoir management, this quantification of uncertainty must be seen as important, and effort should be focused on reducing the uncertainty of the most influential factors. Early production data provide insight into the range of models that may match the historical information.

In terms of the significance of individual parameters, fault density is the single, most influential factor and varies the recovery factor by as much as 9, although this is a gas field, and the faults do not completely seal. This also points to the importance of including near-seismic resolution faults and subseismic faults in this type of reservoir. The second, most influential factor is the choice of fault permeability, which, in this case, can affect production by as much as 3%. By explicitly modeling fault seal potential, the recovery factor drops by as much as 7.

However, the importance of this work does not only lie in the figures but in the recognition that structural heterogeneity should not be subordinated. Naturally, different results would be expected in other fields whose reservoirs were formed in different environments of deposition. This is also indicated by Lescoffit and Townsend (2005) and highlights the need for complimentary studies where the effect of faults in various sedimentary environments is examined. The value of this type of study should, however, not be underestimated in helping to determine where further work should be concentrated.

ACKNOWLEDGMENTS

Thanks to geophysicist Marit Moxnes, who classified the faults. Also, thanks to all the great colleagues in the Reservoir and Uncertainty Modeling project in Statoil-Trondheim, in particular, Alfhild Lien Eide, who did the 3-D facies modeling for this project, Jan Ole

Aasen, who set up the experimental design, and Guillaume Lescoffit, whose technical support has been invaluable. Thanks also to Kevin Keogh, who helped us with the written language. Thanks to Rock Deformation Research Group, Leeds University for their enthusiastic analysis of the cored faults, giving input on fault permeability and deformation processes, and to the Norwegian Computing Center, Jon Gjerde, Lars Holden, and Knut Hollund, who designed the HAVANA Structural Uncertainty Modeling software to our specifications and made this study a possibility. Also thanks to Statoil for the permission to publish and to the referees K. Hill, D. Graules, and P. Boult, whose valuable comments made the chapter more attractive to the reader.

REFERENCES CITED

Abrahamsen, P., 1992, Bayesian kriging for seismic depth conversion of a multi-layer reservoir, in S. Amilcar, ed., Proceedings from the Fourth International Geostatistical Congress, Troia, Portugal, September 13–18: Geostatistics Troia '92: Dordrecht, The Netherlands, Kluwer Academic Publishers, v. 1, p. 385–398.

Bailey, W. R., T. Manzocchi, J. J. Walch, K. Keogh, D. Hodgetts, J. Rippon, P. A. R. Nell, S. Flint, and J. A. Strand, 2003, The effect of faults on the 3-D connectivity of reservoir bodies: A case study from the east Pennine coal field, U.K.: Petroleum Geoscience, v. 8, p. 263–277.

Damsleth, E., V. Sangolt, and G. Aamodt, 1998, Subseismic faults can seriously affect fluid flow in the Njord field off western Norway— A stochastic fault modeling case study: Society of Petroleum Engineers Paper 49024, p. 295–304.

England, W. A., and C. Townsend, 1998, The effects of faulting on production from a shallow marine reservoir— A study of the relative importance of fault parameters: Society of Petroleum Engineers Paper 49023, p. 279–294.

Evans, J. P., 1990, Thickness-displacement relationships for fault zones: Journal of Structural Geology, v. 12, no. 8, p. 1061–1090.

Fisher, Q. J., and R. J. Knipe, 1998, Fault sealing processes in silisiclastic sediments, in G. Jones, Q. J. Fisher, and R. J. Knipe, eds., Faulting, fault sealing and fluid flow in hydrocarbon reservoirs: Geological Society (London) Special Publication 147, p. 117–134.

Fisher, Q. J., and R. J. Knipe, 2001, The permeability of faults within siliciclastic petroleum reservoirs of the North Sea and Norwegian continental shelf: Marine and Petroleum Geology, v. 18, p. 1063–1081.

Foxford, K. A., J. J. Walsh, J. Watterson, I. R. Garden, S. C. Guscott, and Burley, S. D, 1998, Structure and content of the Moab fault zone, Utah, U.S.A., and its implications for fault seal prediction, in G. Jones, Q. J. Fisher, and R. J. Knipe, eds., Faulting, fault sealing and fluid flow in hydrocarbon reservoirs: Geological Society (London) Special Publication 147, p. 269–282.

Hesthammer, J., and H. Fossen, 2000, Uncertainties associated with fault sealing analysis: Petroleum Geoscience, v. 6, p. 37–45.

Holden, L., P. Mostad, B. F. Nielsen, J. Gjerde, C. Townsend, and S. Ottesen, 2003, Stochastic structural modeling: Mathematical Geology, v. 35, no. 8, p. 899–914.

Hollund, K., P. Mostad, B. F. Nielsen, L. Holden, J. Gjerde, M. G. Contursi, A. J. McCann, C. Townsend, and E. Sverdrup, 2001, HAVANA— A fault modeling tool, in Hydrocarbon seal quantification: Norwegian Petroleum Society Special Publication 11, p. 157–171.

Jacobsen, T., H. Augustsson, J. Alvestad, P. Digranes, I. Kaas, S. T. Opdal, 2000, Modelling and identification of remaining reserves in the Gullfaks field: Society of Petroleum Engineers Paper 65412, 28 p.

Knai, T. A., and R. J. Knipe, 1998, The impact of faults on fluid flow in the Heidrun field, in G. Jones, Q. J. Fisher, and R. J. Knipe, eds., Faulting, fault sealing and fluid flow in hydrocarbon reservoirs: Geological Society (London) Special Publication 147, p. 269–282.

Knipe, J. J. 1997, Juxtaposition and seal diagrams to help analyze fault seals in hydrocarbon reservoirs: AAPG Bulletin, v. 81, no. 2, p. 187–195.

Lescoffit, G., and C. Townsend, 2005, Quantifying the impact of the fault modeling parameters on production forecasting for clastic reservoirs, in P. Boult and J. Kaldi, eds., Evaluating fault and cap rock seals: AAPG Hedberg Series, no. 2, p. 137–149.

Lia, O., H. Omre, H. Tjelmeland, L. Holden, and T. Egeland, 1997, Uncertainties in reservoir production forecasts: AAPG Bulletin, v. 81, p. 775–802.

Lindsay, N. G., F. C. Murphy, J. J. Walsh, and J. Watterson, 1993, Outcrop studies of shale smears on fault surfaces, in S. Flint and A. D. Bryant, eds., The geological modelling of hydrocarbon reservoirs and outcrop analogues: International Association of Sedimentologists Publication, v. 15, p. 113–123.

Manzocchi, T., A. E. Heath, J. J. Walsh, and C. Childs, 2002, The representation of two-phase fault-rock properties in flow simulation models: Petroleum Geoscience, v. 8, p. 119–132.

Ottesen Ellevset, S., R. J. Knipe, T. Svava Olsen, Q. J. Fisher, and G. Jones, 1998, Fault controlled communication in the Sleipner Vest field, Norwegian continental shelf; detailed, quantitative input for reservoir simulation and well planning, in G. Jones, Q. J. Fisher, and R. J. Knipe, eds., Faulting, fault sealing and fluid flow in hydrocarbon reservoirs: Geological Society (London) Special Publication 147, p. 283–297.

Thore, P., A. Shtuka, M. Lecour, T. Ait-Ettajer, and R. Cognot, 2002, Structural uncertainties: Determination, management, and applications: Geophysics, v. 67, no. 3, p. 840–852.

Yielding, G., J. J. Walsh, and J. Watterson, 1992, The prediction of small-scale faulting in reservoirs: First Break, v. 10, p. 449–460.

Yielding, G., B. Freeman, and D. T. Needham, 1997, Quantitative fault seal prediction: AAPG Bulletin, v. 81, p. 897–917.

Quantifying the Impact of Fault Modeling Parameters on Production Forecasting for Clastic Reservoirs

Guillaume Lescoffit
Statoil Research and Technology, Trondheim, Norway

Chris Townsend[1]
Statoil Research and Technology, Trondheim, Norway

ABSTRACT

Recent developments in structural modeling allow the detailed representation of the fault geometries, and their impact on reservoir behavior can be fully assessed. However, it is unclear how much the most advanced developments, such as subseismic fault and fault-seal modeling, contribute to a better understanding of a field. This study was designed to set up a workflow for assessing the impact of structural uncertainty and to test the workflow on a synthetic model to analyze the production response to this uncertainty.

A synthetic oil reservoir model was built, using a simplified tilted block structure, which was then modified with different groups of faults, and three different sedimentary models for building the facies and property model. Two experiments were conducted, where both sedimentary model and structural settings would vary. Experiment 1 included the sedimentary model as a varied input, whereas experiment 2 focused on the structural factors in each defined sedimentary model.

The models were built using a combination of RMS™ and Havana software systems, and then the results were imported into a reservoir simulator, Eclipse™. Production results were extracted and analyzed using statistical methods, to determine the impact of each input variable.

The results pointed out that apart from varying reservoir properties, some of the structural factors had a significant impact on production history (in particular fault-seal model and fault pattern). In addition, experiment 2 showed that different sedimentary settings do not respond in the same way to different structural settings, thus demonstrating the potential value of conducting this type of study in the early phases of field development.

[1]*Present address*: Shell EP Nederlanse Aardolie Maatschappij, Assen, The Netherlands.

Copyright ©2005 by The American Association of Petroleum Geologists.
DOI:10.1306/1060761H23166

INTRODUCTION

Over the past decade or so, it has been realized that including more realistic geological scenarios in reservoir models leads to better reservoir performance prediction and history matching. Much of this effort in geological modeling has concentrated on sedimentologic and stratigraphic reservoir description (e.g., Henriquez and Jourdan, 1995). This has been the focus of effort because a large knowledge base already existed, and there was a common belief that these elements would have the dominant effect on reservoir performance. Several studies have been published that have investigated the effects of the various modeling parameters on reservoir simulation results (Kjønsvik et al., 1994; Jones et al., 1995; Brandsæter et al., 2001). These studies have attempted to build realistic geological models and tested different scenarios using experimental design to minimize the number of simulations and maximize the confidence in the results. Considerably less effort has been put into determining the effects of fault description parameters. Walsh et al. (1998) looked at the effects of subseismic-scale faults and concluded that they had little impact on upscaled permeabilities, whereas England and Townsend (1998) added faults of a subseismic scale to the models of Kjønsvik et al. (1994). Relative effect of the added faults was compared to the sedimentary parameters, and a small number of the fault description parameters were examined.

The effect of faults have sparsely been modeled in a realistic manner because of a lack of good modeling tools and a poor understanding of fault systems. During the past few years, it has been increasingly recognized that faults commonly have a significant effect on reservoir simulations (e.g., Knai and Knipe, 1998). The handling of faults in past reservoir modeling typically would have involved the inclusion of a few of the larger structures, and their properties were modeled as either fully open or totally sealing. Smaller, seismically mapped faults and concepts of subseismic-scale faults have sparsely been considered. During the early 1990s, several fault property studies were carried out (e.g., Knipe, 1992), which led to the suggestion that it is too simplistic to treat faults as only totally open or sealing, and that fault plane permeability needs to be included. Measurements of fault permeabilities indicated a general reduction in permeability in relation to the host rock by a factor of between 0.1 and 0.0001 for sandstone reservoirs. Other fault-sealing studies investigated the distribution of fault rock types along fault planes, leading to suggestions that fault lithologies were highly heterogeneous, and that this variation in fault rock properties might be related to the sedimentary rocks that are faulted (e.g., Lindsay et al., 1993; Yielding et al., 1997; Manzocchi et al., 1999). Knai and Knipe (1998) and Ottesen et al. (1998) were among the first to show how these fault-seal analysis techniques can be applied to reservoir simulation models and can significantly improve results. More recent work has concentrated on the automation of fault-seal quantification and the development of tools that can include this in the form of transmissibility multipliers ready for use in dynamic models (Manzocchi et al., 1999; Hollund et al., 2002).

In addition to fault-seal modeling, three-dimensional (3-D) modeling tools have advanced in their capabilities and can now build reliable 3-D structural models. These contain faults as integral elements where cells adjacent to fault surfaces are separated from cells of the same layer, but on the opposite side of the fault, by a distance representing the fault displacement. Moreover a tool called Havana, which was originally built to model subseismic-scale faults (Munthe et al., 1993), has been developed to model seismic-scale faults using its unique capabilities for displacing horizons and 3-D grids according to a parameterized displacement operator.

The aim of this study is to analyze the relative effect of the different input factors that are required to realistically model both fault seal and geometry. This analysis has been carried out in a similar manner to the earlier sedimentological studies of Jones et al. (1995) and Kjønsvik et al. (1994), whereby the effects of each factor have been measured by running several reservoir simulations. To achieve statistically significant results, experimental design has been implemented to extract the optimum number of scenarios. The structural model used in this study represents a fairly simple, tilted fault block cut by several seismic-scale faults. In the experiments, fault displacement, sealing characteristics, and density of faulting are varied. To have the most realistic geological property model, real oil field stratigraphic-sedimentary parameters were included. Three separate sedimentary environments have been modeled to assess how much the results of the structural factor analyses are relevant to a diverse range of geological environment.

GEOLOGICAL MODEL

The geological model aims to represent a typical North Sea tilted fault block (Figure 1). The model comprises five zones and has a maximum depth of 2225 m (7299 ft), rising to 2000 m (6600 ft) in depth at the top of the structure. Two of the horizons were modeled as erosive surfaces, and the others formed conformable surfaces. The geological model covers a region of 1.5 km (0.9 mi) in the dip direction by 1 km (0.6 mi) in the strike direction. The reservoir interval has an average thickness of 80 m (262 ft). It contains several seismic- and subseismic-scale faults, which are fairly typical in size,

FIGURE 1. Structural map. The model represents a south-dipping block with 17 major faults. Subseismic faults, in white, may be added.

geometry, and density for a North Sea field. As much as possible, the basic structure of the model was kept constant between scenarios, especially the horizons, grid topology, and fault locations.

The reservoir properties were modeled using object-based stochastic techniques. The basic input data and geological concepts are based on real field examples. The aspects of the model, which were varied for the sensitivity study, include the conceptual sedimentary model and the reservoir properties. The structural factors that were varied include fault displacement, fault zone thickness, fault-seal model, fault permeability, and fault pattern (i.e., fault density).

The geological model has been built using the following workflow (see Figure 2; Table A in this chapter's Appendix):

1) Five smooth but undulating and south-dipping horizons were generated.
2) The largest faults were generated as zero-displacement faults. These were used to control the geometry of the 3-D grid and were given a zero displacement to make the grid building process less cumbersome. The faults were initially generated as Pixat format faults and then converted into RMS™ format faults comprising fault lines and surfaces.
3) A 3-D grid was built around the horizons and zero-displacement faults. The grid was built in RMS and used the fault lines to control fault offset (zero displacement in this case), location, and slope.
4) Reservoir properties were modeled in the 3-D grid. Modeling properties in an unfaulted grid simplifies some of the techniques and ensures that sedimentary bodies, which cross faults, are correctly modeled.
5) The 3-D grid was altered to consider fault displacements along the fault traces defined originally. This approach ensured that the displaced cells were all located along the grid lines defined by the zero-displacement faults. The displacements for each fault were defined in a parameter file, and these were applied using the displacement options in Havana. The smaller faults were also added as displacement structures during this stage.
6) Seal calculations were made for the faults, and these were converted into fault transmissibility multipliers.

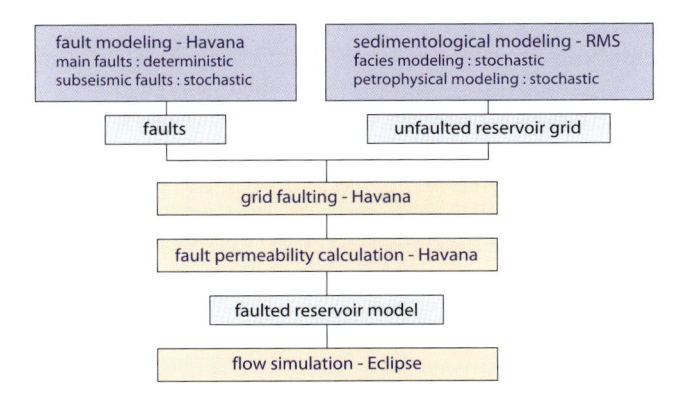

FIGURE 2. Experimental workflow. A combination of Havana and RMS is used to build the geomodel and the dynamic model.

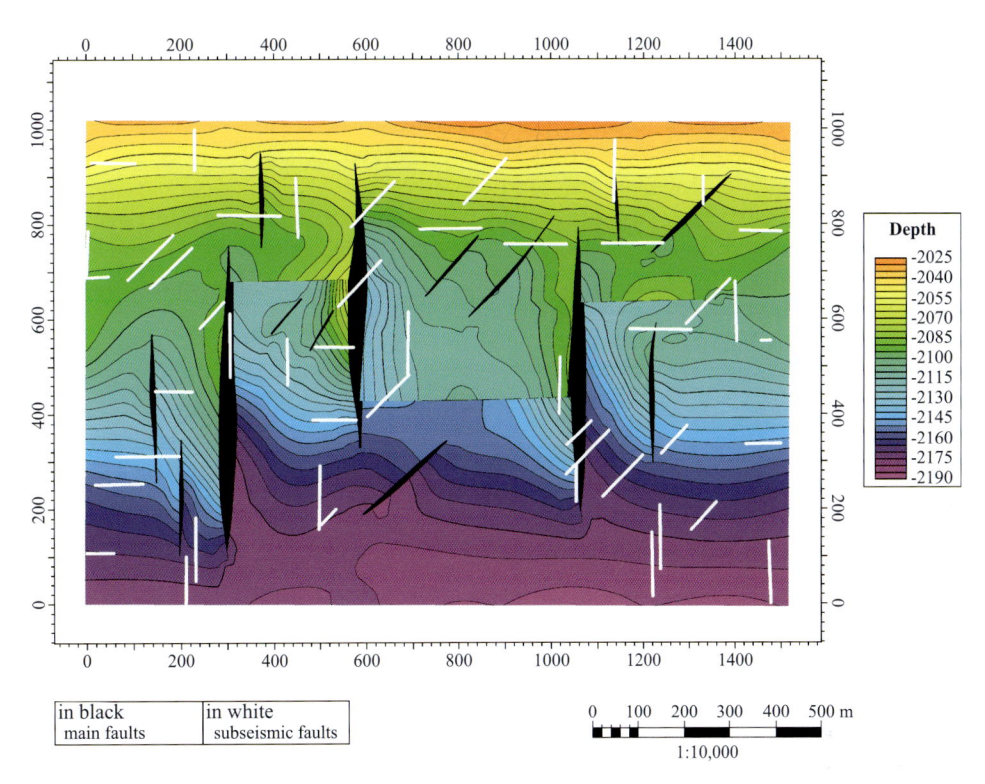

GEOLOGICAL MODEL DESCRIPTION

The geological property modeling was based on a grid cell size of 50 × 50 m (164 × 164 ft) with variable cell thickness, totaling approximately 110,000 cells

($33 \times 19 \times 231$ cells). The 3-D geological grid was used for both property modeling and reservoir simulation. This avoided any need for upscaling of properties and allowed the modeled heterogeneities to be preserved at the modeling scale. However, this led to a fairly large dynamic model with long simulation run times.

The modeling of reservoir properties was carried out using a combination of RMS and STORM software. Each of the four reservoir zones where modeled individually, and the general workflow was as follows:

- modeling of geological objects (for most zones)
- modeling of facies types in objects
- modeling of petrophysical properties for each facies type: porosity, horizontal permeability, vertical permeability, and clay content

All 3-D property modeling was undertaken with no conditioning to well data. This could, of course, have been done by generating synthetic well data and then using this to condition further realizations. The reason for doing unconditioned modeling was to generate realizations with a high degree of variability. Considering the relatively small size of the modeled volume and the large scale of the objects being modeled, well conditioning would have resulted in a low variability between realizations.

Sedimentary Models

The sedimentary models were based on real field cases. Three different environments were investigated: deltaic and tidal deltaic from the Åre and Tilje formations, respectively, of the Heidrun field (Haltenbanken, mid-Norway), fluviodeltaic from the Tarbert Formation of the Gullfaks field (North Viking Graben, Norway), and fluvial from the Ness Formation of the Gullfaks field.

The four zones in the 3-D model used in this study have been filled with representative sediments from overlying (commonly continuous) zones. To ensure consistency between the different reservoir scenarios, zone 3 was chosen to be a relatively poor reservoir zone, which restricted vertical flow. The geological objects, facies types, and modeling procedures used are summarized in Table B in this chapter's Appendix.

Petrophysical Model

Porosity and horizontal and vertical permeability were modeled stochastically as part of the petrophysical property modeling. For the purpose of fault modeling, the clay content was also modeled using similar methods. The vertical permeability was defined for each facies type using a kv/kh (vertical permeability/horizontal permeability) relationship based on the method developed by Ringrose et al. (2001).

Structural Modeling

Framework Building

Fault location and geometry were modeled deterministically, whereas stochastic techniques were used to model fault displacement and sealing properties. The fault pattern consists of three fault types: large (typically mapped from seismic data and included in the reservoir grid), intermediate (typically identified on seismic data but not commonly included in a reservoir grid), and small (typically subseismic-scale faults) (see Table C in this chapter's Appendix).

The fault pattern factor examined in this study was essentially the presence or absence of the smallest set of faults. These faults were modeled such that when present, they had a significant impact on flow, and they increased the possibility of vertical communication between layers.

Fault position and geometry could also have been modeled using stochastic techniques. This would have required a new grid for each structural realization and may have resulted in topological differences between grids. Because this was likely to cause larger variations in the results than those created by the factors that were under investigation, this sensitivity was not attempted in the models built for this study.

An important consideration in the modeling of faults is how they will be incorporated into a dynamic model. Faults typically displace geological layers, causing alternative fluid pathways; they commonly contain a thick zone of low-permeability fault rock, which commonly inhibits cross-fault flow. Fracturing can be common and lead to reduced permeability damage zones or high-permeability streaks in the vicinity of fault structures. The techniques that are currently available include considering juxtaposition of different strata in a 3-D corner-point grid with fault-seal effects, represented by transmissibility multipliers. Along-fault flow is difficult to model for a large number of faults using current technology.

Fault geometric modeling, 3-D grid building, and fault incorporation in the grid were carried out using a combination of the fault modeling tools RMS and Havana. An initial set of fault centerlines were generated for the top and bottom surfaces. These data were read into Havana and first converted to Havana's parametric fault model (PFM) format and then into RMS format faults (i.e., fault surfaces and fault lines). These zero-displacement faults were imported into RMS and then used in the 3-D corner-point grid building process to control where fault displacements would occur in the grid. This allowed the z-coordinate lines in the 3-D grid to slope along the future location of the fault surfaces and fan away from the faults to their more normal vertical orientation. This unfaulted grid was then imported into Havana, and the same PFM format faults were used to displace the grid. The PFM faults were added using their correct displacements; they followed the already pre-defined fault locations and displaced the grid precisely along the sloping z-coordinate lines.

One of the main advantages of modeling faults in a 3-D grid in this manner is that displacement can be altered between realizations, because their values exist as a parameter set in the PFM fault format. Other benefits include the initial grid building being simplified, because no displacements need to be modeled, and the displacements, when added later, are considerably smoother and cleaner (i.e., simpler) than when included as part of the grid building process.

After the large faults were included as displacements in the grid, both the intermediate and small fault sets were added as PFM format faults. These smaller faults were also constructed from line data, but they were modeled as zigzag structures through the grid, and it was therefore not necessary to modify the grid to consider displacements. Again, because the faults are added to the grid, displacements could be changed.

Fault displacement was one of the key factors examined in this study. The base case set of faults was assigned displacements of 80 m (262 ft) for the large faults, 50 m (164 ft) for the intermediate faults, and 30 m (98 ft) for the smallest faults. The displacement on all faults was reduced by 15% to obtain a low value. Typically, in any seismic interpretation, considerable uncertainty exists in the actual displacement measured. A value of 15% is probably on the low side of any realistic uncertainty.

Havana has a feature for truncating faults against other faults. This was used to generate realistic fault patterns. The truncation order was set up so that the larger faults truncated the smaller faults.

Fault-seal Modeling

In reservoir grids, faults are implemented as displaced cell boundaries with no volume or property. The effect of their displacement is considered in the initialization of the dynamic model when juxtaposition of non-neighboring cells is brought into the transmissibility calculations. The effect of the fault-seal transmissibility multipliers must be estimated to consider the effects of both fault thickness and permeability.

Transmissibility between two adjacent cells is defined as

$$T = \frac{C_{\text{darcy}}}{\frac{1}{T_1} + \frac{1}{T_2}}$$

where $T_i = K_i R_i D_i S_i$. T is the transmissibility value; C_{darcy} is a constant; T_1 and T_2 (T_i) are the cells' specific transmissibility values; K is the cell permeability; R is the net-to-gross ratio; D is the distance between the cell center and the interface; and S is the contact area. Transmissibility between any two cells can be modified using a multiplier, T_m. In the case of faulted cells, this multiplier is used to consider fault seal and is calculated as the product of the faulted and unfaulted transmissibilities. Where the faulted transmissibility, T_f is

$$T_f = T \times T_m$$

Several ways of adding transmissibility multipliers into a dynamic model are available in Eclipse™: simple multipliers (MULTX Eclipse keyword), two-way multipliers (MULTX and MULTX- keywords), and connections multipliers (EDITNNC and MULTIPLYTRANX keywords). It has commonly been argued that using cell face transmissibilities would increase both simulation time and memory requirements. A preliminary comparison between the three methods showed that no significant difference (<1%) was observed between the initialization times or the simulation times. The memory requirements are equivalent, because in this case, the number of connections is always the same. However, to use the cell face transmissibilities, Eclipse needs a significantly larger memory allocation, although this is not often used.

Havana has the functionality to read in a fault-displaced reservoir grid, estimate the fault seal, and write out the effect as a transmissibility multiplier. Several options exist for how fault seal is estimated, but the basis to all of these involves estimating fault permeability and thickness. Fault permeability can be estimated using techniques such as shale gouge ratio (SGR) and shale smear and clay smear potential, or it can be modeled more simply by estimating the permeability directly for each fault. Thickness can also be modeled as a specific value or as a function of displacement. Havana has the option of modeling both parameters using stochastic techniques, which allows for the effects of uncertainty in the fault-seal input parameters and fault heterogeneity to be considered.

Two alternative methods for estimating fault permeability have been used to determine the fault-seal model:

- simple permeability function: This assigns a permeability reduction factor based on the matrix permeability of fault-adjacent grid blocks. This factor is constant across a single fault surface. An additional multiplier has been added based on the facies juxtaposition across a fault (Table 1).
- complex permeability function: This consisted of calculating both shale smear and SGR across fault surfaces. Shale smear is modeled as zero permeability when present and no change from the matrix when absent. The shale gouge ratio is then used to

TABLE 1. Parameters for the simple permeability function.

Permeability Reduction Factor		
Fault factor	Permeability (md), low case	Permeability (md), high case
	1×10^{-4}	3×10^{-4}
Facies Combination Multiplier		
Shale-shale	0.1	
Shale-sand	0.1	
Shale-silt	0.1	
Silt-silt	1	
Silt-sand	1	
Sand-sand	66	

estimate the permeability of the fault plane where shale smear is absent. Both methods are dependent on a detailed stratigraphic description taken from the property modeling. Because the property modeling has been carried out using stochastic techniques, each realization results in its own unique fault permeability estimates (Table 2).

The second part of any fault-seal calculation, fault thickness, has also been studied to determine its effect on dynamic models. The fault thickness was calculated according to the equation

$$t = \alpha \times d$$

where t = fault zone thickness, and d = fault displacement. High and low values for the parameter α were chosen as representative values for faulted clastic sequences, taken from both published and unpublished sources. The effect of increasing this parameter will essentially tighten the fault seal by increasing the damage zone thickness (Table 3).

DYNAMIC MODELS

From the geological model, a dynamic model was built that attempted to represent the geology as accurately as possible. This dynamic model was then used to determine the effects of the chosen factors. The same sized grid was used for all the simulations (33 × 19 × 231 cells). Eclipse 100 was used for reservoir simulation.

The input to the dynamic models is a grid including reservoir properties and transmissibility multipliers. The grid and reservoir properties were prepared in RMS and Havana and exported in the Eclipse format ready for direct use in the reservoir simulator. The transmissibility multipliers were calculated in Havana as Eclipse format files. No other data manipulations were necessary. The fluid properties at reservoir conditions are given in Table 4.

The oil-water contact was positioned at 2225 m (7299 ft), and the pressure datum at 2000 m (6600 ft) was set at 222 bar (219 atm). For calculating the initial fluid saturation conditions, an input table of capillary pressure corresponding to the Heidrun reservoir conditions was used. The capillary pressure is computed first, according to the formula $P_c = \Delta\rho g h$, where $\Delta\rho$ is the fluid density difference, and h is the height above the oil-water contact. The simulator then computes the corresponding water and oil saturation.

The relative permeability and water saturation were given in interpolation tables, arranged in five classes of local permeability: 0–25, 25–75, 75–200, 200–900, and >900 (all are in millidarcys). They were derived from the capillary pressure calculated by the simulator.

The production strategy was set so that there was a general updip water drive. Five wells were used, two injectors located in the deepest part of the structure and three producers at the top of the structure (see the well locations in Figure 3). The well positions were the same in all the models.

The two water injectors were controlled by a 600-bar (592-atm) bottomhole pressure and a maximum water flow threshold of 2500 Sm^3/day (standard cubic meters/day). Water injection was allowed over the whole reservoir thickness, partly because of vertical permeability barriers being present, and also because the estimated vertical permeability values are not known with confidence. By injecting over the whole thickness, the influence of variation in this property is reduced.

The three production wells were controlled at 180-bar (177-atm) bottomhole pressure. Each well was perforated at specific layers, which were located based on early-stage simulation tests. The perforations were supposed to create an even pressure distribution in the field and a balanced production between all of the layers.

The end of production for each realization was set at 95% water cut for the field as a whole.

TABLE 2. Parameters for the complex permeability function.

Shale Gouge Ratio vs. Permeability		
Shale Gouge Ratio	Permeability, Low Case (md)	Permeability, High Case (md)
0–0.1	0.11	0.3
0.1–0.15	0.1	0.12
0.15–0.5	0.02	0.07
0.5–1	0.001	0.003

TABLE 3. Summary table of the different input factors and levels.

Input Factors	Levels	
Fault pattern	low connectivity, 17 faults	high connectivity, 17 faults + 50 subseismic faults
Fault throw	low displacement = 85% of base displacement	high displacement = 100% of base displacement
Fault-seal model	homogeneous permeability	smear gauge ratio and shale smear factor
Average fault permeability	low	high
Fault zone thickness	low, $\alpha = 0.0133$	high, $\alpha = 0.04$

SENSITIVITY ANALYSIS AND RESULTS

Experimental Design

An experimental design approach was used to set up the experiments using the least number of reservoir simulations, generate a statistically valid set of results, and achieve the objectives of the study. The reason for using this type of approach was that a large number of model realizations would have been built and reservoir simulations run on them if all possible factor combinations had been used. A total number of X^n realizations are required for a full factorial study, where n is the number of factors analyzed, and X is the number of levels for each factor. This study consisted of two separate experiments totaling 192 simulations. In addition to these realizations, several replicates would be required to test the noise levels. The computational time taken to build each realization was approximately 12 CPU hr, and the simulation run times were 20–30 CPU hr. Clearly, the amount of central processing unit (CPU) time required to have run full factorial experiments plus several replicates would have been prohibitive. Therefore, the experimental design approach was used.

Experimental design is basically a method used for reducing the number of required experiments (in this case, reservoir simulations). The basic premise is to choose a selected number of experiments with specific factor value combinations that will represent the range of possible combinations available. This should lead to a representative reflection of the range of variability in the experiment as a whole. Furthermore, it should allow the individual factors to be assessed for their significance to the experiment and allow their effects to be compared relative to one another. The most important part of an experimental design is that the results can be analyzed to determine their statistical significance, the main point being that the overall results reflect the expected results if the full experiment had been carried out. Ideally, if additional reservoir simulations were run and their results added to those of the experimental design, their influence on the overall results of the study would be negligible.

Setting up the experimental design required a significant amount of preparation time. However, it helped our understanding of the influence of the various factors. The main difficulty was defining the input factors and handling a combination of both quantitative and qualitative factor types.

The type of experimental design chosen for this study was a 2^k fractional factorial design, which was reduced using the D-optimization method. The two main experiments were set up in the following manner:

- A global comparison is made, whereby the influence of structural factors could be compared to the sedimentary models. In this experiment, all the fault modeling factors were varied as described in Table D in this chapter's Appendix, and the sedimentary model was also allowed to vary. Several of the modeling parameters were generated stochastically to introduce geological variability between the realizations. The noise introduced by the stochastic parameters was a concern, in that its variability could have been greater than the effects of the parameters themselves. This was controlled by running five replicates for each experiment, which aimed to reduce the overall noise level. A total of seven factors were examined in this experiment. The experimental design, which dictates the parameter values to be used in each realization, is presented in Table E in this chapter's Appendix.
- A second experiment was carried out, whereby the structural factors were varied using each of the

TABLE 4. Fluid properties.

Fluid Properties		
	μ (cp)	ρ (kg/m³)
Oil	2.313–2.94, in relation to the local reservoir pressure	919
Water	0.38	1033

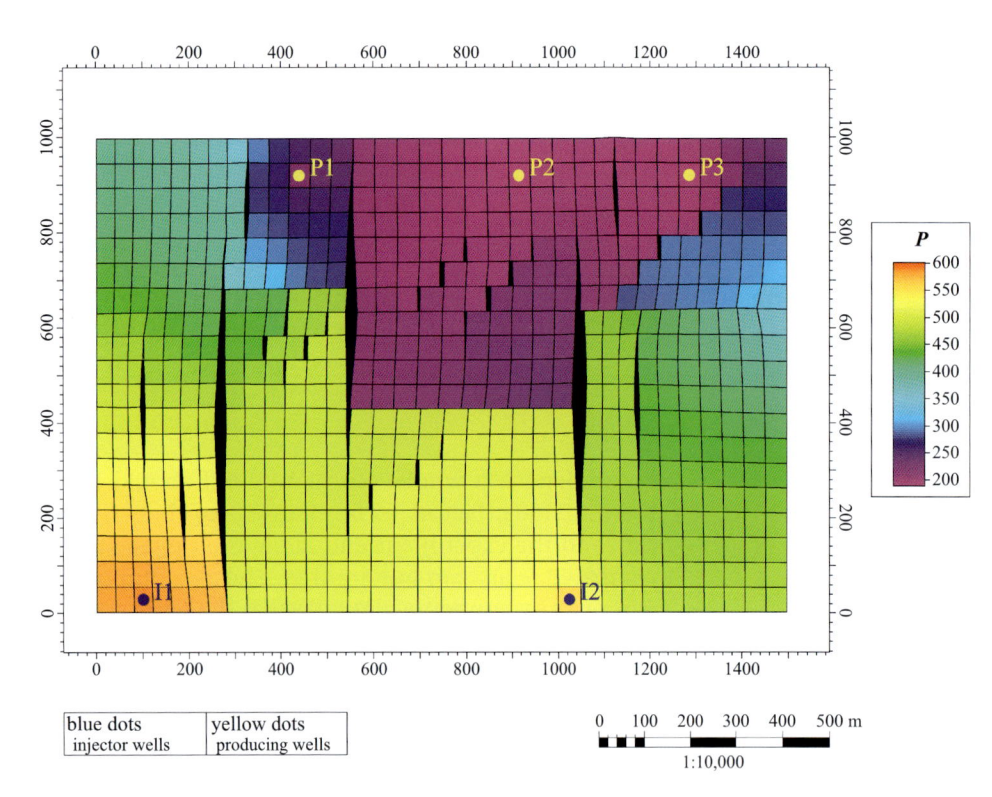

FIGURE 3. Reservoir pressure (bar) after 20 yr of production, case 10. The well locations are indicated: Two injectors in the deepest part (I1 and I2) and three producers in the upper part (P1, P2, and P3) are shown.

were selected that reflected the effects on both earlier and later production responses. Elapsed time and recovered oil volume were the key measures, each made at three different time steps:

- water breakthrough (10% water cut)
- one moveable pore volume injected
- end of production (95% water cut)

sedimentary model individually. This experiment was carried out not only to examine the effect of the structural factors in the absence of the highly variable sedimentary models, but also to see if their effects changed when applied to different depositional environments. The experimental design for this experiment is presented in Table E in this chapter's Appendix. To represent the variability of facies simulation, five replicates of each realization were run.

For the factors that were to be analyzed, a high and low value had to be selected. These values were chosen with the following in mind: They formed realistic values backed up by geological data, and the high-low values tried to reflect either the true geological variability or the general uncertainty that is commonly present in oil field data and its interpretation. The resulting weight is not the absolute weight of one factor but two extreme values defining the realistic range of influence. In this study, a mix of qualitative (sedimentary model and fault-seal model) and quantitative factors (fault pattern, fault displacement, fault permeability, and fault zone thickness) has been used. All the factors and possible values are presented in Table D in this chapter's Appendix.

Result Extraction

To assess the relative importance of the various factors, the results of each Eclipse simulation had to be measured numerically. For this, six different measures

All the measured simulation results were then compiled to give

1) the relative weight of each input factor in each measurement, measured by the average deviation formula:

$$A_d = \frac{1}{N}\sum_{i=1}^{N} |X_i - X|, \quad \text{where} \quad X_i = \frac{1}{K_i}\sum_{j=1}^{K_i} Y_{ij}$$

Y is the response variable; i and N are, respectively, the level and the number of levels for the input factor; and K is the number of simulation runs with the same level for this specific input factor.

2) the significance level of each input factor in each measurement, using analysis of variance (ANOVA).

The analysis of the simulation results is summarized in Figure 3, where the relative influences of each input factor for each experiment are plotted as stacked bar charts. These charts plot the variability only brought about by the studied variables. These do not show the total variability in each experiment. Plotting the data this way allows the relative influence of each factor to be directly compared between different measurements or experiments.

The input factors that are not significant are not plotted onto the graphs.

FIGURE 4. Experiment 1 results. The relative influence of each input factor is plotted for each measurement: elapsed time and recovery factor at three time steps: water breakthrough (WBT), one moveable pore volume injected (1 mpvi), and end of production (95% water cut). Factors that are not significant are not plotted. This experiment shows the high influence of the sedimentary model factor and lower but significant influence for two structural factors, fault-seal model, and fault pattern.

Result Analysis

Before analyzing the results, it is important to redefine the context of these experiments. Experiment 1 used unquantifiable input variables (e.g., the sedimentary models), which means they cannot be ranked. To do so, it would be necessary to describe them completely as a set of quantifiable factors (e.g., permeability and porosity), which was beyond the scope of this study. In experiment 2, results are strongly linked to one sedimentary model. The focus was set on a limited number of input factors, mainly to reduce the computation time. It should be made clear that it is not possible to generalize the conclusions; they are strongly limited to these reservoirs and experimental settings and have no global predictive value. However, some of the conclusions prove how interesting the approach can be.

Experiment 1

Experiment 1 was designed to measure the impact of structural factors relative to different sedimentological environments. It is clear from the results (Figure 4) that the variability caused by the sedimentary model has a much higher impact on the dynamic behavior than the structural factors, because it considers 50–75% of the variability.

However, despite the dominance of the sedimentary model influence, the impact of the structural variables is also significant, in particular for the fault-seal model and the fault pattern.

It is also interesting to note that the influence of the factors vary differently for each type of measurement (recovery and time), as well as through production time. This second observation can be of high interest when assessing the uncertainties. For instance, here, the fault pattern has no influence on the recovery measurements, whereas it is a significant factor for the time-related measurements. Additionally, a clear diminution of its impact through production time exists. In these settings, although the predictions would be strongly influenced by the fault pattern at the beginning, its impact would be less significant at the end of the field's life.

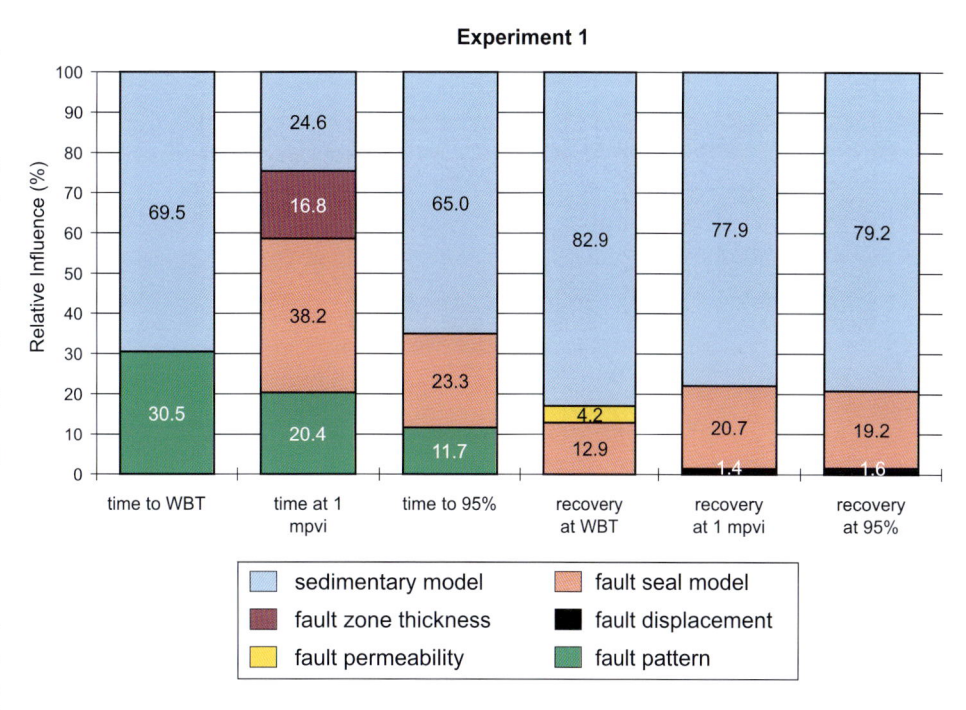

Finally, the other input factors have no significant influence on the dynamic response. Fault throw, fault zone thickness, and fault average permeability have low impact. It does not mean that they do not matter, but that their variability, or the uncertainty attached to them, does not strongly affect the results.

Experiment 2

Experiment 2 was designed to remove the influence of varying the depositional environment and to get an idea of the response to structural factors in the different sedimentary model. Results are presented in Figure 5. Here, the variability brought by the property models is much lower; each experiment uses one single sedimentary model, the only variability is brought by the stochastic approach (use of replicates).

Two classes of input factors can easily be identified, based on how sensitive the models are to them:

- primary factors: In all models, sedimentary model, fault-seal model, and fault pattern are significant for almost all measurements.
- secondary factors: Secondary factors, such as fault permeability, fault throw, and fault zone thickness, can have different behavior between the models.

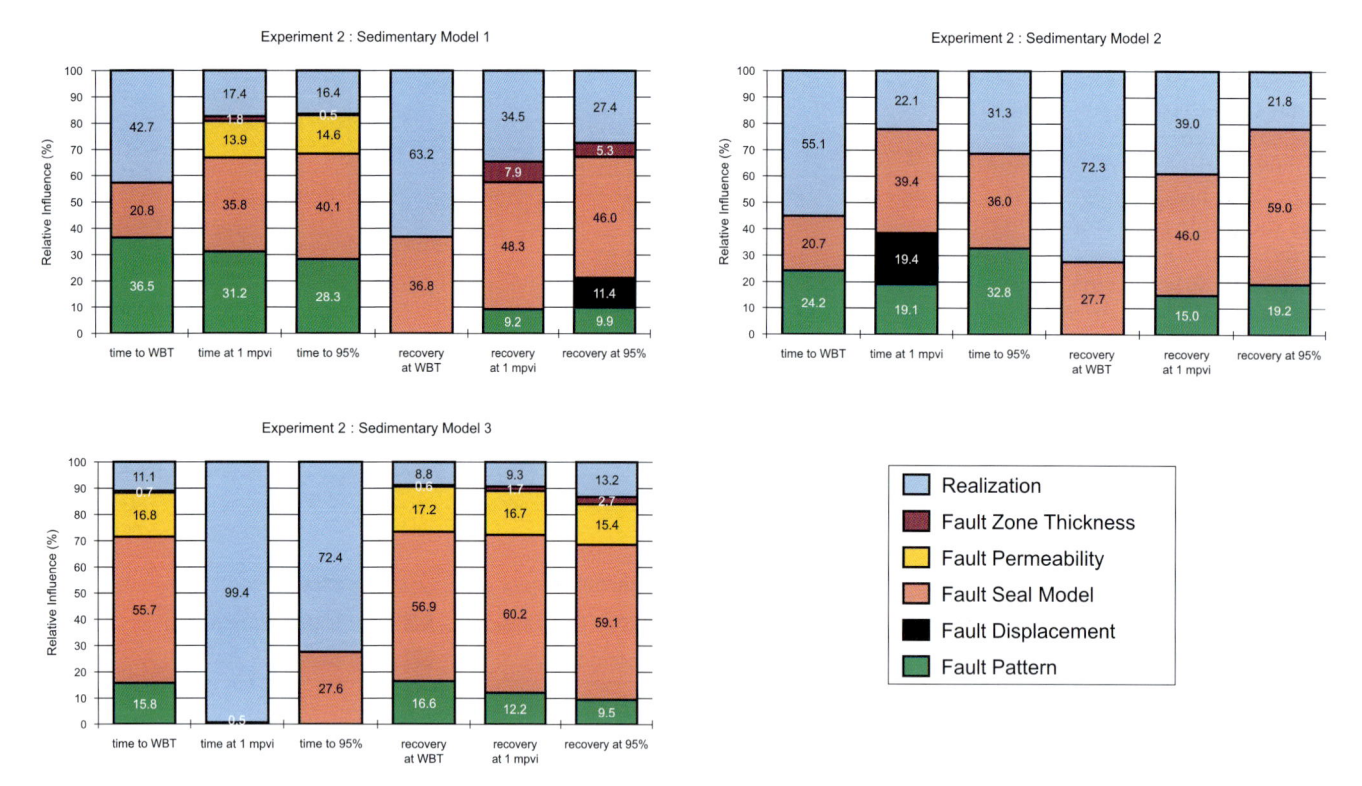

FIGURE 5. Experiment 2 results. The relative influence of each input factor is plotted for each measurement: elapsed time and recovery factor at three time steps: water breakthrough (WBT), one moveable pore volume injected (1 mpvi), and end of production (95% water cut). Factors that are not significant are not plotted. The results show how variable the response to structural factors can be in different sedimentological settings.

It is clear that the sedimentary models have a strong influence on how much the secondary factors impact the production history. With less complex sedimentary models, it would be possible to link the variations in sensitivity to the model characteristics. Here, the three sedimentary models are too complex and, at the same time, not different enough to interpret with confidence the variations in the responses. However, these results show that the impact of the secondary factors is difficult to predict and highly dependent on the sedimentological settings.

CONCLUSION

This study showed that beside the variability caused by the sedimentological and petrophysical property modeling, structural factors can have a very significant effect on reservoir performance. Here, the fault-seal modeling approach and the subseismic fault pattern were the most significant structural factors. Although 3-D stratigraphic modeling has been in use for more than a decade and is commonly integrated into the reservoir modeling workflows, advanced structural modeling is just breaking through. The influence of the structural factors, as well as the variations between experiments, highlight the need for integrating more structural information in the reservoir models.

Finally, we proved here how valuable this type of work can be, especially for a field at the development stage: today, sensitivity studies are not often used for any purpose other than economical uncertainty assessment, whereas they could be used to determine which area of the models should be studied in more detail so as to predict potential problems in the reservoir management of a field, that is, not to panic when a problem occurs, because it was expected as a possible outcome at the onset of production. The tools used here are now available to help the geoscientists and engineers run such evaluations in a relatively short time frame.

ACKNOWLEDGMENTS

The authors thank Statoil for permission to publish this chapter.

We also thank our colleagues for their significant contribution through discussions and expertise: Inge Brandsæter, Philip Ringrose, Jan Ole Aasen, Signe Ottesen, Arve Næss, and Maria Grazia Contursi.

Jamie Burgess, Gary Couples, Richard Suttill, and Alastair Welbon are thanked for their helpful review of the manuscript.

TABLE A. GRID AND ZONES GEOMETRY. THE 3-D GRID CONTAINS ABOUT 100,000 ACTIVE CELLS

Grid Features	X	Y	Z
Dimensions X, Y, Z	1500 m	1000 m	~80 m
Number of layers	33	19	231

Zones features	Zone number	Number of layers	Top surface
Top	1	66	Normal
	2	28	Erosion surface
	3	107	Erosion surface
Bottom	4	30	Normal

TABLE B. DESCRIPTION OF THE SEDIMENTARY MODELS AND OF THE MODELING PROCEDURES USED

Sedimentary Models	Model 1		Model 2		Model 3	
Geological sequence Zone 1	Åre 2B → Tilje 2 shoreface	facies modeling Gmpp	NER-1-3 → Ness-3B flood plain, with coal and fluvial-channel deposits	facies modeling FF/FG + Gmpp	Tarbert-1B → Tarbert-3A fluvial and tidal channels	facies modeling Gmpp
Zone 2	shoreface	Gmpp	tidal channel overlying marine deposits	FF/FG	fluvial and tidal channels, coal, and cement	Gmpp + TgSim
Zone 3	shoreface to tidal (channels and mud flats)	Gmpp	marine deposits and calcite	petrophysics only	tidal and lagoonal deposits	Gmpp
Zone 4	tidal channels and mudflat to shoreface	Gmpp	lower shoreface to flood-plain deposits (fluvial channels and coal); calcite-cemented layers	Gmpp + TgSim	sequence from offshore transition to fluvial channels	TgSim + Gmpp

Gmpp = general market point process; TgSim = truncated Gaussian simulation; FF/FG = fluvial facies/fluvial grid.

TABLE C. FAULTS PARAMETERS

Fault Set	Length Range (m)	Strike (degrees)	Dip (degrees)	Maximum Displacement (m)
Large: 3 faults	600–800	0	60	80
Intermediate: 3 faults	500–650	90	90	50
Small: 11 faults	180–400	45	60	30
Extra Fault Set 50 subseismic faults	150	0–45–90	90	10

TABLE D. INPUT FACTORS. THE TABLE SUMMARIZES THE PRINCIPAL CHARACTERISTIC OF EACH OF THE INPUT FACTORS, AS WELL AS THE DIFFERENT VALUES/TYPES USED. EXPERIMENT 1 AND 2 DIFFER BY THE USE OF THE SEDIMENTARY MODEL AS AN INPUT FACTOR

Input factors	Levels		
Sedimentary model	Model 1 Heidrun Tilje and Åre	Model 2 Gullfaks Ness	Model 3 Gullfaks Tarbert
Fault pattern	low connectivity (0) 17 faults	high connectivity (1) 17 faults + 50 subseismic faults	
Fault throw	low displacement (0) = 85% of base displacement	high displacement (1) = 100% of base displacement	
Fault seal modeling method	homogeneous permeability (0)	smear gauge ratio and shale smear factor (1)	
Average permeability	low (0)	high (1)	
Fault zone thickness	low (0)	high (1)	

Experiment 1 covers all rows; Experiment 2 excludes the Sedimentary model row.

TABLE E. EXPERIMENTAL DESIGN TEMPLATES. THE TABLE OUTLINES THE PARAMETER FACTORS THAT VARIED IN EACH REALIZATION.

Experiment 1

Realization Number	Sedimentary Model	Fault Pattern	Fault Displacement (m)	Fault-seal Model	Fault Permeability (md)	Fault Zone Thickness (m)
1	2	0	1	1	1	1
2	2	1	0	0	0	1
3	2	0	0	1	1	0
4	2	1	1	0	1	0
5	2	1	0	0	0	1
7	3	1	1	0	0	0
8	3	1	0	1	0	0
9	3	0	0	0	1	1
10	3	0	1	1	1	1
11	1	1	1	1	1	1
12	1	0	1	0	0	0
13	1	0	0	1	0	0
14	1	1	1	0	1	0
15	1	1	0	0	0	1

Experiment 2

Realization Number	Fault Pattern	Fault Displacement (m)	Fault-seal Model	Fault Permeability (md)	Fault Zone Thickness (m)
1	0	0	1	1	1
2	0	1	1	0	0
3	1	1	0	0	0
4	1	0	1	0	0
5	1	1	0	1	1
6	0	0	0	1	0
7	0	1	0	0	1
8	1	0	1	0	1
9	1	1	1	1	0

Each factor can take a low (0) or high (1) value, which is based on realistic reservoir characteristics and uncertainty range. Running a full experiment (with all possible combination) would require too much computation time. Therefore, the experimental designs are reduced and optimized. Experiment 1 includes the sedimentary model as one of the variable factors (with three levels). Experiment 2 is run in each of the sedimentary model.

REFERENCES CITED

Brandsæter, I., H. T. Wist, A. Næss, O. Lia, O. J. Arntzen, P. Ringrose, A. Martinius, and T. R. Lerdahl, 2001, Ranking of stochastic realizations of complex tidal reservoirs using streamline simulation criteria: Petroleum Geoscience, v. 7, special issue, p. 53–63.

England, W. A., and C. Townsend, 1998, The effects of faulting on production from a shallow marine reservoir—A study of relative importance of fault parameters: Society of Petroleum Engineers Paper 49023, p. 279–294.

Henriquez, A., and C. Jourdan, 1995, Challenges in modelling reservoirs in the North Sea and on the Norwegian Shelf: Petroleum Geoscience, v. 1, no. 4, p. 327–336.

Hollund, K., P. Mostad, B. F. Nielsen, L. Holden, J. Gjerde, M. G. Contursi, A. J. McCann, C. Townsend, and E. Sverdrup, 2002, Havana—A fault modelling tool, in A. G. Koestler and R. Hunsdale, eds., Hydrocarbon seal quantification: Norwegian Petroleum Society Special Publication 11, p. 157–171.

Jones, A., J. Doyle, T. Jacobsen, and D. Kjønsvik, 1995, Which subseismic heterogeneities influence waterflood performance? A case study of a low net-to-gross fluvial reservoir: Geological Society (London) Special Publication 84, p. 5–18.

Kjønsvik, D., J. Doyle, T. Jacobsen, and A. Jones, 1994, The effects of sedimentary heterogeneities on production from a shallow marine reservoir—What really matters?: Society of Petroleum Engineers Paper 28445, p. 27–40.

Knai, T. A., and R. J. Knipe, 1998, The impact of faults on fluid flow in the Heidrun Field, in G. Jones, Q. J. Fisher, and R. J. Knipe, eds., Faulting, fault sealing and fluid flow in hydrocarbon reservoirs: Geological Society (London) Special Publication 147, p. 269–282.

Knipe, R. J., 1992, Faulting processes and fault seal, in R. M. Larson, H. Brekke, B. T. Larsen, and E. Talleraas eds., Structural and tectonic modelling and its application to petroleum geology: Norwegian Petroleum Society Special Publication 1, p. 325–342.

Lindsay, N. G., F. C. Murphy, J. J. Walsh, and J. Watterson, 1993, Outcrop studies of shale smears on fault surfaces, in S. Flint and A. D. Bryant, eds., The geological modelling of hydrocarbon reservoirs and outcrop analogues: International Association of Sedimentologists Special Publication 15, p. 113–123.

Manzocchi, T., J. J. Walsh, P. Nell, and G. Yielding, 1999, Fault transmissibility multipliers for flow simulation models: Petroleum Geoscience, v. 5, p. 53–63.

Munthe, K. L., H. Omre, L. Holden, E. Damsleth, K. Heffer, T. S. Olsen, and J. Watterson, 1993, Subseismic faults in reservoir description and simulation, in Proceedings of the 68th Annual Technical Conference and Exhibition, Society of Petroleum Engineers, Houston, Texas: SPE Paper 26500, p. 843–850.

Ottesen, S., S. O. Ellevset, R. J. Knipe, T. Svava Olsen, Q. J. Fisher, and G. Jones, 1998, Fault controlled communication in the Sleipner Vest field, Norwegian continental shelf; detailed, quantitative input for reservoir simulation and well planning, in G. Jones, Q. J. Fisher, and R. J. Knipe, eds., Faulting, fault sealing and fluid flow in hydrocarbon reservoirs: Geological Society (London) Special Publication 147, p. 283–297.

Ringrose, P., R. Wen, and O. Lia, 2001, Effective flow and connectivity properties of mathematical cubes, in G. Yielding, B. Freeman, and D. T. Needham, eds., Conference digest of the Third Institute for Mathematics and Its Applications Conference on Modelling Permeable Rock, March 27–29, 2001: Churchill College, Cambridge, United Kingdom, 3 p.

Walsh, J. J., J. Watterson, A. Heath, P. A. Gillespie, and C. Childs, 1998, Assessment of the effects of sub-seismic faults on bulk permeabilities of reservoir properties: Geological Society (London) Special Publication 127, p. 99–114.

Yielding, G., B. Freeman, and D. T. Needham, 1997, Quantitative fault seal prediction: AAPG Bulletin, v. 81, p. 897–917.

Using Buoyancy Pressure Profiles to Assess Uncertainty in Fault Seal Calibration

Peter Bretan
Badley Geoscience Limited, Lincolnshire, United Kingdom

Graham Yielding
Badley Geoscience Limited, Lincolnshire, United Kingdom

ABSTRACT

Effective fault seal model calibration is dependent on the quality of the available data. Buoyancy pressure profiles provide a method to assess the potential uncertainty involved in deriving key input data, such as V_{clay} (volumetric clay fraction), and in the empirical equations used to derive seal-failure criteria. For a membrane-sealing fault, seal failure occurs when the buoyancy pressure exerted by the hydrocarbon column is equal to the minimum capillary entry pressure of the fault zone. If the seal is intact, the predicted fault zone capillary entry pressure value must be higher than the buoyancy pressure and, on a buoyancy pressure profile, plot to the right of the buoyancy pressure trend line. Predicted values that are lower than the buoyancy pressure indicate either that the fault is leaking or that one or more of the input data are in error.

Buoyancy pressure profiles are used postmortem to identify which data analysis techniques and seal-failure criteria best predict the observed hydrocarbon contacts in a given area. They can also be used to verify the threshold shale gouge ratio values that represent the onset of fault sealing.

The effect of varying key input parameters can be rapidly checked using buoyancy pressure profiles without resorting to more time-consuming probabilistic approaches. The analysis can identify which data preparation techniques provide appropriate estimates of the fault zone capillary entry pressure, relative to the observed buoyancy pressure, thereby significantly reducing potential uncertainty in the fault seal calibration.

Copyright ©2005 by The American Association of Petroleum Geologists.
DOI:10.1306/1060762H23167

INTRODUCTION

A standard method for evaluating fault seal in mixed clastic (sand-shale) sequences involves mapping the juxtaposition pattern of nonreservoir against reservoir intervals on a three-dimensional (3-D) fault surface and predicting the likely fault rock composition where the reservoir is in contact with another reservoir (e.g., Yielding et al., 1997). A principal aim of this method is to derive an empirical relationship between fault rock composition and the capillary entry pressure for seal failure (Yielding et al., 1997; Fisher et al., 2001; Childs et al., 2002; Bretan et al., 2003). Two different approaches can be used to derive this empirical relationship. The first approach relies on algorithms, such as shale gouge ratio or SGR (Yielding et al., 1997; Freeman et al., 1998; Yielding, 2002), to predict the distribution of fault rock types through the analysis of the sequence geometry and the volumetric clay fraction (V_{clay}) of the stratigraphy adjacent to the fault (see Appendix for definition). The predicted fault rock property or seal attribute is then calibrated at faults in hydrocarbon traps where the sealing behavior can be demonstrated using in-situ pressure data from wells (e.g., Fristad et al., 1997; Fisher et al., 2001; Yielding, 2002). The SGR-based approach, first developed in the North Sea, has become the standard method of assessing the seal potential in many hydrocarbon provinces worldwide (Yielding, 2002). Uncertainty in the structural interpretation (e.g., Childs et al., 1997; Hesthammer and Fossen, 2000) and in the reservoir stratigraphy can result in different juxtaposition patterns of sand-against-sand reservoirs at the fault. A major source of uncertainty is the V_{clay} parameter used in the SGR algorithm. Different estimates for V_{clay} give rise to significantly different estimates for the onset of seal (Bretan et al., 2003).

The second approach for estimating seal potential relies on empirical relationships between fault rock permeability, capillary entry pressure, and clay content from laboratory-derived measurements on fault gouge samples (e.g., Gibson, 1994, 1998; Fisher and Knipe, 2001; Sperrevik et al., 2002). Hydraulic properties obtained through experimental studies are typically correlated against fault rock classification (Knipe et al., 1998; Ottesen Ellevset et al., 1998). Samples used in laboratory-based analysis are biased to low-displacement faults (less than 1 m [3.3 ft] throw) and seldom originate from faults that are known to be sealing to hydrocarbons. Furthermore, there is uncertainty converting capillary entry pressures derived using mercury-air tests under laboratory-based conditions to equivalent pressures at reservoir conditions in the subsurface (see also Sperrevik et al., 2002, for more discussion on the uncertainties).

The relationship between capillary entry pressure for seal failure and fault rock composition is equivocal. Calibration plots of seal attribute, typically SGR, against maximum across-fault pressure differential (Fristad et al., 1997; Yielding et al., 1997; Yielding, 2002) exhibit a continuous relationship for seal failure. The maximum capillary entry pressure that the fault can support systematically increases as the SGR value increases (Figure 1a). A similar, continuous relationship

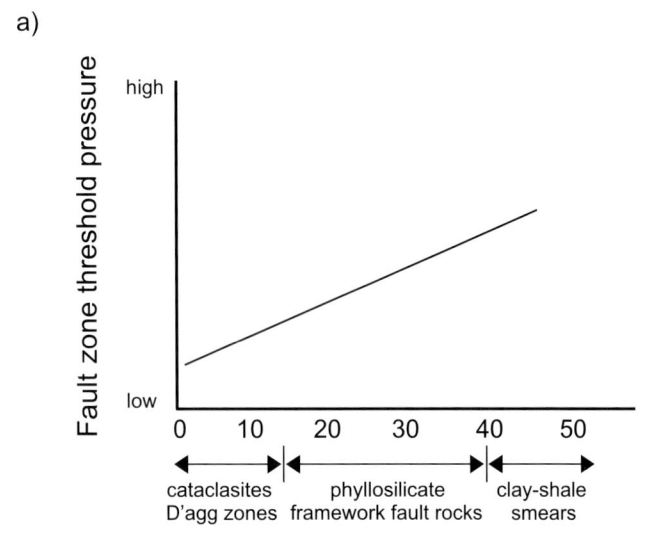

FIGURE 1. Schematic diagram showing (a) continuous and (b) discontinuous relationship between fault zone capillary entry pressure for seal failure and fault rock classification based on fault zone clay content (modified after Fisher and Knipe, 1998). Current methods for predicting capillary entry pressure are typically based on a continuous function. D'agg zones = disaggregation zones.

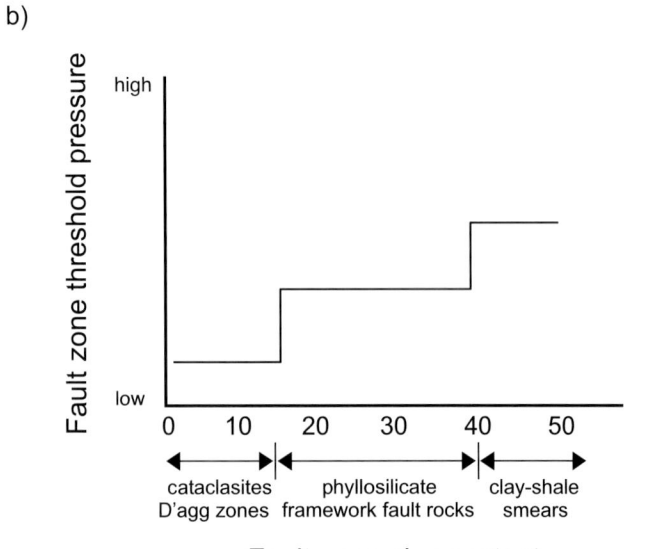

is evident in the equations derived from laboratory-based entry pressure measurements of fault rock samples (Sperrevik et al., 2002). In contrast, several subsurface (e.g., Fulljames et al., 1997; Ottesen Ellevset et al., 1998; Yielding, 2002; Gibson and Bentham, 2003) and microstructure studies (e.g., Fisher and Knipe, 1998, 2001; Takahashi, 2003) tend to show a more discontinuous relationship between fault zone composition and seal failure (Figure 1b). A general trend of decreasing fault rock permeability, and, hence, increasing capillary entry pressures, with increasing clay content exists. However, this trend is punctuated with relatively uniform permeability values over a range of clay contents. More abrupt changes in permeability occur at the transition between the main fault rock types (Fisher and Knipe, 1998, 2001). Current methods for predicting fault zone capillary entry pressure using fault rock composition are commonly based on a continuous function (e.g., Yielding, 2002; Bretan et al., 2003).

The aim of this contribution is to describe a method to assess the veracity of the key input parameters used to establish seal-failure criteria in the SGR-based approach for fault seal assessment. Uncertainty associated with the geological model and stratigraphic sequence is not considered here. The method is based on buoyancy pressure-depth profiles (Childs et al., 2002). Using buoyancy pressure profiles as a point of reference by which other input variables can be assessed provides a quick-look check on the validity of the input data and/or predictive algorithm. Here, the profiles are not used to estimate seal potential or column heights. Instead, the method is used to establish which data analysis techniques and seal calibrations best predict the observed hydrocarbon contacts in a given area, thereby significantly reducing potential uncertainty in the seal analysis.

BUOYANCY PRESSURE PROFILES

Buoyancy pressure, caused by buoyancy forces exerted by the hydrocarbon column, is the difference in pressure between the hydrocarbon phase and the water phase measured at the same depth (Figure 2) (e.g., Schowalter, 1979; Watts, 1987; Brown, 2003). The primary source of data for buoyancy pressure profiles is pressure-depth points obtained from repeat formation tests (RFTs). Pressure gradients are typically derived by linear regression of the pressure-depth data points. Few fault seal studies report the accuracy of the pressure data used in calibrating fault seal attributes. Yielding et al. (1999) report a measurement accuracy of 1 bar (about 14 psi) for pressure data from the Gullfaks field, North Sea. Given the available data, the range of uncertainty associated with buoyancy pressure data is considered to be relatively narrow compared to the larger uncertainties that may exist in the structural model, juxtaposition pattern of reservoirs at the fault, and the estimate of V_{clay}.

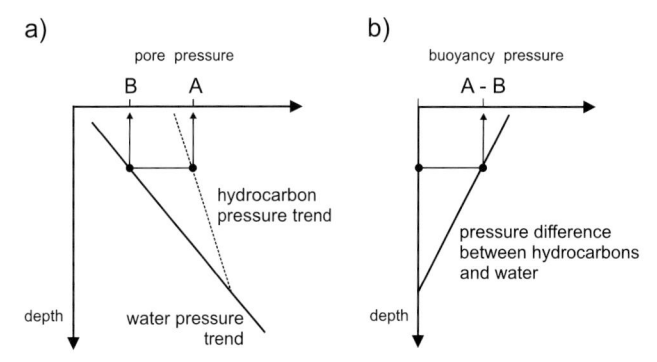

FIGURE 2. Pressure data shown as (a) pore pressure-depth profile and (b) buoyancy pressure-depth profile. Buoyancy pressure is the difference in pressure between the hydrocarbon (A) and water phase (B) measured at the same depth.

Fisher et al. (2001) have described capillary effects on fault sealing based on homogeneous fault zone properties. Here, we describe effects associated with heterogeneous faults. Figure 3 shows buoyancy pressure profiles incorporating heterogeneous fault zone properties for a fault shown as a two-dimensional section (Figure 3a). Fault zone composition, estimated using the SGR algorithm, is derived from the modeled fault surface at points where sand is juxtaposed against sand across the fault and plotted against depth (Figure 3b). The SGR values are transformed to capillary entry pressure using the empirical equations derived from calibrations with in-situ pressure data or from laboratory-based measurements. When the SGR values are transformed to capillary entry pressure using a continuous varying function, the overall shape of the capillary entry pressure data distribution (Figure 3c) is similar to the fault zone composition (Figure 3b). Seal failure occurs when the buoyancy pressure trend line intersects the left-hand edge of the cloud of capillary entry pressure data (C in Figure 3c). The maximum height of the hydrocarbon column that can be supported by the critical point without leaking is the difference in depth between the intersection point of the buoyancy pressure trend line with the minimum capillary entry pressure of the fault and the point where the buoyancy pressure equals zero, that is, at the oil-water or gas-water contact (Childs et al., 2002; Bretan et al., 2003). The depth on the fault where leakage occurs is a function of the buoyancy pressure gradient and the heterogeneous distribution of fault zone clay content. If the fault is sealing to hydrocarbons, the predicted capillary entry pressure of the fault zone must exceed the buoyancy

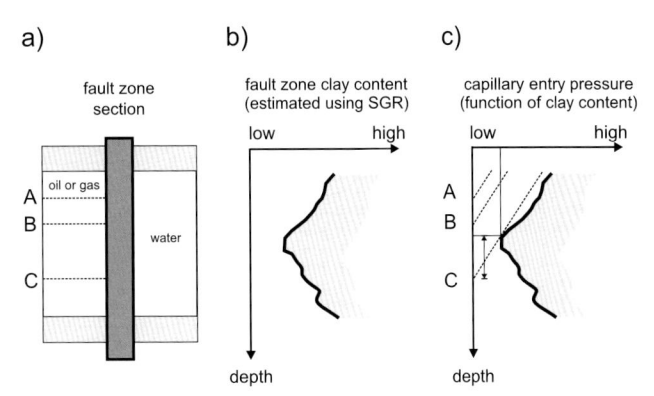

FIGURE 3. A schematic diagram showing the relationship between (a) a 2-D fault zone section, (b) fault zone composition (estimated using SGR) from sand-on-sand juxtapositions plotted against depth, and (c) predicted capillary entry pressure and buoyancy pressure trend lines. Shaded region in (b) and (c) is the distribution of data points derived from a heterogeneous fault surface. Fault zone composition is transformed to capillary entry pressure using continuous transformations. Lines labeled A, B, and C in (a) and (c) represent different levels of hydrocarbon fill and buoyancy pressure trend lines, respectively. Leakage occurs when the buoyancy pressure trend line intersects the minimum fault zone capillary entry pressure. Maximum hydrocarbon column height supported at critical point on fault is shown by a vertical arrow. If the fault is sealing to hydrocarbons, the derived capillary entry pressure of the fault zone must exceed the buoyancy pressure. Data points representing the capillary entry pressure must plot to the right of the buoyancy pressure trend line.

pressure for a given depth (Childs et al., 2002; Bretan et al., 2003); that is, data points representing the capillary entry pressure plot to the right of the buoyancy pressure trend line. If the capillary entry pressure values plot to the left of the buoyancy pressure trend line for part of the fault surface that is known to be sealing to hydrocarbons from in-situ pressure data, then it is likely that one or more of the data input variables or predictive algorithms are in error.

APPLICATIONS

In this section, we demonstrate the practical application of the method using three examples from the North Sea (Figure 4) where the SGR-based methodology for predicting fault seal has been widely used (e.g., Fristad et al., 1997; Ottesen Ellevset et al., 1998). The first example is a fault from the Oseberg Syd field located in block 30/9, northern North Sea. This fault was chosen to illustrate the effect of changing various input parameters on estimating seal potential for a fault with proven hydrocarbon contacts. The second example is from the Gullfaks Sør field located in block 34/10, northern North Sea, and is used to illustrate an example of a leaking fault. Finally, we show an example of a fault from Field C, located in the central North Sea, that is leaking and sealing to the same hydrocarbon phase. Details of the methodology for calculating fault seal attributes on 3-D gridded fault surfaces are described elsewhere (e.g., Yielding et al., 1997, 1999). The SGR attribute was calculated using the standard algorithm (Freeman et al., 1998; Yielding, 2002) except where otherwise mentioned. The transformation from SGR to capillary entry pressure uses the empirical relationship between SGR and across-fault pressure difference (equation 1 in Bretan et al., 2003).

Example: Sealing Fault

The fault used for this example is fault 1 in Fristad et al. (1997), which is located in the southwestern part of the Oseberg Syd field (Figure 5a) (see Fristad et al., 1997, for a complete description). Wells are located in the footwall (well 30/9-13S) and in the hanging wall (well 30/9-14) of fault 1 and have different hydrocarbon columns (Figure 5b). Pressure data (Figure 5c) indicate that the fault is sealing to hydrocarbons because the

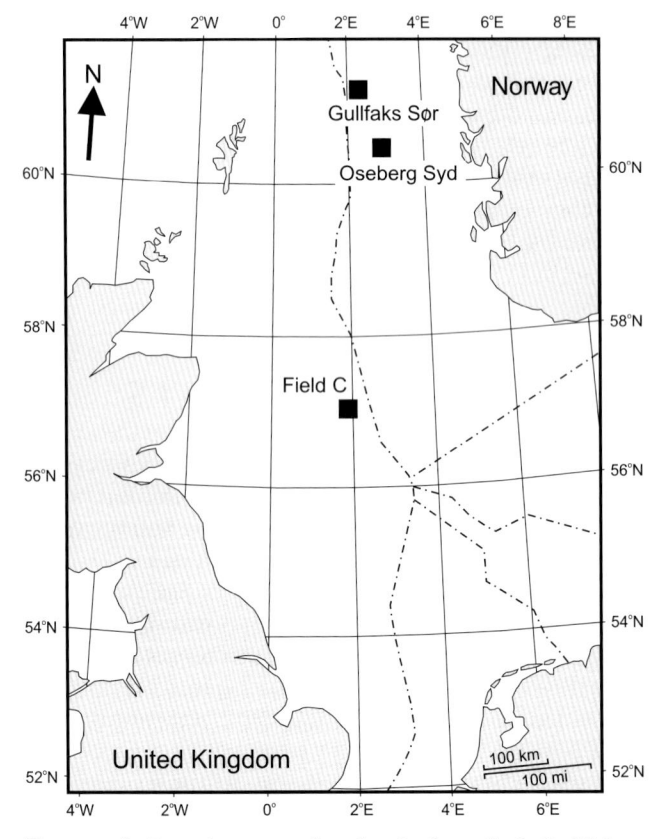

FIGURE 4. Location map for the Oseberg Syd, Gullfaks Sør, and Field C Fields, North Sea.

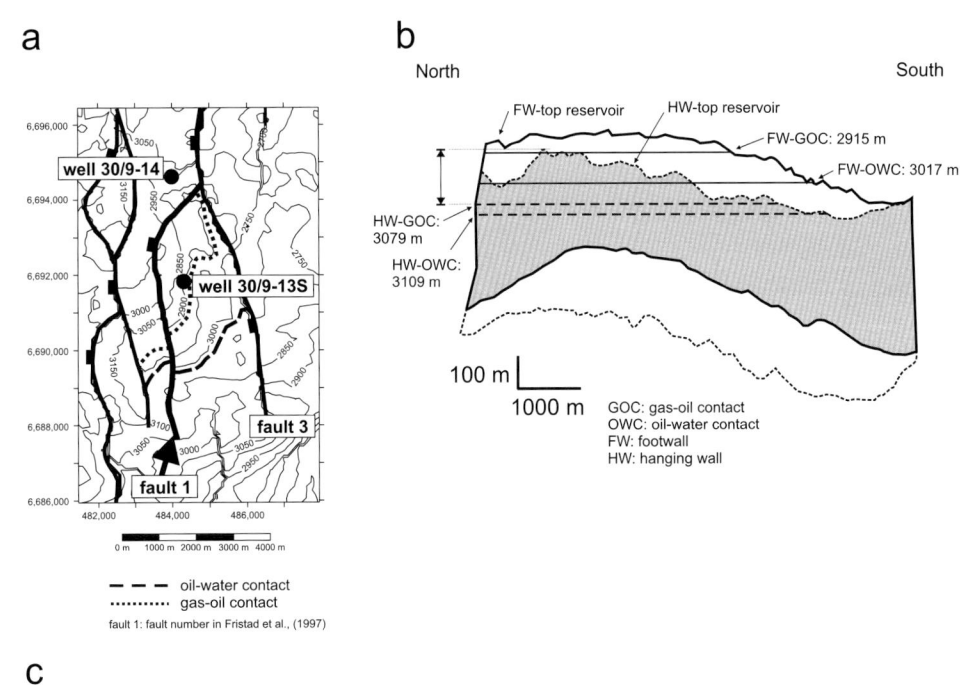

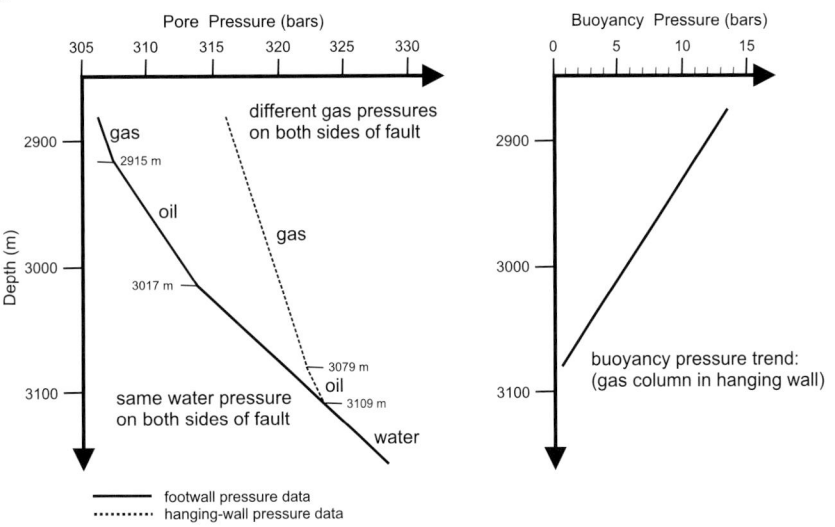

FIGURE 5. (a) Top reservoir structure map showing location of fault 1 (arrowed) and wells in the southwest part of the Oseberg Syd field; (b) fault plane diagram for fault 1, viewed from the west, showing reservoir juxtaposition (shaded) and hydrocarbon contacts. Shale gouge ratio values derived from reservoir juxtapositions over the depth range are shown by vertical arrow. (c) Pressure-depth and buoyancy-depth profiles showing pressure trends and hydrocarbon contacts from wells on either side of fault 1. The buoyancy pressure trend line was calculated from the gas column in the hanging wall.

gas pressures have not equalized across the fault (Fristad et al., 1997). In the footwall of fault 1, the gas-oil and oil-water contact are controlled by structural spill on fault 3 situated to the east of fault 1. In the hanging wall, the oil-water contact is controlled by a structural spillpoint along the southern part of fault 1 (Fristad et al., 1997). The gas-oil contact in the hanging wall is the only contact controlled by fault seal on fault 1. Fristad et al. (1997) derived a pressure difference of 9.5 bar (about 138 psi) measured across the fault, where gas-bearing intervals in the footwall are juxtaposed against gas-bearing intervals in the hanging wall. Brown (2003) pointed out that seal capacity should be estimated from the buoyancy pressure calculated from the tallest petroleum column instead of measuring the pressure difference across the fault (see also Bretan et al., 2003). The buoyancy pressure calculated using the gas column in the hanging wall is 13.5 bar (about 196 psi) at the gas-against-gas juxtaposition (Figure 5c). For fault 1 to be sealing, the predicted fault zone capillary entry pressure must be greater than 13.5 bar (about 196 psi).

Variability in V_{clay} data is one of the major uncertainties in fault seal analysis. Bretan et al. (2003) have shown that different vintages of V_{clay} logs can give rise to substantially different estimates of fault zone entry pressure and, hence, column heights. Figure 6a shows two different estimates for the V_{clay} content in the Brent reservoir sequence for well 30/9-14, which is located in the hanging wall of fault 1. Curve 1 was an early estimate of V_{clay} derived from an initial petrophysical interpretation of wire-line logs (Fristad et al., 1997). Curve 2 is a revised estimate incorporating kaolin and mica (S. Sperrevik, Norsk Hydro, 2000, personal communication). The different V_{clay} logs, when used as inputs into the SGR algorithm, produce markedly different SGR distributions (Figure 6b), clearly illustrating the effect of variability in the V_{clay} parameter on the calculated SGR values. Capillary entry pressures predicted using the SGR distributions 1 and 2 in Figure 6b are shown in Figure 6c and d, respectively. The predicted fault zone capillary entry pressure values using SGR distribution 1

FIGURE 6. (a) Different V_{clay} logs for the Brent reservoir sequence in well 30/9-14 used to derive SGR; (b) SGR distributions from reservoir juxtapositions using the V_{clay} logs shown in (a) over the depth range shown by vertical arrow in Figure 5b. Diagrams c–f are buoyancy pressure profiles illustrating the effect of changing input parameters on predicting the capillary entry pressure for fault 1. The parameters are different V_{clay} logs (c, d), equations derived from laboratory-based analysis of fault gouge samples (e), and a distance-weighting SGR algorithm (f). The SGR values were transformed to capillary entry pressure using the empirical equation in Bretan et al. (2003) for moderate burial depths (depth constant $C = 0.25$) except in (e). The buoyancy pressure trend line was calculated from the gas column in the hanging wall. Dashed line in (e) and (f) is the buoyancy pressure trend used to predict the hydrocarbon contact.

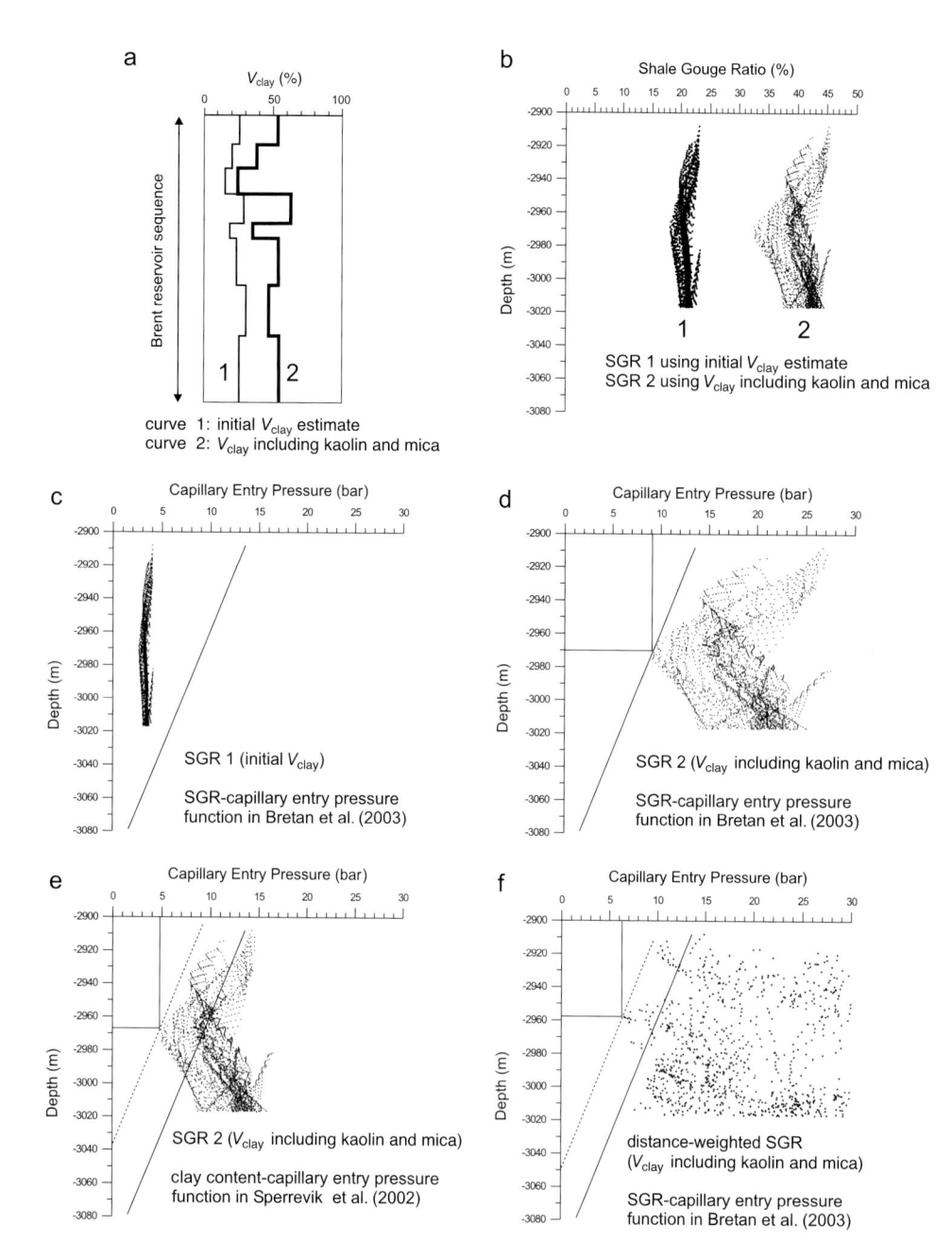

(initial V_{clay} estimate) are lower than the observed buoyancy pressure trend exerted by the gas column in the hanging wall (Figure 6c). The data plot to the left of the buoyancy pressure trend line, contradicting the observation of seal in the subsurface. In contrast, the capillary entry pressure values derived using SGR distribution 2 (revised estimate for V_{clay}) are higher than the buoyancy pressure, plot to the right of the buoyancy pressure trend (Figure 6d), and are consistent with the observation of seal in the subsurface. The intersection of the buoyancy pressure trend with the minimum capillary entry pressure at a depth of 2970 m (9741 ft) in Figure 6d gives a predicted value similar to the observed gas-oil contact (3079 m; 10,101 ft). Constructing diagrams like the examples in Figure 6c and d can quickly alert the explorationist to inadequacies in the estimate of the V_{clay} input parameter.

Figure 6e illustrates the effect of changing the function used to transform the SGR value to capillary entry pressure. In Figure 6e, the equations relating fault rock clay fraction to laboratory-measured capillary entry pressure (Sperrevik et al., 2002) were used to transform the SGR values. For fault 1, the burial depth at the time of faulting was shallower than approximately 500 m (1600 ft) (Fristad et al., 1997), with the maximum burial depth equal to the present-day depth. A value of 200 m (660 ft), representing the original thickness of the Brent reservoir sequence, was used for the burial depth at the time of faulting. Shale gouge ratio distribution 2 shown in Figure 6b was transformed into capillary entry pressures using equations 10 and 11 of Sperrevik et al. (2002). The Sperrevik equations used to derive the capillary entry pressure underestimate the observed buoyancy pressure by about 5 bar (70 psi). The predicted depth of the gas-oil contact is about 3035 m (9954 ft) compared to the observed depth of 3079 m (10,099 ft). Uncertainties involved in deriving representative transformations from laboratory-based analysis are described by Sperrevik et al. (2002) and also in Yielding (2002). It is likely that one or more of the uncertainties are contributing to the lower predicted capillary entry pressure and shallower predicted contact in Figure 6e. An additional bias is that the multivariable equations derived from the fault gouge samples are shown as best fit trends through the available data. In terms of estimating seal strength using fault gouge samples, it is the lowest capillary entry pressures for any given clay content that are important (Yielding, 2002).

The effect of changing the SGR algorithm on the predicted capillary entry pressure is shown in Figure 6f. The implicit assumption in the standard SGR method is that, on average, wall rock is entrained in the fault zone in the same proportion as it occurs in the slipped interval (Fristad et al., 1997; Yielding et al., 1997; Freeman et al., 1998; Yielding, 2002). Shale beds farthest from the point of calculation in the slipped interval are assumed to make the same amount of clay contribution to the fault zone as shale beds closest to the point. The average V_{clay} content of the fault zone is equal to the average V_{clay} content of the wall rock in the slipped interval (Figure 7). Other algorithms incorporate a distance-weighting dependence for the incorporation of clay into the fault zone and are generically termed smear factors (SF) (Yielding et al., 1997), e.g., the clay smear potential algorithm of Bouvier et al. (1989), Jev et al. (1993), and Fulljames et al. (1997). The smear factor is inversely related to the distance from the calculation point, reflecting the observation that more clay material will be entrained into the fault zone when the calculation point is closer to the shale beds (Figure 7). This same method of weighting can be applied in the SGR algorithm, giving a distance-weighted SGR (see Appendix of this chapter). The distance-weighted SGR value was derived using V_{clay} curve 2 in Figure 6a. The minimum capillary entry pressure predicted using distance-weighted SGR (Figure 6f) is lower than the observed buoyancy pressure trend. The predicted gas-oil contact (about 3050 m [10,004 ft]) derived using a distance-weighted SGR algorithm is about 30 m (100 ft) shallower than the observed contact. The effect of applying a distance-weighting function to the standard SGR algorithm is to increase the dispersion in the range of SGR values (compare Figure 6d with f). Therefore, the lowest SGR values are lower than computed using the standard SGR algorithm, thereby reducing the predicted capillary entry pressure.

Example: Leaking Fault

This section describes an example of a leaking fault, Figure 8a from the Gullfaks Sør field (see Linjordet et al., 1997, for details on fault seal analysis and pressure communication in Gullfaks Sør). Capillary entry pressures for a leaking fault should be lower than the buoyancy pressure and plot to the left of the buoyancy pressure trend line on a buoyancy pressure profile. The fault described herein separates fault block 2 from fault block 3 in Linjordet et al. (1997) (Figure 8a). Wells are located in the footwall (well 34/10-36) and in the hanging wall (well 34/10-33). The fault is characterized by a complex hydrocarbon-pressure distribution (Figure 8b, c). In the footwall, the gas-water contact occurs at a depth of 3331 m (10,925 ft). In the hanging wall, gas-oil and oil-water contacts in the Brent reservoir section occur at a depth of 3305 and 3363 m (10,840 and 11,030 ft), respectively. Where the gas-bearing reservoirs are juxtaposed across the fault, the pressures in the gas phase occur on the same trend, indicating that gas is probably in communication across this fault. The SGR values derived from the fault zone separating the juxtaposed gas-bearing reservoirs (Figure 8d) range from 16 to 34%. Although most of the SGR values in Figure 8d are higher than the generally accepted threshold for seal (about 15–20%; Yielding, 2002), the predicted capillary entry pressures at depths shallower than about 3270 m (10,725 ft) plot to the left of the buoyancy pressure trend calculated from the gas column in the footwall (Figure 8e). The distribution of data points is consistent with the interpretation that this part of the fault is leaking to gas. The example in Figure 8 clearly illustrates the importance of incorporating buoyancy pressure data from hydrocarbons (if present) to establish the threshold SGR values for sealing (see also Fisher et al.,

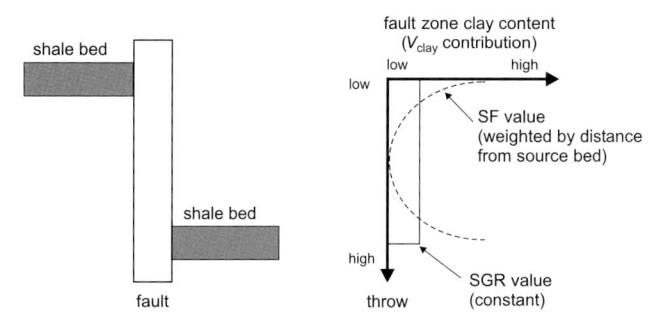

FIGURE 7. Schematic representation of the V_{clay} contribution to fault zone composition in the SGR and SF algorithms. SGR algorithm assumes constant V_{clay} contribution in the throw window, whereas SF algorithms model a nonlinear increase in V_{clay} contribution with decreasing distance to shale beds. A distance-weighting function can be incorporated into the standard SGR method by assuming that the total V_{clay} contribution made by both algorithms is the same (that is, the area under the curves is balanced).

FIGURE 8. (a) Top reservoir structural map showing location of the fault (arrowed) and wells in the Gullfaks Sør field (redrawn from figure 2 in Linjordet et al., 1997); (b) schematic fault plane diagram, viewed from the east, of the fault arrowed in (a) showing reservoir juxtapositions (shaded) and hydrocarbon contacts. Shale gouge ratio data were derived from reservoir juxtapositions over the depth range shown by vertical arrow. (c) Pressure-depth and buoyancy-depth profiles showing pressure trends. The buoyancy pressure was calculated from the gas column in the hanging wall. (d) Shale gouge ratio values from reservoir juxtaposition, and (e) predicted capillary entry pressure and buoyancy pressure trend lines. Most of the capillary entry pressure values plot to the left of the buoyancy pressure trend line and are consistent with the interpretation that the upper part of this fault is leaking to gas. The SGR values were transformed to capillary entry pressure using the empirical equation in Bretan et al. (2003) for moderate burial depths (depth constant $C = 0.25$).

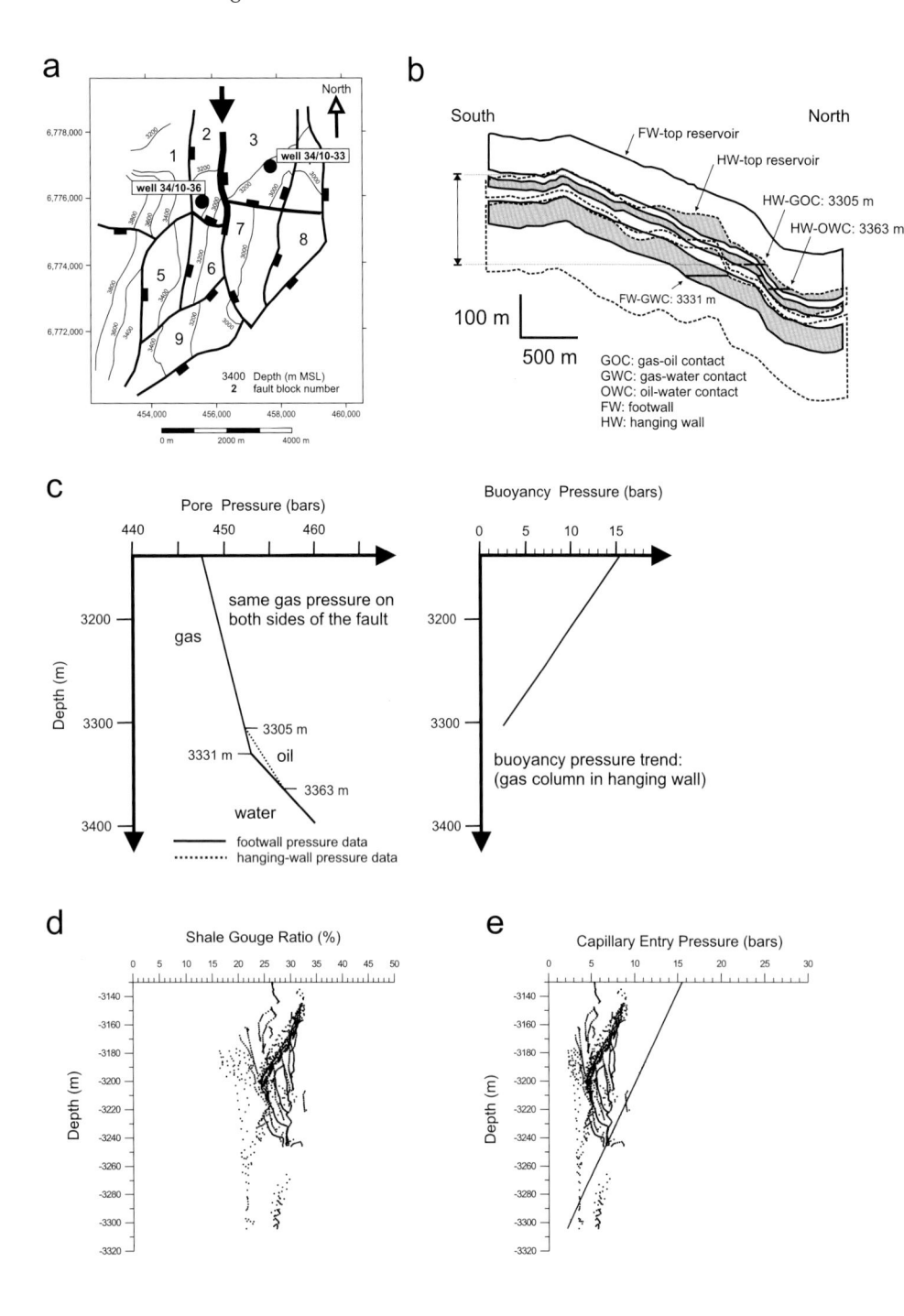

2001). The implication of this analysis is that evaluating the sealing potential of a fault based on the SGR values alone can be misleading. Where hydrocarbon columns are tall enough, their buoyancy pressure can overcome the initial seal.

Example: Complex Sealing and Leaking Fault

The final example, Figure 9, is derived from a fault located in Field C in the central North Sea (Figure 4). Gas leakage and pressure equalization occurs across the upper part of the fault (Figure 9a, b). At a lower structural level, the fault separates high-pressure aquifer from low-pressure gas. Below the gas-water contact in the footwall, the fault separates differently pressured aquifers. An SGR-buoyancy pressure calibration plot for this fault shows no obvious threshold for the onset of sealing (figure 11 of Bretan et al., 2003). We have replotted the calibration data in terms of SGR against depth below the top reservoir (Figure 9c), predicted capillary entry pressure, and observed buoyancy pressure trends (Figure 9d). The predicted capillary entry pressures for the fault zone above 214 m (701 ft) occur to the left of the buoyancy pressure trend calculated using the hanging-wall gas column. This is consistent with pressure communication in the gas leg on the upper part of the fault. Between depths 214 and 330 m (701 and 1082 ft), the capillary entry

FIGURE 9. (a) Schematic fault section across a fault in Field C, central North Sea, showing hydrocarbon contacts. The fault is shown as a vertical slab showing part of the fault that is leaking (horizontal lines) and sealing to gas (solid fill). (b) Pressure-depth and buoyancy-depth profiles showing pressure trends. Buoyancy pressure trend lines calculated from the gas columns in the footwall and in the hanging wall; (c) SGR values from sand-on-sand reservoir juxtaposition over the depth range shown by vertical arrow in (a); and (d) capillary entry pressures and buoyancy pressure trends. Depth values are depth below top of reservoir. The SGR values were transformed to capillary entry pressure using the empirical equation in Bretan et al. (2003) for deep burial depths (depth constant $C = 0$). HP = high water pressure; LP = low water pressure.

pressures plot to the right of the hanging-wall buoyancy pressure trend (dashed line in Figure 9d) but to the left of the buoyancy pressure trend calculated using the footwall gas column (solid line in Figure 9d). If the tallest gas column, that is, the footwall gas column, was used to evaluate the sealing potential for this fault, one interpretation would be that the fault is leaking to gas upward from about 260 m (852 ft) because the predicted capillary entry pressure data plot to the left of the footwall buoyancy pressure trend (Figure 9d). Although the buoyancy pressure in the footwall is greater than the capillary entry pressure of the fault zone, gas leakage into the hanging wall is prevented because the observed pressure in the footwall gas leg at a depth of 260 m (852 ft) below the top reservoir is lower than the aquifer pressure in the hanging wall (Figure 9b).

The buoyancy pressure profile in Figure 9d can be used to help understand the sealing behavior of this fault. One plausible filling history is shown in Figure 10. It is assumed that the difference in aquifer pressure at the time of charge was negligible. Regional studies show that gas migration and water flow were from the hanging wall into the footwall block. Initially, at a small degree of trap fill, the fault is sealing to gas (Figure 10a).

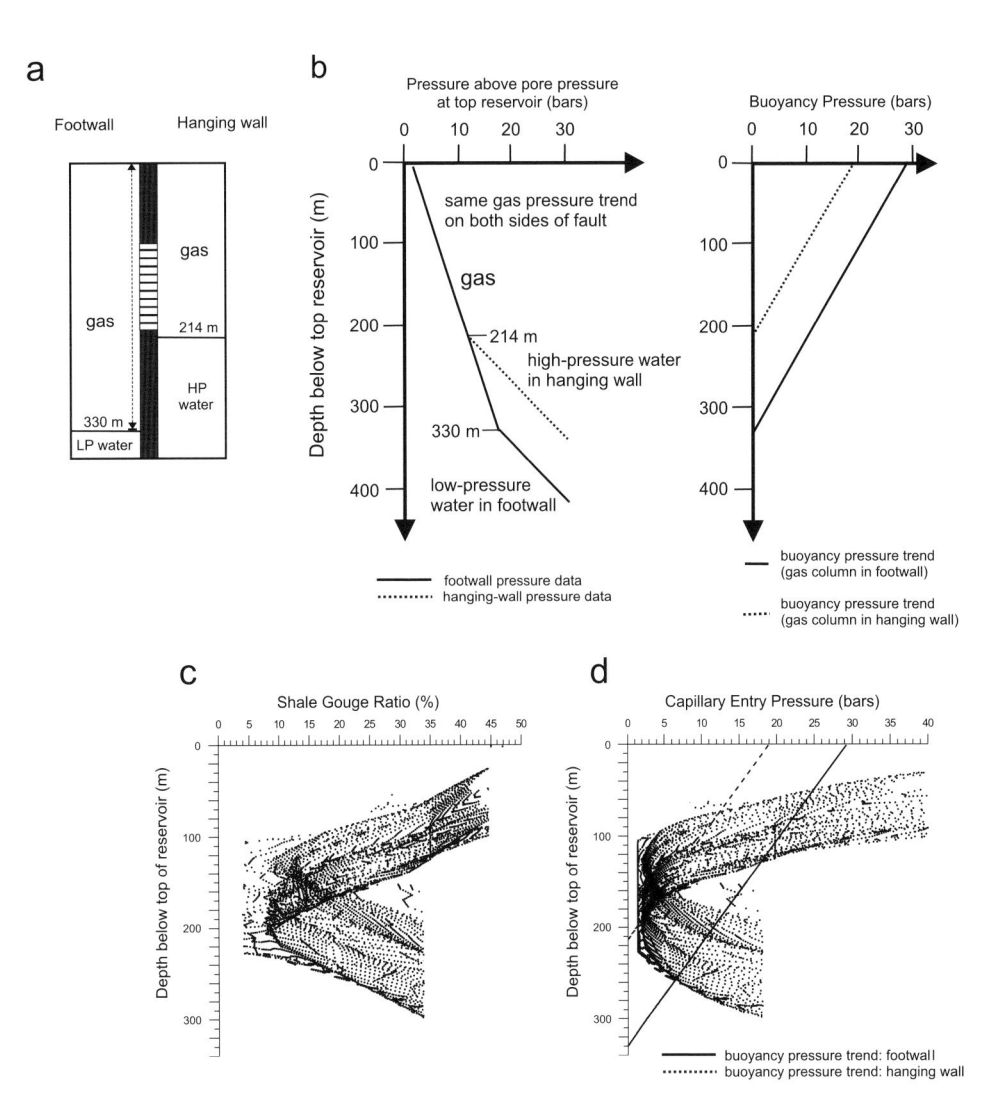

Gas charge increases the height of the gas column and, hence, buoyancy pressure in the hanging wall. The increase in buoyancy pressure moves the buoyancy pressure trend toward the right until the pressure trend intersects the lowest predicted capillary entry pressure of the fault (line ii in Figure 10a). Gas begins to leak across the fault at this point. Leakage across the fault begins to fill the footwall with gas, with leakage restricted to the depth range where the capillary entry pressure of the fault zone is lower than the buoyancy pressure exerted by the hanging-wall gas column (Figure 10b). Gas pressures equalize across the fault when the depth of the gas-water contact in the footwall equals that in the hanging wall. Continued charge increases the height of both gas columns until the gas-water contact in the hanging wall is close to the base of the low-seal window on the fault. This window will have relatively high fault zone permeability as well as low capillary entry pressure. As the gas-water contacts push below the high-permeability window (Figure 10c), the flow of water across the fault is reduced, causing the aquifer pressure to increase in the hanging-wall block. Gas continues to

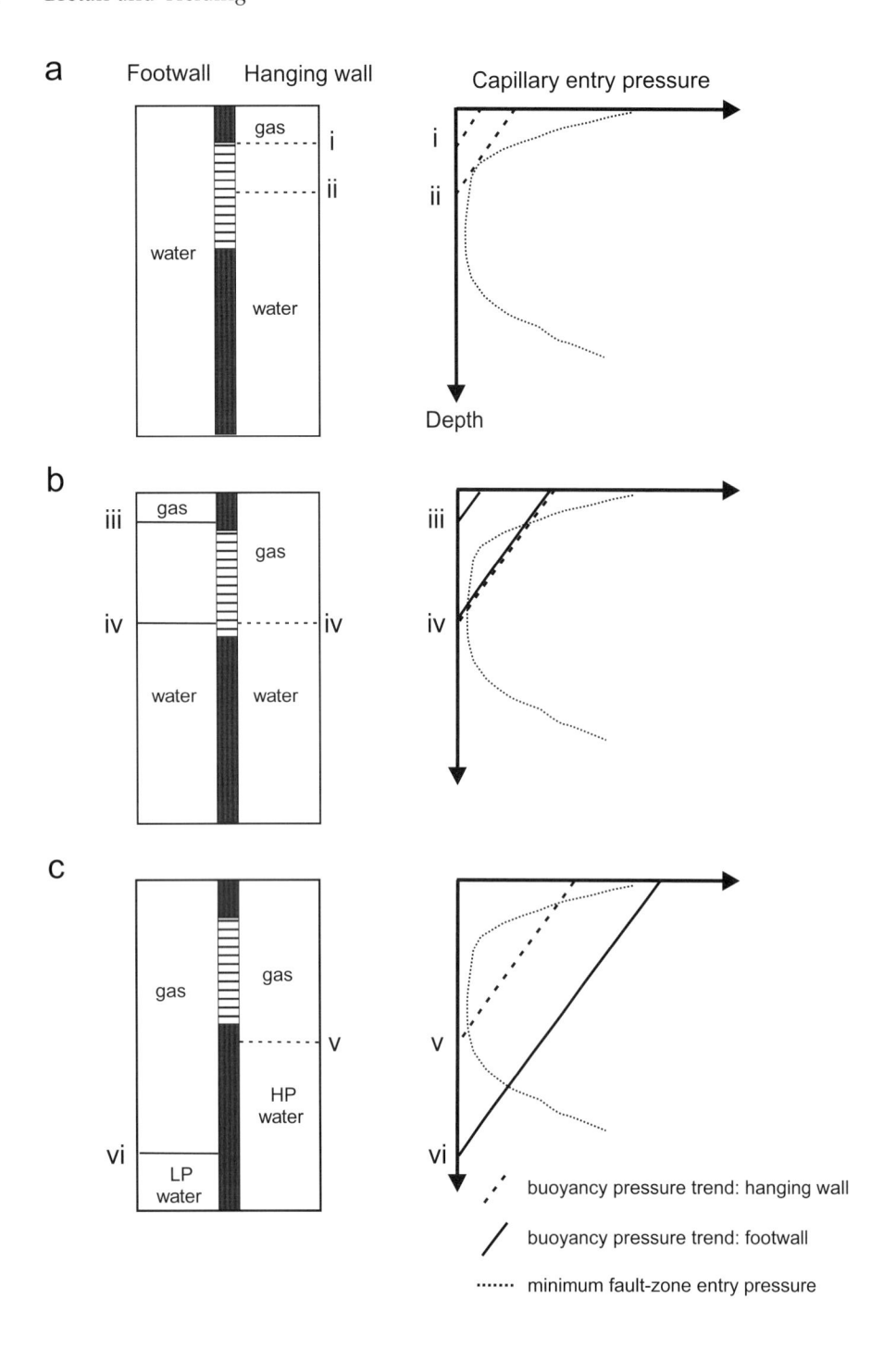

FIGURE 10. Fault zone sections and schematic buoyancy pressure profiles illustrating one possible filling history for the fault in Figure 9. Filling of the hanging-wall block is shown in (a), seal failure and across-fault leakage in (b), and present-day situation in (c). Lines labeled i to vi represent different levels of hydrocarbon fill. During initial migration (a), the height of the gas column (i) increases until the buoyancy pressure equals minimum capillary entry pressure of fault zone (ii). Gas begins to leak across the fault to fill the footwall block (iii in b). Gas pressure equalization occurs when the gas-water contacts are at the same depth on both sides of the fault (iv). Continuing gas charge (c) moves the gas-water contact in the hanging wall below the low-permeability window (v), reducing the flow of water across the fault. Further gas migration across the fault increases the height of the gas column in the footwall (vi). HP = high-pressure water; LP = low-pressure water.

leak into the footwall, causing the gas-water contact to move farther down in the footwall block. The gas-water contact in the hanging-wall side is a form of perched contact.

CONCLUSIONS

- Buoyancy pressure profiles can be used to evaluate uncertainty in fault seal calibration studies. The method can be used to quickly identify which seal analysis methodologies best predict the depth of observed hydrocarbon contacts.
- For a sealing fault, the predicted fault zone capillary entry pressure must be higher than the observed buoyancy pressure. If the predicted capillary entry pressure is lower than the buoyancy pressure, either the fault is leaking or one or more of the data input parameters and/or predictive algorithms are in error.
- Shale gouge ratio values that represent the onset of fault sealing should be verified using buoyancy pressure data from observed hydrocarbons columns.
- Observed buoyancy pressure trends show that gas leakage across a fault in the Gullfaks Sør field may have occurred at SGR values as high as 32%. High buoyancy pressures associated with a tall gas column have overcome the sealing capacity of the fault zone.

- Using buoyancy pressure profiles as a point of reference by which other input variables can be assessed provides a quick-look check on the validity of the input data.

ACKNOWLEDGMENTS

The authors are grateful to Rob Weeden, Peter Boult, and an anonymous reviewer for their thoughtful and constructive comments.

APPENDIX

Original definitions of the SGR and smear factor (SF) algorithms are as follows:

$$\mathrm{SGR} = \sum [(V_{\mathrm{clay}} \times \mathrm{thickness})/\mathrm{throw}] \times 100\% \quad (1)$$

$$\mathrm{SF} = \sum (\mathrm{thickness}/\mathrm{distance}) \quad (2)$$

where V_{clay} = volumetric clay fraction of the interval; thickness = thickness of the interval at the fault; distance = distance from the center of the interval to the calculation point.

A hybrid equation incorporating equations 1 and 2 can be expressed as

$$\sum [(V_{\mathrm{clay}} \times \mathrm{thickness})/\mathrm{distance}] \times 100\% \quad (3)$$

We then normalize so that the average distance-weighted SGR value should be the same as the average SGR value derived using the standard algorithm to reflect the equal contribution of sand- and clay-rich rocks to fault zone composition, although the distribution would be heterogeneous (more clay is entrained closer to the source beds). We define a distance-weighted SGR (dwSGR) algorithm as

$$\mathrm{dwSGR} = C \times \sum [(V_{\mathrm{clay}} \times \mathrm{thickness})/\mathrm{distance}] \times 100\% \quad (4)$$

where C = average SGR/average SF.

Equation 4 is one of many possible different ways of weighting the SGR value by the distance from the source bed.

REFERENCES CITED

Bouvier, J. D., C. H. Kaars-Sijpesteijn, D. F. Kluesner, C. C. Onyejekwe, and R. C. Van Der Pal, 1989, Three-dimensional seismic interpretation and fault sealing investigations, Nun River field, Nigeria: AAPG Bulletin, v. 73, p. 1397–1414.

Bretan, P., G. Yielding, and H. Jones, 2003, Using calibrated shale gouge ratio to estimate hydrocarbon column heights: AAPG Bulletin, v. 87, p. 397–413.

Brown, A., 2003, Capillary pressure effects on fault-fill sealing: AAPG Bulletin, v. 87, p. 381–395.

Childs, C., J. Watterson, and J. J. Walsh, 1997, Complexity in fault zone structure and implications for fault seal prediction, in P. Møller-Pedersen and A. G. Koestler, eds., Hydrocarbon seals: Importance for exploration and production: Norwegian Petroleum Society Special Publication 7, p. 61–72.

Childs, C., O. Sylta, S. Moriya, J. J. Walsh, and T. Manzocchi, 2002, A method for including the capillary properties of faults in hydrocarbon migration models, in A. G. Koestler and R. Hunsdale, eds., Hydrocarbon seal quantification: Norwegian Petroleum Society Special Publication 11, p. 127–139.

Fisher, Q. J., and R. J. Knipe, 1998, Fault sealing processes in siliciclastic sediments, in G. Jones, Q. J. Fisher, and R. J. Knipe, eds., Faulting, fault sealing and fluid flow in hydrocarbon reservoirs: Geological Society (London) Special Publication 147, p. 117–134.

Fisher, Q. J., and R. J. Knipe, 2001, The permeability of faults within siliciclastic petroleum reservoirs of the North Sea and Norwegian continental shelf: Marine and Petroleum Geology, v. 18, p. 1063–1081.

Fisher, Q. J., S. D. Harris, E. McAllister, R. J. Knipe, and A. J. Bolton, 2001, Hydrocarbon flow across faults by capillary leakage revisited: Marine and Petroleum Geology, v. 18, p. 251–257.

Freeman, B., G. Yielding, D. T. Needham, and M. E. Badley, 1998, Fault seal prediction: The gouge ratio method, in M. P. Coward, T. S. Daltaban, and H. Johnson, eds., Structural geology in reservoir characterization: Geological Society (London) Special Publication 127, p. 19–25.

Fristad, T., A. Groth, G. Yielding, and B. Freeman, 1997, Quantitative fault seal prediction: A case study from Oseberg Syd, in P. Møller-Pedersen and A. G. Koestler, eds., Hydrocarbon seals: Importance for exploration and production: Norwegian Petroleum Society Special Publication 7, p. 107–124.

Fulljames, J. R., L. J. J. Zijerveld, and R. C. M. W. Franssen, 1997, Fault seal processes: Systematic analyses of fault seals over geological and production time, in P. Møller-Pedersen and A G. Koestler, eds., Hydrocarbon seals: Importance for exploration and production: Norwegian Petroleum Society Special Publication 7, p. 51–59.

Gibson, R. G., 1994, Fault-zone seals in siliciclastic strata of the Columbus basin, offshore Trinidad: AAPG Bulletin, v. 78, p. 1372–1385.

Gibson, R. G., 1998, Physical character and fluid-flow properties of sandstone-derived fault gouge, in M. P. Coward, T. S. Daltaban, and H. Johnson, eds., Structural geology in reservoir characterization: Geological Society (London) Special Publication 127, p. 83–97.

Gibson, R. G., and P. A. Bentham, 2003, Use of fault-seal analysis in understanding petroleum migration in a complexly faulted anticlinal trap, Columbus basin, offshore Trinidad: AAPG Bulletin, v. 87, p. 465–478.

Hesthammer, J., and H. Fossen, 2000, Uncertainties associated with fault sealing analysis: Petroleum Geoscience, v. 6, p. 37–45.

Jev, B. I., C. H. Kaars-Sijpesteijn, M. P. A. M. Peters, N. L. Watts, and J. T. Wilkie, 1993, Akaso field, Nigeria: Use of integrated 3-D seismic, fault slicing, clay smearing, and RFT pressure data on fault trapping and dynamic leakage: AAPG Bulletin, v. 77, p. 1389–1404.

Knipe, R. J., G. Jones, and Q. J. Fisher, 1998, Faulting, fault sealing and fluid flow in hydrocarbon reservoirs: An introduction, in G. Jones, Q. J. Fisher, and R. J. Knipe, eds., Faulting, fault sealing and fluid flow in hydrocarbon reservoirs: Geological Society (London) Special Publication 147, p. vii–xxi.

Linjordet, A., P. E. Nielsen, and E. Siring, 1997, Heterogeneities modeling and uncertainty quantification of the Gullfaks Sør Brent Formation in-place hydrocarbon volumes: Society of Petroleum Engineers Formation Evaluation, v. 12, p. 202–207.

Ottesen Ellevset, S., R. J. Knipe, T. S. Olsen, Q. Fisher, and G. Jones, 1998, Fault controlled communication in the Sleipner Vest field, Norwegian continental shelf: Detailed, quantitative input for reservoir simulation and well planning, in G. Jones, Q. J. Fisher, and R. J. Knipe, eds., Faulting, fault sealing and fluid flow in hydrocarbon reservoirs: Geological Society (London) Special Publication 147, p. 283–297.

Schowalter, T. T., 1979, Mechanics of secondary hydrocarbon migration and entrapment: AAPG Bulletin v. 63, p. 723–760.

Sperrevik, S., P. A. Gillespie, Q. J. Fisher, T. Halvorsen, and R. J. Knipe, 2002, Empirical estimation of fault rock properties, in P. Møller-Pedersen and A. G. Koestler, eds., Hydrocarbon seals: Importance for exploration and production: Norwegian Petroleum Society Special Publication 11, p. 109–125.

Takahashi, M., 2003, Permeability change during experimental fault smearing: Journal of Geophysical Research, v. 108, p. 2235–2250.

Watts, N., 1987, Theoretical aspects of cap-rock and fault seals for single- and two-phase hydrocarbon columns: Marine and Petroleum Geology, v. 4, p. 274–307.

Yielding, G., 2002, Shale gouge ratio—Calibration by geohistory, in A. G. Koestler and R. Hunsdale, eds., Hydrocarbon seal quantification: Norwegian Petroleum Society Special Publication 11, p. 1–15.

Yielding, G., B. Freeman, and T. Needham, 1997, Quantitative fault seal prediction: AAPG Bulletin, v. 81, p. 897–917.

Yielding, G., J. A. Overland, and G. Byberg, 1999, Characterization of fault zones for reservoir modeling: An example from the Gullfaks field, northern North Sea: AAPG Bulletin, v. 83, p. 925–951.

Evaluation of Late Cap Rock Failure and Hydrocarbon Trapping Using a Linked Pressure and Stress Simulator

A. E. Lothe
Sintef Petroleum Research, Trondheim, Norway

H. Borge
Sintef Petroleum Research, Trondheim, Norway

Ø. Sylta
Sintef Petroleum Research, Trondheim, Norway

ABSTRACT

Hydraulic fracturing and leakage can be controlling factors for hydrocarbon leakage in overpressured sedimentary basins over geological time. Knowledge of the lateral flow properties of major faults is needed to simulate how pressure generation and dissipation can influence the sealing potential of cap rocks. The hydraulic fracture processes in the cap rock need to be evaluated to quantify timing and the amount of hydraulic leakage.

To address these issues, we use a single-phase simulator, which calculates pressure generation resulting from mechanisms, such as shale compaction and drainage, and mechanical and chemical compaction in sandstones. Pressure dissipation and lateral flow are simulated between different pressure and stress compartments defined by major fault patterns at the top reservoir level. An empirical model for the minimum horizontal stress is applied to the Griffith–Coulomb failure criterion and the sliding criterion to estimate hydraulic fracturing.

Only minor changes, if any at all, in the amount and timing of hydraulic fracturing and leakage in the modeled pressure compartments are present when the coefficient of internal friction and frictional sliding are varied. When varying fault permeability, low fault permeabilities give early leakage, whereas high permeabilities result in late or no hydraulic fracturing and leakage. Our simulations also suggest that leakage in one pressure compartment influences the neighboring compartments, and large compartments control the leakage pattern in surrounding areas. The amount of cumulative leakage depends on the timing of leakage and size of the compartment. Uncertainties of timing and leakage for different compartments can be estimated using the pressure measured in the wells today as calibration. The uncertainty in the estimates can be used as guidelines for possible hydrocarbon leakage risks.

INTRODUCTION

Knowledge of the timing and distribution of hydraulic fractures is essential for evaluating hydrocarbon potential in many overpressured sedimentary basins (Grauls, 1998; Hermanrud and Bolas, 2002). Overpressure can be generated by several mechanisms, including rapid sedimentation and compaction disequilibrium of sediments, hydrocarbon diagenesis, sediment diagenesis (Swarbrick and Osborne, 1998), aquathermal pressuring (Chapman, 1980; Miller and Luk, 1993,) and lateral compression (Bour et al., 1995). Field data indicate that overpressuring of sedimentary units can cause episodic fluid expulsion into overlying layers during basin subsidence (Hunt, 1990). The understanding of hydraulic fracturing and leakage has been limited by a lack of good field exposures and also a lack of cores from fractured cap rocks (Cosgrove, 1998). One way to achieve new knowledge of which parameters influence hydraulic failure of cap rocks leading to leakage is to use basin simulators that couple pressure (Borge and Sylta, 1998; Borge, 2000) and time-related stresses (Lothe et al., 2004).

The relation between hydraulic fracturing and leakage in sedimentary basins is not well known. It is recognized that large volumes of fluid can become trapped during significant overpressure buildup in pressure compartments (Darby et al., 1996). A rapid fluid pressure release should be expected when the cap rock fails (Cosgrove, 1998). Capillary leakage of hydrocarbons is not considered in this work, because only the water-phase flow is modeled. The failure mode will depend on the burial depth, stress regime, and pressure. The simplest scenario assumes leakage along one fault (Verbeek and Grout, 1993). Overpressures will, in this case, be highest at the top of the pressure compartment, resulting in fault nucleation and growth from the reservoir into and through the cap rock. Alternatively, fracture swarms may develop as shear failures in the deeper parts of the cap rock and tensile failures in the shallower parts (Cosgrove, 1998). This is illustrated in Figure 1, where high overpressures have been building up in the deepest part of the reservoir. Low transmissibilities across nonjuxtaposed reservoir faults decrease the lateral fluid flow, and hydraulic failure and leakage will occur. Reactivation of preexisting faults is a third possibility (Wiprut and Zoback, 2002). However, mechanisms involving slip along preexisting fault zones facilitated by overpressure buildup or tectonism are not considered. We will analyze possible overpressure-initiated failure at the highest point of the compartments along faults with shear failures at deep burial depths and mixed or tensile failures at shallower depths (Figure 1).

Lateral fluid flow between pressure compartments is an important and sometimes controlling factor on pressure buildup in an overpressured basin. Lateral fluid flow in faulted reservoir systems is predominantly controlled by the permeability of the major bounding faults. Different attempts have been made to measure fault permeabilities in the laboratory (Morrow et al., 1984; Zhang et al., 2001; Sperrevik et al., 2002). Sperrevik et al. (2002) observed a relationship between the fault permeability and the mineralogy of the faulted rock, the effective stress conditions, and the history to the reservoir prior to, during, and after deformation. They measured very low permeabilities ($<10^{-7}$ md) in faults with clay smears at depth (>3.5 km; >2.1 mi). The fault rock permeability with a phyllosilicate framework lies in the order of less than 10^{-4} md at greater than 3.5-km (2.1-mi) depth. Revil and Cathles (1999) present examples of measurements in shaly sands and demonstrated a rapid decrease in permeability with increasing clay content. In the present work, we do not discuss the

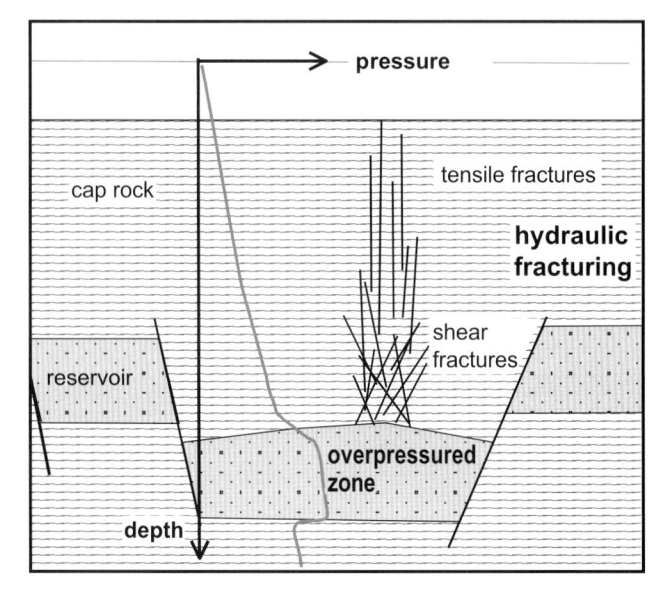

FIGURE 1. Sketch showing hydraulic fracturing and leakage of fluids from overpressured reservoir units.

mechanisms of how lateral flow across a major fault could occur, but instead consider how changes in the fault permeability may influence hydraulic fracturing and leakage. The analysis is based on the fault-permeability model of Borge and Sylta (1998).

Hydrocarbon buoyancy contributes generally little to pressure buildup (Bjørkum et al., 1998; Swarbrick and Osborne, 1998). Hydrocarbon leakage, however, is expected to be sensitive to hydraulic fracturing and water fluid leakage in an overpressured area. Hydrocarbon fill history is also time dependent and relies on the charge, trapping, hydraulic fracturing, and leakage histories. In this study, we aim to predict and quantify uncertainties with respect to the timing and amounts of leakage from overpressured areas. The results from these single-phase simulations may therefore be used as input to the secondary hydrocarbon migration modeling for estimating hydrocarbon volumes in prospects.

The simulations in this study are carried out on a data set from the Halten Terrace, offshore mid-Norway (Figure 2). The pressure distribution in this area is described in, e.g., Skar et al. (1999), Teige et al. (1999), and Berg et al. (2000), and pressure simulations have been carried out by Borge (2000, 2002) and Lothe et al. (2004). The geology in the area has been extensively described; see Blystad et al. (1995) and Koch and Heum (1995) for further information. The study area is selected because of two main reasons. (1) None of the eight wells penetrating Jurassic reservoir rocks at Haltenbanken prior to 1996 found hydrocarbons, and cap rock leakage seemed to be the main explanation for these results. Later drilling has, however, resulted in hydrocarbon discoveries (Hermanrud and Bolås, 2002). (2) A pressure compartmentalization is described in the area (e.g., Berg et al., 2000). The main focus of this chapter, however, will be on the method and not on the specific study area.

METHODS AND DATA

In the present study, the fault traces mapped at the top reservoir level delineate the lateral extent of the pressure and stress compartments that are used in a pressure and stress simulator (Pressim; Borge, 2000; Lothe et al., 2004). The lateral Darcy flow of formation

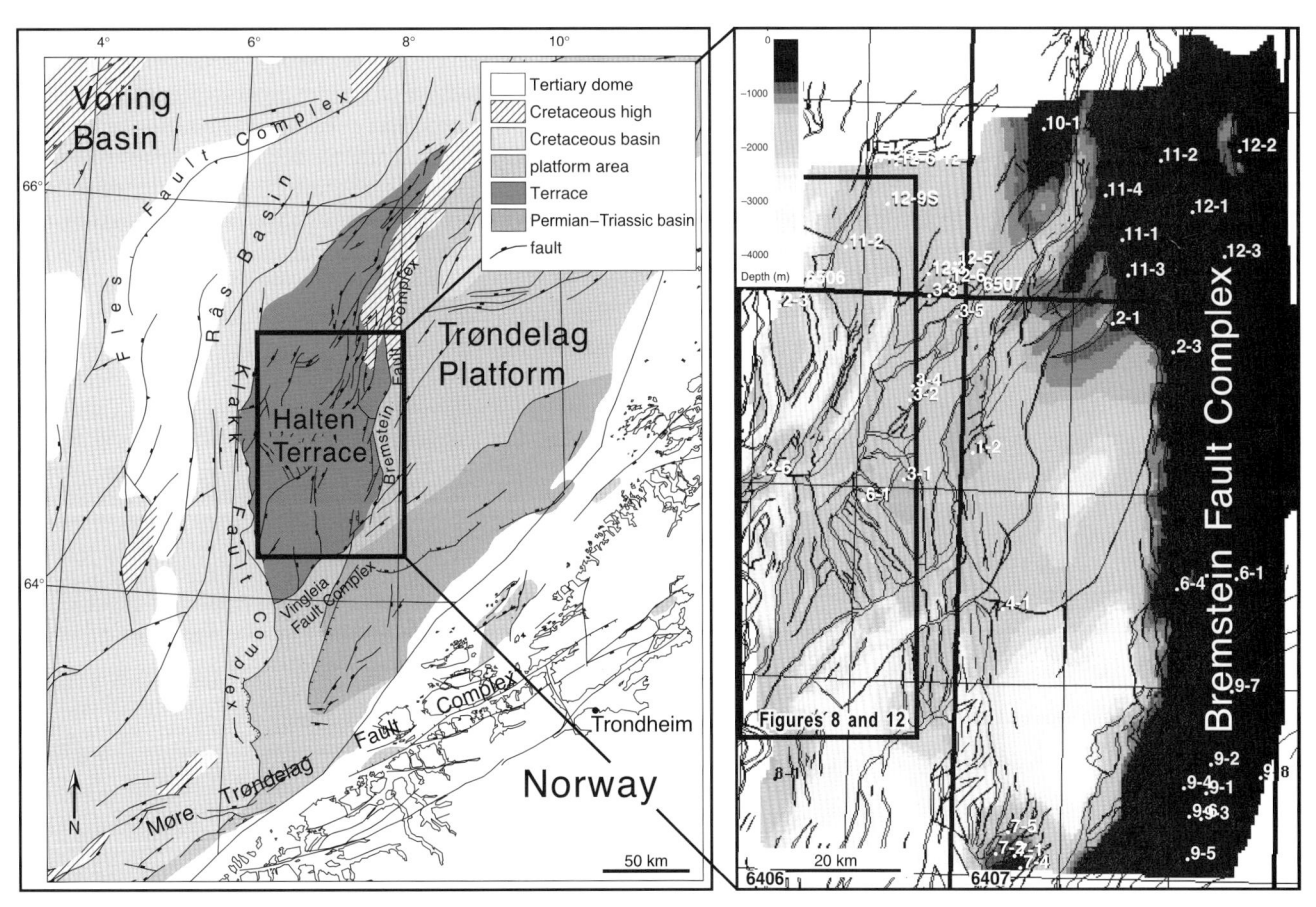

FIGURE 2. Left map shows the mid-Norwegian region, where study area is marked. Modified from Blystad et al. (1995). Right map shows present-day depth map of the top reservoir (top Garn Formation). Calibration wells and major faults at the top reservoir are marked. Frame shows area presented in Figures 8 and 12.

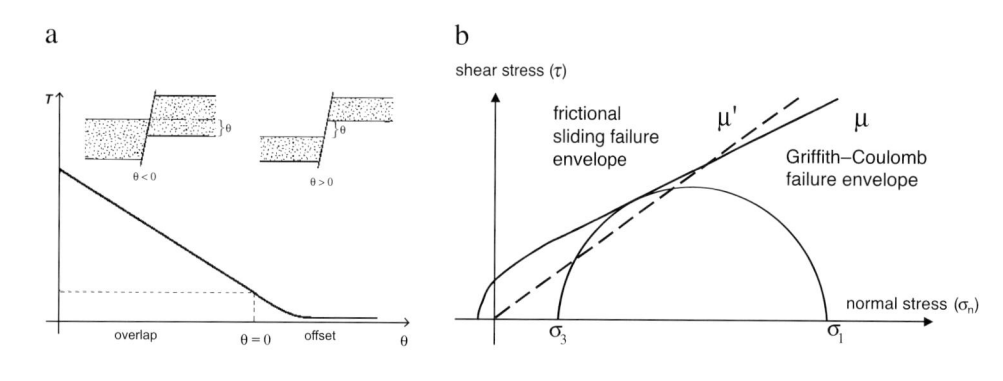

FIGURE 3. (a) Lateral fault transmissibilites (T) is defined by the overlap and offset (θ) of the reservoir unit. (b) Failure criteria used in the simulations.

water across low-permeable faults is calculated using an explicit forward Euler solution technique (Borge, 2002). Effective top seals stop flows out of compartments. Depth-converted maps of the overlying sediments are used to construct the burial history that is adjusted for decompaction. The development of pressure and stresses are reported for a series of time steps. Time steps are correlated to the depositional ages of the stratigraphic horizons that are used to build the model. A porosity-depth relationship is used to model mechanical compaction in the shales, and a kinetic model for quartz cementation (Walderhaug, 1996) is used to model chemical compaction of the sands. Hydrocarbon migration is not considered in this work that present one-phase simulation.

The fault transmissibility depends on the burial depth, the length, width and the dip-slip displacement of the faults, the thickness of the reservoir layers, and the permeability inside the fault zone (Borge and Sylta, 1998). Juxtaposition faults (faults where the reservoir is self-juxtaposed) have high transmissibilites, whereas faults with no overlap have lower transmissibilites (Figure 3a). An exponential decrease in fault permeabilities is used with increasing separation of reservoir units. Therefore, all faults will be modeled to have at least some capacity for flow. The fault-permeability function used in the simulator is empirical based (Borge and Sylta, 1998). It decreases logarithmically with increasing burial depth.

The geomechanical properties for the cap rock are allowed to vary through time with changing burial depths (Lothe et al., 2004). Isotropic horizontal stresses are assumed, and the minimum horizontal stress is estimated using an empirical formula (D. Grauls, 1996, personal communication). The vertical stress varies vs. time, depending on sediment loading. No strain is included, and a passive sedimentary margin is assumed. The Griffith–Coulomb failure criterion and the frictional sliding criterion are used to simulate hydraulic fracturing from the overpressured compartments (Figure 3b).

The data used are a present-day depth-converted map of the top reservoir unit (top Garn Formation), the present-day thickness of the reservoir, and a fault map at the top reservoir level (Figure 2; Table 1). Drillstem tests and formation tests are used for the pressure measurement in the different wells. The pressure data from 43 wells in the area are available at the top reservoir level and are used to calibrate the model (Table 1). The mean deviation (mean deviation = $\frac{1}{n}\sum_{i=1}^{n}|P_{i,\text{measured}} - P_{i,\text{modeled}}|$) is calculated between modeled and measured overpressure in the pressure compartments containing pressure observations from wells. A constant paleowater depth of 200 m (660 ft) is used in the study area. Poisson's ratio and Young's modulus are varied with depth, based on calibration with laboratory measurement of North Sea shale (Horsrud et al., 1998; Lothe et al., 2004).

TABLE 1. Input data used in the simulations.

Input Data	Pressim-water-flow Simulations
Depth-converted seismic horizons	seven time steps; 90, 80, 65, 20, 5, and 2 Ma, and present
Isopach map reservoir	Garn Formation (185–160 Ma)
Fault maps	fault map at top Garn Formation level
Pressure data	pressure measurements in Garn Formation from 43 wells used to calibrate the simulations. The wells are 6407/8-2, 6407/7-3, 6407/7-5, 6407/9-1, 6407/9-2, 6407/9-3, 6407/9-4, 6407/9-5, 6407/9-6, 6407/4-1, 6407/9-7, 6406/8-1, 6407/2-1, 6407/6-4, 6406/3-1, 6407/2-3, 6406/6-1, 6406/2-6, 6506/12-1, 6506/12-3, 6506/12-5, 6506/12-8, 6407/1-1, 6407/1-2, 6407/1-3, 6406/3-2, 6406/3-4, 6406/2-3, 6507/11-4, 6507/12-1, 6507/12-2, 6507/12-3, 6407/2-2, 6507/11-2 (see also Lothe et al., 2004).

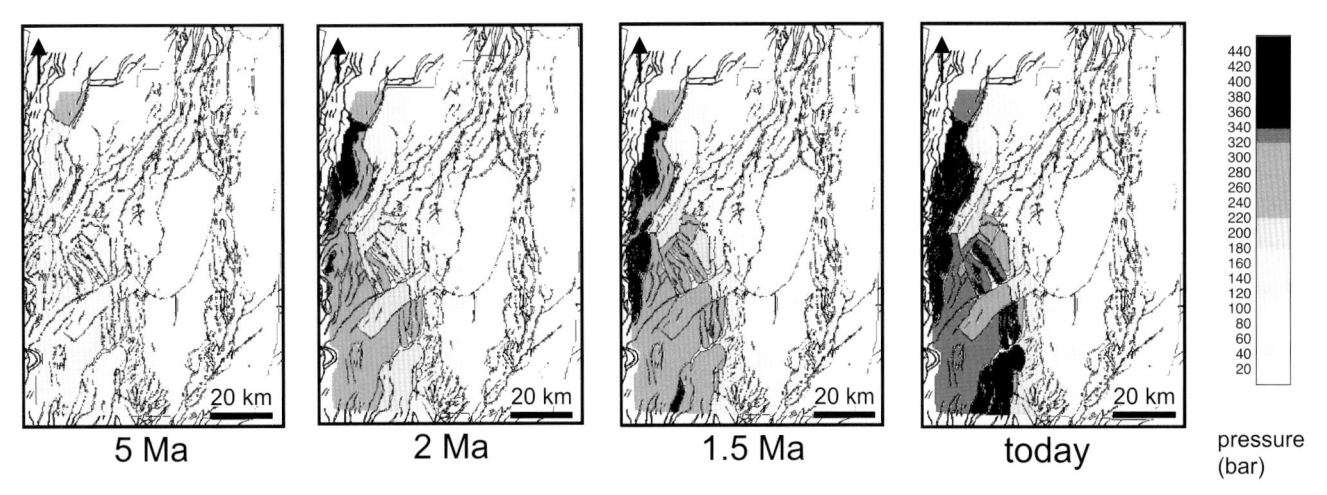

FIGURE 4. The development of simulated overpressure using the base case during the last 5 m.y. in the western part of the study area. Gray scale shows overpressures in bar.

The simulations account for the flow and stress development for the last 90 m.y. The input parameters used in the base case are listed in the Appendix. The parameters will be varied in the sensitivity runs.

RESULTS

The aim of the simulations was to determine which input parameters will have a major impact on the timing and extent of hydraulic leakage in an overpressured sedimentary basin. A base case was run to use as a reference for simulation sensitivities. The base case illustrates the variations in the modeled pressure and hydraulic leakage over the study area. Different simulations were carried out by varying the coefficient of internal friction (μ) and the coefficient of frictional sliding (μ'; see also Lothe et al., 2004). The time-dependent sensitivity of late hydraulic fracturing and amounts of leakage were studied by varying the permeability and, thereby, the transmissibility of major faults that control the lateral extent of the pressure compartments.

The pressure buildup over the last 90 m.y. was simulated for the base case. Figure 4 illustrates the overpressure distribution in the area from 5 Ma to the present day. Hydrostatic pressures prevail in nearly the whole basin at 5 Ma. Between 5 and 2 Ma, a rapid pressure increase occurs in the western part of the area. This trend continues until today, with an increase in overpressure in the most deeply buried units and a spreading of the overpressured part of the simulated area (Figures 4, 5). The high overpressures lead to hydraulic fracturing and leakage through the cap rock above the reservoir between 2 and 1.5 Ma (Figure 5). The modeling shows that the failure strength is exceeded in some of the compartments with the high overpressures (Figures 4, 5). One of the compartments where hydraulic fracturing occurs is compartment K (Figures 6, 7, located in Figure 8a). In the simulation, this compartment reaches its limits for hydraulic fracturing at 1.9 Ma,

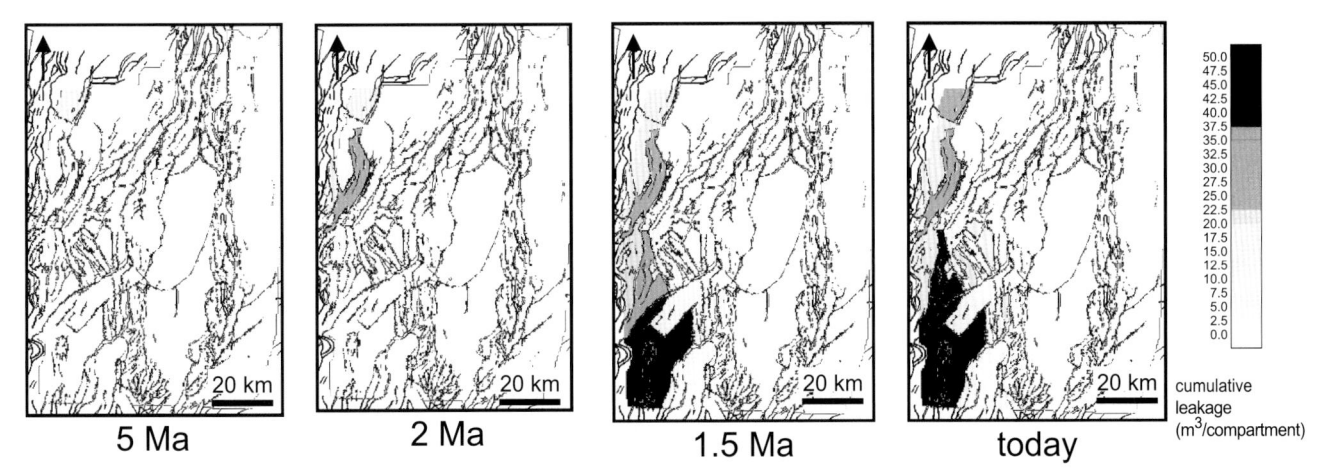

FIGURE 5. Development of hydraulic fracturing and leakage during the last 5 m.y. in western part of study area for the base case. Gray scale indicates cumulative leakage in cubic meters per compartment.

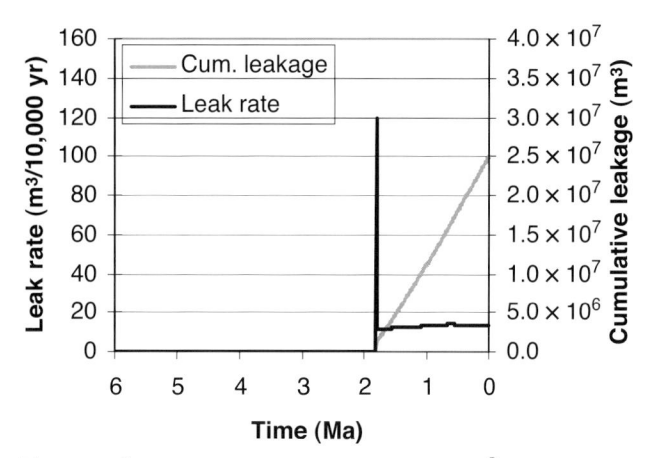

FIGURE 6. Modeled cumulative leakage (m³) and leakage rate (m³/10,000 yr) for compartment K during the last 6 m.y. Note that leakage occurs in one major pulse.

which is marked with a peak in the leakage rate (Figure 6). When we study the changes in effective stress through time in this compartment, we observe a rapid decrease when the pressure increases (Figure 7). The first rapid increase in overpressure starts at about 5 Ma. At about 3.2 Ma, the overpressure decreases, before a new peak builds up at about 1.9 Ma.

From the evaluation of the overpressure curves, one would perhaps expect hydraulic leakage to occur at about 3 Ma in compartment K. However, the model suggests that this does not happen until 1.9 Ma (Figure 6) because the neighboring compartments to K, WK, and EK start to leak at about 3.2 Ma (Figure 8). This leads to a loss in accumulated overpressure in K at about 3 Ma and illustrates that leakage from one compartment may have a critical effect on leakage from neighboring compartments. The same effect is observed in the south, where early leakage in compartment P controls the timing of later leakage in compartment WP.

The Significance of Coefficient of Internal Friction and Frictional Sliding

Pressure generation processes started to become effective during the last 5 m.y. in the basin. This is mainly caused by rapid late burial and the associated quartz cementation in the reservoir unit (Dalland et al., 1988; Teige et al., 1999). In the simulator, two failure criteria were used: first, the Griffith–Coulomb, and second, the frictional sliding criterion (Figure 3). To test which parameters control the amount and timing of hydraulic fracturing and leakage, the coefficients of internal friction (μ) and frictional sliding (μ') were varied (Figure 9). In the base case, $\mu = 0.6$, and $\mu' = 0.7$ (Figure 6). Figure 10 shows results from such simulations, where the coefficients are varied from $\mu = 0.3$ and $\mu' = 0.7$, to $\mu = 0.6$ and $\mu' = 1.0$. These simulations gave minor changes in the

leak rate, cumulative leakage (2.2–2.5 × 10⁷ m³) and the timing (<150 ka) for the failure (see also Lothe et al., 2004). The largest changes were found for very high values of μ' (1.0) in the peak leakage rates at failure. This resulted in lower peak leakage rates (Figure 10b).

Significance of the Permeability Across Major Faults

Sensitivities of the timing of failure and amount of leakage to fault permeability was also tested (Figure 11). The fault-permeability curve used is discussed in Borge and Sylta (1998), where kinks in curves are supported by Loosveld and Franssen (1992). The fault-permeability curves are based on empirical data (see Discussion section below) and also on the calibration between simulated pressures and measured pressures in the basin. The permeability values were applied in a low, base, and high case for all the master faults that define pressure compartments (Figure 12). Using low fault permeabilities resulted in pressure buildups that were modeled to occur at earlier times. More compartments, therefore, suffer from hydraulic fracturing and leakage compared to runs using higher permeabilities (base and high case; Figure 12). Using high permeabilities resulted in some compartments that did not fail at all. The timing of failure in compartments WK, EK, and K in the northern part of the study area differed with only a few hundred thousand years, when the low and the base cases were compared (Figure 12b, d). Comparing the base case and the high case, however, a marked delay in the timing of the hydraulic fracturing for compartments WK and EK, from 3.3 to 1.8 Ma, is observed, and no leakage from compartment K is observed (Figure 12d, f). In the southern area, compartment P fails approximately 1 m.y. earlier in the low case compared to the high case. For

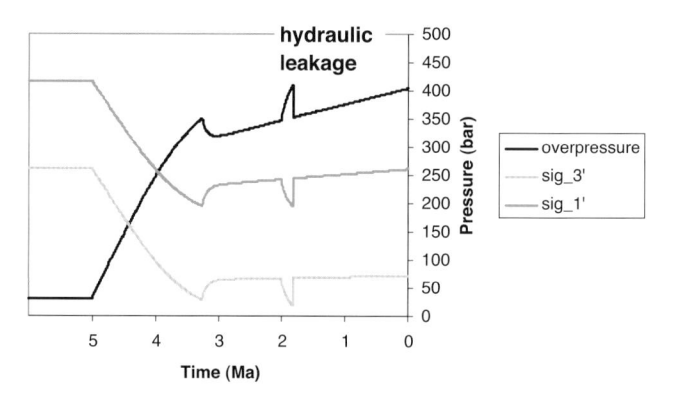

FIGURE 7. Modeled overpressure (bar), effective minimum horizontal stress (sig_3′; bar), and effective vertical stress (sig_1′; bar) for compartment K. Note the rapid pressure buildup at about 5 Ma. Hydraulic leakage is modeled to occur at about 1.9 Ma (see Figure 6).

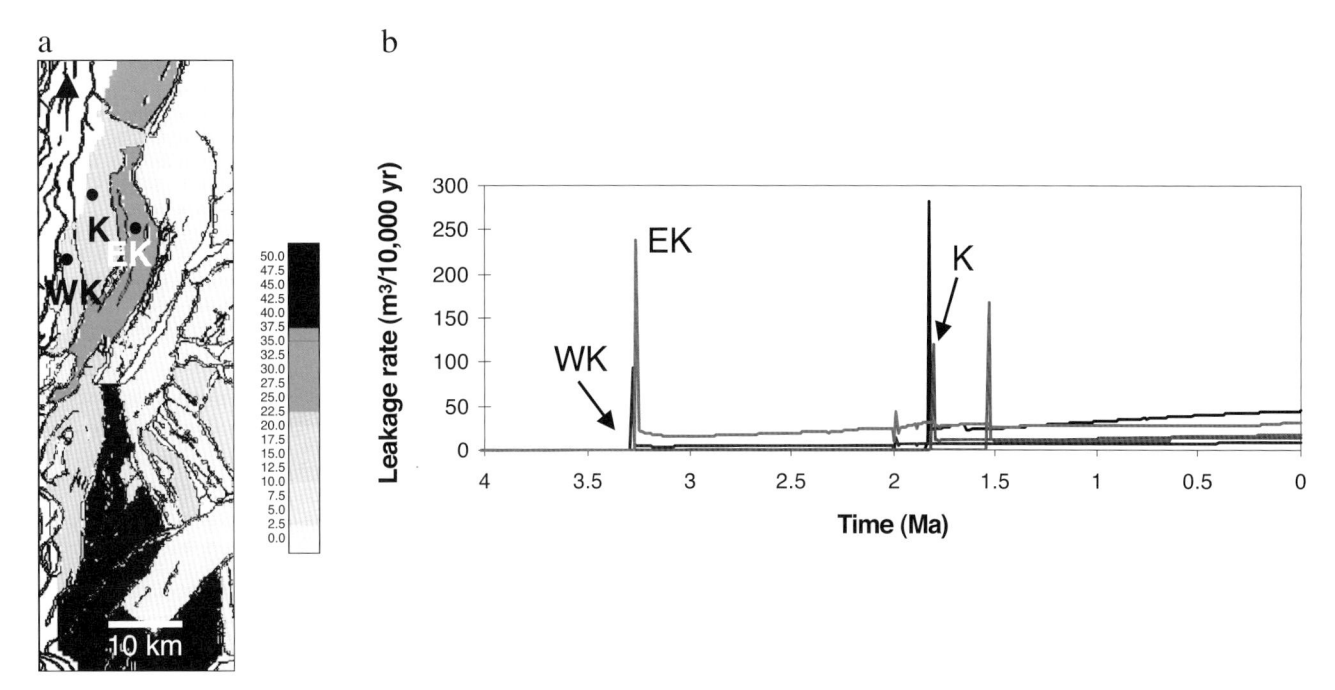

FIGURE 8. (a) Leakage volumes in different pressure compartments. Gray scale shows cumulative leakage to present day (m³). (b) Timing of leakage from the different compartments in the study area. The earliest leakage event from compartments WK, EK, and K is marked.

the neighboring compartment WP, no leakage is observed in the high case.

For the same three cases, the cumulative leakages for the respective compartments have been calculated (Figure 13). The results for the base case show that the highest cumulative leakage occurred from compartment EK in the northern area and compartment P in the southern area. Because the amounts of leaked fluid from the two compartments were much larger than from the neighboring compartments, the two would control the amount and also the timing for the leakage in the neighboring cells. Compartment K is shown as an example of this effect (Figures 7, 8). A progressive reduction is observed in the total amount of cumulative leakage because of the time dependency when comparing the leakage from compartment EK in the low, base, and high cases (Figure 13). This is also observed for other compartments, except for compartment P, which displays an increase in cumulative leakage in the high case compared to that of the base case. This is because the neighboring compartment to the south (compartment S) does not fail in the high case (Figures 12e, 13c). Accordingly, higher fluid pressures are developed in compartment P, and the cumulative leakage is much higher relative to those in the low and base cases. This effect overrules the effect of timing on the cumulative leakage in this example.

A strong dependence on fault permeability for both the timing of hydraulic failure and hydraulic leakage rates is noted from the previous descriptions. To quantify the uncertainty in the simulations, fault permeabilities have been systematically varied to investigate their effect on the timing and amount of leakage in selected overpressured compartments. As shown in Figure 11, the permeability across the faults varies according to the depth of burial. To be able to compare the different cases, the permeability typical for a burial depth of 4 km (2.4 mi) was used. Figure 14 shows how the timing of leakage varies vs. fault permeability for results for compartment K, which was analyzed in isolation. Figure 15 shows the same sensitivity for other overpressured compartments in the study area. The data points on the curves show at which time a particular compartment fails in different permeability runs when the permeabilities of the fault zones are varied.

The simulations show a strong dependency of the permeability across the major faults on the timing of

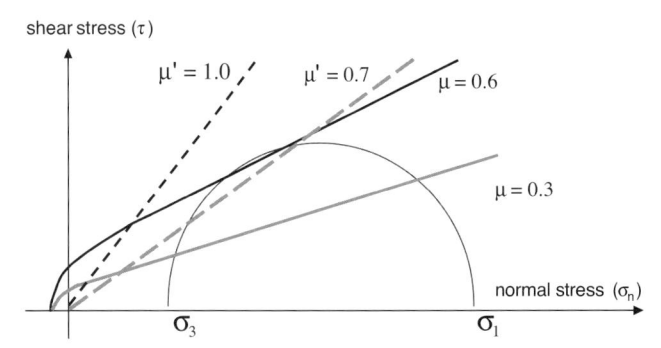

FIGURE 9. The coefficients of internal friction and frictional sliding used in two different simulations ($\mu = 0.3$ and $\mu' = 0.7$ and as $\mu = 0.6$ and $\mu' = 1.0$).

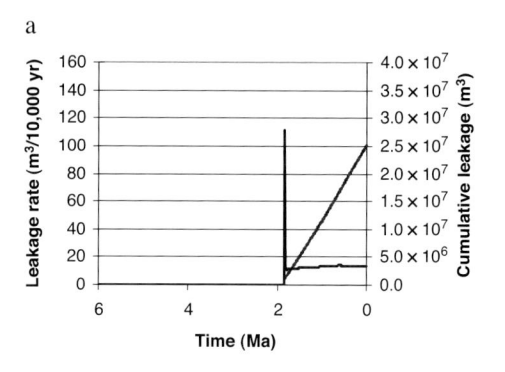

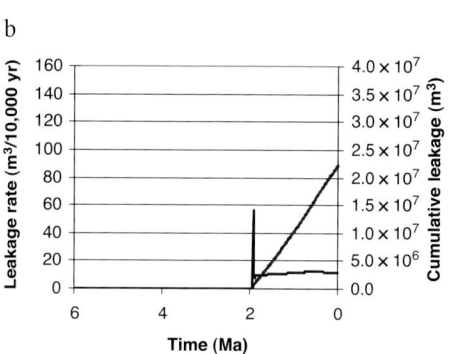

FIGURE 10. Cumulative leakage (m³) and leakage rate (peak) (m³/10,000 yr) in compartment K in runs with (a) $\mu = 0.3$ and $\mu' = 0.7$ and (b) $\mu = 0.6$ and $\mu' = 1.0$.

hydraulic fracturing in overpressured areas. An early and significant buildup of overpressure was displayed in the simulation runs, where a low permeability was applied. However, deviations between the simulated pressures and pressures measured in wells are large (Figure 15). This is because the early hydraulic fracturing gives too high leakage rates and too low simulated pressures for the present situation. When high fault permeabilities are used, only few compartments and, in the extreme case, no compartment will fail. A marked jump in the timing (from 5 Ma back to before 20 Ma) of the hydraulic failure in the different compartments can be observed for varying fault permeability models (Figure 15). The timing changes from a very early failure using low permeability input to late failure using higher permeabilities. We observe less change in the timing of leakage from around 4 to 3.5 Ma and from 1.9 to 1.7 Ma (Figure 16) when the changes during the last 5 m.y. are studied in more detail. This is typical for all of the studied compartments, except for compartment P, which fails in all cases.

When cumulative leakage vs. permeability is studied, two compartments with higher leakage rates are observed; compartment EK in the north and compartment P in the south (Figure 17). These two compartments show high cumulative leakage in the runs when they go to failure using low-permeability faults and a rapid decrease in leakage approaching no failure with high-permeability faults. The deviations between simulated pressures and those measured in the calibration wells for the varying permeability runs increase either when the permeability is very low and an early leakage occurs or when permeabilities are very high and no failure occurs (Figure 17).

In the northern area, the first leakage is observed for compartments EK and WK followed by compartment K for all simulated fault permeabilities (Figure 18a). Cumulative leakage vs. permeability curves are used to distinguish between compartments EK and WK (Figure 18c). Leakage is markedly higher for compartment EK at low fault permeabilities, because a larger fluid pressure has built up in this compartment because of its deep burial and the relative size of the compartment. This compartment controls the leakage in the smaller, neighboring compartments. A relative decrease in the leakage also occurred from compartment EK in the runs where compartment K has a larger leakage at a permeability of 1×10^{-8} md (Figure 18c). Studying the cumulative leakage in the three different compartments and the uncertainty in the runs, we see that compartment WK will most likely go to failure but at low leakage rates (Figure 18a, c). Compartment K will have a low leakage rate if the compartment goes to failure using low permeabilities ($<1 \times 10^{-7}$ md). Using high permeabilities, the compartment will not fail (Figure 18c).

In the southern area, the timing of leakage does not show such a clear trend for the different permeability runs as in the north. Compartment WP is an exception, in that it has the same plateaus in the timing of failure as the northern compartments (between 1.8 and 2 Ma, and between approximately 4.5 and 3.6 Ma; Figure 18a, b). For compartments P and S, the timing is more linearly linked to the permeability (Figure 18b). This is because S is the largest compartment in this area, with high hydraulic leakage rates at low fault permeabilities (Figure 18d). Therefore, it influences the leakage from compartment P, where an increase

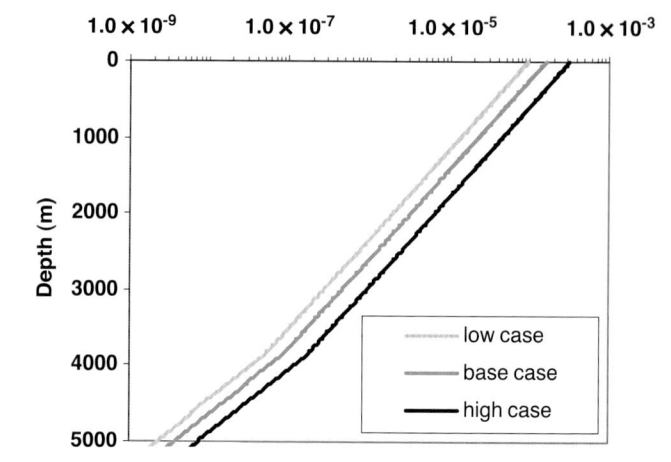

FIGURE 11. Depth-dependent fault permeabilities used for three different cases, low, base, and high, used as modeling input for all faults in the study area.

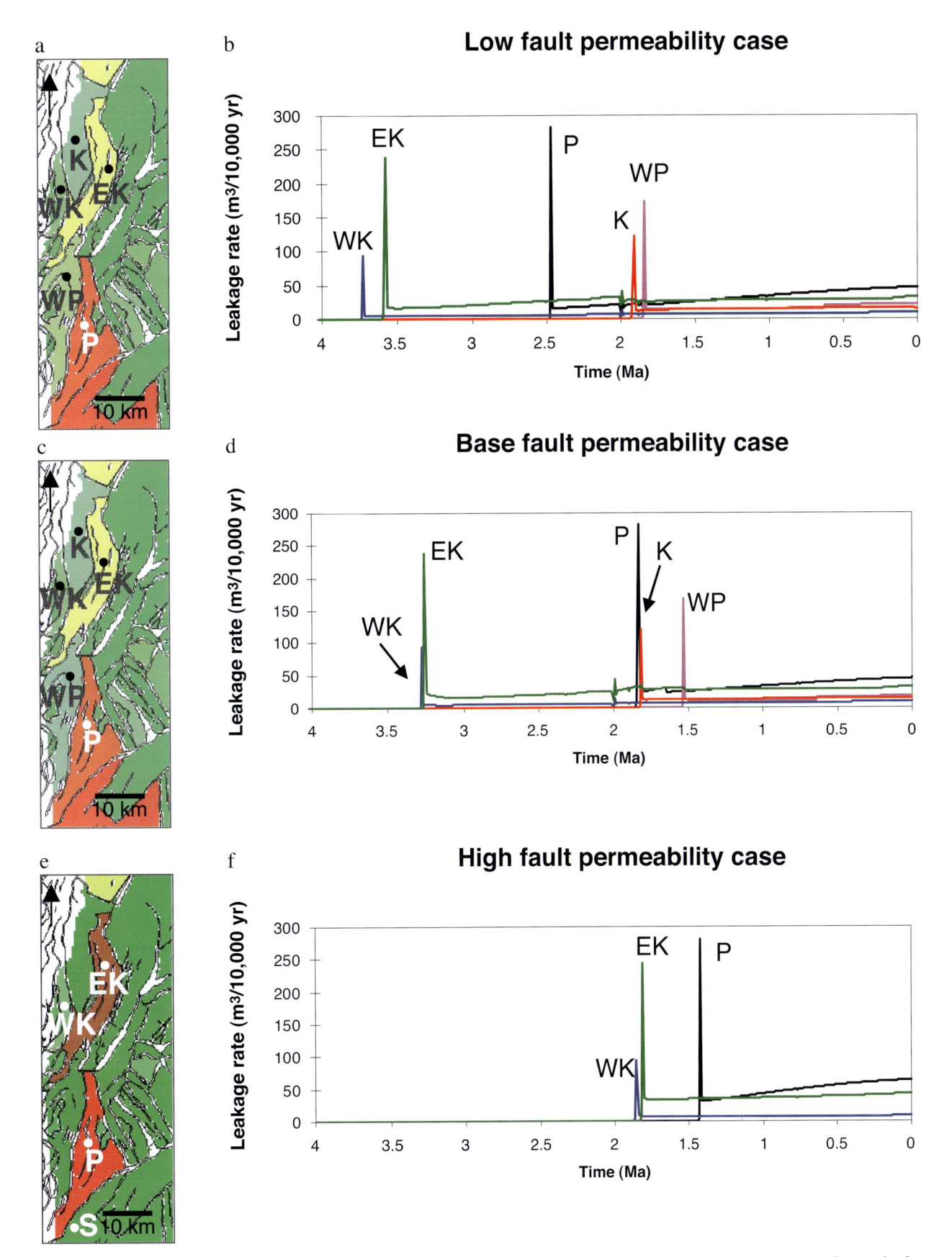

FIGURE 12. Fault permeabilities varied in the low, base, and high cases. Maps (a, c, and e) present cumulative leakage in the different compartments. (Color scale as in Figure 8; m³.) Graphs (b, d, and f) show the variation leakage throughout the last 4 m.y. Note that enhanced permeability causes delay in the timing of the hydraulic fracturing.

a) Low fault permeability case

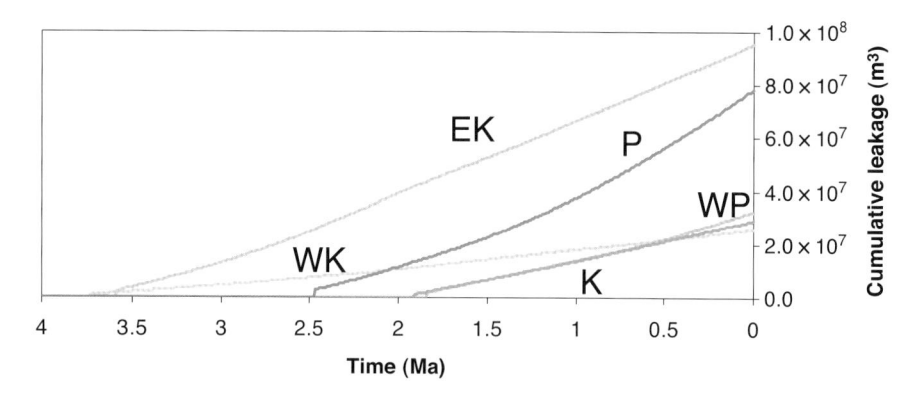

b) Base fault permeability case

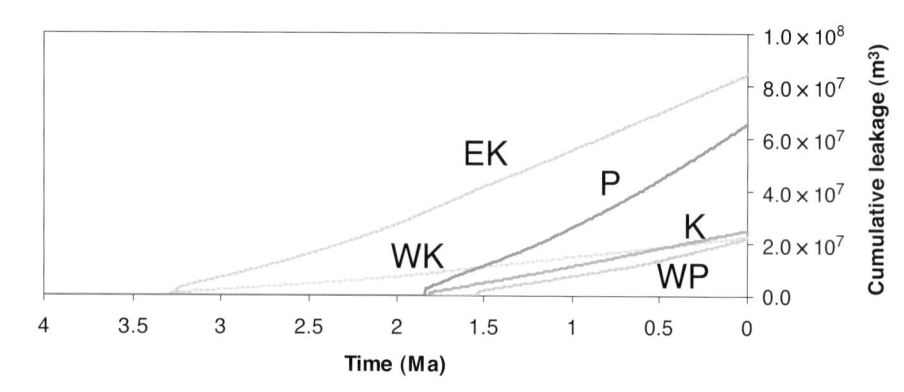

c) High fault permeability case

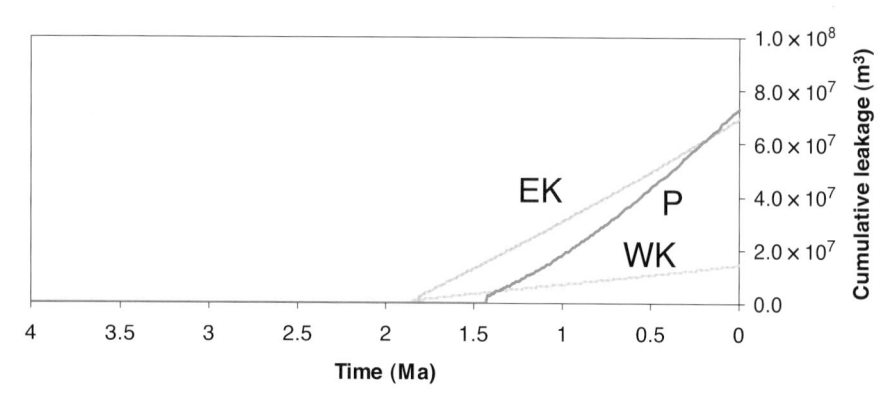

FIGURE 13. The cumulative leakage with time in the three fault permeability cases for different compartments.

mean deviations (>15 bar; Figure 15). A linear trend between timing and amount is observed for the small compartments during the last 5 m.y. (Figure 19b). The two larger compartments (P and EK) are characterized by a more complex relationship between timing and cumulative leakage.

DISCUSSION

Uncertainty is an important issue to address to evaluate the quality of the simulations. Therefore, well measurements of present reservoir pressures were used to calibrate the simulations. Unfortunately, pressure data are not available for the cap rock. Furthermore, not only the pressure measurement would be of interest to estimate, but also the possibilities for hydrocarbon finds. However, many of the highly overpressured compartments that have been simulated in this study have not yet been drilled, and calibration is therefore not possible. Using the predictions from the present simulations, however, better estimates may be obtained.

in the cumulative leakage is observed for lower permeability runs when compartment S is still intact (Figure 18b, d). The amounts of hydraulic leakage vary significantly for the low, base, and high cases, especially for the two largest compartments in the northern and southern areas, compartment EK and compartment P, respectively (Figure 18c, d).

When plotting the timing of hydraulic fracturing vs. cumulative leakage for all the compartments, a linear trend in each compartment between 5 Ma and the present emerges (Figure 19a). When failure occurs before 5 Ma, extreme values for cumulative leakage are observed. A similar situation is seen for runs with large

We start out assuming that hydrocarbons have accumulated in the highly overpressured compartments. The preferred situation from an exploration point of view would be that the cap rocks for these compartments did not fail. However, if one or more of the compartments did fail, the timing and the amount of leakage are important factors to evaluate. One scenario is that the pressure buildup may have led to very early fracturing, but that the cap rock subsequently became chemically healed, and the compartment could be refilled. This, however, does not appear to be the case in the study area, because hydraulic fracturing and leakage is predicted during the last 5 m.y. Late leakage may be

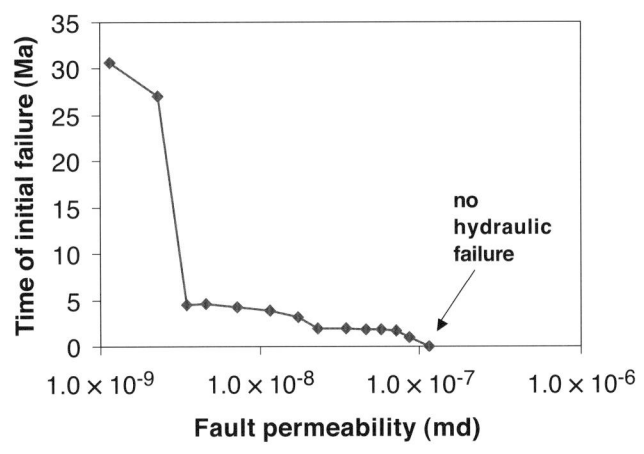

FIGURE 14. Time of hydraulic failure for compartment K vs. fault permeability at 4 km (2.4 mi) depth. The data points show results from different runs. The highest fault permeability gives no pressure buildup and, hence, no hydraulic fracturing in the compartment.

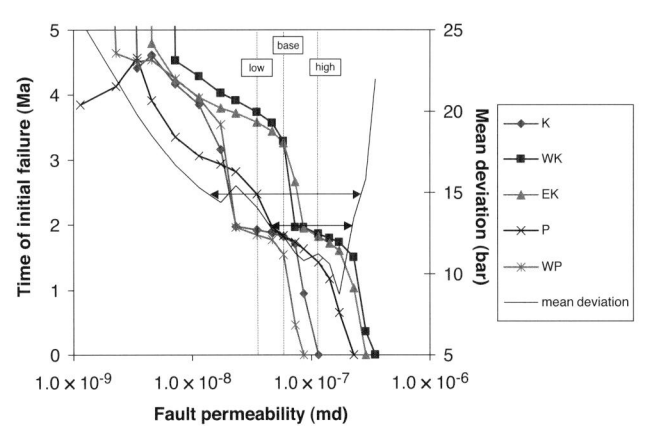

FIGURE 16. Illustration of the fault permeability at 4 km (2.4 mi) depth vs. timing of hydraulic leakage in the different compartments for the last 5 m.y. The right y-axis shows the mean deviations in the different runs. The fault permeabilities used in the low, base, and high cases are marked.

optimal for accumulations of hydrocarbons, because only small amounts of fluid would have time to escape.

To rank the possibilities of hydraulic fracturing and leakage in the compartments, we have studied three important factors: the timing, the amount of leakage, and the size of the accumulation. As an example, we can compare the timing and amount of leakage in compartments WP and WK (Figure 16). In this data set, WP is the first compartment that does not fail when the permeability is increased in the simulations (Figure 16). However, using lower permeabilities, the cumulative leakage for compartment WP increases rapidly because of earlier failure (Figure 17). Compartment WK may therefore contain a more preferable prospect for hydrocarbon exploration, although modeling suggests leakage at high permeabilities (Figure 16). More important perhaps is that the cumulative leakage in compartment WK seems to be small when low permeabilities are used (Figure 17). In such a case, the size of the field will be the controlling factor. For a large field, it would be best for the hydrocarbon accumulation that a low but steady leakage occurs. As seen directly from the cumulative leakage-vs.-permeability plot (Figure 17), the large compartments, in this case, EK and P, would have too large cumulative leakages to be good exploration targets. Thus, the smaller compartments with

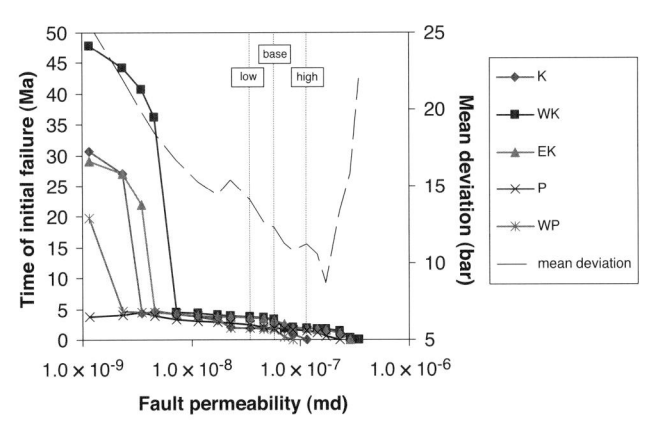

FIGURE 15. Timing of failure in different compartments depending on the fault permeability at 4 km (2.4 mi) depth. The right y-axis shows the mean deviations between measured and simulated pressures in the whole basin in the different runs. The fault permeabilities used in the low, base, and high cases are marked.

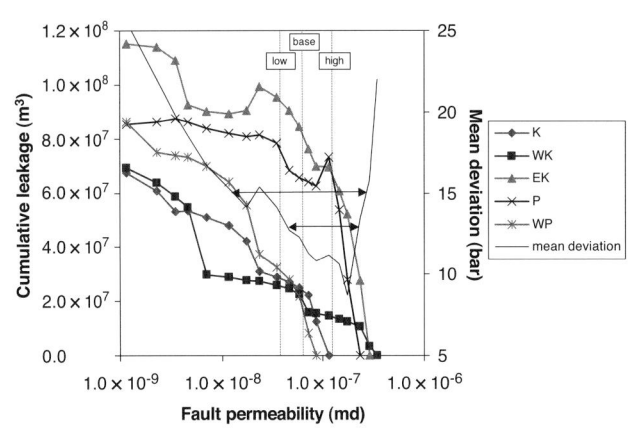

FIGURE 17. Cumulative leakage (m^3) vs. permeability (md) for the simulated fault at 4 km (2.4 mi) burial depth. Note the high cumulative leakage simulated for compartments EK and P. The right y-axis shows the mean deviations in the different runs. The fault permeabilities used in the low, base, and high cases are marked. Note that the base and high cases give mean deviation less than 13 bar.

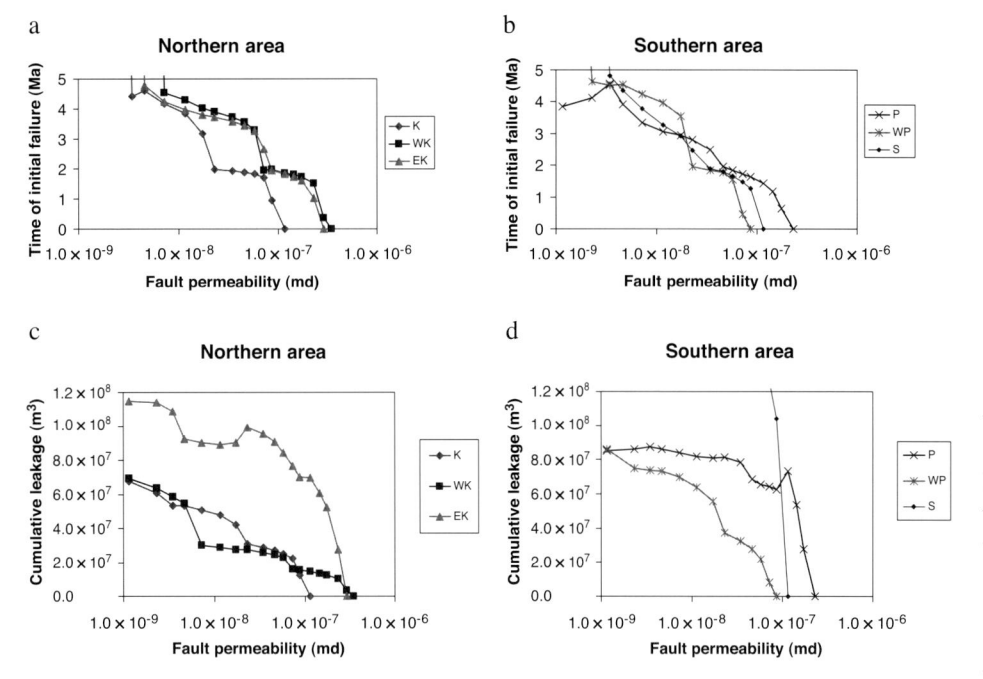

FIGURE 18. Timing of hydraulic fracturing vs. permeability across faults in (a) the northern study area and (b) the southern study area. Cumulative leakage vs. fault permeability in (c) the northern and (d) southern areas. The largest amounts of leakage are simulated for EK and S.

either late fracturing or early but low leakage rates are more favorable.

The uncertainty assessment can be used to evaluate the predictability of the simulations. Figures 16 and 17 show the calculated mean deviation between measured and simulated pressures in all compartments where wells exists. Large mean deviations are observed when very high or low fault permeabilities are used in the simulations. If a mean deviation of 13 bar is considered acceptable, then the uncertainties in the simulations of the timing and cumulative leakage for one compartment such as K can be considered. If a mean deviation of 13 bar is acceptable, then late leakage from compartment K is predicted from 1.9 Ma to the present day. The cumulative leakage is estimated not to exceed 29×10^6 m^3 (1.024×10^9 ft^3). The runs with lower deviations do not fail (Figure 16). Compartment P most likely fails, except when using high permeability values. The timing of the failure in compartment P is simulated between 2.5 and 0.6 Ma, with cumulative leakage in the scale of $280-78 \times 10^6$ m^3 ($9.888-2.754 \times 10^9$ ft^3), respectively. An uncertainty limit of 15 bar gives more or less the same results for most compartments (Figure 17). Differences are observed when low permeabilities are used, and the rate of leakage increases rapidly. A relatively large change in the modeled timing of hydraulic leakage and cumulative leakage is seen for compartments K and WP, when the acceptable deviation is changed from 13 to 15 bar. Both fail from 1 to 1.5 m.y. earlier, with higher leakage rates resulting (Figures 16, 17).

Effects on Hydrocarbon Migration

Different effects have to be considered when the probability of hydrocarbon leakage is evaluated. In the simulations, we assume leakage from the highest point in the structure, with less or no leakage downdip. Still, hydrocarbons may be trapped downdip in compartments that have experienced hydraulic leakage at the crest. To be able to evaluate the probability for hydrocarbon reserves, more detailed studies of smaller areas, incorporating more faults, need to be carried out. The presence of internal minor faults in the compartments will influence fluid flow.

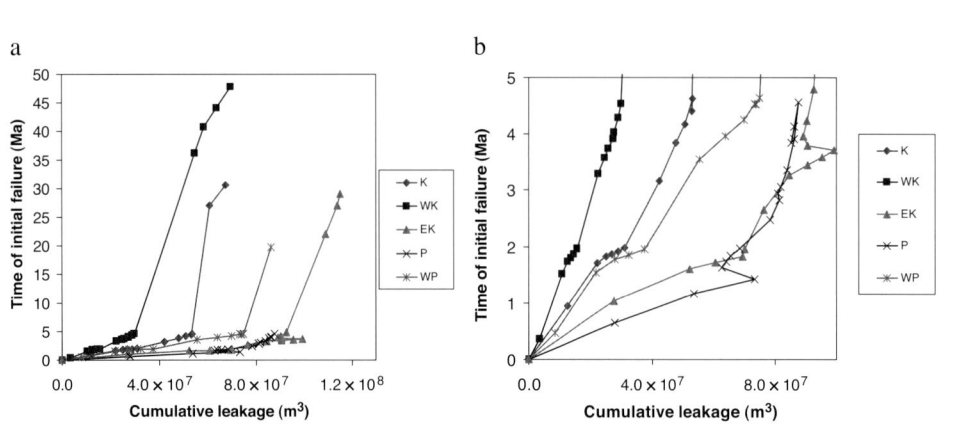

FIGURE 19. Time of hydraulic failure vs. cumulative leakage for all the compartments during (a) the last 50 m.y. and (b) the last 5 m.y. The smallest compartments (WK, K, and EK) show a linear relationship in the last 5 m.y.

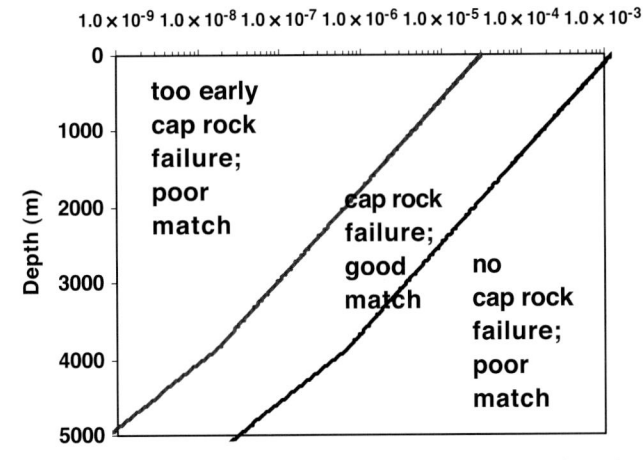

FIGURE 20. Variations in fault permeability used in the simulations giving hydraulic fracturing and leakage. Matches to the pressure measurements with wells are outlined.

Furthermore, assuming leakage from the highest point of the structure, time delays in the leakage should be considered.

Comparison of Permeabilities Used in Modeling with Measured Permeabilities

Faults in clastic sequences commonly represent significant barriers to fluid flow. Fisher and Knipe (2001) measured the permeability of faults in siliciclastic rocks in the North Sea and the Norwegian continental shelf. Faults typical in impure sandstones (V_{shale} [volume of shale] = 15–40%) have permeabilities of about 0.001 md at depths greater than 4 km (2.4 mi), whereas clay-rich sediments (V_{shale} > 40%) deform to produce clay smears with very low permeabilities (<0.001 md; Fisher and Knipe, 2001). Sperrevik et al. (2002) observed a relationship between the permeability and the mineralogical composition of the faulted rock, the effective stress conditions, and the history of the reservoir prior to, during, and after deformation. They measured very low permeabilities (<10^{-7} md) in faults with clay smears at depths (>3.5 km; >2.1 mi). This is in accordance with the permeability values used in our simulations, which resulted in hydraulic failure in the simulations (Figure 20). With less clay content in the fault zones, the data from Sperrevik et al. (2002) gave too high permeabilities in the fault, and the pressure barriers are not effective enough to simulate hydraulic fracturing. According to this data set, lower permeabilities should be used when simulating the shallow part of the faults (<3 km; <1.8 mi). However, because overpressure is unlikely to occur at this shallow depth of burial, this is probably not critical for the timing and amount of the hydraulic leakage modeled. Clay smear is probably not the controlling sealing factor in all the faults with deeper burial depths in the study area. Diagenesis and quartz cementation in the fault zones are processes that contribute to fault sealing at large depths (>3 km; >1.8 mi). Therefore, better control on the lateral leakage would be gained by integrating the effect of the clay content of the faults (shale gouge ratio) during subsidence into the simulator.

In addition to the initiation of hydraulic fracturing, the continuation of fluid flow and leakage should be considered. Gutierrez et al. (2000) present results from laboratory tests of extensional fractures in shales. The tests show a decrease in permeability during increasing normal stress across a fracture and after shearing of the fracture under constant high normal stress. However, the fractures never completely closed. This indicates that fractures once created are difficult to close by mechanical loading. Then, as long as sufficient hydraulic gradients are obtained, fluid can still flow along fractures even in the absence of large overpressures. The results from Gutierrez et al. (2000) are in accordance with the simulation of continuous leakage we worked on, as long as the pressure compartment is fractured and is at the leakage pressure gradient.

In this chapter, we have presented a type of analysis that can be carried out using a coupled pressure and stress simulator to evaluate hydraulic fracturing and leakage. The simulator itself is described in Borge and Sylta (1998), Borge (2002), and Lothe et al. (2004). Other input parameters than the ones we discussed can influence the timing of fracturing. Lothe et al. (2004) shows a large influence of the geomechanical parameters used for the cap rock (Poisson's ratio and Young's modulus). The handling of stresses in the simulator has room for significant improvements. Nevertheless, by varying the input parameters in different runs, the uncertainty in the simulations can be found with a larger confidence than we could before.

CONCLUSIONS

The main conclusion from this work is that the transmissibilites across faults between overpressured compartments will have a major impact on the timing and amount of hydraulic fracturing and leakage, and this can be simulated. Low fault permeabilities cause restricted lateral flow, overpressure buildup, and early hydraulic fracturing, and the resultant simulated pressures give large deviations compared to the measured pressures in wells. Using high fault permeabilities in the simulations gives a delay in the hydraulic failure of some compartments, and in some cases, no failure will occur. Hydraulic leakage in one compartment will influence the amount and timing of leakage in the neighboring pressure compartments. The largest compartments in

APPENDIX: VALUES OF PARAMETERS USED IN THE SIMULATIONS
(SEE BORGE, 2000 FOR EXPLANATION OF PARAMETERS AND NOMENCLATURE)

Description	Symbol	Value	Unit
Pressure Generation and Accumulation			
Accumulating depth	z_A	2500	m
Generating depth	z_S	4100.0	m
Salinity	s	50,000	ppm
Accumulating exponent	A	3.45	
Shale drainage thickness	γ	100.0	m
Minimum reservoir thickness	z_{min}	0.10	m
Minimum net/gross ratio	N/G_{min}	0.050	
Maximum shale compaction depth	z_{shale}	10,000.0	m
Hydrostatic gradient	$\rho_w g$	0.1030	bar/m
Lithostatic gradient	$\bar{\rho}g$	0.220	bar/m
Time step	Δt	10,000	yr
Diameter of quartz grain size	D	0.0003	m
Fraction of detrital quartz	f	0.65	
Molar mass of quartz	M	0.06009	kg/mol
Density of quartz	ρ_{quartz}	2650.0	kg m^{-3}
Temperature at which quartz cementation starts	T_{C0}	80.0	°C
Temperature at which quartz cementation is completed	T_{C1}	175.0	°C
Quartz precipitation rate factor	r_1	1.98×10^{-18}	mol/m² s
Quartz precipitation rate exponent	r_2	0.022	°C
Sand porosity at seabed	ϕ_{S0}	0.45	
Sand porosity constant 1	η_1	2400	m
Sand porosity constant 2	η_2	0.50	
Temperature at seabed	T_0	4.0	°C
Temperature gradient	$\partial T/\partial z$	0.035	°C m^{-1}
Irredusible water saturation (Garn Formation)	ϕ_{C1}	0.040	
Clay coating factor (Garn Formation)	C	0.50	
Minimum dissipation volume	V_{min}	1.0×10^6	m³
Hydraulic Leakage			
Poisson's ratio (shales) at surface	ν_{z_0}	0.40	
Poisson's ratio (shales) at accumulating depth	ν_{z_A}	0.27	
Poisson's ratio (shales) at sealing depth	ν_{z_S}	0.20	
Poisson's ratio (shales) at maximum shale compaction depth	$\nu_{z_{shale}}$	0.02	
Young's modulus (shales) at surface	E_{z_0}	600	bar
Young's modulus (shales) at accumulating depth	E_{z_A}	20,000	bar
Young's modulus (shales) at sealing depth	E_{z_S}	40,000	bar
Young's modulus (shales) at maximum shale compaction depth	$E_{z_{shale}}$	90,000	bar
Coefficient of thermal expansion	α_T	1.00×10^{-5}	
Bulk modulus (shale)	K_s	170,000.0	bar
Coefficient of internal friction	μ	0.6	
Coefficient of sliding friction	μ'	0.7	
Lateral Transmissibility			
Lateral transmissibility		0.00069	
Percent transmissibility remaining at no overlap	p	0.05	
Width of fault blocks	b	20.0	m
Porosity at seabed	ϕ_0	0.45	
Rate of change in porosity versus depth	c	0.00039	m^{-1}
Porosity where the $K - \phi$ curve changes between deep and shallow relationships	ϕ_b	0.1	
Permeability where the $K - \phi$ curve changes between deep and shallow relationships	K_b	0.00001	md
Rate of change in fault zone permeability (log) vs. depth (log) for shallow faults	δ_{sh}	5.0	md m^{-1}
Rate of change in fault zone permeability (log) vs. depth (log) for deep faults	δ_{de}	7.0	md m^{-1}

an area will control the timing and amount of leakage in the smaller ones. The uncertainty in the simulations can be estimated by comparing the simulated pressures with the measured pressures in the wells, and therefore, one may be able to improve risk predictions of hydrocarbons in undrilled traps.

ACKNOWLEDGMENTS

We thank Norsk Hydro ASA for their funding to Lothe's Ph.D. thesis on hydraulic fracturing, providing data, and giving permission to publish. Colleagues at Sintef Petroleum Research are thanked for useful discussions. Roy H. Gabrielsen has kindly corrected an early version of the manuscript.

REFERENCES CITED

Berg, T., K. Berge, M. Cecchi, C. Ekeli, L. Hansen, O. Norvik, and D. Waters, 2000, Fault seal and overpressure analysis of Halten Terrace (abs.): Norwegian Petroleum Society Conference on Hydrocarbon Seal Quantification, Stavanger, Norway, October 16–18.

Bjørkum, P. A., O. Walderhaug, and P. H. Nadeau, 1998, Physical constrains on hydrocarbon leakage and trapping revisited: Petroleum Geoscience, v. 4, p. 237–239.

Blystad, P., R. B. Færseth, B. T. Larsen, J. Skogseid, and B. Tørudbakken, 1995, Structural elements of the Norwegian continental shelf: Part II. The Norwegian sea region: Norwegian Petroleum Directorate Bulletin, v. 8, 42 p.

Borge, H., 2000, Fault controlled pressure modelling in sedimentary basins: Doktor Ingenør Thesis: Norwegian University of Science and Technology, Trondheim, Norway, 148 p.

Borge, H., 2002, Modelling generation and dissipation of overpressure in sedimentary basins: An example from the Halten Terrace, offshore Norway: Marine and Petroleum Geology, v. 19, p. 377–388.

Borge, H., and Ø. Sylta, 1998, 3D modelling of fault bounded pressure compartments in the North Viking Graben: Energy, Exploration and Exploitation, v. 16, p. 301–323.

Bour, O., I. Lerche, and D. Grauls, 1995, Quantitative models of very high fluid pressures: The possible role of lateral stresses: Terra Nova, v. 7, p. 68–79.

Chapman, R. E., 1980, Mechanical versus thermal causes of abnormally high pore pressures in shales: AAPG Bulletin, v. 64, p. 179–183.

Cosgrove, J. W., 1998, The role of structural geology in reservoir characterization, in M. P. Coward, T. S. Daltaban, and H. Johnson, eds., Structural geology in reservoir characterization: Geological Society (London) Special Publication 127, p. 1–13.

Dalland, A. G., D. Worsley, and K. Ofstad, 1988, A lithostratigraphic scheme for the Mesozoic and Cenozoic succession offshore mid- and northern Norway: Norwegian Petroleum Directorate Bulletin, v. 4, p. 1–65.

Darby, D., R. S. Haszeldine, and G. D. Couples, 1996, Pressure cells and pressure seals in the UK Central Graben: Marine and Petroleum Geology, v. 13, p. 865–878.

Fisher, Q., and R. J. Knipe, 2001, The permeability of faults within siliciclastic petroleum reservoirs of the North Sea and Norwegian continental shelf: Marine and Petroleum Geology, v. 18, p. 1063–1081.

Grauls, D., 1998, Overpressure assessment using minimum principal stress approach: Overpressure in petroleum exploration, in A. Mitchell and D. Grauls, eds., Workshop Proceedings, Pau, France, April 1998: Bulletin Centre Recherche, Elf Exploration and Production Memoir 22, p. 137–147.

Gutierrez, M., L. E. Øino, and R. Nygård, 2000, Stress-dependent permeability of a de-mineralised fracture in shale: Marine and Petroleum Geology, v. 17, p. 895–907.

Hermanrud, C., and H. M. N. Bolås, 2002, Leakage from overpressured hydrocarbon reservoirs at Haltenbanken and in the northern North Sea, in A. G. Koestler and R. Hunsdale, eds., Hydrocarbon seal quantification: Norwegian Petroleum Society Special Publication 11, p. 221–231.

Horsrud, P., E. F. Sønstebø, and R. Bøe, 1998, Mechanical and petrophysical properties of North Sea shales: International Journal of Rock Mechanical Mining Science, v. 35, p. 1009–1020.

Hunt, J. M., 1990, Generation and migration of petroleum from abnormally pressured fluid compartments: AAPG Bulletin, v. 78, p. 1811–1819.

Koch, J.-O., and O. R. Heum, 1995, Exploration trends of the Halten Terrace, in S. Hanslien, ed., Petroleum exploration and exploitation in Norway: Norwegian Petroleum Society Special Publication 4, p. 235–251.

Loosveld, R. J. H., and R. C. M. W. Franssen, 1992, Extensional vs. shear fractures: Implications for reservoir characterisation: Presented at the European Petroleum Conference held in Cannes, France, November 16–18, Society of Petroleum Engineers: SPE Paper 25017, p. 65–66.

Lothe, A. E., H. Borge, and R. H. Gabrielsen, 2004, Modeling of hydraulic leakage by pressure and stress simulations and implications for Biot's constant: An example from the Halten Terrace, offshore mid-Norway: Petroleum Geoscience, no. 10, p. 199–213.

Miller, T. W., and C. H. Luk, 1993, Contributions of compaction and aquathermal pressuring to geopressure and the influence of environmental conditions: Discussion: AAPG Bulletin, v. 77, p. 6–10.

Morrow, C. A., L. Q. Shi, and J. D. Byerlee, 1984, Permeability of fault gouge under confining pressure and shear stress: Journal of Geophysical Research, v. 89, p. 3193–3200.

Revil, A., and L. M. Cathles, 1999, Permeability of shaly sands: Water Resources Research, v. 35, p. 651–662.

Skar, T., R. T. V. Balen, L. Arnesen, and S. Cloetingh, 1999, Origin of overpressures on the Halten Terrace, offshore mid-Norway: The potential role of mechanical compaction, pressure transfer and stress, *in* A. C. Aplin, A. J. Fleet, and J. H. Macquaker, eds., Mud and mudstones: Physical and fluid flow properties: Geological Society (London) Special Publication 158, p. 137–156.

Sperrevik, S., P. A. Gillespie, Q. J. Fisher, and R. J. Knipe, 2002, Empirical estimation of fault rock properties, *in* A. G. Koestler and R. Hunsdale, eds., Hydrocarbon seal quantification: Norwegian Petroleum Society Special Publication 11, p. 109–125.

Swarbrick, R. E., and M. J. Osborne, 1998, Mechanisms that generate abnormal pressures: An overview, *in* B. E. Law, G. F. Ulmishek, and V. I. Slavin, eds., Abnormal pressures in hydrocarbon environments: AAPG Memoir 70, p. 13–34.

Teige, G. M. G., C. Hermanrud, L. Wensaas, and H. M. N. Bolås, 1999, The lack of relationship between overpressure and porosity in North Sea and Haltenbanken shales: Marine and Petroleum Geology, v. 16, p. 321–335.

Verbeek, E. R., and M. A. Grout, 1993, Geometry and structural evolution of gilsonite dikes in the eastern Uinta basin, Utah: U.S. Geological Survey Bulletin, v. 1787-HH, 42 p.

Walderhaug, O., 1996, Kinetic modelling of quartz cementation and porosity loss in deeply buried sandstone reservoirs: AAPG Bulletin, v. 80, p. 731–745.

Wiprut, D., and M. D. Zoback, 2002, Fault reactivation, leakage potential, and hydrocarbon column heights in the northern North Sea, *in* A. G. Koestler and R. Hundale, eds., Hydrocarbon seal quantification: Norwegian Petroleum Society Special Publication 11, p. 203–219.

Zhang, S., T. E. Tullis, and V. J. Scruggs, 2001, Implications of permeability and its anisotropy in a mica gouge for pore pressures in fault zones: Tectonophysics, v. 335, p. 37–50.

Sealing by Shale Gouge and Subsequent Seal Breach by Reactivation: A Case Study of the Zema Prospect, Otway Basin

Paul J. Lyon
Australian School of Petroleum, University of Adelaide, Australia

Peter J. Boult
Australian School of Petroleum, University of Adelaide, Australia and also Department of Primary Industries and Resources South Australia, Adelaide, Australia

Richard R. Hillis
Australian School of Petroleum, University of Adelaide, Australia

Scott D. Mildren[1]
Australian School of Petroleum, University of Adelaide, Australia

ABSTRACT

The Zema prospect, located in the Otway Basin of South Australia, hosts an interpreted 69-m (226-ft) paleohydrocarbon column. Two faults are significant to prospect integrity. The main prospect-bounding fault (Zema fault) shows a significant change in orientation along strike, with some parts of the fault trending northwest–southeast and other parts trending east–west, all at a consistent dip of about 70°. The fault shows a complex splay and associated relay zone at its western tip. An overlying fault shows a similar northwest–southeast trend.

Shale volume (V_{shale}) derived from the gamma-ray log was tied to seismic horizon data in order to model across-fault juxtaposition and shale gouge ratio on the Zema fault. Shale volumes of greater than 40% correspond with paleosol shale lithotypes identified in the core that are characterized by high mercury injection capillary entry pressures of 55 MPa (8000 psi), capable of supporting gas columns far beyond the

---
[1]*Present address:* JRS Petroleum Research Pty. Ltd., Magill, Adelaide, Australia.

Copyright ©2005 by The American Association of Petroleum Geologists.
DOI:10.1306/1060764H23169

structural spillpoint of the trap. V_{shale} values of 20–40% correspond to silty shale lithotypes characterized by mercury capillary entry pressures equivalent to gas column heights of less than 30 m (100 ft). Sands correspond with V_{shale} values of less than 20%.

Juxtaposition modeling of the Pretty Hill reservoir interval that is displaced across the Zema fault against the Laira Formation seal demonstrates the existence of both sand-on-sand juxtaposition and sand-on-silty shale juxtaposition above the paleofree-water level. Hence, juxtaposition alone cannot explain the observed paleocolumn. It is therefore likely that fault damage processes on the fault plane were responsible for holding back the original 69-m (226-ft) column. Shale gouge ratio values show a gradual decrease from 32% at the top of the fault trap to less than 14% at the structural spillpoint. The fault damage zone is likely to consist of phyllosilicate framework rock types. Because the Zema trap was not filled to structural spillpoint, it is likely that the percentage of shale gouge in the fault zone not only provided the original sealing mechanism but also limited the original column height. This is supported by fault zone capillary entry pressures calculated from shale gouge ratio values, which indicate that the fault zone is only capable of supporting a maximum column height of 72 m (236 ft), just 3 m (10 ft) more than the interpreted column height of 69 m (226 ft).

Geomechanical analysis shows that the northwest–southeast-trending parts of the faults are optimally orientated in the in-situ stress field for reactivation. A spontaneous potential (SP) anomaly in the Zema-1 well, which was recorded in a northwest–southeast-striking fault damage zone through the seal, confirms the existence of open, permeable fracture networks. These are likely to have been generated by recent reactivation that caused the breach and subsequent leakage of the entire original hydrocarbon column.

INTRODUCTION

The Zema prospect is one of seven fault-bound prospects drilled in the Penola Trough in the Otway Basin of South Australia that have been interpreted as hosting a paleohydrocarbon column, i.e., a column that was once present but has since leaked away (George et al., 1998; Lisk et al., 2000). These paleohydrocarbon columns occur in the same play-type as five commercially producing gas fields: Katnook, Ladbroke Grove, Redman, Haselgrove, and Haselgrove South (Figure 1). The occurrence of both paleohydrocarbon columns and live columns in traps that are both full to spill and, in some cases, partially filled makes the Penola Trough an ideal location for testing and calibration of existing seal analysis methods. This case study of the Zema prospect is the first of a series of fault seal analyses to be undertaken in this area as part of a 3-yr project.

The Zema-1 well intersected a 65-m (213-ft) zone of paleohydrocarbons interpreted from drillstem test data, amplitude-vs.-offset seismic anomalies, ditch gas response, wire-line logs, and core (Boult et al., 2004) in the informally named Pretty Hill Sandstone reservoir of the Pretty Hill Formation. This paleohydrocarbon zone comprises a 50-m (160-ft) gas cap and a 15-m (49-ft) oil leg (Lovibond et al., 1995; Boult, 1997). The Pretty Hill Sandstone is regionally sealed by the overlying Laira Formation, which shows an average thickness of 400–500 m (1300–1600 ft) over the Zema prospect. The seal potential of the Laira Formation has been extensively studied. Boult (1997) deduced that fine shale paleosols in this formation form effective capillary top seals to hydrocarbon migration. These sealing units are abundantly distributed in the Laira Formation and show a high probability of being laterally continuous over prospects (Boult, 1997). For these reasons, initial top-seal risk in the Penola Trough is considered to be minimal. The presence of many breached columns in this area has thus been interpreted to be a fault-related seal issue (Hillis et al., 1995; Jones et al., 2000; Willink and Lovibond, 2001; Boult et al., 2002a).

Conventional fault seal analysis involves a first-order assessment of the seal potential caused by across-fault juxtaposition of different lithology types (Smith, 1966; Allan, 1989; Yielding et al., 1997), followed by a second-order analysis of the seal potential of the fault itself (Knipe, 1997; Yielding et al., 1997; Yielding, 2002). Shale volume estimates determined from well-log data are tied to seismic horizon data and projected onto fault plane surfaces in the modeling of across-fault juxtaposition (Downey, 1984; Watts, 1987; Jev et al., 1993). Predictive algorithms, such as shale gouge ratio (SGR) (Yielding et al., 1997) and shale smear factor (Lindsay et al., 1993; Gibson, 1994), are used to determine the

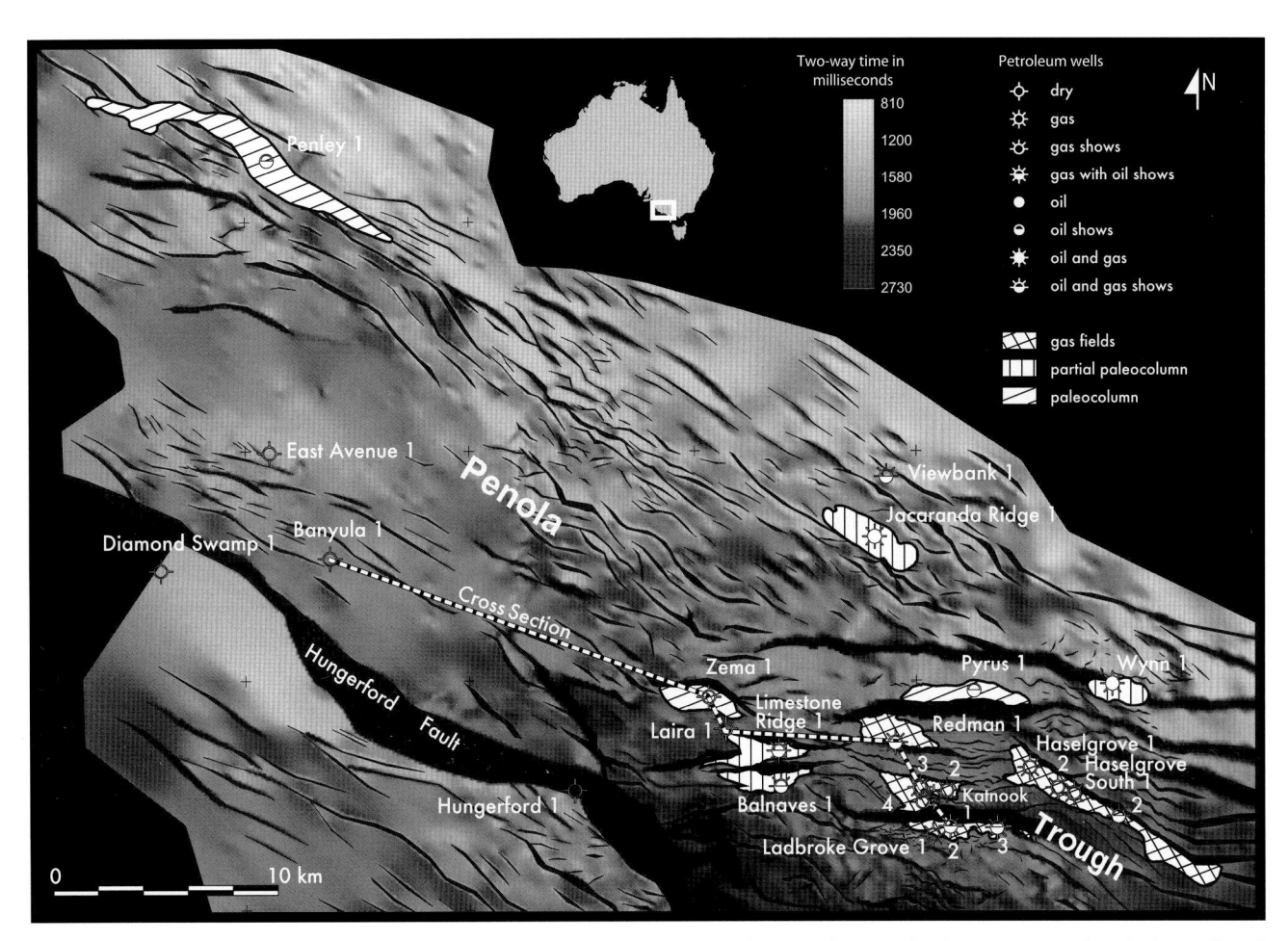

FIGURE 1. Location and status of drilled hydrocarbon prospects in the Penola trough, Otway Basin, South Australia. A near top Pretty Hill Formation horizon is displayed. The dashed white line indicates the location of Figure 3.

seal potential of a fault by estimating the amount of clay material entrained into the fault zone by mechanical processes. Predictive shale damage and smear algorithms are dependent on the accuracy of the V_{shale} determination (Yielding, 2002). Predictive algorithms are commonly compared and calibrated with pressure data and known column heights (Jev et al., 1993; Yielding, 2002; Bretan et al., 2003) or laboratory-based measurements (Sperrevik et al., 2002). Bretan et al. (2003) have calibrated SGR values to across-fault pressure data for a worldwide data set to define depth-dependent seal failure envelopes. These seal failure envelopes define the fault zone capillary entry pressure (FZP) for a given SGR by the equation FZP (bar) = $10^{(SGR/27 - C)}$, where $C = 0.5$ for burial depths less than 3 km (1.8 mi), $C = 0.25$ for burial depths between 3 and 3.5 km (1.8 and 2.1 mi), and $C = 0$ where burial depth is greater than 3.5 km (2.1 mi). Fault leakage occurs when the minimum FZP is exceeded by the buoyancy pressure of the hydrocarbon column (Bretan et al., 2003; Bretan and Yielding, 2005). This allows the maximum sustainable buoyancy pressure and, thus, the hydrocarbon column height to be predicted (Bretan et al., 2003).

An assessment of the likelihood of subsequent breach of viable prospects through postcharge reactivation of faults is required in certain circumstances (Sibson, 1996; Mildren et al., 2002; Wiprut and Zoback, 2002; Jones and Hillis, 2003). Recently reactivated faults are known from outcrop studies to be associated with permeable fracture networks (Barton et al., 1995). Recent reactivation of faults in the Otway Basin, like many other Australian basins, is considered a key exploration risk (Boult and Jones, 2000; Mildren et al. 2002).

Jones et al. (2000) assessed the probability of fault reactivation in the Penola Trough using two-dimensional fault analysis seals technology (2-D FAST). Fault analysis seals technology is a technique that quantifies the relative risk of fault reactivation in terms of in-situ stress and rock properties (Mildren et al., 2002, 2005). Although FAST predictions of fault reactivation have been successfully calibrated to the present-day distribution of hydrocarbon columns in the Timor Sea (Mildren et al., 2002), 2-D FAST predictions of fault reactivation did not fully explain the distribution of paleocolumns in the Penola Trough (Boult et al., 2002a). This suggested that mechanisms other than recent fault reactivation may be significant to hydrocarbon leakage. Boult et al. (2002a) proposed a mechanism of breach by fracturing of intact top seal caused by stress perturbations that occur locally around faults (Ramsey, 1967). Areas of high differential stress and, thus, high risk of cap seal breach can be predicted through stress modeling (Hunt and Boult, 2005).

We believe that to unequivocally calibrate the relatively new predictive methods of fault-related seal breach analysis in the Penola Trough, a thorough fault seal analysis is required, consisting of first-order fault geometry and juxtaposition analysis, a second-order analysis of fault plane properties, and an analysis of fault reactivation, using quantitative methods. A full and quantitative assessment of these parameters will allow us to fully constrain both the original sealing mechanisms in place and the influence of fault seal breach caused by recent fault reactivation. We present here the methodology and results of a uniquely customized fault seal analysis for the Zema prospect that should be applicable elsewhere in the Otway Basin.

Detailed structural interpretation using both seismic time data and pseudodepth-converted data (Lyon et al., 2004) was undertaken. A first-order analysis of juxtaposition (Allan, 1989; Yielding et al., 1997) is presented, together with a methodology of calibrating V_{shale} predictions from the gamma-ray and spectral gamma-ray log to capillary entry pressures and lithological interpretation from core analysis. The SGR method (Yielding et al., 1997) was used to assess the likelihood of fault zone sealing caused by entrainment of shale gouge in the fault zone (Knipe, 1992). The minimum fault zone capillary pressure was calculated from SGR values computed on the fault plane for comparison with predicted buoyancy pressure gradients for the original column prior to breaching (Bretan et al., 2003). Reactivation potential was determined using three-dimensional (3-D) FAST to assess the risk of recent reactivation (Mildren et al., 2002, 2005).

REGIONAL GEOLOGY

The Otway Basin formed as a result of rifting and continental breakup of Australia and Antarctica. Rifting was initiated along the presently onshore part of the basin in the Late Jurassic (Lovibond et al., 1993) (Figure 2). The Penola Trough is one of a series of onshore half-graben structures that formed as a result of this initial rifting (Perincek et al., 1994). It is a northwest–southeast-trending structure and is bound to the southwest by a large northeast-dipping listric fault complex known as the Hungerford–Kalangadoo fault system (Finlayson et al., 1993) (Figure 1).

Initial sedimentation in the Penola Trough was dominated by lacustrine sediments of the Casterton Formation (Morton, 1990) (Figure 2). This was followed by the deposition of the Crayfish Group, which includes the Pretty Hill Formation and the overlying Laira Formation, under predominantly fluvial to lacustrine conditions (Kopsen and Scholefield, 1990; Padley et al., 1995). The Pretty Hill Formation has been informally subdivided in to several members, namely, the Lower Sawpit shale member, the Sawpit Sandstone member, the Upper Sawpit shale member, and the main target reservoir, the Pretty Hill Sandstone Member at the top of the sequence (Boult et al., 2002b) (Figure 2). The Pretty Hill Sandstone consists of massive, slumped, and cross-bedded sand packages, which were deposited in a braided fluvial environment, and it has been lithologically classified as a

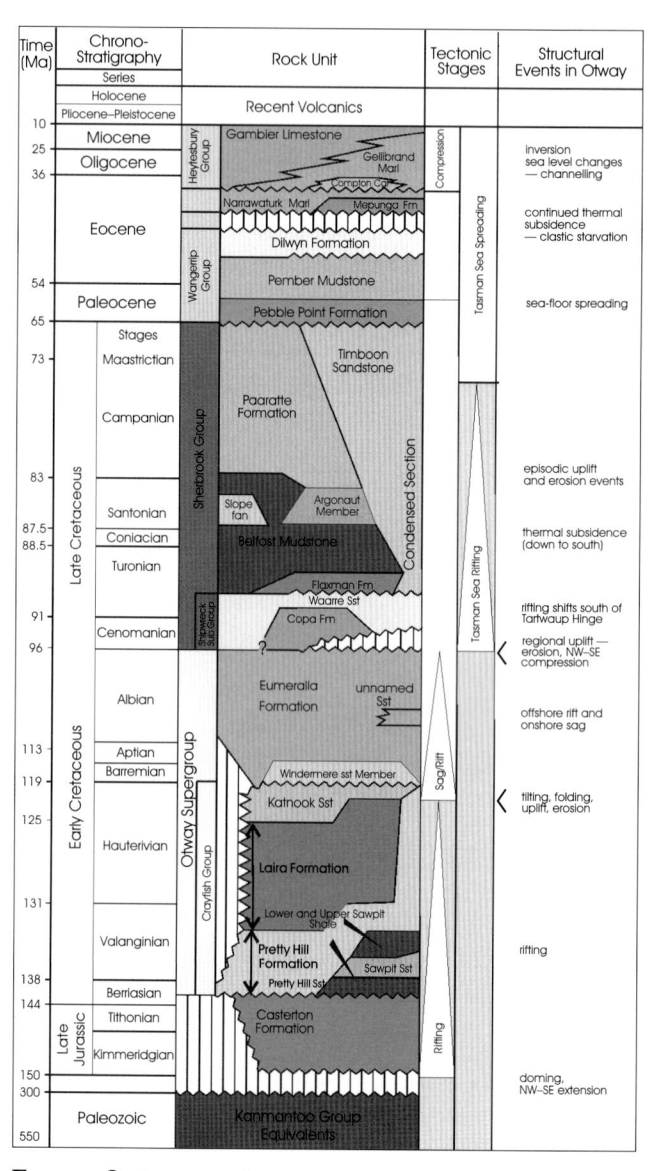

FIGURE 2. Stratigraphy and detailed tectonic history of the Penola Trough, Otway Basin.

litharenite to feldspathic litharenite (Alexander, 1992; Little, 1996). The Laira Formation, which comprises interbedded siltstones, shales, and occasional sands, was deposited in an overbank to lacustrine environment (Kopsen and Scholefield, 1990). The Laira Formation has been subdivided into five units based on well correlation of algal content (Lovibond et al., 1993) (Figure 3).

The Casterton Formation and Crayfish Group show synsedimentary growth into many of the Early Cretaceous intrarift faults formed during the initial rifting phase. These intrarift faults strike east–west in the central part of the Penola Trough, but at the margins of the trough, the strike is predominantly northwest–southeast (Figure 1).

Cessation of rifting was coincident with a period of erosion on uplifted footwall escarpments marked by the locally angular unconformity surface at the top of the Crayfish Group. This was followed by thermal sag with widespread deposition of the Eumeralla Formation of as much as 2 km (1.2 mi) thickness (Figure 2) (Hill

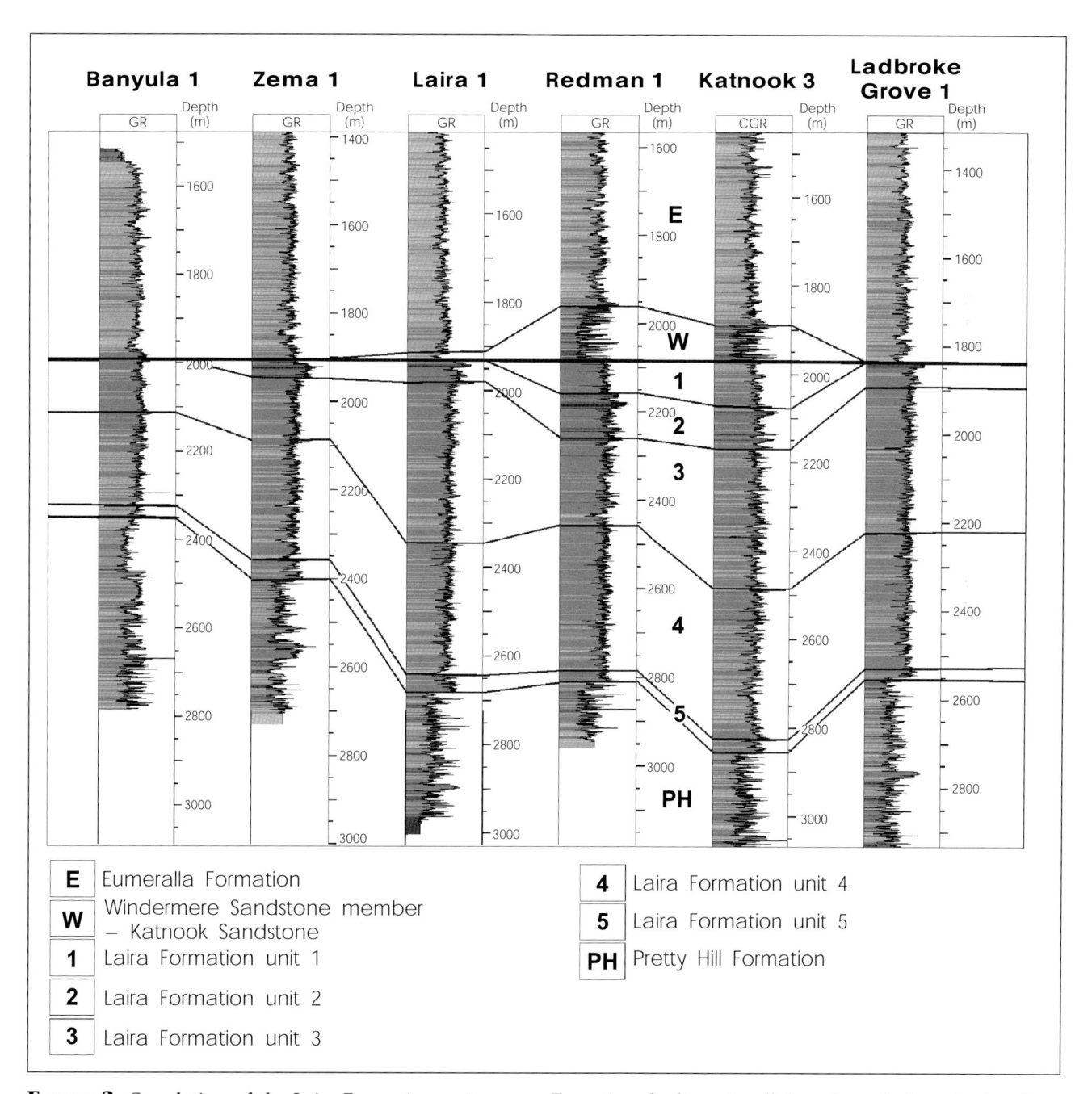

FIGURE 3. Correlation of the Laira Formation units across Zema-1 and adjacent wells based on algal content and showing anomalously high gamma-ray values associated with unit 2 (information supplied in Lovibond et al., 1993). See Figure 1 for location.

and Durrand, 1993). Compaction of relatively thick Crayfish Group sediments up against the hanging walls of large half-graben-bounding faults, under the weight of the Eumeralla Formation, allowed synsedimentary faulting to penetrate the base of the Eumeralla Formation, despite the cessation of continental extension. Rifting between Australia and Antarctica was reinitiated in the Late Cretaceous, but the rifting depocenter shifted 50 km (30 mi) to the southwest of the Penola Trough, thus leaving the Penola Trough and similar depocenters as failed rift structures (Finlayson et al., 1993; Lovibond et al., 1993). The Eumeralla Formation was deposited in a fluviolacustrine environment and consists of interbedded volcanogenic lithic sandstones, siltstones, coals, and claystones (Cockshell et al., 1995; Morton et al., 1995). Sediments of the overlying Sherbrook Group were deposited in the Late Cretaceous under fluviodeltaic conditions and consist of frequent mudstone units and massive sandstone sediment packages (Morton et al., 1995). The Tertiary succession can be subdivided into the Wangeripp, Nirranda, and Heytesbury groups and generally shows a transition to a dominantly marine environment of deposition showing frequent occurrences of massive limestone sequences.

Postrift faults of the Penola Trough, which show a more consistent northwest–southeast orientation than the synrift faults of the Early Cretaceous, commonly show displacement through the top of the Eumeralla Formation and the Sherbrook Group, and in some cases, displacement of reflectors only 200 m (660 ft) below the surface is discernable on seismic data.

Substantial hydrocarbon charge in the Penola Trough is indicated by the filling of most of the traps in the region to structural spillpoint. Thermal maturity modeling suggests that its peak was contemporaneous with the deposition of the Eumeralla Formation (Lovibond et al., 1995; Duddy, 1997).

THE ZEMA PROSPECT

Previous structural interpretation in the area has commonly failed to recognize the complex kinematic relationships between older, Early Cretaceous synsedimentary faulting and younger, postrift faulting (Aburas and Boult, 2001; Boult et al., 2002a). Because of the structural complexity of fault geometries in the area (Chantraprasert et al., 2001), the authors consider it important to understand the true depth geometries of both the main Zema prospect-bounding fault and any associated faulting. In the absence of prestack depth migration seismic data for the area, a laterally invariant velocity function was applied to the time-migrated seismic data to convert it to seismic data scaled by depth. The method describing the determination of the velocity function used in the depth conversion is beyond the scope of this chapter; for a full discussion of the technique, the reader is referred to Lyon et al. (2004). Both time-based and depth-based seismic profiles were used interactively in the workstation interpretation, together with variance slices that were computed for both seismic data sets. Interpretation made in the time domain was cross-checked with the depth interpretation to ensure that it was structurally valid. Further quality control of fault interpretation was undertaken using 3-D visualization and modeling software to eliminate inherent interpretation error.

Interpretation of four prominent seismic reflectors was undertaken after the fault geometries had been accurately mapped. The horizons interpreted were the unconformity surface at the top of the Laira Formation, the top of the Pretty Hill Formation, the top of the Eumeralla Formation, and a prominent reflector in the Sherbrook Group. The stratigraphic assignment of each horizon pick was validated by the creation of a synthetic seismogram derived from the Zema-1 sonic and checkshot log. The fault and horizon interpretation made on the depth-scaled seismic sections could be tied to the known depths of formation tops and fault interpretation in the Zema-1 well. Hence, the interpretation made in the depth domain was used in the subsequent fault seal analysis without the need to apply a more complex depth conversion to the time interpretation (e.g., Birmingham et al., 1985; Blackburn, 1986).

The depth interpretation model of the Zema prospect shows two major faults, which are significant to its trap integrity (Figure 4). Maximum displacements are found at the central parts of both faults, with a gradual decrease in displacement toward the fault tips. The abrupt change in throw observed at the western end of the main prospecting bounding Zema fault plane (referred to as the Zema fault herein) is caused by a fault splay (Figure 5). Although relay and fault splay intersections are generally considered to be critical to fault trap integrity (Childs et al., 1997), this fault splay occurs beyond the extent of the paleohydrocarbon column. It is therefore not thought to be significant to the trap geometry and is thus excluded from subsequent fault seal analysis.

From east to west, the Zema fault shows a variation in strike from east–west to northwest–southeast then to east–west (Figures 4, 5). Fault dips are fairly consistent, lying within a range of 60–70° NNE. The maximum throw for the top Pretty Hill Formation and top Laira Formation observed on the fault is 270 and 204 m (885 and 669 ft), respectively. A second fault directly overlies the northwest–southeast segment of the Zema fault, where it mimics the underlying Zema fault geometry (Figure 4). At the eastern end of this fault, however, it clearly splays away from the main Zema fault trend to a more east-southeast–west-northwest strike. The overlying fault has displaced both the top Eumeralla Formation

FIGURE 4. Three-dimensional display of the fault geometry of the main Zema prospect-bounding fault (Zema fault) and the overlying fault. Intersections of the footwall and hanging-wall horizons with the faults are indicated. Darker shading indicates northwest–southeast-trending fault segments; lighter shading indicates more east–west-trending fault segments.

and Sherbrook Group horizons. The maximum throw for the top Eumeralla Formation and Sherbrook Group horizons is 87 and 50 m (285 and 164 ft), respectively.

The height of the paleohydrocarbon column was determined from the difference in depth between the pick of the paleofree-water level (PFWL) in the well and the top of the structure in the depth model, which was found to be 69 m (226 ft). The 15-m (49-ft) thickness of the paleo-oil leg is constrained by the identification of the top and bottom of the oil zone in the Zema-1 well, but the implied 54-m (177-ft) height of the overlying gas cap is subject to the uncertainty associated with the depth conversion (see data limitations section). The PFWL lies 180 m (590 ft) above the structural spillpoint of the trap in the depth model.

SHALE VOLUME

Prediction of whether or not all or part of a fault plane is sealing in terms of juxtaposition and SGR requires an accurate determination of shale volume from log data. An estimate of the shale volume (V_{shale}) can be derived directly from the gamma-ray log by determining the maximum average gamma-ray values, assumed to be 100% shale (shale line), and the average minimum gamma-ray values, assumed to be 0% shale (sand line). All gamma-ray values can thus be assigned a percentage of V_{shale} by assuming either a linear relationship or a nonlinear relationship of increasing V_{shale} with increasing gamma-ray values (Rider, 2000). The relationship between gamma-ray values and V_{shale} is dependent on the age of the units. For well-consolidated rocks, the relationship is more accurately described by a nonlinear function (Rider, 2000).

The anomalously high gamma-ray values that occur in unit 2 of the Laira Formation (Figure 3) were investigated on the basis that these elevated gamma-ray values may be caused by high quantities of organic matter instead of high clay content (Boult, 1997). The frequent adsorption of uranium by organic matter commonly results in anomalously high gamma-ray values that are not representative of true shaliness and therefore cannot be used to define a shale line. All units in the Crayfish Group in the Zema-1 well, including unit 2, could be correlated across to the Laira-1 well (Figure 3), where a spectral gamma-ray log was run. A crossplot of total gamma-ray values and thorium values from the spectral gamma ray over the Laira Formation interval in Laira-1 shows a good correlation (correlation coefficient = 0.8; Figure 6). Because the thorium log is a better indicator of shaliness than the total gamma-ray log (Rider, 2000), it can be reasonably inferred that the elevated values of the gamma-ray log in unit 2, which are associated with equally high Th values, are indeed caused by a genuinely high abundance of shale. Inspection of the neutron-density log in Zema-1 also shows that unit 2 is associated with the largest positive separation observed on the neutron-density log.

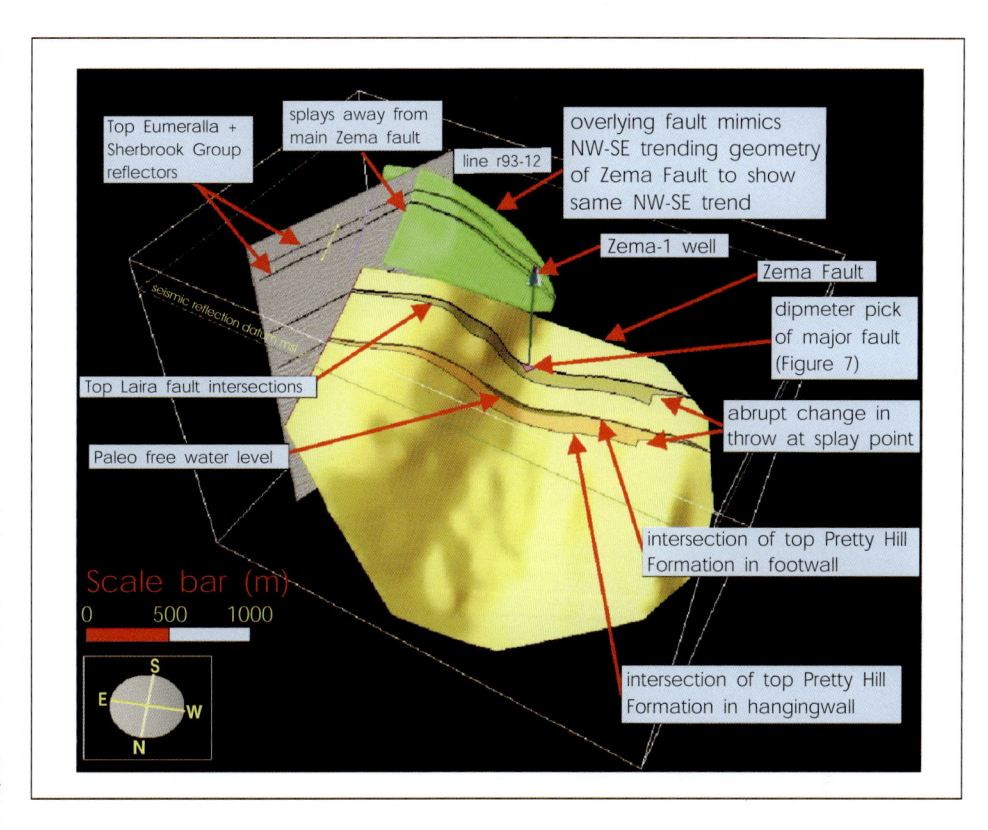

These observations confirm unit 2 as the most shale-rich unit. Unit 2 was therefore used to define the shale line at 156° API. The sand line was picked in the Pretty Hill Formation at 28° API. We can justify what may appear to be a rather high gamma-ray value for the sand line by extensive petrological work done by Little (1996), which had already shown that between 20 and 40% of the rock matrix of the Pretty Hill Sandstone in Zema-1 comprises feldspars and igneous rock

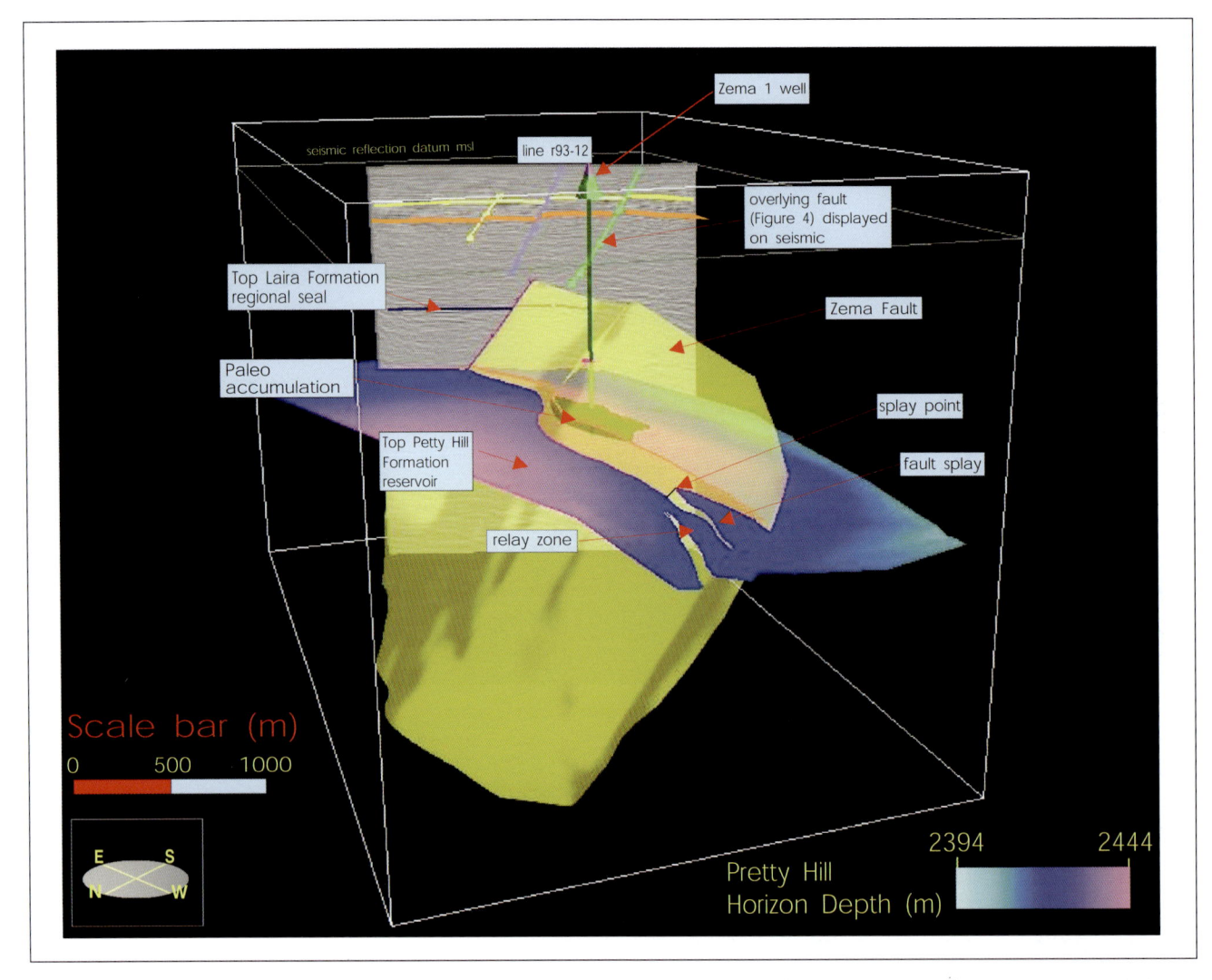

FIGURE 5. Three-dimensional display of the Zema trap geometry showing the top Pretty Hill Formation reservoir horizon, main bounding fault with splay point, and relay zone. The 3-D extent of the interpreted paleohydrocarbon accumulation is also shown. Darker shading on the fault indicates northwest–southeast-trending fault segments; lighter shading on the fault indicates more east–west-trending fault segments.

fragments, and that generally less than 5% is detrital clay matrix. Thus, the relatively high gamma-ray value for this sand is caused by the nonclay matrix grains, which do not contribute to the formation of shale gouge. Furthermore, the good correlation between the gamma-ray and thorium values show that the gamma ray is a good indicator of shaliness for these lithology types.

The nonlinear conversion of gamma ray to V_{shale} for consolidated rocks (Rider, 2000) was used to derive a V_{shale} log:

$$V_{shale} = 0.33(2^{2GRI} - 1),$$

where Gamma-ray Index (GRI)

$$= \frac{GR(\text{log value}) - GR(\min)}{GR(\max) - GR(\min)} \qquad (1)$$

GR(min) is the minimum average gamma-ray value (sand line); GR(max) is the maximum average gamma-ray value (shale line); and GR(log) is the actual gamma-ray log value.

CORE CALIBRATION

The V_{shale} log was used to subdivide the Laira and the Pretty Hill formations into 45 intraformational beds or isochores (Bouvier et al., 1989). This detailed lithological interpretation was necessary for a thorough juxtaposition analysis, because even a 1-m (3.3-ft) sand that was unfavorably juxtaposed across the fault could be crucial to initial fault seal integrity.

Each intraformational bed was also assigned a lithology type based on calibration with core. Figure 7 shows

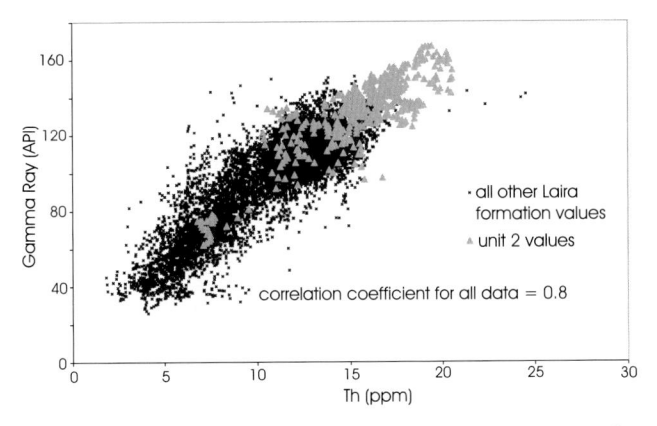

FIGURE 6. Gamma-ray (API) vs. thorium (ppm) crossplot for the Laira Formation in Laira-1. A good correlation between the two data sets indicates that the gamma-ray curve is a good indicator of shale content in this region.

the interpretation of a cored interval through the base of the Laira Formation. An intra-Laira sand occurs at the top of the core. Below this sand, the core consists of predominantly silty shale with occasional thin interbeds of relatively clay-rich paleosol units. Mercury injection capillary pressure curves show that the dominant silty shale lithotype has an entry pressure of 2 MPa (300 psi), suggesting that it is only capable of holding back a total gas column height of 21 m (69 ft) (Figure 7). The clay-rich paleosol beds show an entry pressure of 55 MPa (8000 psi) and are thus capable of holding back a gas column far beyond the structural spillpoint of the trap. A comparison of the core with the V_{shale} log shows that the sandstone and silty shale lithologies have average V_{shale} values of 15 and 33%, respectively. Two thick paleosol units that are resolvable with the gamma-ray tool show average V_{shale} values of 46 and 48%. Thinner paleosol horizons identified in the core, in addition to those highlighted in Figure 7, are less than 80 cm (31 in.) thick and are thus beyond the resolution of the gamma-ray tool.

Intraformational beds of the Laira and Pretty Hill formations were classified into three broad lithology types for the purposes of qualitative juxtaposition analysis on the basis of this cross-comparison with the core. All beds with an average V_{shale} of less than 20% were assigned to a sand lithology type. Values between 20 and 40% were assigned to a silty shale lithology type with expected capillary pressures equivalent to just a few tens of meters of gas column height capacity. Values above 40% were assigned to a shale lithology type expected to have high capillary pressures capable of holding back a gas column to the structural spillpoint. The Laira Formation consists of a roughly equal mixture of both silty shale and shale, with four prominent thin interbeds of sandstone near its base. The Pretty Hill Formation consists of mostly sandstone with sparse interbeds of 10–15-m (33–49-ft)-thick silty shale.

JUXTAPOSITION

The interpreted internal stratigraphy of the Laira and Pretty Hill formations in the Zema-1 well was isopached proportionally between the top Laira Formation and the top Pretty Hill Formation seismic horizons in the depth model, thus providing a detailed stratigraphic interpretation for the footwall and slightly thicker section in the hanging wall. The lateral continuity of sands is discussed under the data limitations section.

Figure 8 shows the mapped juxtaposition relationships for the slipped interval on the fault plane where the top Pretty Hill Formation horizon has been displaced. Of key significance is the occurrence of an intra-Laira Formation sand juxtaposed against Pretty Hill Formation reservoir sand above the PFWL. Furthermore, several occurrences of Laira Formation silty shale are also juxtaposed against Pretty Hill Formation sands that occur above the PFWL. These silty shale units show an average V_{shale} of approximately 32%, which is equivalent to the silty shale lithology types identified in the core through the base of the Laira Formation (Figure 7). It is therefore inferred that these silty shales are likely to have similar entry pressures and are thus unable to hold back a gas column greater than about 20 m (66 ft).

Fault juxtaposition relationships alone cannot explain the interpreted 69-m (226-ft) paleocolumn at Zema. Accordingly, the mechanical processes occurring in the fault damage zone were considered as a mechanism for sealing.

SHALE GOUGE RATIO

Shale gouge ratio was computed along the fault plane using throw and average V_{shale} from the hanging wall and footwall (Yielding et al., 1997; Yielding, 2002). Figure 9a shows the shale gouge computation results for the slipped interval on the fault plane where the top Pretty Hill Formation has been displaced. The SGR values over sand-on-sand, silty shale-on-sand, and silty shale-on-silty shale intervals show a gradual decrease from values as high as 34% at the top of the trap to values as low as 15% at the structural spillpoint.

Shale gouge ratio values from the top of the trap to the depth of the interpreted PFWL, sampled at 5-m (16-ft) intervals, were converted to fault zone capillary entry pressure (FZP) using the equation $FZP = 10^{(SGR/27-C)}$ (Bretan et al., 2003) (Figure 9b). A C value of 0.5 was used, because the burial depth is less than 3 km (1.8 mi). If it is assumed that leakage results when the buoyancy pressure of the hydrocarbon column exceeds the minimum FZP on the fault plane (Bretan and Yielding, 2005), then these data can be used to predict the maximum

188 Lyon et al.

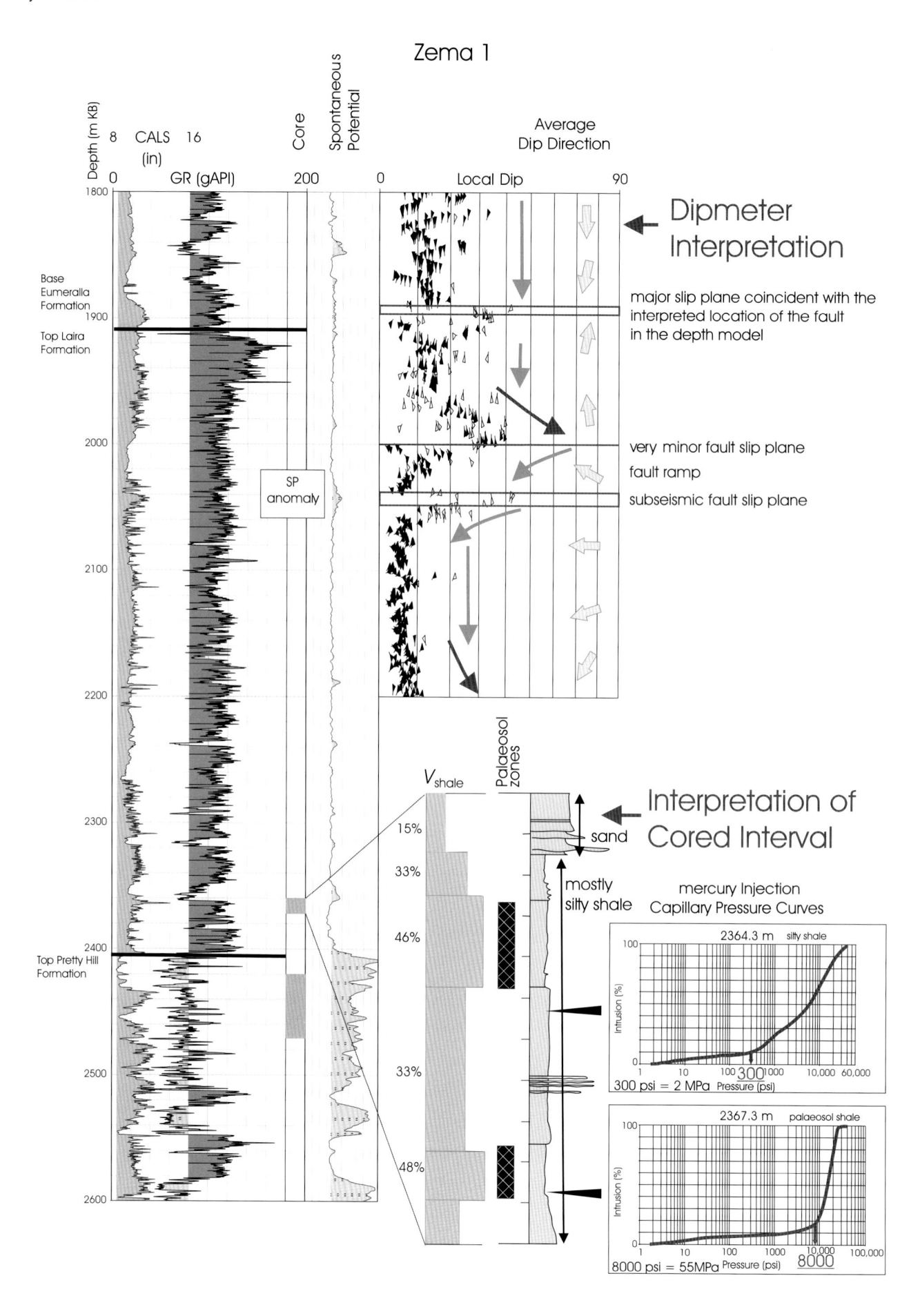

column height sustainable by the fault zone. The interpreted buoyancy gradients for both gas (8 kPa/m) and oil (4.3 kPa/m) were calculated using subsurface density measurements made in adjacent wells (oil density = 0.53 g/cm^3; gas density = 0.143 g/cm^3; and water density = 0.97 g/cm^3). The buoyancy gradients were used to predict the maximum column height for a gas cap associated with the known 15-m (49-ft) oil leg based on the observed FZP values (Bretan et al., 2003). The predicted total maximum column height is 72 m (236 ft) (Figure 9b). The predicted point of leakage is near the top of the fault where SGR values are greater than 30% (Figure 9b). This predicted column height interpreted from FZP values is only 3 m (10 ft) greater than the interpreted paleocolumn height of 69 m (226 ft).

Therefore, shale gouge developed on the fault plane is a plausible mechanism to explain the primary entrapment of the hydrocarbon column at Zema. Furthermore, the filling of the trap down to the PFWL may have been controlled by capillary leakage related to the amount of shale gouge in the fault zone.

FAULT CHARACTERIZATION

A dipmeter interpretation was used to determine the position of the Zema fault interpreted on seismic data in the Zema-1 well, so that the fault rock properties could be investigated from wire-line logs (Figure 7). Distinct zones of abrupt change in dip and strike are present that indicate that the fault plane, as identified on seismic data, consists of multiple slip planes at the resolution of the dipmeter. The main slip plane interpreted on the dipmeter occurs at 1890 m (6200 ft), near the base of the Eumeralla Formation, consistent with the interpreted location of the fault in the depth seismic interpretation (Figure 4). The 8-m (26-ft) zone of faulting that is interpreted from the dipmeter at 2040 m (6692 ft) corresponds to a distinct spontaneous potential anomaly and represents the lower fault of a relay ramp that dips northeast and has a strike of 300° (northwest–southeast). The spontaneous potential anomaly suggests that this fault surface and slip plane (beyond the resolution of the seismic data) is permeable. However, spontaneous potential effects, which are associated with the main fault plane, are disguised by the more frequent sandstone beds at the base of the Eumeralla Formation.

REACTIVATION ASSESSMENT (FAST)

The reactivation potential of the Zema fault analysis seals technology and the overlying fault was assessed using Fault Analysis Seals Technology (FAST) (Mildren et al., 2002, 2005). A brief overview of the technique is given here, but for a more detailed discussion, the reader is referred to Mildren et al. (2005). FAST provides an assessment of the likelihood of a fault to reactivate in the present-day stress field. The reactivation potential is quantified in terms of the increase in pore pressure (ΔP) required for the resolved shear and effective normal stress acting on a fault plane of a given dip and strike to exceed a specified failure envelope. Three inputs are required for FAST analysis: the failure envelope, the in-situ stress field, and the fault orientation data (Figure 10a, b).

The following stress gradients in the Penola Trough were determined by Jones et al. (2000) for a depth range of 2500–3000 m (8200–10,000 ft) and were used to determine the 3-D Mohr circle used in the FAST analysis herein:

minimum horizontal stress (σ_h) = 16.1 MPa/km.
overburden stress (σ_v) = 22.4 MPa/km.
maximum horizontal stress (σ_H) = 28.7 MPa/km.
pore pressure (P) = 9.8 MPa/km.
maximum horizontal stress (σ_H) orientation = 156°N.

The failure envelope was determined using the results from triaxial testing of a core sample through a cataclasite fault zone in the Pretty Hill Formation in the Banyula-1 well (Dewhurst and Jones, 2002) (see Figure 1 for location):

$$\tau = 5.40 + 0.78\sigma'_n \quad (2)$$

where τ is shear stress at failure, and σ'_n = effective normal stress (i.e., $\sigma_n - P$), with all pressure units in megapascals.

The above criteria were then used to assign a reactivation risk (ΔP) to the 3-D fault surface elements of the Zema fault and overlying fault (Figure 10c). The northwest–southeast-trending part of the Zema fault is associated with ΔP values in the range of 6–7 MPa/km (red colors) and is critically orientated in the in-situ

FIGURE 7. Zema-1 caliper (CALS), gamma-ray (GR), spontaneous potential (SP), and dipmeter logs. Interpretation of the cored interval through the base of the Laira Formation, together with the interpreted V_{shale}, the main paleosol zones and mercury injection capillary pressure (MICP) curves for the core (bottom). The dipmeter interpretation through the fault zone is also shown (top). A spontaneous potential anomaly associated with a subseismic fault slip plane is highlighted. CALS = caliper logs; KB = Kelly Bushing.

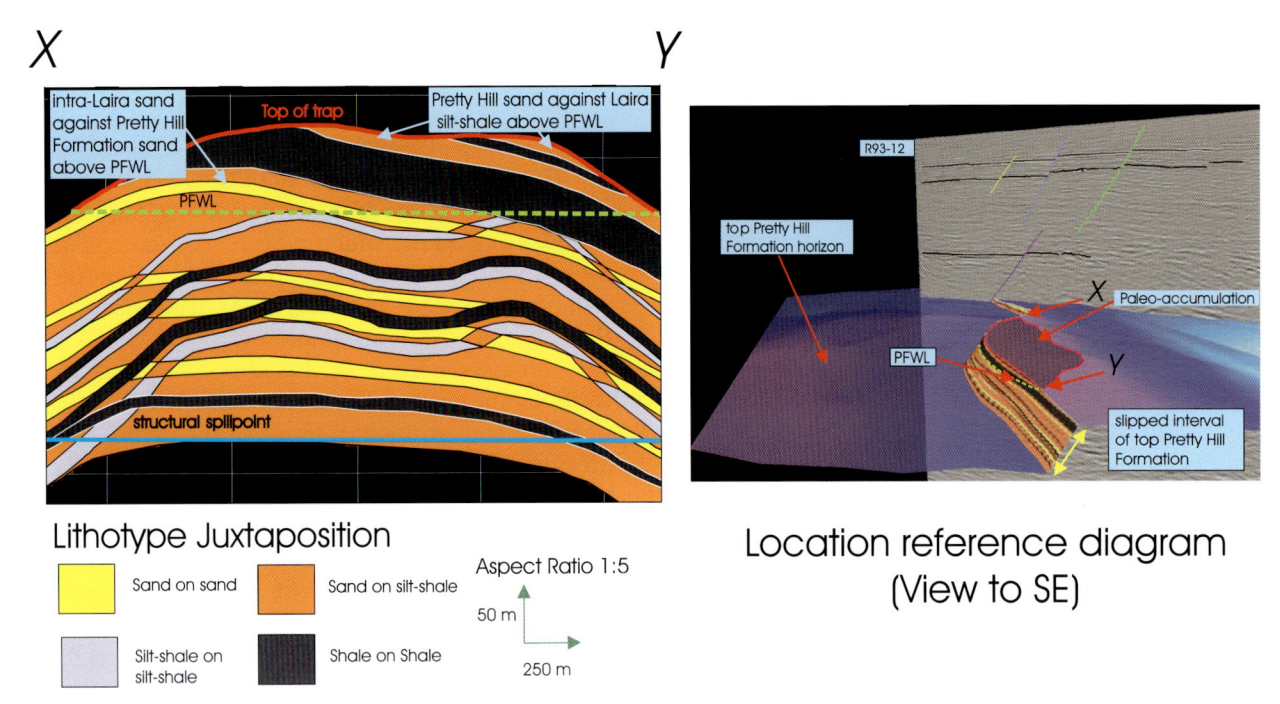

FIGURE 8. View from the hanging wall toward the footwall of juxtaposition modeling results for the slipped interval on the fault plane, where the top Pretty Hill Formation reservoir has been displaced (left). The position of points X and Y are indicated on the location reference diagram (right). PFWL = paleofree-water level.

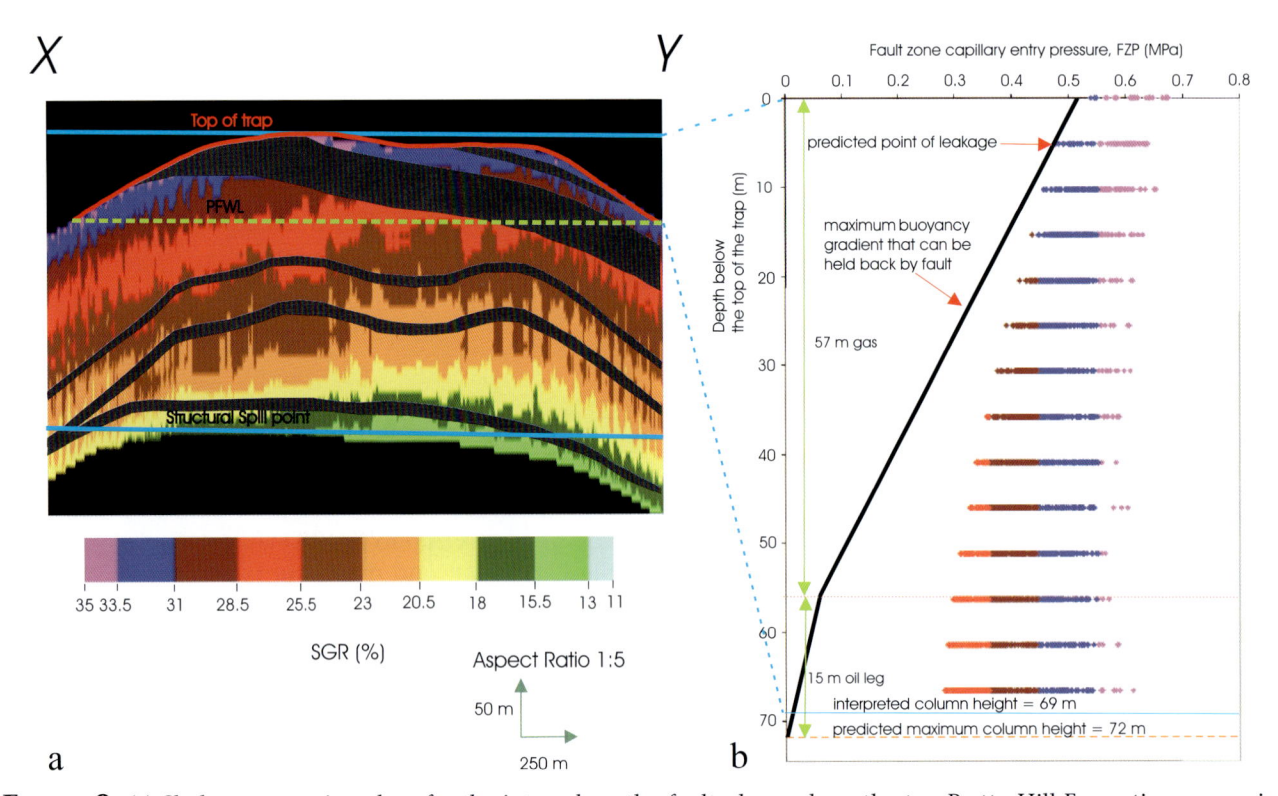

FIGURE 9. (a) Shale gouge ratio values for the interval on the fault plane where the top Pretty Hill Formation reservoir has been displaced. The position of points X and Y are indicated in Figure 8. (b) The fault zone capillary entry pressure (FZP) profile, calculated from SGR values sampled at 5-m (16-ft) increments from the top of the trap to the PFWL. The predicted maximum buoyancy profile and, thus, predicted maximum column height are shown.

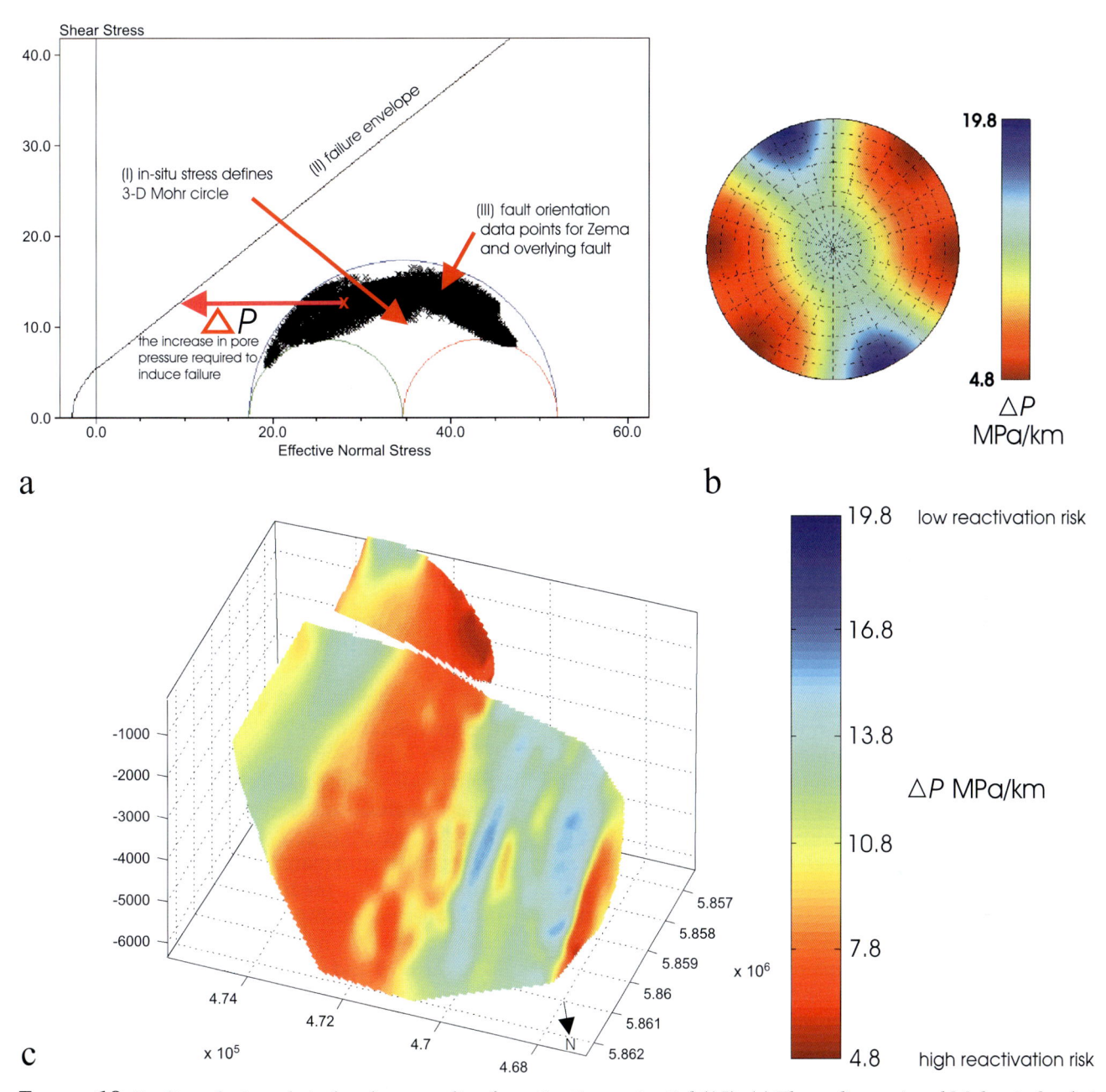

FIGURE 10. Fault analysis seals technology results of reactivation potential (ΔP). (a) Three-dimensional Mohr circle plot for a depth of 2750 m (9022 ft) with explanatory notes showing how ΔP is calculated and the three key inputs used in the FAST analysis: (I) in-situ stress, (II) failure envelope, and (III) fault orientation data points. (b) Stereographic projection of ΔP values showing the full range in ΔP for all possible fault orientations. (c) The Zema fault and overlying fault surfaces colored by their respective ΔP attributes.

stress field for reactivation. The more east–west-trending part of the fault generally shows values greater than 12 MPa/km and are thus less critically oriented for reactivation (blue colors). The overlying fault shows the lowest ΔP values that range from 5 to 5.5 MPa/km over the northwest–southeast-striking zone, where it directly overlies the northwest–southeast-striking part of the Zema fault.

DATA LIMITATIONS

Unavoidable limitations are imposed on any fault seal analysis by the available data. Structural uncertainty arises from the limited resolution of seismic data, which prevents the incorporation of subseismic structural elements that make up the fault zone into the fault zone model (Childs et al., 1996, 1997; Doughty, 2003;

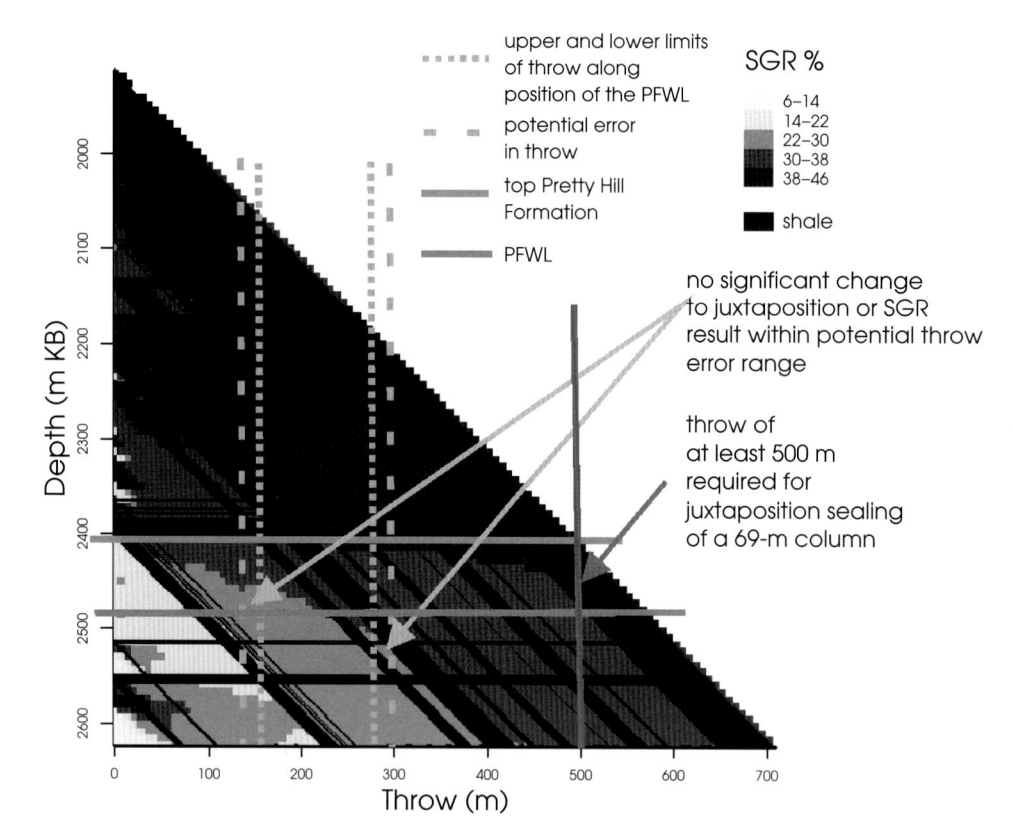

FIGURE 11. Triangle diagram showing an estimate of the expected juxtaposition and SGR distribution for a given throw magnitude. Shale gouge ratio values for the full range of throw values for the top Pretty Hill Formation along the position of the PFWL (i.e., the range in throw values between points X and Y in Figure 8) are indicated together with potential error bars. The top of the Pretty Hill Formation and PFWL are also indicated on the diagram. KB = Kelly Bushing.

Koledoye et al., 2003), such as the relay zone and multiple slip planes, which were identified on the dipmeter log. Further uncertainty exists in determining the scale of heterogeneity in terms of fault zone composition and thickness (Fisher and Knipe, 1998; Sperrevik et al., 2002; Yielding, 2002). Although algorithms such as SGR can provide a good estimate of the average shale content of fault zones at the seismic scale, it remains difficult to upscale the heterogeneities of fault zones as observed in both log data and core samples into a meaningful and representative fault seal model. Furthermore, the fluid-flow properties of fault zones (Gibson, 1998; Weher et al., 2000), which may be controlled by thermal history and effective stresses at the time of deposition and postdeposition (Fisher and Knipe, 1998; Fisher et al., 2003), must also be considered. In addition to these generic limitations of fault seal analysis, several limitations that are specific to the data quality and methodology adopted in this analysis of the Zema prospect exist.

1) There is incomplete 3-D seismic data coverage over the prospect. The 3-D survey spans much of the east–west-trending part of the fault, to the west of the Zema-1 well (Figures 4, 5). Furthermore, the 3-D coverage of the prospect is at the edge of the 3-D St. George survey and is therefore not optimally migrated. The fault has been interpreted using 2-D data to the east of the Zema-1 well. The existence of additional relay zones and fault splay structures, similar to the one identified in the 3-D survey (Figure 5), are not discernable from the 2-D seismic data.

2) Potential error in the magnitude of throw on the Zema fault is highly probable because of the nature of the depth conversion used (Lyon et al., 2004) and also because of possible interpretation error limited by the relatively poor seismic resolution at this depth. A sensitivity analysis was undertaken by constructing a triangle diagram (e.g., Knipe, 1997) to assess how a conservative error margin of ±25 m (±82 ft) in throw would affect the predictions of across-fault juxtaposition relationships and SGR (Figure 11). The results show that potentially leaky juxtaposition of sand-on-sand and sand-on-silty shale would still be expected in the 25-m (82-ft) throw error margins. Figure 11 clearly shows that a throw in the order of 500 m (1600 ft) magnitude is needed to result in a viable juxtaposition seal of shale-on-sand for a 69-m (226-ft) column. No significant change to the SGR values in the specified 25-m (82-ft) throw error margin exist.

3) The sand line picked in the Pretty Hill Formation may have led to a slight overestimation in the V_{shale} calculation, because petrological analysis of the Pretty Hill Sandstone typically shows 5% interstitial clays (Little, 1996). It is therefore postulated that even the lowest gamma-ray values in the Pretty Hill Formation may not truly represent a 0% shale line.

4) The V_{shale} curve requires more mercury injection capillary pressure data points to be confident of precise calibration of V_{shale} to capillary entry pressure. However, relating the observations made in core and mercury injection capillary pressure data

to the V_{shale} curve demonstrated in this chapter does provide a useful reference for qualitative assessment of juxtaposition relationships.
5) The degree to which there is lateral variation in lithology type along the fault plane and away from well control limits the analysis. The sands that were cored at the base of the Laira Formation (Figure 7) are correlatable as far as the Laira-1 well; however, more detailed stratigraphic data or analyses are required to determine the lateral continuity of other units.
6) Uncertainty in upscaling laboratory-derived geomechanical fault rock properties to the seismic scale of the FAST analysis exists. In this case, we have used data from a cataclasite in the Pretty Hill Formation in Banyula-1, which is not necessarily representative of the predicted phyllosilicate framework fault rock predicted for the Zema fault above the PFWL.

A conservative range has been defined for the potential error in throw. However, the contribution in terms of quantitative error of other uncertainties to the cumulative error of the analysis is difficult to determine. The fact that the predicted column height obtained from SGR values (Figure 9b) is only 3 m (10 ft) larger than the interpreted paleocolumn height from the structural interpretation suggests an overall reliability in the first- and second-order fault seal analysis.

The sensitivity analyses described in Mildren et al. (2005) show significant changes to absolute ΔP values from FAST analysis when varying the failure envelope used. There is, however, no change to the fact that the northwest–southeast parts of the fault are critically oriented in the in-situ stress field for reactivation.

DISCUSSION

Our structural interpretation shows the existence of two distinct fault planes associated with the Zema prospect. The slightly thicker Crayfish Group in the hanging wall of the Zema fault confirms it formed as a synsedimentary fault in the Early Cretaceous. The overlying fault shows no synsedimentary deposition, and the initiation age of the fault is thus difficult to constrain. However, the geometry of the overlying fault mimics the geometry of the underlying Zema fault over its northwest–southeast-trending section. This may be caused by stress perturbation in the cover, because of the Zema fault, which, in turn, may be related to its reactivation.

The results of juxtaposition analysis show that both sand-on-sand and sand-on-silty shale zones of juxtaposition occur above the interpreted PFWL. The sand-on-sand juxtaposition zone has a high potential for leakage (Allan, 1989). The silty shale lithology shows a similar V_{shale} value to the silty shale lithology identified in the core and is therefore likely to be only capable of holding back a gas column of just over 20 m (66 ft). It is therefore suggested that juxtaposition relationships alone cannot explain the retention of a 69-m (226-ft) column. However, it must be stressed that the simple assumption that sand-on-sand juxtaposition is nonsealing is not always correct. Fault damage zones in clean sandstone reservoirs can be associated with quartz-cemented cataclasites (Fisher et al., 2000, 2003) that are capable of supporting large columns. Indeed, an example of such a cataclasite zone was intersected in the Pretty Hill Formation at Banyula-1 (Jones et al., 2000; Dewhurst and Jones, 2002). Cataclasis and quartz cementation may be the dominant fault rock deformation processes in the Pretty Hill Formation, where sandstones have slipped past each other. However, where the Laira Formation has been downthrown against the Pretty Hill Formation (the critical part of the fault zone to trap integrity), the fault zone is unlikely to consist of well-developed zones of cataclasite over sand-on-sand intervals. The fault rock types will instead be dominated by phyllosilicate framework rock types based on the observed SGR range of 18–32% (Fisher and Knipe, 1998). This viewpoint is further supported by the recognition of a phyllosilicate framework fault rock in a core from Jacaranda Ridge-1, where the Pretty Hill Formation is more shaly (see Figure 1 for location).

We suggest that the 32–18% shale gouge that developed along the slipped interval between the Laira and Pretty Hill formations was responsible for holding back the original 69-m (226-ft) hydrocarbon column. Furthermore, it is possible that the extent of shale gouge development in the fault zone limited the height of the column to only 69 m (226 ft). This hypothesis is supported by Figure 9b, which shows that predicted fault zone capillary entry pressures calculated from SGR values predict a maximum hydrocarbon column height of 72 m (236 ft) (for a 15-m [49-ft] oil leg) based on the buoyancy gradients of the gas and oil phases calculated in nearby wells (Boult, 1997). Leakage is predicted to occur near the top of the fault, where the SGR values are greater than 30%. The fact that there is just a 3-m (10-ft) difference in the predicted column height from SGR values and the interpreted paleocolumn height suggests that it is shale gouge in the fault zone that limited the column height. Lack of charge may also have been the limiting factor. However, this is considered unlikely, because the live hydrocarbon columns at Redman, Katnook, Haselgrove, and Haselgrove South (Figure 1) all lie within 15 km (9 mi) of Zema and are full to structural spillpoint. Considering the evidence of abundant charge in the area and maturity modeling (Boult et al., 2004), it is suggested that the trap may have experienced more than one episode of capillary leakage of part of the gas column, where added buoyancy pressure from additional hydrocarbon charge was sufficient to exceed the fault zone capillary pressure of the fault (Schowalter, 1979; Fisher et al., 2001).

Fault analysis seals technology indicates that the prospect-bounding northwest–southeast part of the Zema fault is critically oriented in the in-situ stress field for reactivation, associated with relatively low ΔP values of 6–7 MPa/km. Furthermore, the recognition of a spontaneous potential anomaly within part of a similarly oriented fault zone through the Laira Formation seal in Zema-1 indicates that at least part or parts of the fault zone are likely to be presently permeable (Figure 7). Recent reactivation is therefore likely to have breached the original shale gouge fault seal, causing the entire 69-m (226-ft) column to leak away. Although fault reactivation is the most likely explanation for the breach of the trap based on these observations, the authors do not dismiss the possibility of intact cap seal fracturing as a breach mechanism (Boult et al., 2002b) discussed in more detail in Mildren et al. (2005) and Hunt and Boult (2005).

Interestingly, the lowest ΔP values occur on the overlying fault and not the main Zema fault. Reactivation may have occurred preferentially along this more critically stressed fault, which may also extend down to the top of the reservoir as a subseismic feature. Perhaps growth of these younger faults is soft-linked to the reactivation of the underlying older faults. These younger faults, which are widespread throughout the upper sedimentary succession of the Penola Trough, have, to date, received little attention in the reactivation risking of fault-bound prospects, despite the fact that many are critically orientated for reactivation. The authors therefore stress the importance of a more holistic approach to reactivation risk, which accounts for faults in the upper section that may not necessarily be identified on seismic data at the reservoir level, but which may have propagated downward, causing subseismic fracturing through the cap rock.

A holistic risking method of assessing reactivation in regions such as the Penola Trough therefore demands two considerations: first, a detailed knowledge of both fault geometry and fault kinematics; second, as fault rock properties are strongly controlled by host rock lithology, throw, effective stress, and temperature history. Early Cretaceous synrift faults and shallower postrift faults are likely to be characterized by very different fault rock properties and, thus, different geomechanical properties. Future reactivation risking using FAST in the Otway Basin should attempt to consider these differences through the use of varying failure envelopes (see Mildren et al., 2005).

Further fault seal analysis of breached and intact columns in the Penola Trough will be undertaken to systematically assess juxtaposition, fault damage, and reactivation potential. The establishment of empirical calibrations of SGR to paleo- and present column height and of fault geometry to reactivation potential using FAST across numerous traps will reveal which are the critical risks.

CONCLUSIONS

- The Zema prospect is host to an interpreted paleohydrocarbon column. A paleofree-water level is identified at 69 m (226 ft) below the top of the trap, which is 200 m (660 ft) above the structural spillpoint.
- V_{shale} estimates derived from the gamma-ray log used in this study are calibrated to lithology type and capillary pressure data from core analysis. V_{shale} values of greater than 40% correspond with paleosol shale lithotypes in core that have capillary entry pressures of 55 MPa (8000 psi) and are thus capable of supporting gas columns beyond the structural spillpoint of the trap. V_{shale} values of 20–40% correspond with silty shale lithotypes associated with capillary entry pressures equivalent to gas column heights of less than 30 m (100 ft). Sands correspond with shale volumes of less than 20%. The excellent correlation of the gamma-ray log and thorium log in an adjacent well confirms the reliability of the V_{shale} values as a means of estimating shale content of these rock types.
- Across-fault juxtaposition modeling of the slipped interval of the top Pretty Hill Formation horizon shows several sand-on-sand and sand-on-silty shale windows that occur above the PFWL. Hence, juxtaposition alone cannot explain the observed paleocolumn.
- Shale gouge ratio values show a range of 32–18% over the slipped interval of Pretty Hill Sandstone reservoir, characteristic of a phyllosilicate framework rock-type fault zone (Fisher and Knipe, 1998). Fault zone capillary entry pressures calculated from SGR values predict a maximum hydrocarbon column height of 72 m (236 ft) (for a 15-m [49-ft] oil leg and a 57-m [187-ft] gas cap) based on the buoyancy gradients of the gas and oil phases calculated in nearby wells. The critical leak point is at the top of the prospect-bounding fault zone, where SGR values are greater than 30% (Figure 9b). This predicted column height of 72 m (236 ft) is in close agreement with the paleocolumn height of 69 m (226 ft), which is determined from the structural interpretation and well data. This supports the interpretation that the shale gouge in the fault zone was responsible for entrapment of the original column. It also suggests that the column height may have been limited by capillary leakage related to the amount of shale gouge in the fault zone.

- Geomechanical analysis indicates that the northwest–southeast prospect-bounding part of the Zema fault and the overlying fault are critically oriented in the in-situ stress field for reactivation. It is likely that fracturing generated by recent reactivation caused the entire original column to leak away. This hypothesis is supported by evidence of present-day permeability in a northwest–southeast trending fault damage zone in the top seal as indicated by a spontaneous potential anomaly recorded in the Laira Formation (Figure 7).
- Although a high likelihood of trap breach by fault reactivation exists, we do not dismiss the possibility of intact cap rock failure associated with fault-related stress perturbations as a breach mechanism.
- Many younger faults in the Penola Trough may, as in the case of Zema, be associated with a higher risk of reactivation than the Early Cretaceous prospect-bounding faults. These younger faults may extend down subseismically to the base of the seal and are a significant risk to prospect integrity.

ACKNOWLEDGMENTS

Peter Bretan, Andrew Davids, Richard Suttill, and Paul Theologou are thanked for their constructive reviews. Origin Energy and its joint venture partners are thanked for the provision of 3-D seismic data and regional near top Pretty Hill Formation horizon. All staff and students at the Australian School of Petroleum, University of Adelaide, are thanked for advice, guidance, and sharing of technical expertise. We particularly acknowledge the key contribution made by fellow researchers in the Stress Group (Australian School of Petroleum) and the Australian Petroleum Cooperative Research Center. All staff members at Badleys Geoscience Ltd. are thanked for their excellent software support and guidance and provision of Traptester software. Jerry Meyer of JRS Petroleum Research Pty. Ltd. and Quentin Fisher of Rock Deformation Research, University of Leeds, are also thanked for their informative comments and advice. Primary Industries and Resources of South Australia Publishing Services are thanked for their assistance in the drafting of Figures 1–3 and 7.

REFERENCES CITED

Aburas, A. N., and P. J. Boult, 2001, New insights into the structural development of the onshore Otway Basin, in K. C. Hill and T. Bernecker, eds., Eastern Australian Basins Symposium: Petroleum Exploration Society of Australia Special Publication, p. 447–453.

Alexander, E. M., 1992, Geology and petrophysics of petroleum reservoirs from the Otway Group, Otway Basin: Energy Research and Development Corportation Project 1424, State Energy Research Advisory Committee Project 3/89, end of grant report, 331 p.

Allan, U. S., 1989, Model for hydrocarbon migration and entrapment within faulted structures: AAPG Bulletin, v. 73, p. 803–811.

Barton, C. A., M. D. Zoback, and D. Moos, 1995, Fluid flow along potentially active faults in crystalline rock: Geology, v. 23, p. 683–686.

Birmingham, P. J., K. R. A. Grieves, and D. E. Spring, 1985, Depth conversion techniques in the Gippsland Basin: Exploration Geophysics, v. 16, p. 172–174.

Blackburn, G. J., 1986, Depth conversion: A comparison of methods: Exploration Geophysics, v. 17, p. 67–69.

Boult, P. J., 1997, A review of seal potential in the Penola Trough, western Otway Basin (Petroleum Exploration Licence 32): Boral Energy Resources Ltd., 91 p.

Boult, P. J., and R. M. Jones, 2000, Fault geometry and seal attribute mapping in the Bass and Otway basins, Australia (abs.): AAPG Bulletin, v. 84, p. 1406.

Boult, P. J., B. A. Camac, and A. W. Davids, 2002a, 3D fault modelling and assessment of top seal structural permeability—Penola Trough, onshore Otway Basin: Australian Petroleum Production and Exploration Association Journal, v. 42, p. 151–166.

Boult, P. J., M. R. White, R. Pollock, J. G. G. Morton, E. M. Alexander, and A. J. Hill, 2002b, Lithostratigraphy and environments of deposition, in P. J. Boult, and J. E. Hibburt, eds., The petroleum geology of South Australia: v. 1. Otway Basin, 2d ed.: Department of Primary Industries and Resources, Petroleum Geology of South Australia Series, p. 1–98.

Boult, P. J., P. Lyon, B. Camac, D. Edwards, and D. McKirdy, 2004, Subsurface plumbing of the Crayfish Group in the Penola Trough, in P. J. Boult, D. R. Johns, and S. C. Lang, eds., Eastern Australasian Basins Symposium II: Petroleum Exploration Society of Australia Special Publication, p. 483–498.

Bouvier, J. D., C. H. Kaars-Sijpesteijn, D. F. Kluesner, C. C. Onyejekwe, and R. C. Van der Pal, 1989, Three-dimensional seismic interpretation and fault sealing investigations, Nun River field, Nigeria: AAPG Bulletin, v. 73, p. 1397–1414.

Bretan, P., and G. Yielding, 2005, Using buoyancy pressure profiles to assess uncertainty in fault seal calibration, in P. Boult and J. Kaldi, eds., Evaluating fault and cap rock seals: AAPG Hedberg Series, no. 2, p. 151–162.

Bretan, P., G. Yielding, and H. Jones, 2003, Using calibrated shale gouge ratio to estimate hydrocarbon column heights: AAPG Bulletin, v. 87, p. 397–413.

Chantraprasert, S., K. R. McClay, and C. Elders, 2001, 3D rift fault systems of the western Otway Basin, SE Australia, in K. C. Hill and T. Bernecker, eds., Eastern Australian Basins Symposium: Petroleum Exploration Society of Australia Special Publication, p. 435–445.

Childs, C., J. Watterson, and J. J. Walsh, 1996, A model for the structure and development of fault zones: Journal of the Geological Society, v. 153, p. 337–340.

Childs, C., J. J. Walsh, and J. Watterson, 1997, Complexity

in fault zone structure and implications for fault seal prediction, in P. Moller-Pedersen and A. G. Koestler, eds., Hydrocarbon seals— Importance for exploration and production: Norwegian Petroleum Society Special Publication 7, p. 61–72.

Cockshell, C. D., G. W. O. O'Brien, A. McGee, R. Lovibond, D. Perincek, and R. Higgins, 1995, Western Otway Crayfish Group troughs: Australian Petroleum Exploration Association Journal, v. 35, p. 385–403.

Dewhurst, D. N., and R. M. Jones, 2002, Geomechanical, microstructural and petrophysical evolution in experimentally reactivated cataclasites: Application to fault seal prediction: AAPG Bulletin, v. 86, p. 1383–1405.

Doughty, T. P., 2003, Clay smear seals and fault sealing potential of an exhumed growth fault, Rio Grande Rift, New Mexico: AAPG Bulletin, v. 87, p. 427–444.

Downey, M. W., 1984, Evaluating seals for hydrocarbon accumulations: AAPG Bulletin, v. 68, p. 1752–1763.

Duddy, I. R., 1997, Focussing exploration in the Otway Basin: Understanding timing of source rock maturation: Australian Petroleum Production and Exploration Association Journal, v. 37, p. 178–192.

Finlayson, D. M., B. Finlayson, C. V. Reeves, P. R. Milligan, C. D Cockshell, D. W. Johnstone, and M. P. Morse, 1993, The western Otway Basin— A tectonic framework from new seismic, gravity and aeromagnetic data: Exploration Geophysics, v. 24, p. 493–500.

Fisher, Q. J., and R. J. Knipe, 1998, Fault sealing processes in silliciclastic sediments, in G. Jones, Q. J. Fisher, and R. J. Knipe, eds., Faulting, fault sealing and fluid flow in hydrocarbon reservoirs: Geological Society (London) Special Publication 147, p. 117–134.

Fisher, Q. J., R. J. Knipe, and R. H. Worden, 2000, The microstructure of deformed and undeformed sandstones from the North Sea: Its implications for the origin of quartz cement, in R. H. Worden and S. Morad, eds., Quartz cementation in sandstones: International Association of Sedimentology Special Publication 29, p. 129–146.

Fisher, Q. J., S. D. Harris, E. McAllister, R. J. Knipe, and A. J. Bolton, 2001, Hydrocarbon flow across faults by capillary leakage revisited: Marine and Petroleum Geology, v. 18, p. 251–257.

Fisher, Q. J., M. Casey, S. D Harris, and R. J. Knipe, 2003, Fluid-flow properties of faults in sandstone: The importance of temperature history: Geology, v. 31, p. 965–968.

George, S. C., M. Lisk, E. Summons, and R. A. Quezada, 1998, Constraining the oil charge history of the South Pepper oilfield from the analysis of oil-bearing fluid inclusions: Organic Geochemistry, v. 29, p. 631–648.

Gibson, R. G., 1994, Fault-zone seals in siliciclastic strata of the Columbus Basin, offshore Trinidad: AAPG Bulletin, v. 78, p. 1372–1385.

Gibson, R. G., 1998, Physical character and fluid-flow properties of sandstone-derived fault gouge, in M. P. Coward, T. S. Daltaban, and H. Johnson, eds., Structural geology in reservoir characterisation: Geological Society (London) Special Publication 127, p. 83–97.

Hill, K. A., and C. Durrand, 1993, The western Otway Basin: An overview of the rift and drift history using serial composite seismic profiles: Petroleum Exploration Society of Australia Journal, 21, p. 67–78.

Hillis, R. R., S. A. Monte, C. P. Tan, and D. R. Willoughby, 1995, The contemporary stress field of the Otway Basin, South Australia: Implications for hydrocarbon exploration and production: Australian Petroleum Exploration Association Journal, v. 35, p. 494–506.

Hunt, S. P., and P. J. Boult, 2005, Distinct-element stress modeling in the Penola Trough, Otway Basin, South Australia, in P. Boult and J. Kaldi, eds., Evaluating fault and cap rock seals: AAPG Hedberg Series, no. 2, p. 199–213.

Jev, B. I., C. H. Kaars-Sijpesteijn, P. A. M. Peters, N. L. Watts, and J. T. Wilkie, 1993, Akasco field, Nigeria: Use of integrated 3-D seismic, fault slicing, clay smearing, and RFT pressure data on fault trapping and dynamic leakage: AAPG Bulletin, v. 77, p. 1389–1404.

Jones, R. M., and R. R. Hillis, 2003, An integrated, quantitative approach to assessing fault-seal risk: AAPG Bulletin, v. 87, p. 507–524.

Jones, R. M., P. J. Boult, R. R. Hillis, S. D. Mildren, and J. Kaldi, 2000, Integrated hydrocarbon seal evaluation in the Penola Trough, Otway Basin: Australian Petroleum Production and Exploration Association Journal, v. 40, p. 194–212.

Knipe, R. J., 1992, Faulting processes and fault seal, in R. M. Larsen, H. Brekke, B. T. Larsen, and E. Talleraas, eds., Structural and tectonic modelling and its application to petroleum geology: Norwegian Petroleum Society Special Publication 1, p. 325–342.

Knipe, R. J., 1997, Juxtaposition and seal diagrams to help analyse fault seals in hydrocarbon reservoirs: AAPG Bulletin, v. 1, p. 187–195.

Koledoye, B. A., A. Aydin, and E. May, 2003, A new process-based methodology for analysis of shale smear along normal faults in the Niger Delta: AAPG Bulletin, v. 87, no. 3, p. 445–463.

Kopsen, E., and T. Scholefield, 1990, Prospectivity of the Otway Supergroup in the central and western Otway Basin: Australian Petroleum Exploration Association Journal, v. 30, p. 263–278.

Lindsay, N. G., F. C. Murphy, J. J. Walsh, and J. Watterson, 1993, Outcrop studies of shale smear on fault surfaces: International Association of Sedimentologists Special Publication 15, p. 113–123.

Lisk, M., M. M. Faiz, E. B. Bekele, and T. E. Ruble, 2000, Transient fluid flow in the Timor Sea, Australia: Implications for prediction of fault seal integrity: Journal of Geochemical Exploration, v. 69–70, p. 607–613.

Little, B. M., 1996, The petrology and petrophysics of the Pretty Hill Formation in the Penola Trough, Otway Basin: M.Sc. Thesis, University of South Australia, Adelaide, Australia, 122 p.

Lovibond, R., A. N. Aburas, J. E. Skinner, A. C. Migliucci, R. J. Suttill, and A. J. Buffin, 1993, Petroleum Exploration Licence 32 Permit Assessment Project, onshore Otway Basin South Australia: Boral Energy, ref OT ASSESS6, 39 p.

Lovibond, R., R. J. Suttill, J. E. Skinner, and A. N. Aburas, 1995, The hydrocarbon potential of the Penola

Trough, Otway Basin: Australian Petroleum Production and Exploration Journal, v. 35, p. 358–371.

Lyon, P. J., P. J. Boult, A. Mitchell, and R. R. Hillis, 2004, Improving fault geometry interpretation through "pseudo-depth" conversion to seismic data in the Penola Trough, Otway Basin, in P. J. Boult, D. R. Johns, and S. C. Lang, eds., Eastern Australian Basins Symposium II: Petroleum Exploration Society of Australia Special Publication, p. 695–706.

Mildren, S. D., R. R. Hillis, and J. Kaldi, 2002, Calibrating predictions of fault reactivation in the Timor Sea: Australian Petroleum Production and Exploration Association Journal, v. 42, p. 187–202.

Mildren, S. D., R. R. Hillis, D. N. Dewhurst, P. J. Lyon, J. J. Meyer, and P. J. Boult, 2005, FAST: A new technique for geomechanical assessment of the risk of reactivation-related breach of fault seals, in P. Boult and J. Kaldi, eds., Evaluating fault and cap rock seals: AAPG Hedberg Series, no. 2, p. 73–85.

Morton, J. G. G., 1990, Revisions to stratigraphic nomenclature of the Otway Basin, South Australia: Geological Survey of South Australia, Quarterly Geological Notes, v. 116, p. 2–19.

Morton, J. G. G., E. M. Alexander, A. J. Hill, and M. R. White, 1995, Lithostratigraphy and environments of deposition, in J. G. G. Morton and J. F. Drexal, eds., Petroleum geology of South Australia, Otway Basin: Australia, Mines and Energy South Australia, v. 1, p. 47–94.

Padley, D., D. M. McKirdy, J. E. Skinner, R. E. Summons, and R. P. Morgan, 1995, Crayfish Group hydrocarbons—Implications for palaeoenvironment of Early Cretaceous rift fill in the western Otway Basin: Australian Petroleum Exploration Association Journal, v. 35, p. 517–537.

Perincek, D., B. Simons, and G. R. Pettifer, 1994, The tectonic framework and associated play types of the western Otway Basin, Victoria, Australia: Australian Petroleum Production and Exploration Association Journal, v. 34, p. 460–478.

Ramsey, J. G., 1967, The folding and fracturing of rocks: New York, McGraw-Hill, 568 p.

Rider, M., 2000, The geological interpretation of well logs, 2d ed., chapter 7: Caithness, Scotland, Whittles Publishing, p. 67–90.

Schowalter, T. T., 1979, Mechanics of secondary hydrocarbon migration and entrapment: AAPG Bulletin, v. 63, p. 723–760.

Sibson, R. H., 1996, Structural permeability of fluid-driven fault-fracture meshes: Journal of Structural Geology, v. 18, p. 1031–1042.

Smith, D. A., 1966, Theoretical considerations of sealing and non-sealing faults: AAPG Bulletin, v. 50, p. 363–374.

Sperrevik, S., P. A. Gillespie, Q. J. Fisher, T. Halvorsen, and R. J. Knipe, 2002, Empirical estimation of fault rock properties, in A. G. Koestler and R. Hunsdale, eds., Hydrocarbon seal quantification: Norwegian Petroleum Society Special Publication 11, p. 109–125.

Watts, N. L., 1987, Theoretical aspects of cap-rock and fault seals for single and two phase hydrocarbon columns: Marine and Petroleum Geology, v. 4, p. 274–307.

Weher, F. L., L. H. Fairchild, M. R. Hudec, R. K. Shafto, W. T. Shea, and J. P. White, 2000, Fault seal; contrasts between the exploration and production problem, in Petroleum systems of South Atlantic margins: AAPG Memoir 73, p. 121–132.

Willink, R. J., and R. Lovibond, 2001, Technology, teamwork, respect and persistence: Ingredients of successful exploration in the onshore Otway Basin: Australian Petroleum Production and Exploration Association Journal, v. 41, p. 53–70.

Wiprut, D., and M. D. Zoback, 2002, Fault reactivation, leakage potential, and hydrocarbon column heights in northern North Sea, in A. G. Koestler and R. Hunsdale, eds., Hydrocarbon seal quantification: Norwegian Petroleum Society Special Publication 11, p. 109–125.

Yielding, G., 2002, Shale gouge ratio—Calibration by geohistory, in A. G. Koestler and R. Hunsdale, eds., Hydrocarbon seal quantification: Norwegian Petroleum Society Special Publication 11, p. 109–125.

Yielding, G., B. Freeman, and T. Needham, 1997, Quantitative fault seal prediction: AAPG Bulletin, v. 81, p. 897–917.

Distinct-element Stress Modeling in the Penola Trough, Otway Basin, South Australia

Suzanne P. Hunt

Australian School of Petroleum, Petroleum Engineering Discipline, University of Adelaide, Australia

Peter J. Boult

Australian School of Petroleum, University of Adelaide, Australia and also
Department of Primary Industry and Resources South Australia, Adelaide, Australia

ABSTRACT

The Penola Trough of the Otway Basin, South Australia, is host to five economic gas fields containing an estimated 120 bcf of original gas in place in fault-related traps. However, throughout this trough, many other fault-dependent traps contain paleocolumns or partial paleocolumns. In 2001, the Balnaves 1 well discovered a semibreached structure. This structure was originally thought to be low risk because its associated fault was optimally oriented to seal with respect to the interpreted present-day maximum horizontal stress direction. On subsequent analysis of the wellbore image data, an open conductive fracture network was observed in the seal around the main bounding fault.

We propose that perturbations of the regional stress field around preexisting faults may open a fracture network in the seal. This hypothesis is tested for the Laira Formation (cap seal) using the finite-difference distinct-element method (DEM). To our knowledge, this technique has not previously been used to assess seal integrity.

The DEM has been used before for estimating perturbations around faults. The current work first summarizes and expands previous investigations of the perturbations developed in the two-dimensional (2-D) (horizontal) local maximum (σ_1) and minimum (σ_3) stress magnitudes produced around a single fault, it then uses this understanding to create and assess a 2-D DEM Penola Trough model.

For a single fault, the magnitude of perturbations were examined as a function of $k = \sigma_1/\sigma_3$, θ (the angle between σ_1 direction and the fault strike), friction angle ϕ, and fault stiffness j_{kn} and j_{ks}. The magnitude of stress perturbations are highly sensitive to k,

Copyright ©2005 by The American Association of Petroleum Geologists.
DOI:10.1306/1060765H23157

θ, and φ, but less sensitive to fault stiffness. This insight is applied to horizontal 2-D models to identify areas of potential cap rock failure.

In a 2-D study of the Penola Trough areas of high shear stress are modeled where breached hydrocarbon columns are known to occur. We interpret areas of high shear stress to be zones of fractured rock and possible cap rock failure. Predicting zones of cap rock failure using DEM models could prove to be a very useful exploration tool in locations where cap rocks are known to be brittle and have suffered recent tectonic strain.

INTRODUCTION AND OBJECTIVES

The Otway Basin, South Australia, has been an important, gas-producing region since the late 1980s. The Penola Trough, which is the subject of this study (Figure 1), is an Early Cretaceous, clastic-filled, half graben in the northwestern onshore sector of the Otway Basin. It is host to five economic gas fields containing an estimated 120 bcf of original gas in place (Figure 2). Gas is trapped in braided-fluvial, Pretty Hill Formation or Sawpit Sandstone reservoirs in tilted fault blocks. Cap seal and juxtaposition fault seal is provided by the overbank to lacustrine Laira Formation or Sawpit shale. Four of the gas fields, which occur in the Katnook graben (Figure 3), are filled to structural spill, indicating abundant charge. The fifth gas field, Ladbroke Grove, which is on the footwall of the southern graben-bounding fault, is not full to structural spill. However, Ladbroke Grove field is interpreted to be at the end of the migration pathway for gas and may be charge limited. Other fault-closured structures in the Penola Trough appear to have abundant charge and are interpreted to have been full to either structural or capillary leakage spill (Lyon et al., 2005) but now only contain paleocolumns or partial paleocolumns. In 2001, the Balnaves 1 well, located west of the Ladbroke Grove fault tip (Figures 2, 3), was drilled, resulting in the discovery of a semibreached structure. The structure was thought to be low risk, because the associated fault was optimally oriented to seal with respect to the present maximum horizontal stress direction, determined from borehole fracture orientation and breakout data (Jones et al., 2000). On subsequent analysis of the wellbore image data, an open conductive fracture network was observed in the seal around the main bounding fault (Boult et al., 2002). These and the current authors hypothesized that perturbations of the regional stress field around preexisting faults may have induced fracturing in the seal. This study uses the finite-difference distinct-element method (DEM) (Cundall, 1971) to test this hypothesis. This chapter thus describes a first-pass analysis of the regional seal, the Laira Formation, and a local seal, the Sawpit shale.

The DEM has been used before for estimating perturbations around faults (Su and Stephansson, 1999; Homberg et al., 1997). The current work first summarizes and expands their investigation of the perturbations developed in the two-dimensional (2-D) (horizontal) local maximum (σ_1) and minimum (σ_3) stress magnitudes produced around a single fault; it then uses this understanding to create and assess a 2-D DEM Penola Trough model.

Thus, the key objectives of the work described here are to

- find a method of combining the geomechanics of fault and top seal behavior into the assessment of the causes of hydrocarbon leakage because of structurally induced permeability increases;
- assess (in 2-D) possible mappable relationships between high modeled shear stress and areas of known hydrocarbon leakage using the Penola Trough as a study area.

REVIEW OF NUMERICAL MODELING

Numerical modeling techniques have been developed to help understand and explain observations in complex stress regimes and for prediction of stress away from local points of measurement. Major discontinuities at rock density contrasts, such as faults and mappable seismic horizons, can greatly affect both the magnitude and orientation of the principal stress components. The effect of such heterogeneities contained in the rock mass has been the subject of many field observations (Sonder, 1990; Teufel, 1991; Aleksandrowski et al., 1992; Bell et al., 1992; Mount and Suppe, 1992; Martin and Chandler, 1993; Barton and Zoback, 1994), which numerical modeling attempts to consider.

Several numerical techniques have been proposed for modeling stress and strain in a heterogeneous rock mass. The general applicability of these methods were reviewed by Jing and Hudson (2002) and Jing (2003). The most commonly employed methods are

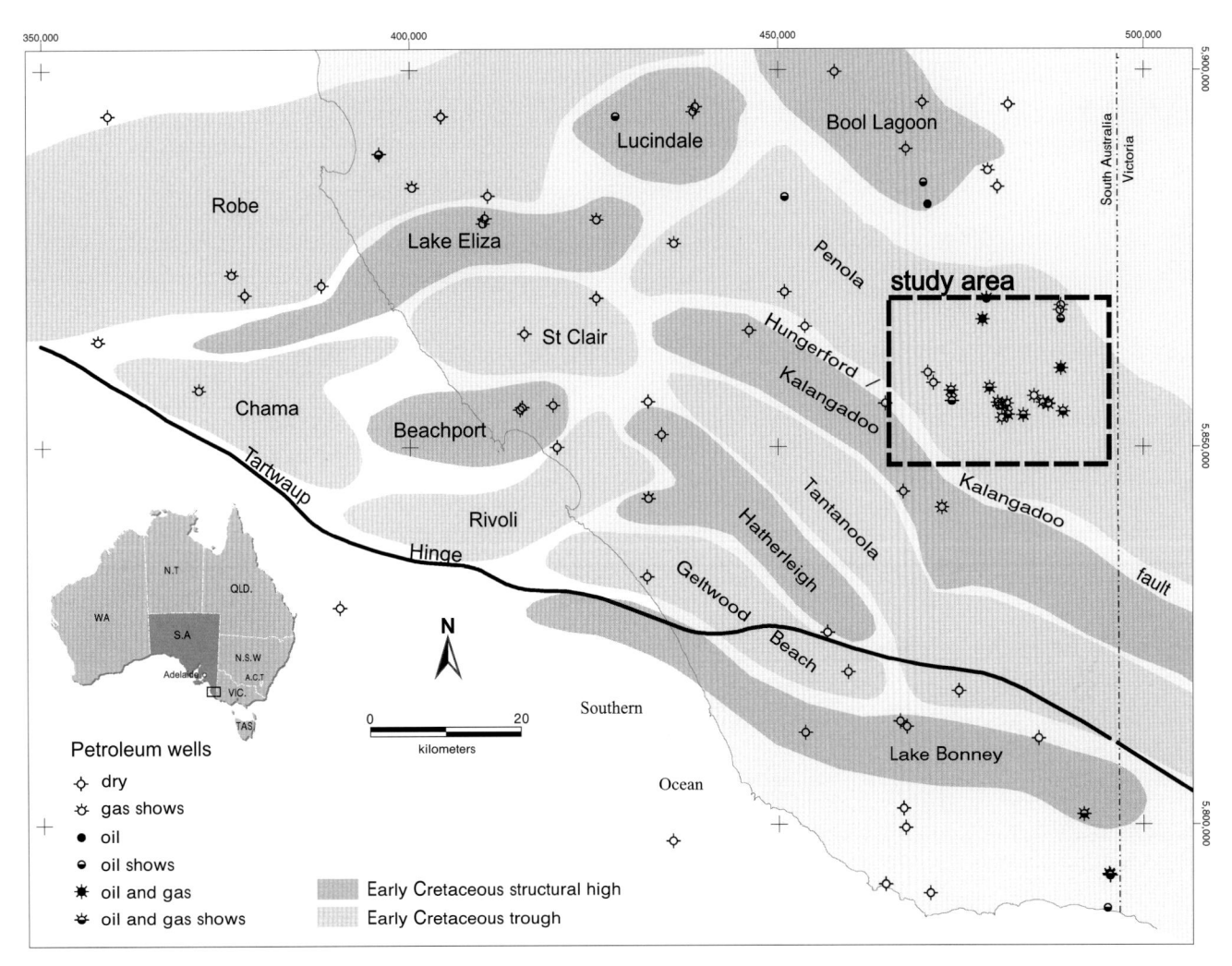

FIGURE 1. Location of the Otway Basin, Australia, showing the main structural features in the region.

the finite-element method (FEM), coupled FEM and boundary-element method (BEM), and the finite-difference method (FDM). The character of these methods makes them difficult to apply when there are numerous fractures. The DEM was developed specifically to be applicable to rock mass fracture-related problems (Cundall, 1971).

The DEM method was developed into 2-D and three-dimensional (3-D) codes, which have recently been used for several geomechanical purposes (Jing, 2003). In particular, these codes and others have been used to assess stress field perturbations caused by discrete fractures and by material property changes at horizon boundaries. A few of these studies are summarized briefly below. The studies were undertaken with DEM, FEM (FEM models that fill a similar role to the DEM method), and BEM software and can be grouped into five main disciplines: (1) tectonics, (2) structural geology, (3) geotechnical and mining engineering, (4) mineral exploration, and (5) petroleum geomechanics.

In the field of tectonics, the plate-scale stress field is based primarily on borehole breakout information and earthquake focal mechanisms. Numerical work involves modeling fault zones and the resultant directional stress perturbations, then comparing these with borehole breakout orientations. Finite-element method models have been used (Golke et al., 1996), and more recently, DEM has been used to successfully match present-day, 2-D horizontal σ_1 orientation patterns in the Cainozoic of the mid-Norwegian margin and northern North Sea with stress orientations from borehole breakouts, focal mechanisms, and in-situ measurements (Pascal and Gabrielsen, 2001). In addition, Homberg et al. (1997) used DEM to model the Morez fault zone in the Jura Mountains and observed that the paleostress trajectories obtained for the Miocene–Pliocene compression can be explained by the perturbation effect of this fault zone.

In structural geology, Kattenhorn et al. (2000) have shown, using BEM, that fault slip can perturb the

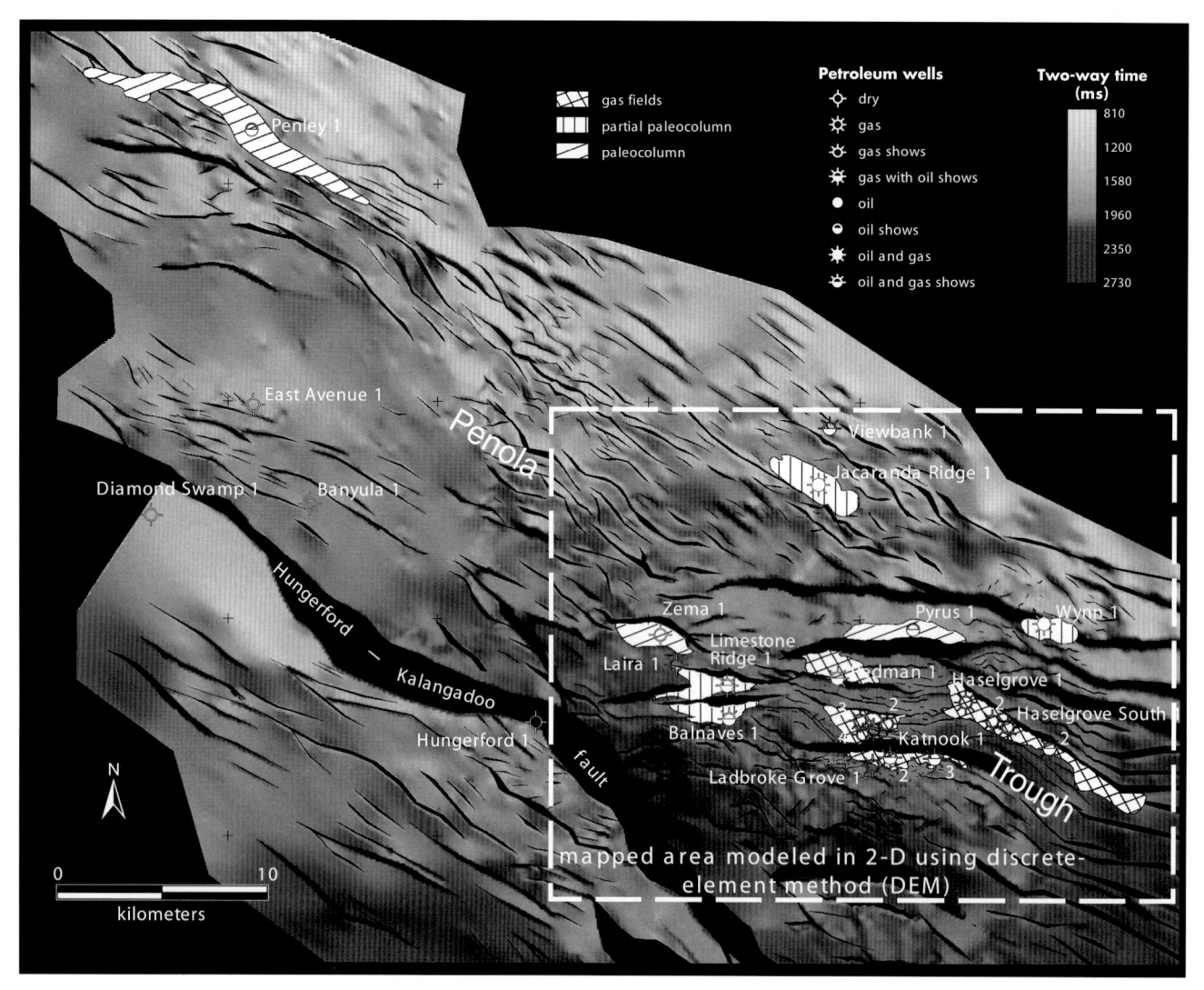

FIGURE 2. Map showing the Penola Trough hydrocarbon field status as of September 2003 and mapped area modeled as a 2-D slice at 2700-m (8858-ft) depth (near base seal) using DEM.

surrounding stress field in a manner that controls the orientations of induced secondary structures. This work has been confirmed by Maerten et al. (2002), who also used a BEM model of normal faulting from a North Sea hydrocarbon reservoir to show that the variability in secondary fault orientations can be attributed to stress perturbation that developed around the larger faults during a single phase of extension.

Recently, in geotechnical and mining engineering, a 3-D DEM has been used as an aid to route selection for tunnels in geologically complex formations in Germany (Konietzky et al., 2001). These modeling results augmented hydraulic fracturing data using the numerical model to predict stress perturbations caused by different geological horizons and by faults with significant throws. In addition, in mining engineering, McKinnon (2001) presented a methodology for using 3-D DEM to calculate a premining stress field. The work was calibrated from stress measurements in a mine containing local stress influences, including topography, mining-induced stresses, and several distinct geological horizons.

In mineral exploration, a geomechanical approach has been used to predict the location of the structurally controlled Mississippi Valley type Pb-Zn and Archean gold mineralization (Holyland and Ojala, 1997; Mair et al., 2000). The models they used were 2-D (horizontal) in the σ_1 and σ_3 plane, where the greatest contrasts in stress occur. This method was applied to epigenetic hydrothermal mineral deposits, where fluid flow is enhanced in dilational sections of structures defined by sites of low minimum principal stress. Hence, DEM predictions of low minimum stress regions or stress lows are used as targets for ore deposition.

This chapter applies knowledge obtained from these studies to modeling a horizontal slice at the mean reservoir-seal contact using DEM. Conditions are chosen to match those in the Penola Trough at a depth of

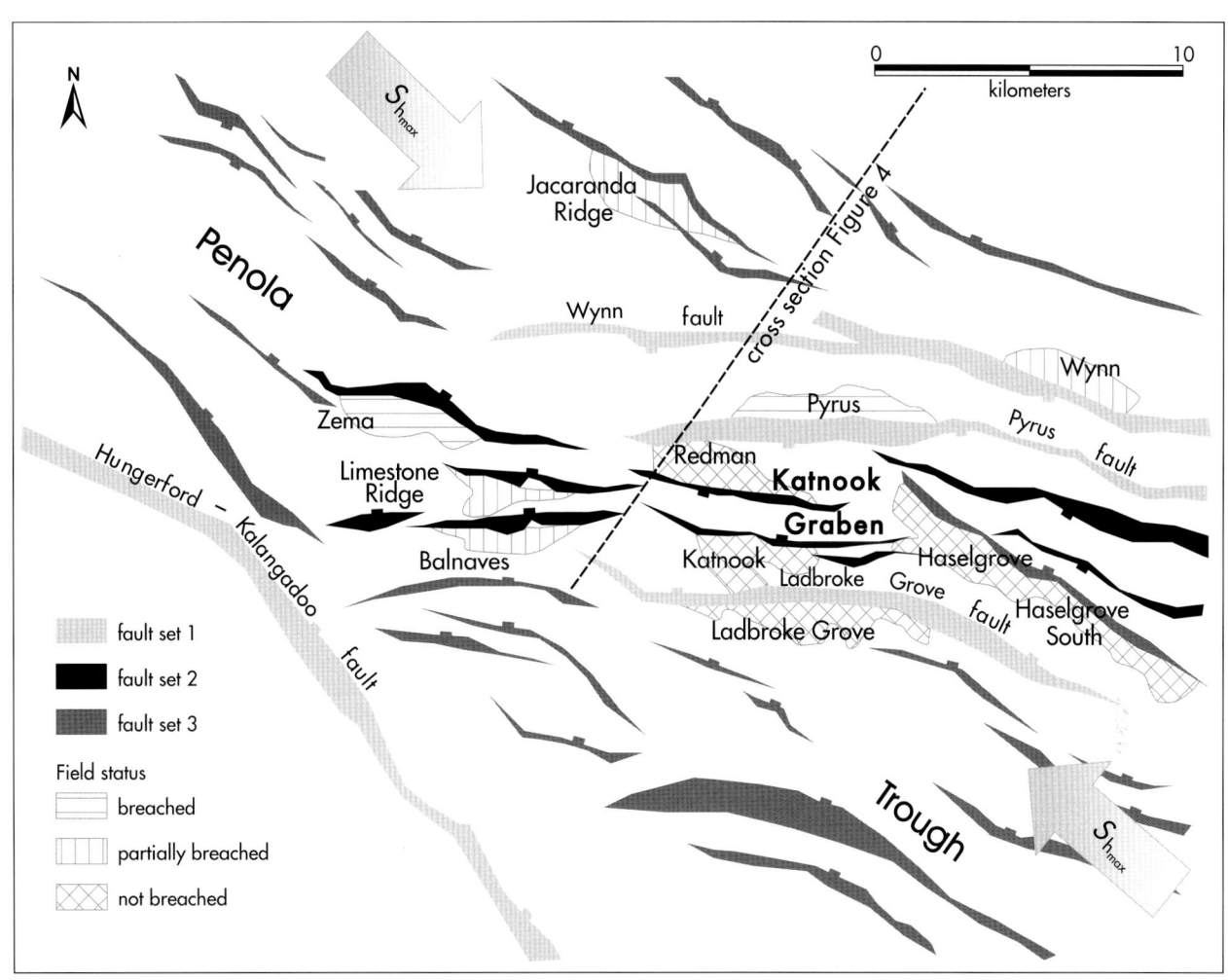

FIGURE 3. Map showing the distribution of fault sets in the central part of the Penola Trough. Fault set 1 is the dominant set that has controlled development of the trough. Fault set 2 is old and only cuts Lower Cretaceous strata. Fault set 3 is relatively young, with throws being higher at shallow levels, indicating downward propagation toward a decollement surface in the main seal.

2700 m (8858 ft) (Figure 4). The aim of the study is to discover whether this geomechanical approach can be used to help predict regions of possible breached top seal. However, the initial work reported here is a sensitivity study of the development of perturbations around faults that are a result of very simple, far-field normal stresses to the DEM model boundaries.

SENSITIVITY ANALYSIS DISCUSSION AND SETUP

The DEM simulates faulting-related deformation using the Coulomb slip model (Figure 5). Below the frictional limit, only elastic deformation (strain) occurs (Figure 5a). At a constant magnitude of applied differential stress, the resultant elastic deformation (strain) is constant and governed by Young's modulus. Elastic deformation is independent of the Coulomb parameters below the frictional limit. Once the fault reaches its frictional limit or shear stress limit, as defined by the strength envelope (Figure 5b), shear displacement occurs along the length of the fault.

To better understand the parameter sensitivity of the basinwide DEM model, stress perturbations around a single fault were modeled in map view. The properties required for description of the intact rock's mechanical behavior were bulk modulus (K), shear modulus (G), and density (d), with the bulk modulus and shear modulus being obtained from the Young's modulus and Poisson's ratio.

Fault properties are conventionally derived from laboratory testing (e.g., triaxial and direct shear tests). Physical properties for joint friction angle, cohesion, dilation angle, and tensile strength, as well as joint normal and shear stiffness, can be derived from these tests. The joint cohesion and friction angle correspond to the parameters in the Coulomb strength criterion. Values for normal and shear stiffness for rock joints

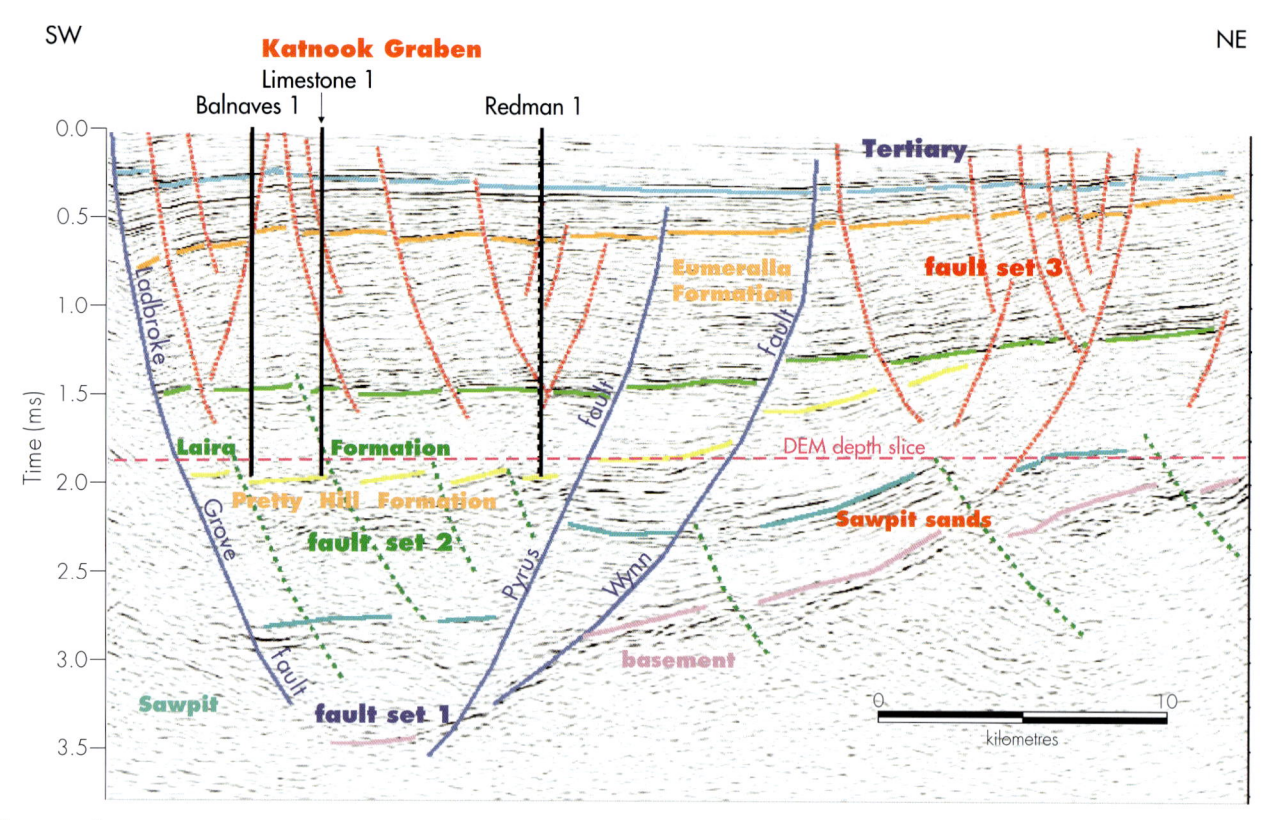

FIGURE 4. Northeast–southwest seismic section showing the Katnook graben bounded by the east–west-trending Pyrus and Ladbroke Grove faults. Three fault sets are also shown. Set 1 is dark blue and thick; set 2 is green and dashed; set 3 is red and dotted. All the wells are projected and appear slightly offstructure. The DEM slice at 2700 m (8858 ft) is equivalent to a two-way traveltime of 1880 ms. Within the Katnook graben, this slice is close to the base of the Laira Formation, whereas on its northern flank, it is close to the Sawpit Sandstone, which is the reservoir for the Jacaranda Ridge and Wynn fields.

typically range from roughly 10–100 MPa/m, for joints with soft clay in-filling, to more than 100 GPa/m, for tight joints in granite and basalt. Published strength properties for discontinuities (which include faults and joints) are more readily available than are stiffness properties. Discontinuity friction angles can range from less than 10° for smooth joints in weak rocks, such as tuff, to more than 50° for rough joints in hard rock, such as granite. Byerlee (1978) indicates that for almost all rock types, for common temperature ranges, the angle of friction is between 30 and 40°. Whether these figures can be extrapolated to the larger scale study of joints in the field with reasonable confidence is not known, but for practical purposes, in rock engineering, upscaling of properties are undertaken using numerical modeling techniques or rock mass classifications systems.

Fracture cohesion can range from zero cohesion to values approaching the compressive strength of the surrounding rock. Lama and Vutukuri (1978) suggest a range between 0 and 75 MPa for clastic sandstones. In this study, for simplicity, cohesion is held at zero, and only the friction angle is varied. This is an adequate approach for the sensitivity analysis herein, because increasing the friction angle has the same effect.

The calibration case that was studied contained a single fault in the center of a 2-D block model, and the finite-difference grid was discretized into approximately 10,000 zones. The fault and block model is shown in Figure 6a.

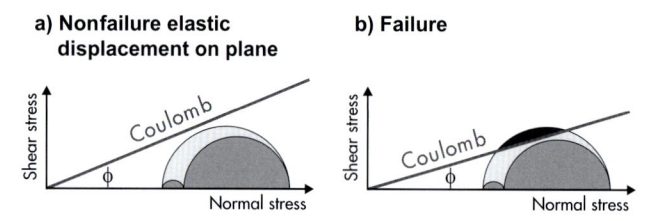

FIGURE 5. (a, b) Schematic diagram of Coulomb friction envelope illustrating failure and nonfailure. For a given state of stress (represented by Mohr circles), low friction angle (ϕ) predicts failure, whereas high ϕ does not (i.e., nonfailure model, purely elastic deformation).

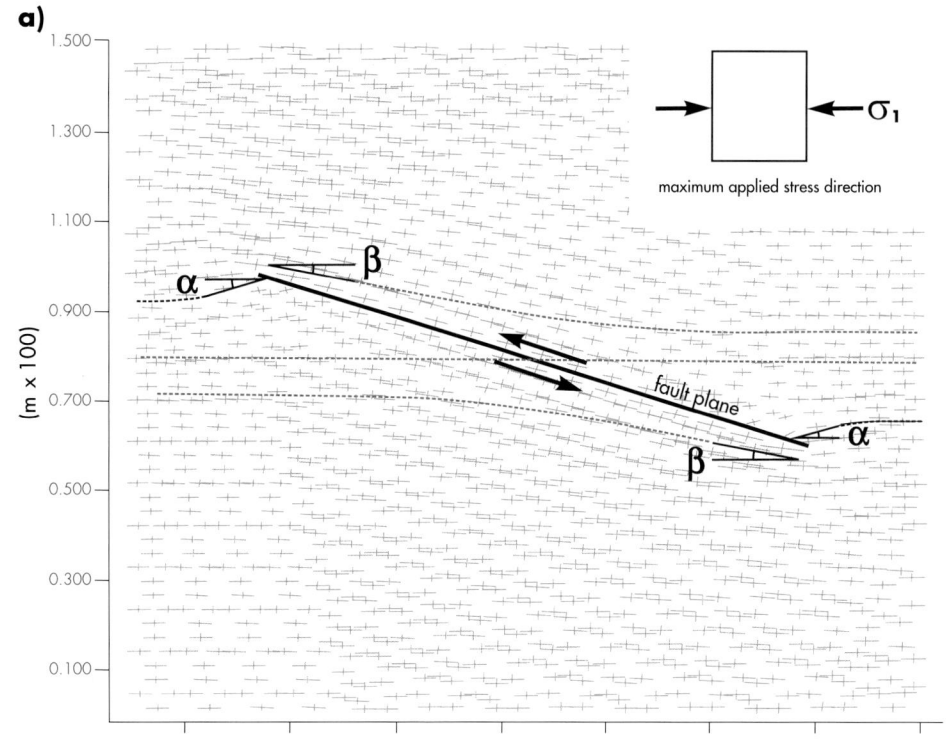

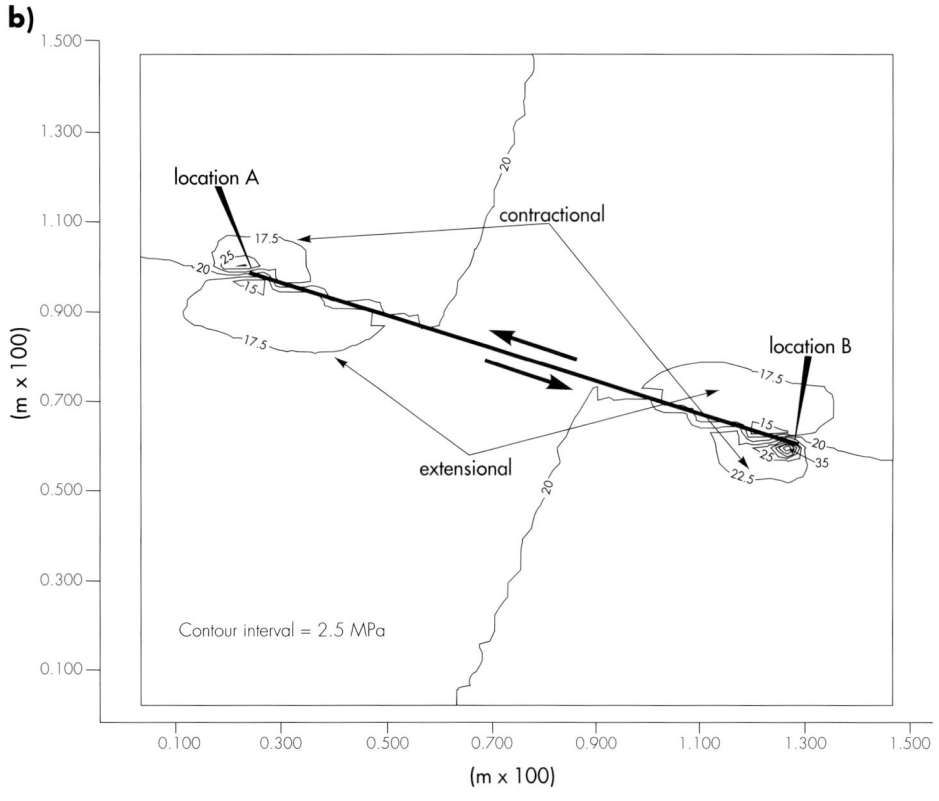

FIGURE 6. Distinct-element model and block model of stress perturbations around a single discontinuity. The figure in (a) shows the principal stress direction for each element of the finite-difference grid perturbed by the fault plane. σ_1 = maximum principal stress. The extensional (α) and contractional (β) angles are the angles between the local trends of σ_1 axes and the direction of σ_1 applied to the boundaries of the model for the largest perturbations occurring in the extensional zone and the contractional zones. The figure in (b) shows the contoured magnitude of the maximum principal stress associated with contraction and extension at the fault plane tips.

SENSITIVITY ANALYSIS RESULTS

Figure 6a shows the orientation and location of the initial fault plane and the loading applied at the model boundaries. A normal traction of 25 MPa was applied to the right and left boundaries, and a normal traction of 10 MPa was applied to the upper and lower boundaries. This model setup was used to study the principal stress perturbation, i.e., σ_1 and σ_3 stress highs and lows around the fault as a function of k, the σ_1/σ_3 ratio, the rotation of the stress field relative to the fault orientation θ, the friction angle ϕ, and the normal and shear joint stiffness constants j_{kn} and j_{ks}.

Figure 6b shows the left-lateral strike-slip fault that developed. The largest deviations of σ_1 are located in the close vicinity of fault tips in a domain, which can be divided for each fault tip into contractional and extensional zones. Near each fault tip, the deformation is contractional on one side of the fault and extensional on the opposite side (Figure 6b). The extreme values, 35 and 15 MPa, are above and below the applied maximum stress value of 25 MPa.

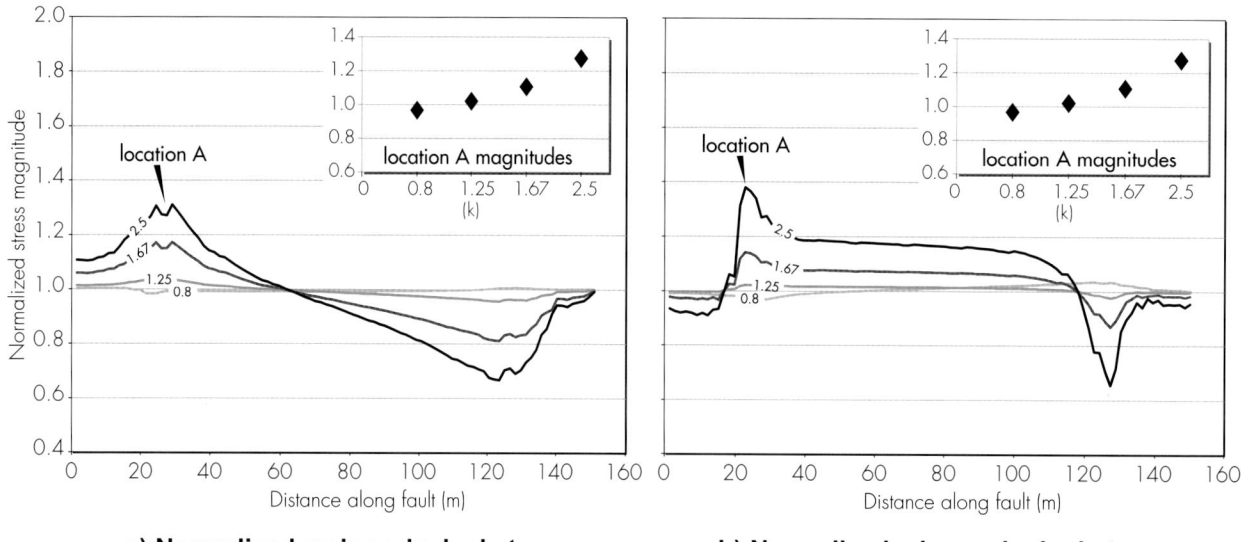

FIGURE 7. (a) Normalized major principal stress as a function of k (σ_1/σ_3) ($\phi = 3°$; $\theta = 25°$; $j_{kn} = 2 \times 10^8$ Pa; $j_{ks} = 1 \times 10^8$ Pa). Inset figure shows magnitude variation of location A. (b) Normalized minor principal stress as a function of k (σ_1/σ_3). Inset figure shows magnitude variation of location A ($\phi = 3°$; $\theta = 25°$; $j_{kn} = 2 \times 10^8$ Pa; $j_{ks} = 1 \times 10^8$ Pa).

Figures 7–9 show stress magnitude at the north side of the fault plane divided by the applied far-field stress (normalized stress). Stress magnitude is plotted against distance along the fault as each parameter varies.

Figure 7a and b shows the effects of varying k (σ_1/σ_3 ratio) on σ_1- and σ_3-normalized magnitudes. As k increases, this perturbation increases in an approximately linear fashion. Location A is in contraction, and location

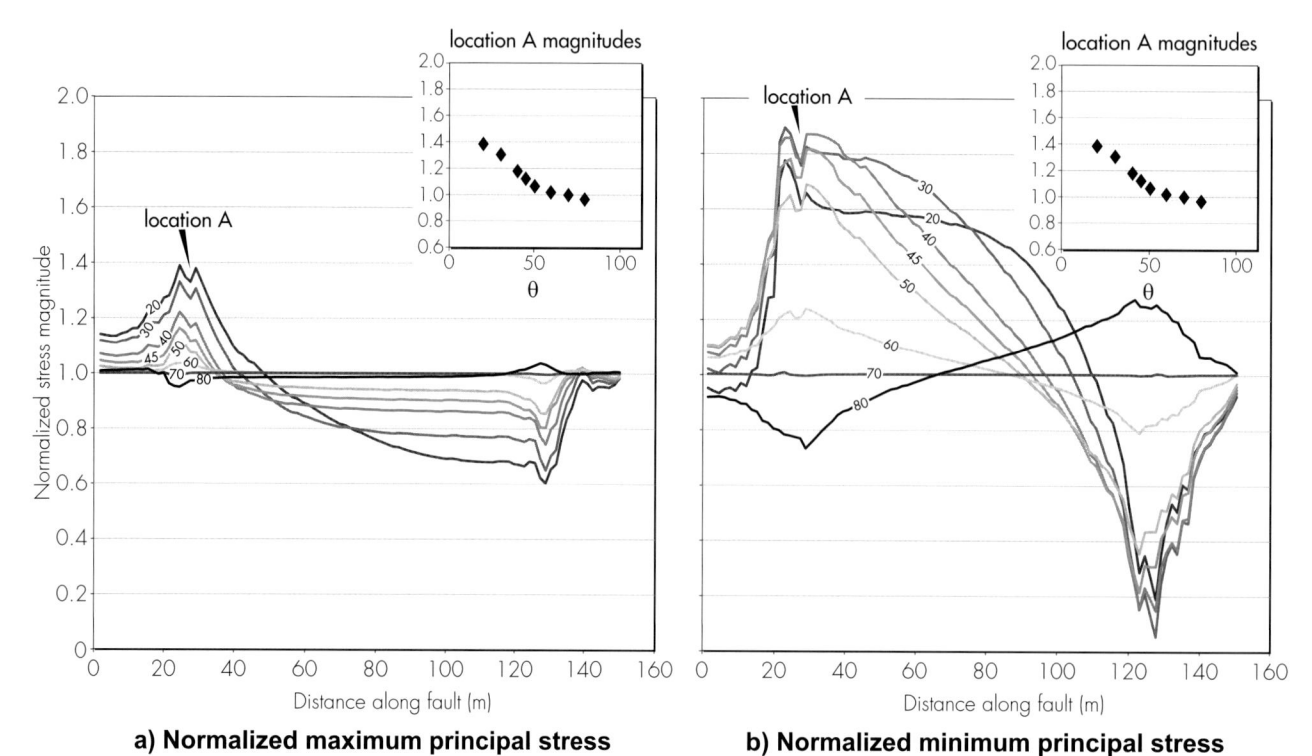

FIGURE 8. (a) Normalized maximum principal stress as a function of the maximum applied principal stress direction (θ) ($k = 2.8$; $\phi = 3°$; $j_{kn} = 2 \times 10^8$ Pa; $j_{ks} = 1 \times 10^8$ Pa). Inset figure shows magnitude variation at location A. (b) Normalized minimum principal stress as a function of the maximum applied principal stress direction (θ) ($k = 2.8$; $\phi = 3°$; $j_{kn} = 2 \times 10^8$ Pa; $j_{ks} = 1 \times 10^8$ Pa). Inset figure shows magnitude variation at location A.

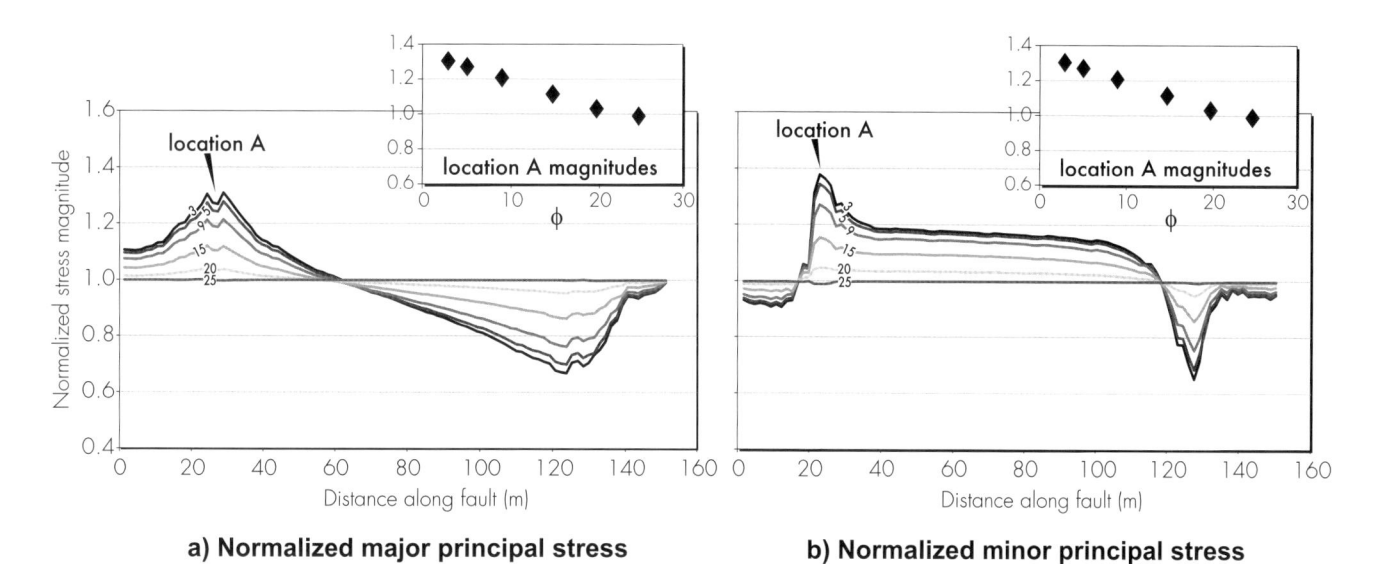

FIGURE 9. (a) Normalized maximum principal stress as a function of friction angle (ϕ) ($k = 2.8$; $\theta = 25°$; $j_{kn} = 2 \times 10^8$ Pa; $j_{ks} = 1 \times 10^8$ Pa). (b) Normalized minimum principal stress as a function of friction angle (ϕ) ($k = 2.8$; $\theta = 25°$; $j_{kn} = 2 \times 10^8$ Pa; $j_{ks} = 1 \times 10^8$ Pa). Inset figures show magnitude variation at location A.

B is in extension. When $k < 1$, the displacement direction on the fault changes, and positive becomes a negative value shown on the graph (at location A) and vice versa (at location B). The change in σ_3 stress magnitude with k has approximately the same magnitude and linear relation to the far-field stress seen for σ_1.

Figure 8a and b shows modeled effects of varying θ (fault orientation relative to the stress field) on σ_1- and σ_3-normalized magnitudes. Fault orientation causes the same degree of perturbation in major principal stress as that caused by magnitude change. At 80°, the maximum applied stress direction is changed, creating a right-lateral displacement and switching extensional and contractional locations. The minor principal stress shows a greater degree of perturbation during rotation than the k ratio perturbation. These values peak to 1.8 times the applied minor stress magnitude.

Figure 9a and b shows modeled effects of friction angle (ϕ) on σ_1- and σ_3-normalized magnitudes. The variation is nonlinear. A rapid decrease in the amount of perturbation occurs as the friction angle increases, followed by a slight decrease in gradient toward the case where nonelastic sliding on the fault plane occurs at $\phi = 25$ (frictional limit).

Previously, Homberg et al. (1997), with a similar model setup, have described local stress orientation perturbation located in the extensional (α) and contractional (β) zones of the fault by two angles α and β. These are the angles between the local trends of σ_1 axes and the direction of σ_1 applied to the boundaries of the model for the largest perturbations occurring in the extensional zone and the contractional zones.

For the ranges of values used in this study, the fracture stiffness parameter generally has an insignificant effect on the stress perturbations (both magnitude and direction) generated around the fault. Despite this, it is worth noting the two different cases that were observed. When $f < 25°$ (where f is the value at which failure (sliding) on the fault plane occurs), the effect of changing fracture stiffness was minimal. Prior to sliding occurring at $f > 25°$, an effect is observed, but the perturbations are two orders of magnitude less than those observed during sliding. As this perturbation (in magnitude) is generated by an elastic deformation on the fault plane, this form of deformation may be physically significant if, for example, a low stiffness material such as a clay smear fills the fault plane.

REGIONAL GEOLOGY OF THE OTWAY BASIN

The Otway Basin (Figure 1) is a divergent passive margin forming a large part of the eastern Australian rift system (Boult et al., 2002), which resulted from the separation of Australia from Antarctica during the Late Jurassic to Early Cretaceous (Lovibond et al., 1995). The northwest- to southeast-trending Penola Trough is one of many half grabens formed along the length of the Otway Basin. A major normal fault, known as the Hungerford–Kalangadoo fault system (Figures 1, 2), bounds a basement high and forms the southern boundary of the Penola Trough. The Katnook graben, within which existing hydrocarbon discoveries are trapped in tilted fault blocks, is an east–west-trending structure bounded by two major faults known as the Pyrus and Ladbroke Grove faults (Figures 3, 4). The main target reservoirs in the Penola Trough are sands of the Early Cretaceous Pretty Hill Formation. Interbedded clays, silts, and fine sands of the Laira Formation form a regionally extensive seal, between 600 and 800 m (2000 and 2600 ft) thick. The Sawpit shale and Sawpit Sandstone form a local reservoir-seal couplet on the northern flank of the trough. For more details, the reader is referred to Lyon et al. (2005).

FAULT SETS

Three distinct sets of faulting have been tentatively recognized in the study area. Further work to more rigorously describe the fault geometries is in progress (P. Lyon, Australian School of Petroleum, 2003, personal communication). However, even if the fault sets presented here are modified by future work, we believe their description is sufficient to demonstrate the potential of the DEM.

Figure 3 shows three distinct genetic fault sets identified by A. Davids (Origin Energy, 2001, personal communication).

- set 1, which cuts Late Jurassic to Holocene strata. Strike is mainly northwest–southeast, but switches to east–west in the Katnook graben area. This is the oldest fault set, formed initially during the Jurassic rifting phase, again active during the Early Cretaceous north–south extension phase and reactivated postcharge during the Miocene to Holocene northwest–southeast compression phase.
- set 2, which cuts Early Cretaceous strata (140–122 Ma). Strike is mainly east–west. This set developed deeper in the Penola Trough as a result of Early Cretaceous north–south extension. This set dominates the Katnook graben and surrounding area and has been interpreted to have propagated upward but has not been active since the Albian.
- set 3, which cuts strata above the Laira Formation to the surface (95 Ma to recent). Fault strikes are mainly northwest–southeast, and they overprint the entire area. These faults are interpreted to have propagated down (with higher throws at shallower levels) toward a decollement surface in the shale-rich Laira Formation and are still active today at the surface (P. Lyon, Australian School of Petroleum, 2003, personal communication).

EXPLORATION HISTORY AND KNOWN PETROLEUM DISTRIBUTION

The exploration history of the South Australian Penola Trough is summarized in Table 1 (B. Camac, 2001, personal communication). The first well to be drilled in the area was Kalangadoo 1 in 1965, which flowed carbon dioxide and methane from a fractured basement. Lack of encouragement, particularly in the South Australian portion of the Otway Basin, resulted in little exploration activity, until Katnook 1 was drilled in 1987–1988 and discovered gas in the Windermere Sandstone. Katnook 2 was drilled shortly after and discovered gas reserves in the Pretty Hill Formation leading to the Pretty Hill Formation becoming a major reservoir target throughout the trough (Table 1).

Varying degrees of exploration success occurred over the following decade with major successes within and proximal to the Katnook graben. The Katnook, Haselgrove, Haselgrove South, and Redman traps were full to their structural spillpoints (Figure 2). Many unsuccessful wells show evidence of possible partial paleocolumns originally full to either structural or capillary spill. Leakage from some of these structures (Limestone Ridge and Balnaves) is interpreted as structurally induced top-seal breach, because faults are not oriented favorably for reactivation and dilation. This is the seal failure mechanism that this study investigates.

THE PENOLA TROUGH NUMERICAL MODEL

Input Parameters for the Penola Trough Model

In summary, it was observed during the calibration studies that the output model result is highly dependent on the following parameters:

- the ratio $k = \sigma_1/\sigma_3$
- θ, the angle between the maximum principal stress, σ_1, and the fault strike
- the fault friction angle

To produce a stress map for the Penola Trough region (Figure 3), we required an accurate definition of the far-field stress magnitude and direction. Previously, S. Mildren (Australian School of Petroleum, 2004, personal communication) gave an estimate of the stress component magnitudes (Figure 10). These estimates were obtained from leak-off pressure data, borehole breakouts, and drilling-induced tensile fracture orientations as interpreted from image logs. They also defined the current stress field as strike slip (with the maximum principal stress acting at a mean angle of 124°N). This stress field is thought to have existed since the Miocene and would have been acting during seal leakage. Hence, it is an appropriate exploration strategy to apply the current stress field to the Penola Trough model to predict areas of high shear stress in the Laira Formation. These areas of high shear stress may represent areas of

Table 1. Summary of historical development and field status of selected wells within the Penola Trough (B. Camac, 2001, personal communication).

Well Name	Date Drilled	Operators	Top of Seal (mSS) L = Laira Formation; S = Sawpit Shale	Top Reservoir (mSS) P = Pretty Hill Formation; S = Sawpit Sandstone	Bounding Fault Orientation	Status
Kalangadoo 1	1965	Alliance	1889 L	none	NA	flowed CO_2 and methane from fractured basement
Katnook 1, 2, and 3	1988	Ultramar	1910 L	2779 P	east–west	gas–full to structural spill
Ladbroke Grove 1	1989	Ultramar	1765 L	2477 P	east–west	gas–very small paleocolumn
Laira 1	1989	Ultramar	1864 L	2614 P	northwest–southeast	dry–off structure
Zema 1	1992	Ultramar	1844 L	2340 P	east–west	gas-oil paleocolumn–was full to capillary spill
Wynn 1	1994	SAGASCO	1846 S	2696 S	northwest–southeast	condensate–partial paleocolumn
Hungerford 1	1994	SAGASCO	1692 L	none	northwest–southeast	no reservoir
Haselgrove 1	1995	SAGASCO	1922 L	2804 P	northwest–southeast	gas–full to structural spill
Haselgrove South 1ST	1995	Boral Energy Resources	1862 L	2878 P	northwest–southeast	gas–full to structural spill
Pyrus 1	1996	Boral Energy Resources	1691 L	2200 P	east–west	possible paleocolumn
Redman 1	1998	Boral Energy Resources	2085 L	2744 P	east–west	gas–full to structural spill
Penley 1	1999	Boral Energy Resources	1399 S	1452 S	northwest–southeast	possible paleocolumn
Jacaranda Ridge 1	1999	Boral Energy Resources	1686 S	2563 S	northwest–southeast	oil–partial paleocolumn
Balnaves 1	2001	Origin Energy Resources	1995 L	2592 P	east–west	gas–possible partial paleocolumn; originally full to structural spill
Limestone Ridge 1	2001	Origin Energy Resources	1991 L	2613 P	east–west	gas paleocolumn–originally full to structural spill

mSS = meters subsea.

potential seal failure caused by the localized development of structural permeability because of open shear-related fracturing.

The fault friction angle parameter was selected on the basis of fault set definition that was described previously. Fault set 2 was given a high friction angle value, high enough to prevent failure on the fault plane in line with its lack of movement since the Early Cretaceous. Set 1, consisting of the larger faults dominating the region, was given a fault friction angle value low enough to produce displacement along the fault plane under the current stress regime. Evidence is also observed validating the selection of this friction angle based on drilling the Ladbroke Grove 1 and Pyrus 1, where abundant calcite (when associated with clay is a relatively weak mineral; Lama and Vutukuri, 1978) was encountered in the Ladbroke Grove and Pyrus faults, and recent movement has been interpreted from seismic data (P. Lyon, Australian School of Petroleum, 2002, personal communication).

Results and Discussion for Penola Trough Distinct-Element Model

The fault geometry used in the model is a close approximation to that shown in Figure 3, and the applied loads and fault parameters used were as described above. The model was run under plain strain conditions, with the maximum and minimum stress magnitudes (both horizontal for this stress regime) at 2700 m (8858 ft)

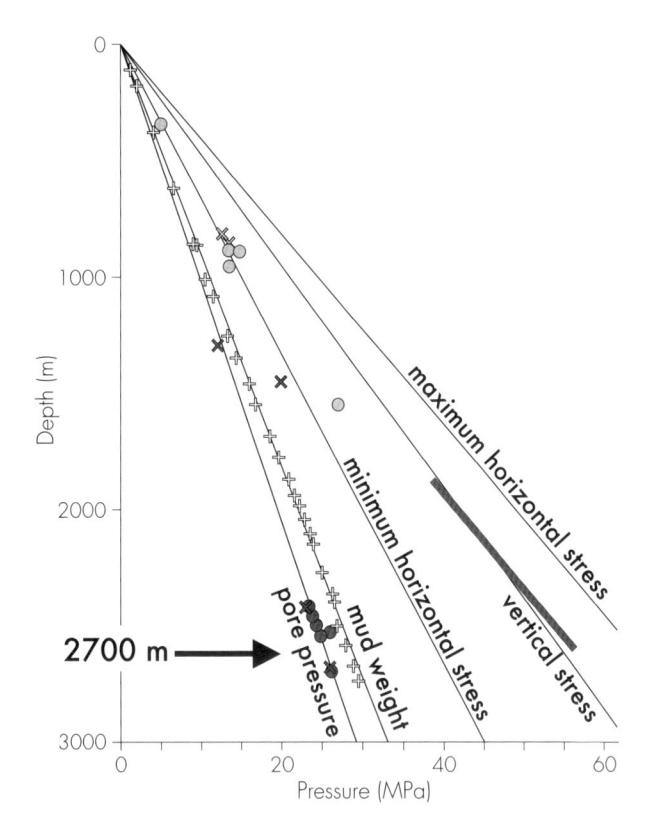

FIGURE 10. Current stress state in Otway Basin obtained from borehole breakout estimates (after S. Mildren, Australian School of Petroleum, 2002, personal communication).

(from Figure 10) applied at the boundaries. The model was discretized to produce around 30,000 zones defining the modeled area. Figure 11 shows the shear stress magnitudes across the region, which were derived from the model. The areas of high shear stress that were generated by the model are interpreted to be the result of fault tip and fault trace curvature of the major Ladbroke Grove, Pyrus, and Wynn faults. In particular, shear stress highs, approximately 10 km (6 mi) in length (east–west) and 3–4 km (1.8–2.4 mi) in width (north–south), were generated at the eastern and western tips of the Wynn and Ladbroke Grove faults, which define the Katnook graben. An extensive region of low shear stress occurs in the graben and adjacent to its northern and southern boundaries (Figure 11).

Bearing in mind the previously discussed parameter sensitivity results, several model runs were undertaken by varying

- the maximum principal stress orientation (σ_H) by ±10°. The locations of the shear stress highs did vary slightly but did not affect the overall location of the lows and highs within and around the main graben structure.
- the friction angle on the three major faults, but this generally only affected the shear stress magnitude values instead of changing the high and low locations. This is consistent with the single fault models described during the sensitivity study.
- the k value, but once again, this affected magnitude instead of the general location of the shear stress highs.

It is thought that very little change was induced in the general location of the shear stress highs and lows because of the relative orientation of the main active faults. The faults run subparallel, and so fault intersections do not affect model output in this case study. Previously, S. P. Hunt and B. Camac (2004, personal communication) have shown that at fault intersections, relative variations in the friction angle, k value, and maximum principal stress directions can significantly alter the locations of the stress highs and lows.

Partial paleocolumns at Wynn and Jacaranda Ridge, which occur stratigraphically lower than the Pretty Hill Formation–Laira Formation reservoir-seal couplet, were trapped in the Sawpit Sandstone and sealed by the Sawpit shale. However, being on the northern flank of the Katnook graben, they are of similar depth to the gas trapped in the Pretty Hill Formation in the Katnook graben. Thus, although Figure 11 is a depth slice of maximum shear stress taken at 2700 m (8858 ft) instead of at the true base of the seal, it is reasonably close in depth to all displayed gas fields and paleocolumns.

The commercial gas fields, Ladbroke Grove Redman, Katnook, Haselgrove, and Haselgrove South gas fields (the last four are all full to spill) are all associated with a relatively low-shear stress zone. However, the breached and probably breached fields of Zema, Balnaves, Limestone Ridge, Jacaranda, and Wynn are all associated with relatively high-shear stress zones.

The only well result that appears not to conform to the numerical model predictions for top-seal breach is Pyrus 1. Here, the base of the seal is 500 m (1600 ft) shallower than the DEM slice in Figure 11, and modeled predictions at the greater depth may not apply here. Furthermore, the throw on the Pyrus fault is so large that a distinct possibility exists that continued postcharge movement on the fault brought the gas-charged reservoir into juxtaposition with the stratigraphically higher Windermere Sandstone on the hanging wall of the fault. Thus, leakage may be juxtaposition leakage instead of top-seal leakage. In addition, the evidence for a paleocolumn in the Pyrus structure is less robust than the other identified paleocolumns in the area, and the trap may be just charge limited. Therefore, this could either be a misidentified paleocolumn or leakage that occurred so long ago that the evidence for charge in the form of residual gas has dissipated. Further work is currently being undertaken to study the Pyrus structure in more detail.

The uncertainty of rock strength and other geometrical parameters in a future 3-D model could be included by conducting a Monte Carlo simulation on

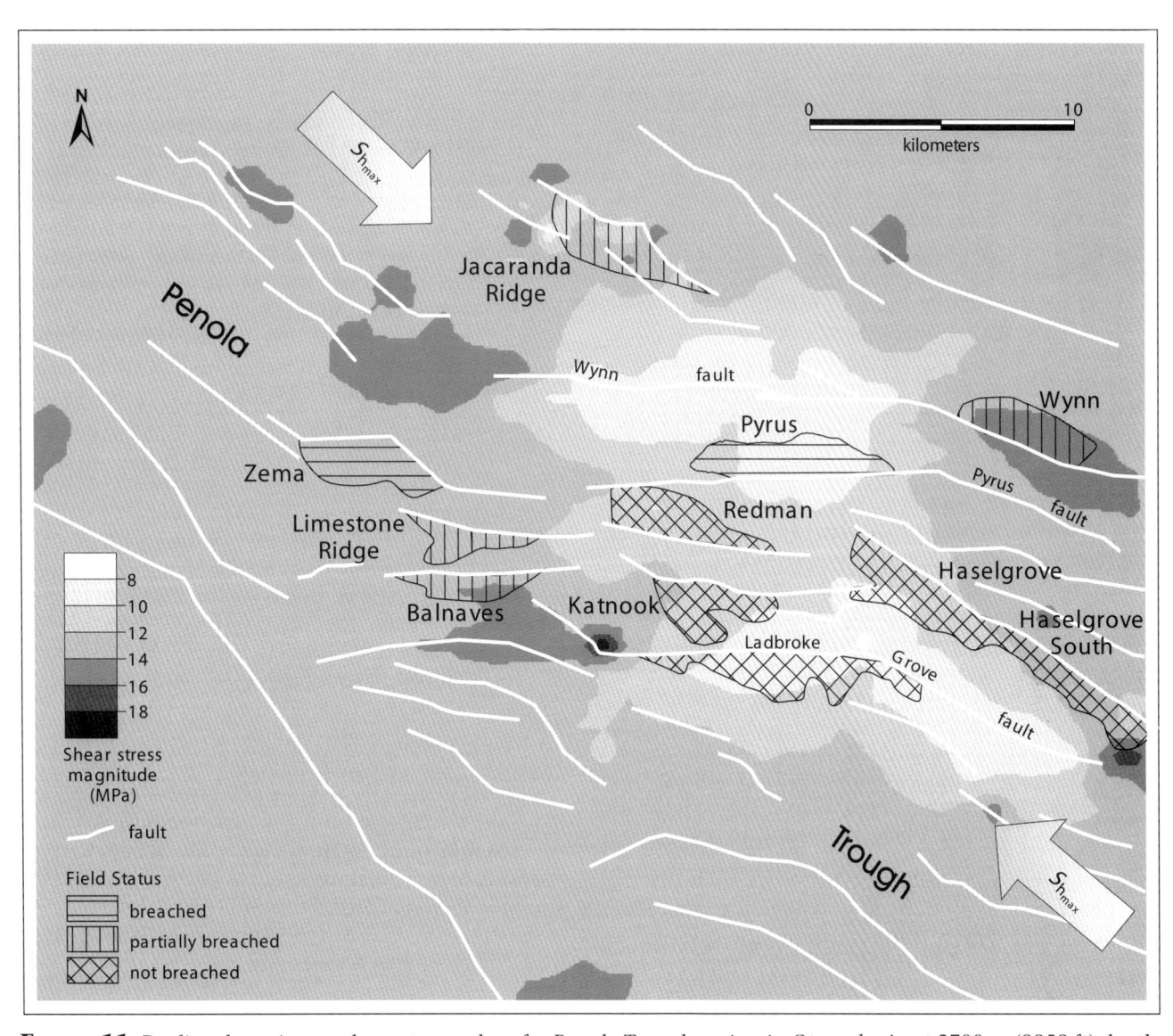

FIGURE 11. Predicted maximum shear stress values for Penola Trough region in Otway basin at 2700 m (8858 ft) depth, compared with breached reservoirs. An overall correlation exists between the occurrence of full-to-spill columns and the location of maximum shear stress lows. The only exception is the Pyrus paleocolumn. This column may have failed by juxtaposition (see text for further explanation). $S_{h_{max}}$ = maximum principal horizontal stress direction.

multiple realizations of the model in much the same way that Ottesen et al. (2005) accounted for structural uncertainty. Four-dimensional modeling or backstepping through time under an evolving stress regime is another uncertainty that could be addressed in the future, but its viability is very dependent on the quality of seismic data, which, at the depth of investigation, is relatively poor in the Penola Trough because of onshore statics.

CONCLUSIONS

Encouraging indications of a direct relationship between the regions of high shear stress that are predicted by our numerical modeling and the locations of the breached reservoirs exist. This suggests that it is possible to use a 2-D DEM model to predict regions of possible high shear stress development in the Laira Formation, which forms a regional Otway Basin top seal. This has implications for future petroleum development in both onshore and offshore Otway exploration. Applying a DEM-type numerical model may provide an additional risking tool for top-seal failure.

In summary this work has

- presented a relationship for k (= σ_1/σ_3), θ (the angle between σ_1 direction and the fault strike), and ϕ (friction angle) between faults and stress perturbation in the surrounding rock mass; all these properties are critical to the magnitude of the stress perturbation;

- presented a mechanism to explain the distribution of breached and unbreached traps in Pretty Hill Formation of the Penola Trough, Otway Basin;
- demonstrated a simple method of combining the mechanics of seal and fault surfaces in 2-D;
- shown that 2-D DEM models can be used to predict areas of possible cap rock failure, thus providing an exploration tool useful where cap rocks are brittle, where maximum stress is horizontal, and where tectonic strain is recent.

Figure 11 is the first 2-D, depth slice, DEM model to be developed for the Katnook graben area of the Penola Trough. The model can be improved, perhaps by expanding to a 3-D model that considers rock strengths (based on stratigraphy). However, unless we can interpret rock strengths from cuttings and/or petrophysical logs, obtaining values for different horizons based on laboratory experiment will prove to be very difficult, unless the following common operational practice of exploration companies is changed. Despite the issue of seal integrity being the primary exploration risk in the Penola Trough and possibly the entire Early Cretaceous of the Otway Basin, not a single core has yet been specifically taken for assessing this risk. Indeed, cores that have been taken and preserved or processed for reservoir assessment, which contain some shale, have not been preserved correctly (D. Dewhurst, Commonwealth Scientific and Industrial Research Organization, 2002, personal communication) for shale rock strength testing.

ACKNOWLEDGMENTS

We thank Origin Energy for their provision of interpreted horizon maps and general discussion on the field status. We also thank Bronwyn Camac for her identification of stress perturbations around some of the major faults in the Penola Trough and also her identification of open tensile fractures in the Laira Formation away from major faults. In addition, we also thank the Stress Group, led by Richard Hillis at the Australian School of Petroleum, for the fruitful discussions we have had. Without them, in particular, Scott Mildren, who contributed Figure 10, which was previously unpublished, we probably would not have done this research. We also thank Alton Brown and Helen Lewis for constructive criticism and suggestions that helped improve the manuscript.

REFERENCES CITED

Aleksandrowski, P., O. H. Inderhaug, and B. Knapstad, 1992, Tectonic structures and wellbore breakout orientation, in J. R. Tillerson and W. R. Wawersik, eds., Proceedings of the 33rd U.S. National Rock Mechanics Symposium: Rotterdam, Balkema, p. 19–37.

Barton, C. A., and M. D. Zoback, 1994, Stress perturbations associated with active faults penetrated by boreholes: Possible evidence for near-complete stress drop and a new technique for stress magnitude measurement: Journal of Geophysical Research, v. 99, no. B5, p. 9373–9390.

Bell, J. S., G. Caillet, and J. Adams, 1992, Attempts to detect open fractures and nonsealing faults with dipmeter logs, in A. Hurst, C. M. Griffiths, and P. F. Worthington, eds., Geological applications of wireline logs II: Geological Society (London) Special Publication 65, p. 211–220.

Boult, P. J., B. A. Camac, and A. W. Davids, 2002, 3-D fault modelling and assessment of top seal structural permeability—Penola Trough Onshore, Otway Basin: Australian Petroleum Production and Exploration Journal, v. 42, no. 1, p. 151–166.

Byerlee, J. D., 1978, Friction of rocks: Pure and Applied Geophysics, v. 116, p. 615–626.

Cundall, P. A., 1971, A computer model for simulating progressive large scale movements in block rock systems, in Proceedings of the Symposium International, Society of Rock Mechanics, Nancy, France, v. 1, p. II-8.

Golke, M., S. Cloetingh, and D. Colentz, 1996, Finite-element modelling of stress patterns along the mid-Norwegian continental margin, 62° to 68°N: Tectonophysics, v. 266, p. 33–53.

Holyland, P. W., and V. J. Ojala, 1997, Computer-aided structural targeting in mineral exploration: Two- and three-dimensional stress mapping: Australian Journal of Earth Sciences, v. 44, p. 421–432.

Homberg, C., J. C. Hu, J. Angelier, F. Bergerat, and O. Lacombe, 1997, Characterisation of stress perturbations near major fault zones: Insights from 2-D distinct-element numerical modelling and field studies (Jura mountains): Journal of Structural Geology, v. 19, no. 5, p. 703–718.

Jing, L., 2003, A review of techniques, advances and outstanding issues in numerical modelling for rock mechanics and rock engineering: International Journal of Rock Mechanics and Mining Sciences, v. 40, p. 283–353.

Jing, L., and J. A. Hudson, 2002, Numerical methods in rock mechanics: International Journal of Rock Mechanics and Mining Sciences, v. 39, p. 409–427.

Jones, R. M., P. J. Boult, R. R. Hillis, S. D. Mildren, and J. Kaldi, 2000, Integrated hydrocarbon seal evaluation in the Penola Trough, Otway Basin: Australian Petroleum Production and Exploration Association Journal, v. 40, no. 1, p. 194–212.

Kattenhorn, S. A., A. Aydin, and D. D. Pollard, 2000, Joints at high angles to normal fault strike: An explanation using 3-D numerical models of fault perturbed stress fields: Journal of Structural Geology, v. 22, p. 1–23.

Konietzky, H., L. te Kamp, H. Hammer, and S. Niedermeyer, 2001, Numerical modeling of in-situ stress conditions as an aid in route selection for rail tunnels in complex geological formations in South Germany: Computers and Geotechnics, v. 28, p. 495–516.

Lama, R. D., and V. S. Vutukuri, 1978, Handbook on mechanical properties of rocks: Series on rock and soil Mechanics: Switzerland, Trans Tech Publications, 481 p.

Lovibond, R., R. J. Suttill, J. E. Skinner, and A. N. Aburas, 1995, The hydrocarbon potential of the Penola Trough, Otway Basin: Australian Petroleum Production and Exploration Journal, v. 35, no. 1, p. 358–371.

Lyon, P. J., P. J. Boult, R. R. Hillis, and S. D. Mildren, 2005, Sealing by shale gouge and subsequent seal breach by reactivation: A case study of the Zema prospect, Otway Basin, in P. Boult and J. Kaldi, eds., Evaluating fault and cap rock seals: AAPG Hedberg Series, no. 2, p. 179–197.

Maerten, L., P. Gilespie, and D. P. Pollard, 2002, Effects of local stress perturbations on secondary fault development: Journal of Structural Geology, v. 24, p. 145–153.

Mair, J. L., V. J. Ojala, B. P. Salier, D. I. Groves, and S. M. Brown, 2000, Application of stress mapping in cross-section to understanding ore geometry, predicting ore zones and development of drilling strategies: Australian Journal of Earth Sciences, v. 47, p. 895–912.

Martin, C. D., and N. A. Chandler, 1993, Stress heterogeneity and geological structures: International Journal of Rock Mechanics and Mining Sciences, v. 30, no. 7, p. 993–999.

McKinnon, S. D., 2001, Analysis of stress measurements using a numerical model methodology: International Journal of Rock Mechanics and Mining Sciences, v. 38, p. 699–709.

Mount, V. S., and J. Suppe, 1992, Present-day stress orientations adjacent to active strike-slip faults: Journal of Geophysical Research, v. 97, no. B8, p. 11,995–12,013.

Ottesen, S., C. Townsend, and K. M. Øverland, 2005, Investigating the effect of varying fault geometry and transmissibility on recovery: Using a new workflow for structural uncertainty modeling in a clastic reservoir, in P. Boult and J. Kaldi, eds., Evaluating fault and cap rock seals: AAPG Hedberg Series, no. 2, p. 125–136.

Pascal, C., and R. H. Gabrielsen, 2001, Numerical modelling of Cenozoic stress patterns in the mid-Norwegian margin and North Sea: Tectonics, v. 20, no. 4, p. 585–599.

Sonder, L. J., 1990, Effects of density contrasts on the orientation of stresses in the lithosphere: Relation to principal stress directions in the Transverse Ranges, California: Tectonics, v. 9, no. 44, p. 761–771.

Su, S., and O. Stephansson, 1999, Effect of a fault on in situ stresses studied by the distinct element method: International Journal of Rock Mechanics and Mining Sciences, v. 36, p. 1051–1056.

Teufel, L. W., 1991, Influence of lithology and geologic structure on in situ stress: Examples of stress heterogeneity in reservoirs, in Reservoir characterisation II: London, UK, Academic Press Incorporated, p. 565–578.

Sedimentology and Petrophysical Character of Cretaceous Marine Shale Sequences in Foreland Basins—Potential Seismic Response Issues

W. R. Almon
Chevron Texaco Inc., Houston, Texas, U.S.A.

Wm. C. Dawson
Chevron Texaco Inc., Houston, Texas, U.S.A.

F. G. Ethridge
Colorado State University, Fort Collins, Colorado, U.S.A.

E. Rietsch
Chevron Texaco Inc., Houston, Texas, U.S.A.

S. J. Sutton
Colorado State University, Fort Collins, Colorado, U.S.A.

B. Castelblanco-Torres
Chevron Texaco Inc., Bakersfield, California, U.S.A.

ABSTRACT

Development of predictive models to estimate the distribution and petrophysical properties of potential mudstone-flow barriers can reduce risks inherent to exploration and exploitation programs. Such a predictive model, founded in sequence stratigraphy, requires calibration with outcrop and subsurface analogs. Detailed sedimentological, petrophysical, and geochemical analyses of Lewis Shale (lower Maastrichtian) samples from southeast Wyoming reveal considerable variability in petrophysically and seismically significant rock properties. Lower Lewis strata represent late-stage transgressive deposits that include a distinctive condensed interval. The overlying progradational Lewis interval consists mostly of interstratified very silty shales and argillaceous siltstones. High-frequency sheet and lenticular sandstone bodies occur in the progradational Lewis package. Sealing capacity, as measured by mercury injection-capillary pressure (MICP) analysis, varies with fabric, texture, and compositional factors that are related to sequence-stratigraphic position. Samples from the Lewis Shale transgressive interval have significantly greater MICP values (average

18,000 psia) and are markedly better seals relative to samples from the overlying Lewis Shale progradational package (average 3000 psia). Transgressive shales with enhanced sealing capacity are characterized by higher total organic carbon and hydrogen index values, lower permeability, and lower detrital silt content. These transgressive shales are enriched in iron-bearing clay minerals and authigenic pyrite. Greater shear wave velocities, larger shear moduli, and higher bulk density also characterize transgressive Lewis Shales. The most promising seal horizons are laterally extensive, silt-poor, pyritic shales occurring in the uppermost transgressive systems tract. Stacking patterns of slow and fast shale horizons can yield seismic responses comparable to those interpreted as hydrocarbon-bearing reservoirs.

INTRODUCTION

This chapter presents an initial effort to use sequence-stratigraphic-controlled variations in shale and mudstone petrophysical and seismic properties to generate predictive models of seal occurrence and to estimate top-seal capacity for application in hydrocarbon exploration and risk analysis. Few systematic studies of seal character and shale sedimentology are available. Consequently, seals remain one of the more poorly understood elements of petroleum systems. The Lewis Shale (Upper Cretaceous, Maastrichtian), which crops out along the eastern margins of the Great Divide and Washakie basins (Figure 1) in south-central Wyoming, provides an interesting analog for understanding stratigraphic architecture of turbidite depositional systems. Previous outcrop and subsurface studies (e.g., Pyles and Slatt, 2000b) established a high-frequency sequence-stratigraphic framework for the Lewis Shale. Witton-Barnes et al. (2000) characterized sandstone lithotypes in the Lewis Shale, and Castelblanco-Torres (2003) completed a detailed study of shale lithotypes from Lewis Shale outcrops and cores. Almon et al. (2002) documented considerable variability in petrophysical properties of shales in the Lewis Shale.

The paucity of information defining systematic variation in shale and mudstone properties reflects the general difficulty of determining the mineralogy and texture of fine-grained lithofacies, as well as the difficulty of accurately measuring the petrophysical properties in rocks having inherently low permeability. This study considers microscale sedimentological aspects of shales, including clay content; percentage of silt-sized grains; degree of bioturbation; organic content; preferred orientation of matrix and larger components; and authigenic components. All of these can be expected to

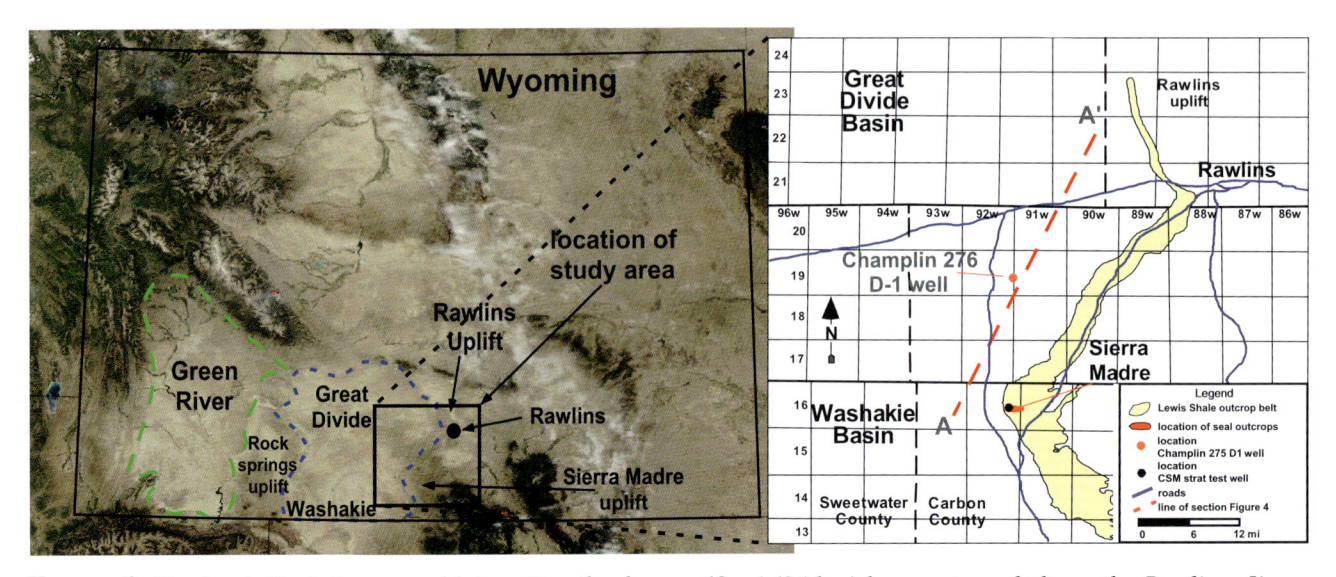

FIGURE 1. The Lewis Shale is exposed intermittently along a 60-mi (96-km)-long outcrop belt on the Rawlins–Sierra Madre uplift west of Cheyenne, Wyoming. The database for Pyles' study consists of outcrops, cores, and well logs. (modified from Pyles, 2000; base map modified after Love and Christiansen, 1985).

influence pore-throat diameter and, consequently, seal character as measured by mercury injection-capillary pressure (MICP) analysis. Additionally, shale facies, shale petrophysical properties, and seal character appear to vary systematically in the context of a sequence stratigraphy.

GEOLOGIC SETTING

The lower Maastrichtian Lewis Shale and Fox Hills Sandstone (Figure 2) were deposited in the final transgression and regression of the Western Interior seaway (Weimer, 1960), which lasted approximately 2.3 m.y. in the study area (Pyles and Slatt, 2000a). The Dad Sandstone occurs in the Lewis Shale and comprises interbedded sandstones and shales (Hale, 1961). These sandstones occur in the middle of the Lewis Shale and are the basis for the informal division of the Lewis Shale into three members: the lower, Dad, and upper members (Gill et al., 1970). The upper member is dominantly dark-gray to olive-gray mudstone, whereas the lower member is comprised of black shale (Gill et al., 1970). On a regional scale, the Lewis Shale interfingers with the overlying Fox Hills Sandstone (Figure 2).

Asquith (1970) constructed stratigraphic cross sections using well logs through the Great Divide and Washakie basins and concluded that the Lewis Shale and Fox Hills Sandstone were deposited as a series of southward-prograding clinoforms that represented a single, progradational event. The progradation was driven by the buildup of a large delta, which comprised well-developed marginal-marine, shelf slope, and basin environments. More recent workers (Winn et al., 1985, 1987; Perman, 1990) have reached similar conclusions.

As seen in the Sierra Madre outcrops of southeastern Wyoming (Figure 3A), the lower Dad Member of the Lewis Shale consists of interbedded dark-gray shales and continuous (thin- to medium-bedded), fine-grained sandstones. These sheetlike sandstones have planar bases marked by *Glossifungities*, which extend into the underlying shales. The middle Dad Member contains thick lenticular bodies of medium-grained sandstone (Figure 3B). These laterally discontinuous sandstones contain an abundance of imbricated, shale, rip-up clasts and have erosional bases (Figure 3C). The flat pebble conglomerate formed by the shale rip-up clasts indicated significant depositional shear in the sand body. Witton-Barnes et al. (2000) interpreted lenticular sandstones in the middle Dad Member of the Lewis Shale as channel-fill lithofacies. Sandstones occurring in the upper Dad interval are fine-grained, thin-bedded, and laterally extensive. They exhibit well-developed internal sedimentary structures, including parallel laminations and climbing ripples (Figure 3D).

HIGH-FREQUENCY SEQUENCE-STRATIGRAPHIC INTERPRETATION

Pyles (2000) generated a high-frequency sequence-stratigraphic framework using gamma-ray and resistivity logs of wells near the Lewis Shale outcrop belt (Figure 4). The datum that forms the base of the framework is the top of an organic-rich black shale unit, which is informally known as the Asquith marker. This organic-rich shale represents an anoxic basinal depositional environment throughout this study area. The Asquith marker is interpreted as a condensed section, representing the maximum transgression of the Lewis seaway (McMillen and Winn, 1991). It is likely that this surface was essentially flat but not necessarily horizontal during deposition of the overlying formations. The Lewis Shale and Fox Hills Sandstone contain only a single, third-order

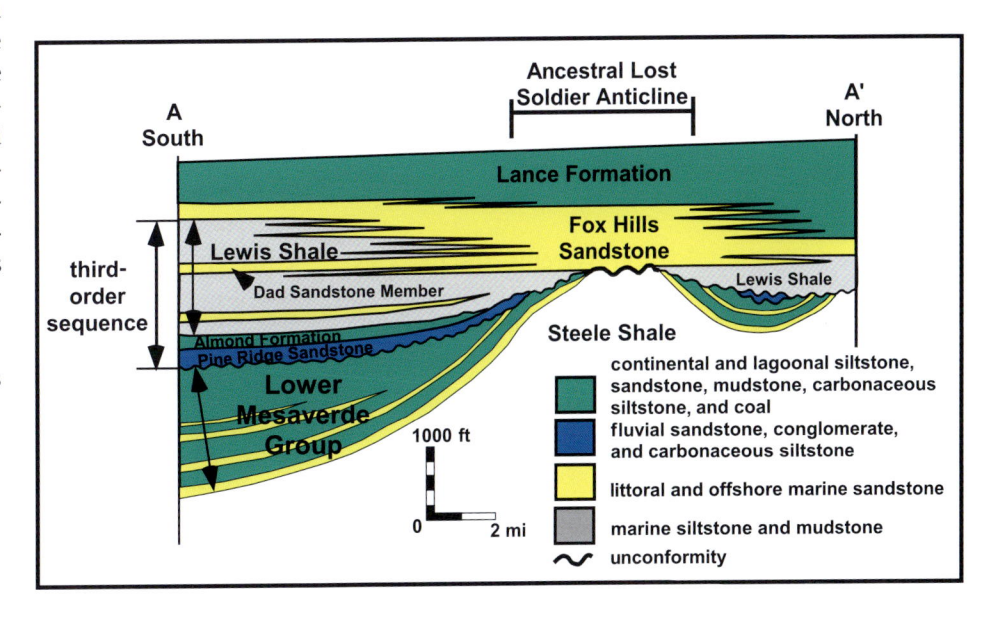

FIGURE 2. Stratigraphic cross section, demonstrating the interfingering nature of the Lewis Shale and Fox Hills Sandstone in the area of the Lost Soldier anticline, south-central Wyoming (modified after Reynolds, 1976).

FIGURE 3. Outcrop photo of lower Dad Sandstone Member of Lewis Shale consisting of (A) thick tan-weathering shales and interstratified thin-bedded sheet sandstones. These sharp-based sandstones are fine-grained and exhibit parallel laminations. The sheet sandstones weather into distinctive rust-colored ledges. (B) The middle Dad Sandstone Member of Lewis Shale consists of thick, tan-weathering, silty shales containing interstratified, thick-bedded, lenticular, sandstone bodies. These sharp-based sandstones are medium grained and range from massive to parallel laminated. The sandstone bodies have irregular (scoured) basal contacts and contain an abundance of shale rip-up clasts. These laterally discontinuous sandstones are resistant to weathering and form distinctive benches. Approximate thickness of sandstone is 25 ft (7.6 m) (gross). Outcrop photo (C) showing base of lenticular sandstone unit in middle Dad Sandstone. Sandstone has irregular (erosional) basal contact with the underlying light-gray claystone. Note the abundance of shale rip-up clasts incorporated into lower part of the irregularly bedded sandstone. Faint laminations are evident in the medium-grained sandstone (15-cm [6-in.] scale). The flat pebble conglomerate formed by the shale rip-up clasts indicated significant depositional shear in the sand body. The thin-bedded (centimeter-scale) sandstones in upper Dad Member of the Lewis Shale (D) are sheetlike and have planar, nonscoured bases and rippled upper surfaces. Detailed examination reveals that parallel-laminated, fine-grained sandstones overlie fine-grained sandstones exhibiting climbing ripples (scale in centimeters and inches).

transgressive and highstand systems tract. Pyles' (2000) high-frequency, sequence-stratigraphic interpretation detected at least 20 high-frequency (probably fourth-order) depositional sequences in the exposed portion of the third-order sequence (Figure 4).

Transgressive Systems Tracts

High-frequency transgressive systems tracts below the Asquith marker are interpreted to record deposition during a period when relative sea level was rising. Thus, they record a deepening succession of depositional environments in which sediment supply is low and probably decreasing. Eight retrogradational to aggradational high-frequency sequences were observed in the Lewis Shale third-order transgressive systems tract (Pyles, 2000). These sequences stack landward, to the north and west (Figure 4). Wire-line log correlations show that the internal stacking pattern is parallel to the upper and lower bounding surfaces (Pyles, 2000). These relatively deep-water shales are excellent seals.

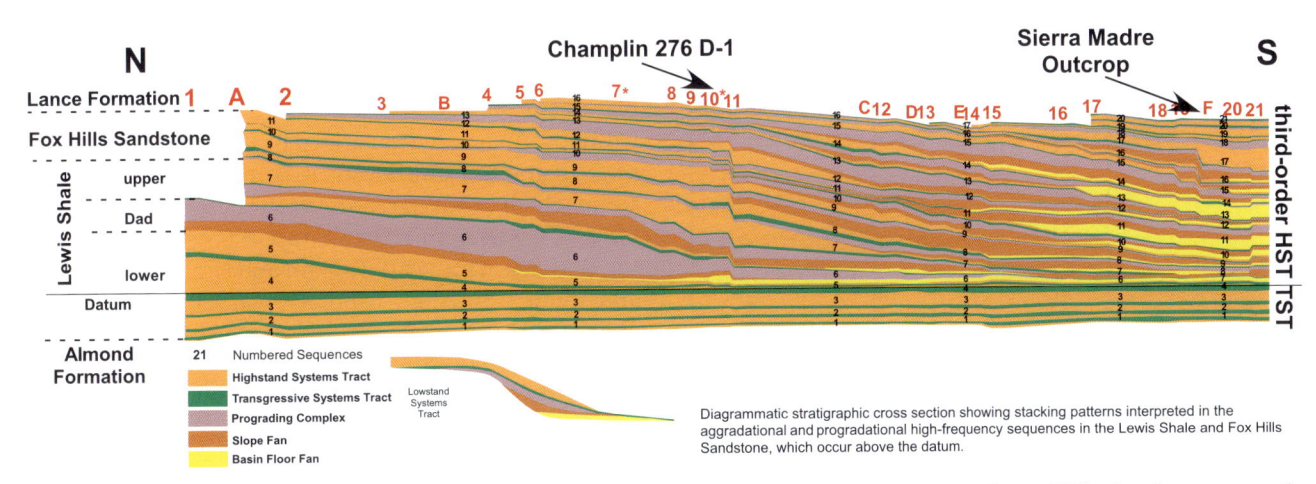

FIGURE 4. The high-frequency sequence-stratigraphic framework for the Lewis Shale and Fox Hills Sandstone reveals that the Lewis Shale consists of at least 20 (probably fourth-order) depositional sequences. Below the Asquith marker, deposition was basically aggradational. The overlying progradational unit consists dominantly of silty shales (third-order highstand) with interstratified fourth-order lowstand sandstones. The section shows systems tracts and significant surfaces. The location of the section is shown in Figure 1 (modified from Pyles, 2000). HST = highstand systems tract; TST = transgressive systems tract.

Highstand Systems Tracts

Thirteen high-frequency [fourth-order(?)] cycles are interpreted to be present in the third-order highstand systems tract (Pyles, 2000), which overlies the maximum flooding surfaces and underlies the sequence boundary. They record several successions of strata deposited in shallowing-upward depositional environments. Well-log correlations show that the internal stacking pattern displays sigmoid, divergent, and convergent geometries (Figure 4). The stratal relationships with respect to the lower bounding surface show onlap in proximal shelf and marginal marine settings and downlap in outer shelf, slope, and basinal environments (Pyles, 2000). These aggradational and progradational sequences contain sandy fourth-order lowstands developed in the third-order highstand systems tract. The high-frequency cycles stack basinward, from north to south (Figure 4). The well-developed, sandy, fourth order lowstand systems tract deposits record below storm wave base deposition from storm-induced gravity flows. Bouma sequences are not evident in these lowstand sandstones. Relatively weak seals (highstand systems tract [HST] shales) are interstratified with the sandstones (potential reservoirs).

Depositional sequences in the Lewis Shale appear to be very similar to those in the Neocomian of West Siberia (Pinous et al., 2001). In the West Siberia basin, the productive sandstone intervals are in a series of clinoforms that have prograded out over organic-rich black, fossiliferous shales (Bazhenov Formation), which are succeeded, in most locations, by dark-gray shales deposited in deep-marine conditions prior to turbidite sedimentation from the approaching clinoforms (Achimov Formation). The Achimov Formation consists of a series of interbedded turbiditic sandstones and hemipelagic shales. The productive sandstones occur in the central parts and toes of the clinoforms. The capping strata are comprised of shales and siltstones, with some minor sandstone, and appear to represent slope deposits. It appears that the turbiditic sandstones in the West Siberia basin, as in the Lewis Shale, were deposited in high-frequency lowstand systems tracts in a third-order highstand systems tract.

METHODS

Representative shales were collected from cores in two wells, the Champlin 276 Amoco D-1 and the Colorado School of Mines (CSM) No. 61 Stratigraphic Test (Figures 5, 6), which gave essentially complete coverage of the Lewis Shale depositional sequence. A petrographic-based microfacies approach was used to develop sedimentologic interpretations of the Lewis shales (Dawson, 2000; O'Brian and Slatt, 1990). The shales were analyzed using quantitative x-ray diffraction and scanning electron microscopy (SEM) techniques. Whole rock mineral identification was based on correspondence of experimental d values with the diagnostic hkl reflections from the International Center for Diffraction Data (1993) reference file and/or other published works. The quantitative analysis method is based on a modification (Srodon et al., 2000) of the matrix-flushing technique by Chung (1974). Minerals that were detected by the presence of a diagnostic reflection but have contents less than 1 wt.% were left as decimal amounts. These and other decimal values are not meant to imply greater precision than indicated from the

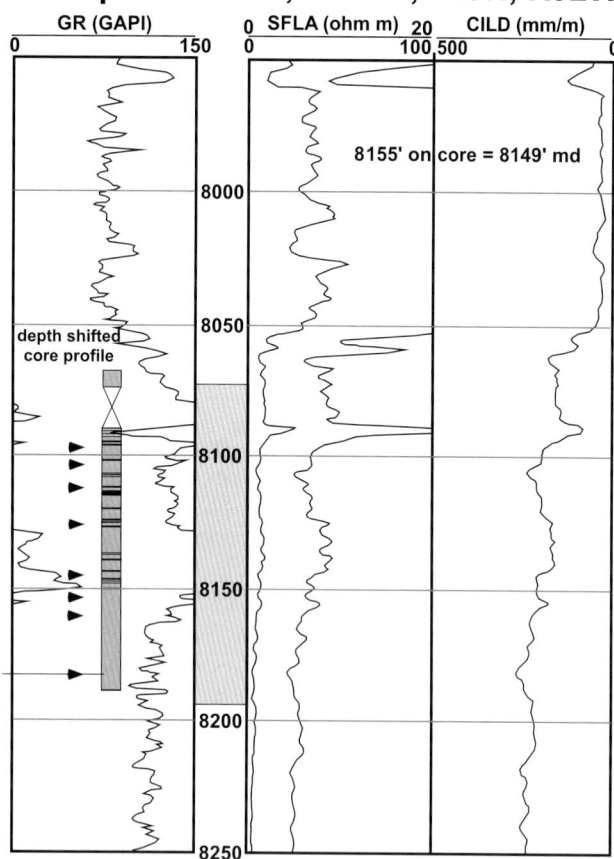

FIGURE 5. Wire-line logs showing the cored interval from the transgressive systems tract in the Champlin 276 Amoco D-1 well. Arrows show the depths of samples examined in this study. GR = gamma ray; SFLA = spherically focused log; CILD = deep induction log.

errors presented in Srodon et al. (2000). Accuracy for major phases ranges from 0.1 to 0.9%, with a mean of 0.5%. Zero values, as used here to indicate that the phase, were below limits of detection. Thirty-one thin sections were studied petrographically and photographed using polarized light microscopy. Percentages of components in thin section were determined by counting (100 points per thin section). Seal capacity was determined by MICP analysis. Physical rock properties were determined by measuring sonic velocities, porosity, permeability, and density at net confining stress from 1000 to 7000 psi.

SEAL CAPACITY ASSESSMENT

The use of these Lewis Shale outcrops as analogs for subsurface turbidite systems (Pyles, 2000; Pyles and Slatt, 2000b) suggested that they should provide a unique opportunity to examine the potential seal characteristics of the shales related to the sandy turbidite deposits (Dad Sandstone Member) in the highstand systems tract of the Lewis Shale. Because any rock type can function as a hydrocarbon seal, provided that the minimum displacement pressure of the potential seal is greater than the buoyant pressure generated by the hydrocarbon column in the accumulation, it is important to understand which lithotypes provide the best potential seals. Lithology is recognized as an important control on sealing capacity of top seals (Ingram and Urai, 1999). A general trend from high capillary seal capacity in fine-grained, clay-rich rocks to low seal capacity in coarse grained, clay-poor rocks is present. If systematic differences exist in fine-grained rock properties and textures in a depositional sequence, it should be possible to predict seal quality from sequence-stratigraphic analysis. High capillary pressure measurements are used to

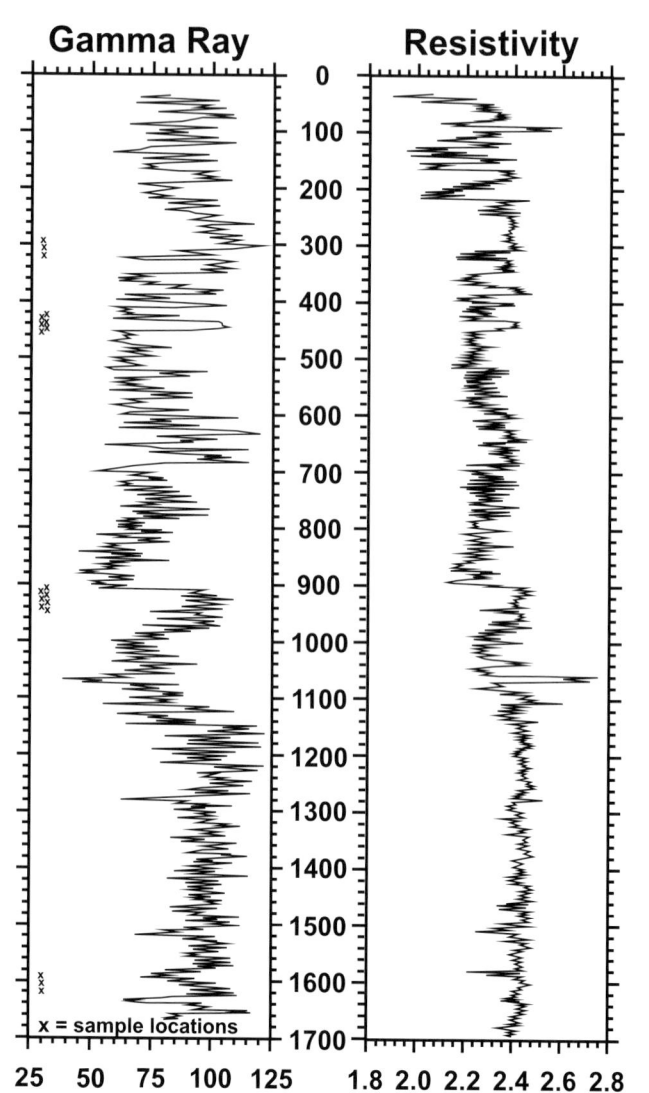

FIGURE 6. Gamma-ray and resistivity log tracks form the highstand systems tract in the CSM Stratigraphic Test No. 61 well in south-central Wyoming. Sample locations are noted to the left of the gamma-ray trace.

evaluate seal capacity. The resulting capillary pressure curves provide quantitative estimates of pore-throat-size distribution, which should define the largest diameter of interconnected, continuous pore throats. These pore throats will determine the ultimate capillary seal capacity. Schowalter (1979) has shown that failure by capillary leakage typically occurs at an average nonwetting phase saturation of 10%. We have adopted the pressure required to achieve 10% nonwetting phase saturation in the seal pore system as the critical value for determining capillary seal capacity. Standard techniques for evaluating seals are discussed in Berg (1975), Schowalter (1979, 1981), and Jennings (1987).

SHALE COMPOSITION

Compositional analyses (XRD) indicate that the Lewis Shale samples belong to a single, moderately variable compositional family (Figure 7). Total clay content ranges from 35 to 71%, with a mean of 52%. Detrital silt (quartz + feldspar mineral) abundance ranges from 24 to 59%, with a mean of 37%. Pyrite (trace to 4%), siderite (trace to 2%), and magnesium calcite (1–4%) occur in all samples as accessory components. The abundance of calcite and dolomite are highly variable. Calcite abundance ranges from 0 to 10%, whereas dolomite abundance ranges between 1 and 8%. With the exception of local bentonite layers, the normalized clay mineral composition is dominated by illite, which accounts for 56–78% (mean 67%) of the clay size fraction. Smectite content ranges from 15 to 36%. Kaolinite (5–10%) and chlorite (0–6%) are minor components. Most illite and smectite are aluminum rich, although iron-rich 2:1 type clays are common in shales from the transgressive systems tract.

FACIES ANALYSIS

Mudstones and shales are essential elements of the petroleum system, serving as both source for and seal on hydrocarbons. Additionally, hydrocarbons must be able to move through the shales and mudstones on a geologic (migration) timescale. Unfortunately, the sedimentologic study of mudstones lags far behind that of sandstones and carbonates. Detailed examinations of sedimentary features are sparse, but Schieber's (1999) work has revealed lateral facies variability in mudstones and has defined several distinct facies types, which are similar to those seen in the Lewis Shale. Five end-member Lewis Shale microfacies are recognizable: (1) massive organic mudstones; (2) organic laminated shales; (3) calcareous-laminated shales; (4) organic bioturbated shales; and (5) massive calcareous shales. These microfacies are described and illustrated below (Figures 8, 9). The average or range of compositional and petrophysical data for each microfacies is given in Tables 1 and 2. Seismically important parameters are listed in Table 3. The distribution of these argillaceous microfacies can be related to the sequence-stratigraphic position. That is, shales from transgressive, highstand, and condensed sequences have distinctive petrographic aspects, seal characteristics, and seismic characteristics.

Late transgressive systems tract deposition (microfacies 1 and 2) in the Lewis Shale is similar to Schieber's (1999) carbonaceous facies associations carbonaceous mudstones, which indicates slow overall deposition in disoxic to anoxic conditions and displays minor bioturbation. Deposits from the lower portions of the highstand systems tract (microfacies 3) are similar to those described in Schieber's (1999) graded mudstone facies (facies association graded mudstones). They show features related to deposition from the waning flow of

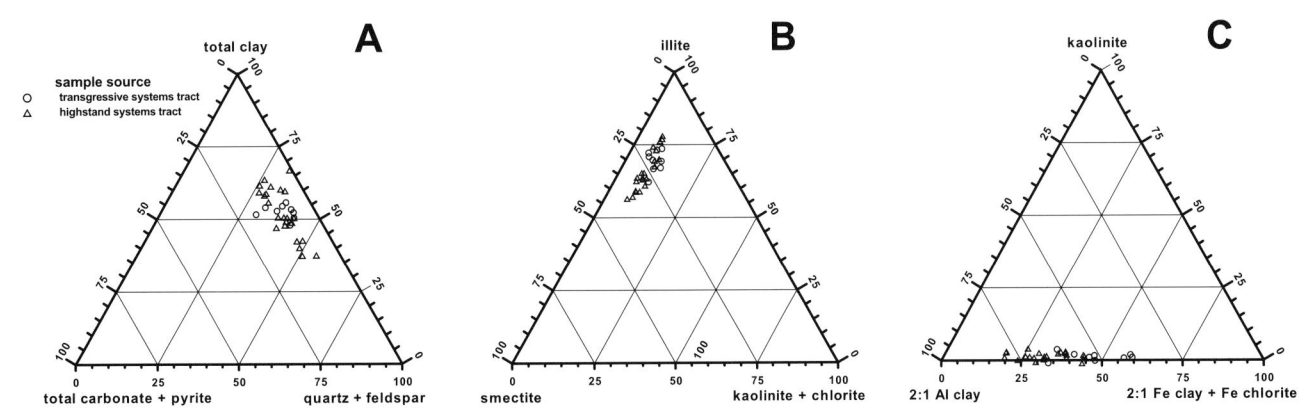

FIGURE 7. Compositional plots for Lewis Shale samples examined in this study. Data are symbol coded by systems tract. (A) Both transgressive systems tract and highstand systems tract samples have relatively constant bulk composition. (B) The clay mineral types do not change between the transgressive and highstand systems tracts. (C) The iron content of the clay mineral suite is greater in the transgressive than in the highstand systems tract.

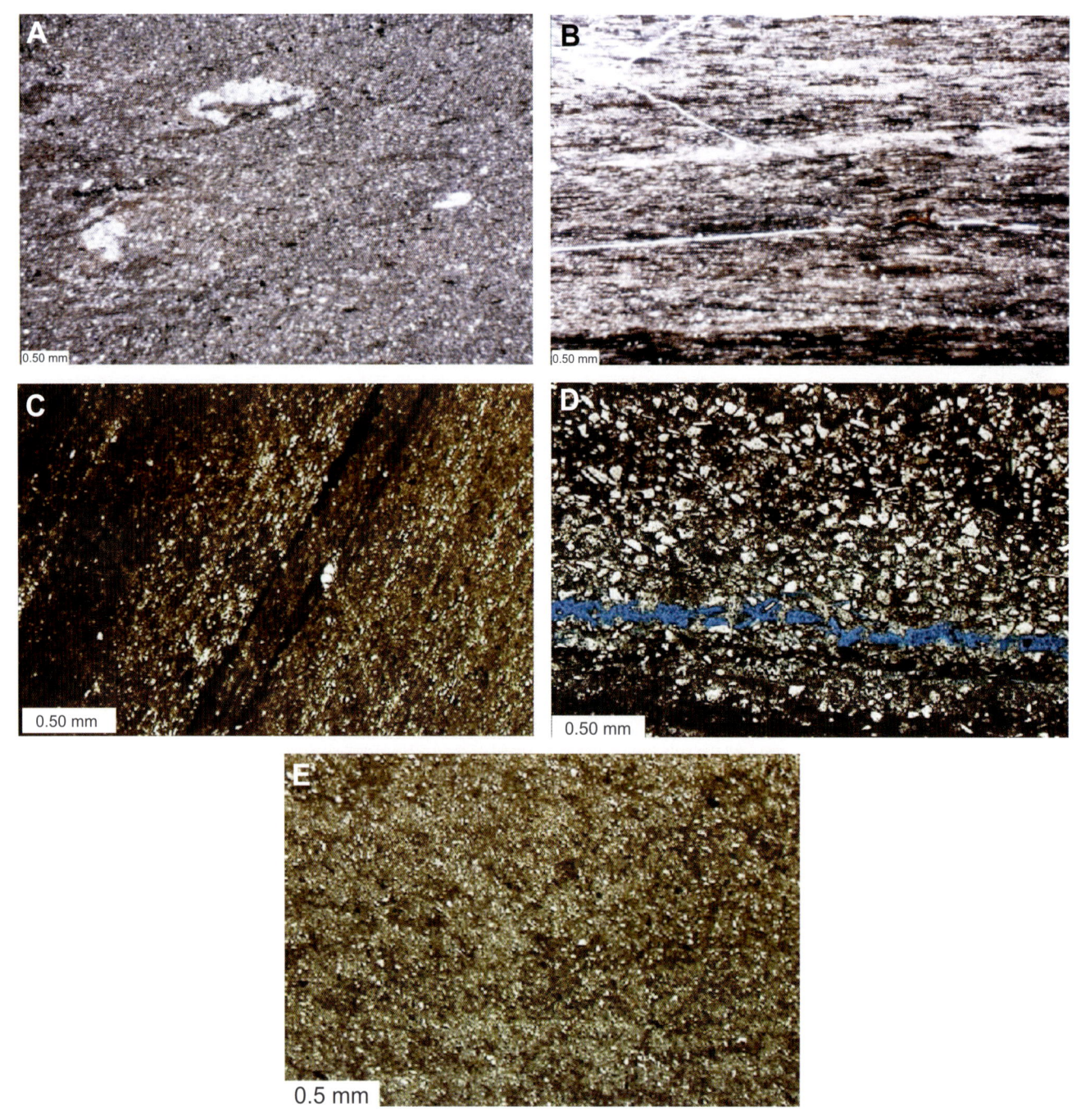

FIGURE 8. Low-magnification, thin-section photomicrographs (plane-polarized light) of the fine-grained Lewis Shale microfacies reveal a wide range of depositional fabrics. (A) The massive organic mudstones of microfacies 1 display wispy laminations and dispersed pyrite. Note foraminifera tests. (B) The organic laminated shales (microfacies 2) are characterized by strong, bedding-parallel organization of clay and organic particles. Note the probable fish bone on one depositional lamination. (C) The calcareous laminated shale (microfacies 3) is marked by the alternation of silty and thin organic-rich laminae and the presence of foraminifera tests on the organic-rich laminations. (D) The organic bioturbated shales (microfacies 4) displays alternating sandy and silty laminae, some of which are disrupted by burrowing. Pelagic deposition is significantly suppressed. (E) In the massive calcareous shale (microfacies 5), lamination is much less pronounced than in microfacies 4. Bioturbation has nearly destroyed any original depositional fabric. Silt is well sorted.

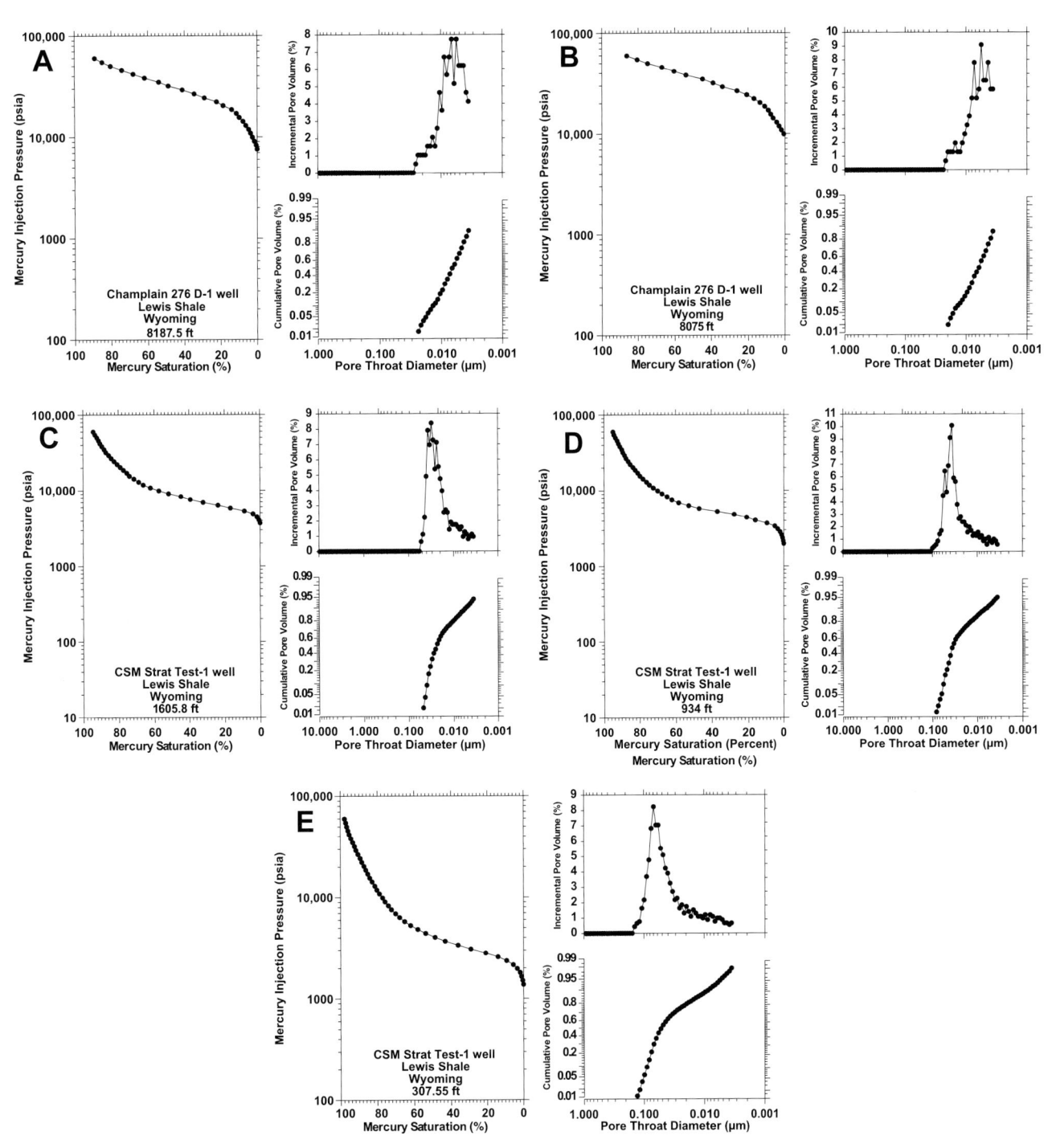

FIGURE 9. High-pressure MICP plots for the Lewis Shale microfacies reveal clear differences in seal character and pore system structure. (A) The pore system of the massive organic mudstones of microfacies 1 is controlled by well-sorted pore throats with a modal diameter of about 0.0065 μm. (B) The organic laminated shales (microfacies 2) also possess a pore system controlled by well-sorted pore throats with a modal diameter of approximately 0.005 μm. These facies have excellent seal capacity. The other microfacies have more complex pore systems. (C) The calcareous laminated shales (microfacies 3) have a pore system containing two populations of pore throats. The dominant population in the silt-rich laminations controls nearly 85% of the pore volume and has a modal diameter around 0.015 μm. (D) In the organic bioturbated shales (microfacies 4), pore system bimodality is more pronounced, and pore throat diameters are slightly larger than in microfacies 3. (E) In the massive calcareous shale (microfacies 5), pore system is controlled by a bimodal, moderately sorted population of pore throats with moderate displacement pressures. The tail of small pore throats is more pronounced than in other microfacies.

TABLE 1. Average or range of compositional data for five microfacies in Upper Cretaceous Lewis Shale, south-central Wyoming.

Microfacies	Detrital Clay (wt.%)	Total Organic Carbon (wt.%)	Framboidal Pyrite (wt.%)	Total Carbonate (wt.%)	Dolomite (wt.%)	Siderite (wt.%)
1	50–53	1.0–1.3	0.5–2.0	5.0–7.0	2.0–3.0	0.5–2.0
2	50–54	1.3–2.8	2.0–4.0	7.0–15	2.0–4.0	1.0–2.0
3	58–69	1.1–1.5	0.6–1.0	1.0–6.0	0.2–1.0	0.0–2.0
4	43–61	0.6–1.8	0.2–3.0	7.0–14	2.0–8.0	0.4–2.0
5	39–44	0.5–0.6	0.2–0.8	8.0–11	4.0–6.0	0.0–1.0

fine-grained turbidites in disoxic conditions. Turbidite influence increases upsection, so that late HST deposits are similar to the burrowed mudstone facies (facies association burrowed mudstones).

Microfacies 1 (Massive Organic Shales)

Microfacies 1 consists of dark-gray and black shales that are thin bedded in core. At the scale of a thin section, observations reveal that the detrital matrix in microfacies 1 has a wispy appearance (Figure 8A). Occasional wavy laminations are present. Local areas in which the long axis of elongate silt grains and thin-shelled pelecypod fragments are aligned parallel with depositional laminae are present in the more massive appearing areas of the matrix. Detrital clay minerals are the dominant (51%) component. Silt-sized grains of detrital quartz and feldspar are conspicuous accessory components. Tests of planktonic foraminifera occur as floating grains in the laminated detrital clay matrix. The chambers of foraminifera tests are cemented with sparry calcite and pyrite. Elongate fragments of terrestrially derived, organic matter comprise an accessory component. Total organic carbon (TOC) content ranges from 1.0 to 1.3 wt.%. Framboidal pyrite (0.5–2.0%) is scattered throughout the compacted clay matrix as a volumetrically minor authigenic component. Total carbonate averages 6.0%, with dolomite being the dominant carbonate species. Siderite (0.5–2.0%) comprises a significant portion of the carbonate fraction. Microfacies 1 is interlaminated with and grades upward into microfacies 2.

Scanning electron microscopy images reveal that microfacies 1 has a strongly compacted appearance. Porosity ranges between 3 and 6%, whereas permeability averages 0.003 md. Shale bulk density ranges from 2.52 to 2.55 g/cm^3. This microfacies provides excellent seal potential. The injection pressure required to achieve 10% nonwetting phase saturation ranges from 15,735 to 18,495 psia (Table 2; Figure 9A). Several seismically significant physical parameters were measured on samples of microfacies 1. The compressional wave velocity of the samples ranges from 14,400 to 15,100 ft/s (4389 to 4571 m/s), whereas the shear velocity ranges from 8070 to 9300 ft/s (2459 to 2834 m/s). Poisson's ratio ranges from 0.145 to 0.275; Young's modulus ranges from 5.60 to 7.22; and the Shear modulus ranges from 2.20 to 3.03.

Interpretation

The abundant framboidal pyrite records bacterial activity near an oxic-anoxic interface (Wilkin and Barnes, 1997). These shales are petrographically comparable to the sparsely fossiliferous transgressive shales described by Abed and Sadaqah (1998). Cretaceous black marine shales are known to coincide with widespread marine transgressions (e.g., Demanison and Moore, 1980; Hallam, 1987; Schlanger et al., 1987). Contrary to the assertion of Heckel (1977), transgressive shales do not necessarily record a deep-marine paleoenvironment of deposition but can represent relatively shallow-marine portions of transgressive depositional sequences (Leckie et al., 1990; Wignall and Newton, 2001). Microfacies 1

TABLE 2. Average or range of petrophysical data for five microfacies in Upper Cretaceous Lewis Shale, south-central Wyoming.

Microfacies	Porosity (%)	Permeability (md)	Mercury Injection Capillary Pressure (psia)
1	3.0–6.0	0.0003–0.0009	15,735–18,495
2	2.0–5.0	0.0001–0.0005	16,685–21,350
3	12.8–20.0	0.001–0.098	2755–5355
4	13.9–18.2	0.005–0.032	1740–7655
5	16.6–17.4	0.007–0.019	1105–2390

TABLE 3. Range of seismically important data for five microfacies in Upper Cretaceous Lewis Shale, south-central Wyoming.

Microfacies	Bulk Density (g/cm³)	Compressional Wave Velocity (ft/s)	Shear Velocity (ft/s)	Poisson Ratio	Young's Modulus	Shear Modulus
1	2.52–2.55	14,400–15,100	8070–9300	0.145–0.275	5.60–7.22	2.20–3.03
2	2.47–2.58	13,700–15,200	7630–10,000	0.114–0.273	5.02–7.92	2.20–3.03
3	2.15–2.26	11,000–13,400	6370–7080	0.198–0.320	2.57–4.10	1.19–1.63
4	2.18–2.31	13,600–15,400	6480–8260	0.264–0.350	4.07–6.03	1.51–2.32
5	2.18–2.25	13,700–13,900	6810–7560	0.281–0.272	3.86–4.98	1.44–1.96

represents a low-energy (below storm wave base), oxygen-depleted paleoenvironment. The juxtaposition of rock textures suggests that the shales have experienced extensive bioturbation. The planktonic foraminifera represent an overlying oxygenated water column. Deposits in this microfacies were probably the result of hemipelagic deposition (Stow et al., 2001), which involved both vertical settling and slow lateral advection through the water column.

Microfacies 2 (Organic Laminated Shales)

Microfacies 2 deposits consist of dark-gray to black shales that display well-developed fissility (Figure 8B) and contain accessory phosphatic nodules, bones, and possible fish teeth. Thin-section examination reveals that microfacies 2 possesses a well-laminated fabric developed by the organization of organic debris, with their long axes parallel to bedding. Detrital clay minerals are the dominant (52%) component. The clay-rich laminae are slightly wavy because of compaction around detrital silt grains and bioclastic components. Thin-shelled pelecypod fragments are aligned parallel with depositional laminae. Most shells have been flattened and broken during compaction. Chambers of foraminifera tests are cemented by calcite and pyrite. Microfacies 2 contains a greater percentage of foraminifera tests than microfacies 1. Framboidal pyrite, a conspicuous accessory component, comprises 2.0–4.0% of the sample mass. These are the highest pyrite abundances in the Lewis Shale. Microfacies 2 contains an average TOC content of 1.3–2.8%, the highest of any microfacies in the Lewis Shale. Total carbonate ranges from 7.0 to 15.0%, with calcite being the dominant carbonate species. Dolomite (2.0–4.0%) and siderite (1.0–2.0%) comprise a significant portion of the carbonate fraction.

Scanning electron microscopy images reveal an abundance of authigenic carbonate, pyrite, and apatite crystals. These components occur as both cements and nodular replacive phases. Microfacies 2 is strongly compacted. Porosity ranges between 2 and 5%, whereas permeability averages 0.001 md. Shale bulk density ranges from 2.47 to 2.58 g/cm³. This microfacies provides excellent seal potential. The injection pressure required to achieve 10% nonwetting phase saturation ranges from 16,685 to 21,435 psia (Table 2; Figure 9B). Among the seismically significant physical parameters, compressional velocity of the microfacies 2 ranges from 13,700 to 15,200 ft/s (4175 to 4632 m/s), whereas the shear velocity is between 7630 and 10,000 ft/s (2325 and 3047 m/s). Poisson's ratio ranges from 0.114 to 0.273; Young's modulus ranges between 5.02 and 7.92; and the shear modulus ranges from 2.20 to 3.03.

Interpretation

These phosphatic shales are comparable to condensed shales discussed by Loutit et al. (1988) and Schutter (1998). Condensed shales are characterized by high-abundance and low-diversity assemblages of planktonic fossils, as well as low abundance of benthic fossils. Moderate to high TOC content and elevated abundances of authigenic minerals are also characteristic. Concentrations of phosphatic bioclasts, elevated abundance of authigenic minerals, and high radioactivity are also common features. The abundance of authigenic pyrite, the lack of benthic fossils, and the scarcity of bioturbation attest to an oxygen-depleted depositional setting for microfacies 2. The abundance of authigenic pyrite, carbonate, and apatite is the end result of a complicated diagenetic history induced by sulfate reduction during early burial diagenesis (Lev et al., 1998). Sedimentologically, condensed shales record highly reduced rates of deposition, which is accompanied by phosphate authigenesis, in a marine setting having minimal current activity (below storm wave base).

Microfacies 3 (Calcareous Laminated Shales)

Microfacies 3 consists of dark-gray to black silty shales that exhibit well-developed fissility (Figure 8C). Thin-section examination reveals that microfacies 3 consists of very finely interlaminated organic-rich black shale and very silty shale. Many of the silt-rich laminae appear graded. Detrital clay minerals are the dominant (61%) component, whereas silt content ranges from 32 to 34%. Planktonic foraminifera tests occur along some of the most clay-rich laminations. These tests are cemented with sparry calcite. Framboidal pyrite, a minor accessory component, comprises 0.6–1.0% of the sample mass. Microfacies 3 contains a TOC content of 1.1–1.5%. Total carbonate ranges from 1.0 to 6.0%, with calcite being the dominant carbonate species. Dolomite (0.2–1.0%) and siderite (0.0–2.0%) comprise a significant portion of the carbonate fraction.

Scanning electron microscopy images reveal an abundance of authigenic carbonate, pyrite, and apatite crystals. These components occur as both cements and nodular replacive phases. Microfacies 3 is not as strongly compacted as microfacies 1 and 2. Porosity ranges from 12.8 to 20%, whereas permeability ranges between 0.001 and 0.098 md. Shale bulk density is between 2.15 and 2.26 g/cm^3. This microfacies provides relatively poor seal potential. The injection pressure required to achieve 10% nonwetting phase saturation ranges from 2755 to 5355 psia (Table 2; Figure 9C). Among the seismically significant physical parameters, compressional velocity of the microfacies 3 ranges from 11,000 to 13,400 ft/s (3352 to 4084 m/s), whereas the shear velocity is between 6370 and 7080 ft/s (1941 and 2157 m/s). Poisson's ratio ranges from 0.198 to 0.320, Young's modulus ranges between 2.57 and 4.10, and the shear modulus ranges from 1.19 to 1.63.

Interpretation

This microfacies reflects the increase in depositional rates produced by the beginning of progradation of the toes of a highstand sediment wedge. Deposition probably resulted from hemiturbidite sedimentation (Stow and Wetzel, 1990), which involved dispersion from dilute turbidity currents during the final stages of deposition or following interaction with positive topographic features. The fine-grained material carried by the turbidity current dispersed beyond the terminal deposit of the turbidite and mixed with any background pelagic or hemipelagic material and deposits formed slowly by vertical settling (Stow et al., 2001). Deposition was episodic. Planktonic foraminifera tests occur along some of the most clay-rich laminations. The lack of bioturbation suggests that microfacies 3 represents a low-energy (below storm wave base), oxygen-depleted paleoenvironment. The planktonic foraminifera were derived from an overlying oxygenated water column.

Microfacies 4 (Organic Bioturbated Shales)

Microfacies 4 consists of dark-gray and black, silty, calcareous shales that are thin bedded in core and break out with a poker chip aspect (Figure 8D). Occasional siltstone and sandstone laminations are present. Slump structures occur in the shale sequence. At the scale of a thin section, the detrital matrix in microfacies 4 has a wispy to massive appearance. Occasional wavy laminations are present. The long axis of elongate silt grains and thin-shelled pelecypod fragments are aligned parallel with depositional laminae. The shales appear to have experienced extensive bioturbation. Detrital clay minerals are the dominant (51%) component. Silt-sized grains of detrital quartz and feldspar are conspicuous accessory components. Silt and sand content ranges from 27 to 41%. In some instances, the coarse detrital grains form a self-supporting framework, but generally, they appear to float in the clay-rich matrix. Framboidal pyrite is a significant accessory component and comprises 2–3% of the sample mass. These values are only slightly lower than those of microfacies 2. Elongate fragments of terrestrially derived, organic matter are an accessory component. Microfacies 4 contains a TOC content of 1.2–1.8%. Total carbonate ranges from 7.0 to 14.0%, with dolomite (2.0–8.0%) being the dominant carbonate species. Siderite (0.5–2.0%) comprises a significant portion of the carbonate fraction.

Scanning electron microscopy images reveal an abundance of authigenic carbonate and pyrite crystals. Microfacies 4 is moderately compacted. Porosity ranges from 13.9 to 18.1%, whereas permeability ranges between 0.005 and 0.007 md. Shale bulk density ranges between 2.25 and 2.31 g/cm^3. This microfacies provides relatively poor seal potential. The injection pressure required to achieve 10% nonwetting phase saturation ranges from 1975 to 3710 psia (Table 2; Figure 9D). These are the lowest values encountered in shales and mudstones of the Lewis Shale. The compressional velocity of the microfacies 4 (14,200–15,400 ft/s; 4328–4693 m/s) and the shear velocity (6480 and 8260 ft/s; 1975 and 2517 m/s) are significantly higher than in microfacies 3. Poisson's ratio (0.264–0.350), Young's modulus (4.07–6.03) and the shear modulus (1.51–2.32) are larger than in microfacies 3.

Interpretation

Microfacies 4 records deposition under variable conditions in terms of current energy and oxygenation levels. This microfacies contains less pyrite and organic matter relative to microfacies 1 and 2 and may record a shift to a somewhat more oxygenated paleoenvironment because the highstand depositional system was firmly established. The large volumes of mud, as well as the significant volumes of silt and sand-sized material, suggest that these sediments may represent deposition from turbidity currents (Stow et al., 2001). Individual flows were discrete events that recurred at irregular intervals. The pulsed nature of deposition generated the depositional laminations.

Microfacies 5 (Massive Calcareous Shales)

Microfacies 5 is comprised of gray laminated mudstones interbedded with thin, black, organic-rich mudstones (Figure 8E). The bedding is parallel and continuous except where interrupted by scour surfaces. Thin-section examination reveals that the individual laminations in microfacies 5 are disrupted and display a wispy to massive aspect. Very fine silt-sized particles of quartz and feldspar (51–59%) are the dominant component. Clay minerals (39–44%) are of secondary importance. The poor organization of the individual detrital laminae may be a result of burrowing. Total organic carbon content ranges from 0.5 to 0.6 wt. %. Framboidal pyrite (0.2–0.8%) is scattered throughout the compacted clay matrix as a volumetrically minor authigenic component. The TOC and pyrite contents are the lowest seen in the Lewis Shale. Total carbonate ranges from 8.0 to 11.0%, with dolomite (4.0–6.0%) being the dominant carbonate species. Siderite (0.0–1.0%) comprises a minor portion of the carbonate fraction.

Scanning electron microscopy images reveal that microfacies 5 is moderately compacted. Small crystals of carbonate cements are scattered throughout the clay-rich matrix. The detrital silt grains do not form self-supporting frameworks. Porosity ranges between 16.6 and 17.4%, whereas permeability ranges from 0.007 to 0.019 md. Shale bulk density ranges from 2.18 to 2.25 g/cm^3. This microfacies provides poor seal potential. The injection pressure required to achieve 10% nonwetting phase saturation ranges from 1105 to 2390 psia (Table 2; Figure 9D). Several seismically significant physical parameters were measured on samples of microfacies 5. The compressional wave velocity of the sample ranges from 13,700 to 13,900 ft/s (4175 to 4236 m/s), whereas the shear velocity is 6810–7560 ft/s (2075–2304 m/s). Poisson's ratio ranges from 0.281 to 0.272; Young's modulus ranges from 3.86 to 4.98; and the shear modulus ranges from 1.44 to 1.96.

Interpretation

The argillaceous siltstones of microfacies 5 mark the top of turbidite deposition (Dad Sandstone Member) in the Lewis Shale depositional sequence. This microfacies contains the smallest volumes of pyrite, siderite, and organic matter in the Lewis Shale. The scarcity of these components suggests that this microfacies was deposited in the most oxygenated setting extant during Lewis Shale deposition. The near equal volumes of mud and silt- and sand-sized material and the poor organization of the individual laminae suggest homogenization of interlaminated shales and siltstones by biologic activity. Microfacies 5 probably represents deposition by hemiturbidite sedimentation (Stow and Wetzel, 1990), which involves dispersion from dilute turbidity currents during the final stages of deposition or following interaction with positive topographic features. The fine-grained material carried by the turbidity current disperses beyond the terminal deposit of the turbidite and mixes with any background pelagic or hemipelagic material and deposits slowly by vertical settling (Stow et al., 2001). Deposition is episodic, but accumulation is commonly sufficiently slow that a restricted ichnofauna is present throughout the microfacies.

Source Rock Character

The average TOC for all analyzed Lewis Shale samples is 1.23 wt.%. However, the bulk of the preserved organic matter is contained in the late stages of the transgressive systems tract (Champlin 276 Amoco D-1 well), where TOC values average 1.68 wt.%. The maximum measured TOC value in this unit is 2.78 wt.%. The TOC values (Figure 10A) in the transgressive systems tract (CSM Stratigraphic Test no. 61 well) range from 0.12 to 1.79 wt.%, with a mean of 1.07 wt.%. In this instance, the organic laminated shales of the condensed section (microfacies 2) are slightly more organic rich than the massive organic mudstones of the latest stage of the transgressive systems tract (microfacies 1). The calcareous laminated shales (microfacies 3), which form the base of the transgressive systems tract, have moderate and highly variable TOC values. The organic bioturbated shales (microfacies 4) and massive calcareous shales (microfacies 5) contain small amounts of organic carbon, and TOC variability is low. The organic matter in the Lewis Shale is uniformly type III (terrestrial organic matter). The organic matter in the transgressive systems tract is significantly more oxidized than that in the highstand systems tract (Figure 10B). Understanding the reason for this difference will require additional study. The temperature of maximum organic generation (Figure 10C) is essentially the same for all samples examined in this study. This fact suggests that the maximum depth of burial has been similar for both cores, although the Champlain 276 D-1 core is still buried to 8100 ft (2468 m) below ground level.

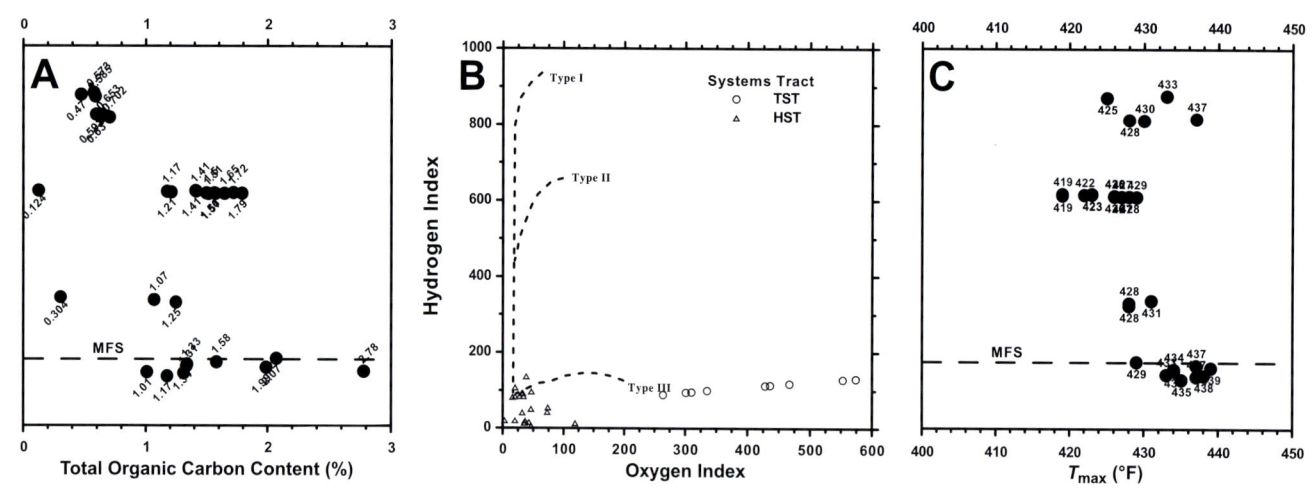

FIGURE 10. Some of the organic properties of the transgressive and highstand systems tracts are significantly different. (A) The transgressive systems tract is more organic rich than the highstand systems tract. (B) The organic matter in the transgressive systems tract is more oxidized than that in the transgressive systems tract. (C) T_{max} values for the two intervals are nearly identical, indicating that they have seen very similar burial histories. TST = transgressive systems tract; HST = highstand systems tract.

DISCUSSION

On the basis of detailed petrographic and petrophysical data, the Lewis Shale can be divided into five microfacies (Figure 11) based on samples from outcrops and continuous cores in portions of the Great Divide and Washakie basins in south-central Wyoming. These facies, listed from best to worst top seals, include organic laminated shales, massive organic mudstones, silty shales, silty calcareous shales, and massive calcareous shale. Transgressive shales and basal transgressive shales are characterized by low species diversity and sparsity of benthic organisms and trace fossils. The relative organic richness and the scarcity of infaunal organisms are consistent with oxygen-depleted bottom conditions (Charvat and Grayson, 1981). Textures observed in the interstratified siltstones of the middle highstand systems tract record a shift to more oxygenated bottom conditions. The upper highstand is marked by increased current activity and the proliferation of bioturbation under significantly more oxygenated bottom conditions (Figure 12).

The best two seals comprise deposits of condensed sections and upper portions of transgressive systems tracts (Figure 13A). Poorer seals comprise deposits from the lower portions of highstand systems tract deposits. These conclusions agree with conclusions reached by Dawson and Almon (2002) for deposits from Gulf Coast-style

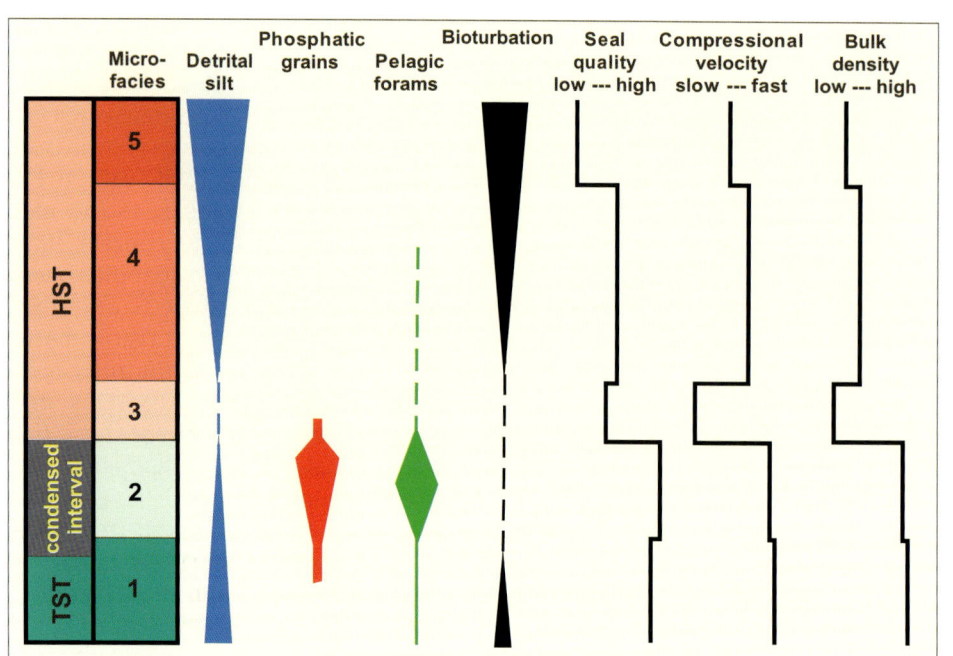

FIGURE 11. Numerous petrographic parameters and seismically important parameters vary systematically through the third-order transgression and highstand recorded by the Lewis Shale. TST = transgressive systems tract; HST = highstand systems tract.

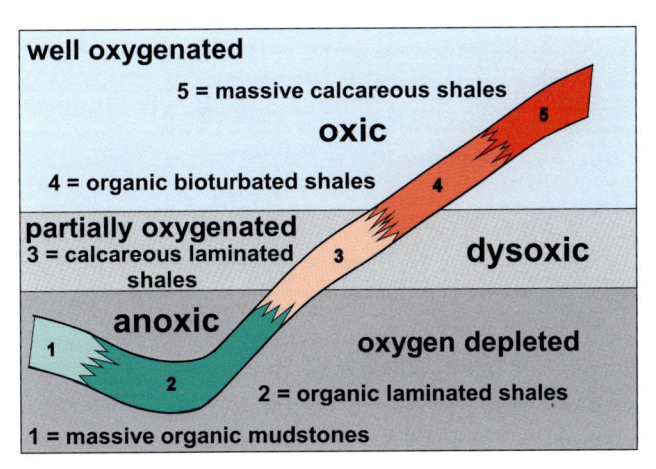

FIGURE 12. The Lewis Shale records deposition under anoxic bottom conditions during the late transgressive systems tract (microfacies 1 and 2) and early highstand (microfacies 3). Bottom conditions become more oxygenated in the middle highstand (microfacies 4), and full oxygenation in the upper highstand systems tract (microfacies 5) leads to an explosion of biologic activity and decrease in the preservation of organic carbon.

the better organization of the detrital clays resulting from suspension settling and associated microcrystalline carbonate cement formed during early submarine diagenesis. No obvious trend in porosity is present with microfacies in the highstand systems tract. This may result from the fact that deposition appears to be dominated by hemiturbiditic and turbiditic processes. Additionally, a strong contrast in shale permeability exists between the transgressive systems tract and the highstand systems tract (Figure 13B). Within the highstand systems tract, a minor tendency for permeability to increase upward is present in the stratigraphic section. This trend appears to be weakly related to increasing silt content, upward in the highstand systems tract. Crossplotting porosity with permeability (Figure 13B) shows that all samples fall along the same trend line, and that a moderate correlation coefficient exists ($R^2 = 0.88$).

Discriminant Function Analysis of the Lewis Shale

Quantitative petrographic and TOC data from the Lewis Shale were used in a discriminant function analysis (DFA) to determine if the five identified microfacies could be discriminated. Data used in this analysis include seven variables: TOC, degree of bioturbation, percent of nonreflecting opaques (including organic

basins and by Sutton et al. (2001) for Upper and Lower Cretaceous deposits of the Denver basin.

Porosity is significantly reduced in the late-stage transgressive systems tract (Figure 13B), relative to all portions of the highstand systems tract. The reduced porosity in microfacies 1 and 2 appears to result from

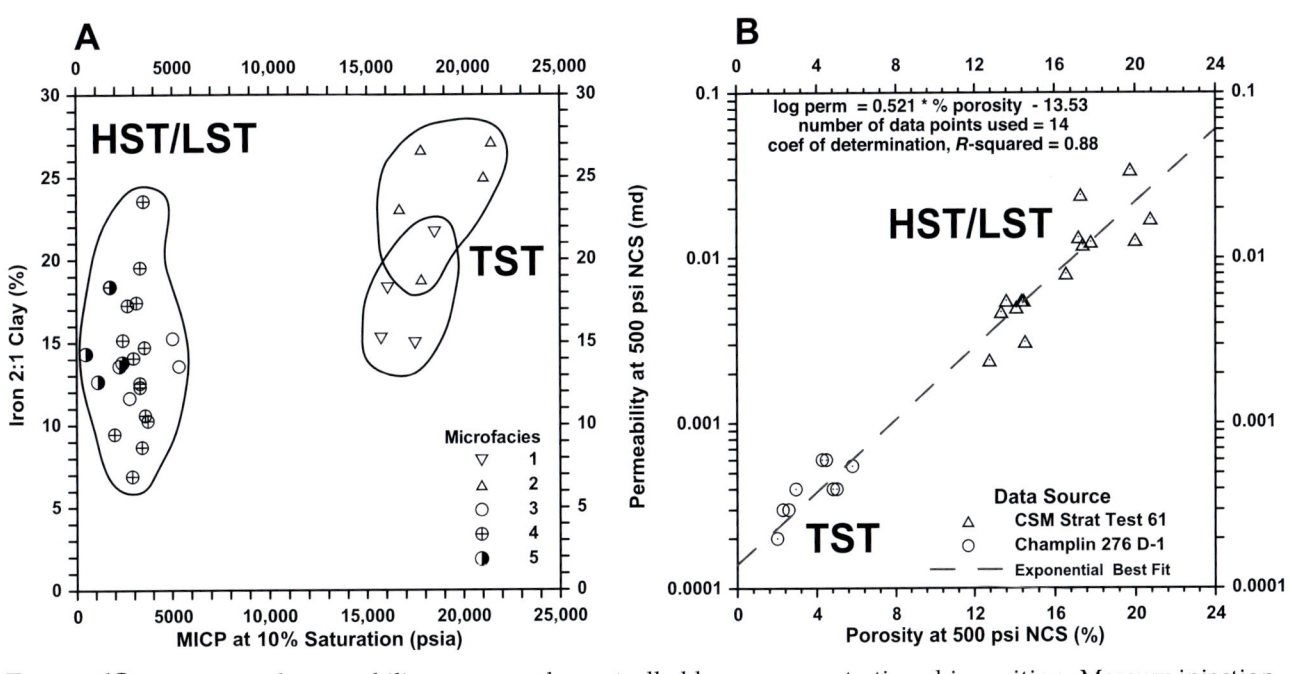

FIGURE 13. Porosity and permeability are strongly controlled by sequence-stratigraphic position. Mercury injection-capillary pressure values and porosity are significantly reduced in the late transgressive systems tract (TST) relative to all parts of the highstand systems tract (HST) interval. (A) Transgressive systems tract shales enriched in iron-bearing clay minerals and pyrite have strongly elevated MICP values. (B) Porosity and permeability in the TST shales is significantly lower than in the HST shales. A strong correlation exists between the two parameters. LST = lowstand systems tract.

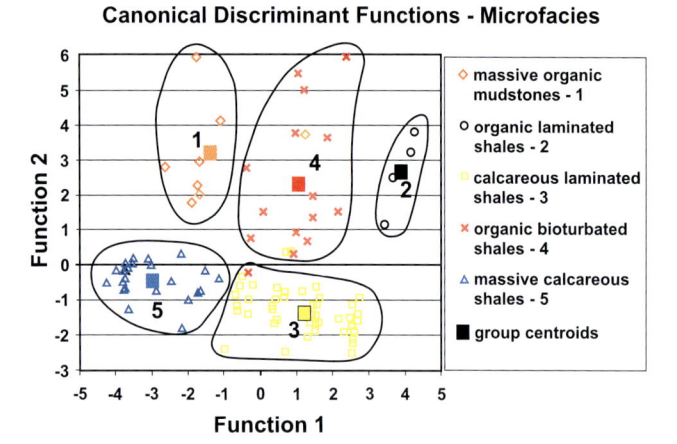

FIGURE 14. Graphic plot of multiple discriminant function analysis, Lewis Shale microfacies. Scores for discriminant functions one and two are plotted for each sample for all microfacies. Note the excellent separation among microfacies groups. Irregular circles enclosing each microfacies group are informally sketched.

matter), average grain size of silt grains, percent silt grains, percent cement, and percent matrix. The TOC was determined on powdered samples as a weight percent of the total sample. The degree of bioturbation was estimated on a scale of 1–6 in thin sections. Average grain size was determined from the measurements of the apparent long axes of 30 grains in thin section. Percent opaques, silt grains, total cement, and matrix were estimated by point counting a minimum of 200 points per thin section. A total of 101 samples was analyzed from all five microfacies (Castelblanco-Torres, 2003). Because the variables are measured on different scales, each variable was transformed to standardized scores before being analyzed statistically.

In DFA, lines of best separation (discriminant functions) are calculated among preestablished groups based on any number of variables. The coefficients of these discriminant functions illustrate the relative importance of the direct contribution of each variable in discriminating among groups. Three points are important concerning DFA. Discriminant functions are linear combinations of all variables. All groups are predetermined on the basis of other criteria, and it is assumed that all samples can be assigned to one of the established groups. Another aspect of DFA, the assignment of unknown samples to one of the established groups, is not considered in this study.

Results of the DFA on the Lewis Shale are given in Figure 14 and Tables 4–7. Figure 14 shows a distinct separation among all five microfacies, with only five samples statistically assigned to a group other than the one in which it originated (Table 4). The first two discriminant functions account for almost 99% of the total variation in the data set (Table 5). Correlations among the variables and the discriminant functions are shown in Table 6. Variables with high correlations are most important in discriminating among microfacies. Thus, samples with relatively high bioturbation indices and

TABLE 4. Predicted microfacies group membership based on counts of all samples over five microfacies groups from the Lewis Shale.

	Predicted Group Membership					
Microfacies	1	2	3	4	5	Total
1	7	0	0	1	0	8
2	0	4	0	0	0	4
3	0	0	48	0	0	48
4	1	0	3	12	0	16
5	0	0	0	0	25	25

TABLE 5. Percentage of variance accounted for by each discriminant functions in separating samples from microfacies 1–5, Lewis Shale.

Function	Eigenvalue	% of Variance	Cumulative Variance (%)
1	4.033	56.9	56.9
2	2.973	41.9	98.8
3	0.630	0.9	99.7
4	0.240	0.3	100.0

TABLE 6. Structure matrix showing pooled within-group correlations between discriminating variables and standardized canonical discriminant functions. Results are shown for all four discriminant functions. Variables are ordered by absolute size of correlation within functions. Only the first two functions are plotted in Figure 14 and considered in this study because they account for 99% of the variation (Table 5).

Variables	Function			
	1	2	3	4
Bioturbation index	−0.924	0.286	−0.112	0.130
Opaques	0.180	0.881	0.098	−0.252
Weight % TOC	0.108	0.444	−0.816	0.004
Average silt grain size	0.089	0.063	−0.213	0.663
% silt	−0.164	−0.131	0.202	0.367
% cement	−0.018	−0.277	−0.241	−0.303
% matrix	0.138	−0.011	0.060	0.272

TABLE 7. Means and standard deviations of raw data from all seven variables used in the DFA and MICP for each of the five microfacies of the Lewis Shale, Wyoming. The best seals are defined as those samples or microfacies with the highest MICP values.

Microfacies	MICP	TOC	Total Silt	Total Matrix	Opaques	Total Cement	Average Size (Phi)	Bioturbation Index
Mean MF 1	12,464.42	1.60	18.10	64.54	13.76	3.55	**4.25**	5.00
Standard deviation MF 1	7876.35	0.56	11.90	13.22	5.62	3.82	1.00	0.93
Mean MF 2	**18,735.52**	1.58	*7.85*	**71.35**	**16.53**	4.38	*4.76*	*0.00*
Standard deviation MF 2	2148.83	0.35	6.28	9.89	5.68	4.37	0.10	0.00
Mean MF 3	1249.04	0.41	20.34	69.56	*1.10*	**9.00**	4.56	1.39
Standard deviation MF 3	418.97	0.66	7.63	11.03	1.03	5.80	0.25	1.11
Mean MF 4	3123.23	1.09	14.31	69.39	12.98	*3.36*	4.68	2.5
Standard deviation MF 4	994.25	0.50	9.47	12.33	7.81	1.97	0.79	0.73
Mean MF 5	*1046.51*	*0.31*	**28.51**	*61.93*	1.48	8.12	4.60	**5.36**
Standard deviation MF 5	497.58	0.29	12.21	14.55	1.02	4.74	0.31	0.86

Bold numbers = highest mean value for all five facies; italic numbers = lowest mean value for all five facies; grain size data in Phi units; largest number = smallest grain size.

high silt content plot to the left, and samples with high percentages of opaques and matrix plot to the right along discriminant function 1 (Figure 14).

Organic laminated shales of microfacies 2 with the highest percentages of opaques, matrix, and TOC and the lowest percentage of silt grains, the lowest mean grain size, and a lack of bioturbation are the best seals (Table 7). Massive calcareous shales of microfacies 5 with the lowest percentages of TOC and matrix and the highest percentage of silt grains and the greatest degree of bioturbation are the poorest seals (Table 7). These two microfacies groups plot on opposite ends of the diagram in Figure 14. Calcareous laminated shales of microfacies 3 with the lowest percentage of opaques, the next to lowest TOC and matrix, and highest percentage of cement are also poor seals (Table 7). Organic bioturbated shales and massive organic mudstones of microfacies 1 and 4 are variable and are moderate to poor seals.

Shale Reflection Modeling Experiment

The presence of a low-velocity and low-density zone (microfacies 3) immediately above a high-velocity, high-density zone (microfacies 2) produces a very strong positive acoustic impedance contrast in the vicinity of the maximum flooding surface (Figure 15). This contrast in acoustic properties, at a shale-on-shale contact, will generate a strong seismic reflection. The arrangement of slow rocks above fast rocks may also generate an amplitude-vs.-offset anomaly similar to those generated by hydrocarbon-filled sandstones.

Seismic data has become the primary tool for exploration. It would be highly beneficial to possess a seismic-based tool for the predrill evaluation of seal capacity. This work has shown that several seismically important parameters display systematic differences and trends in the third-order transgressive and highstand systems tracts of the Lewis Shale (Figure 15). The ability to relate seismic interval velocity to important parameters such as porosity is important in the prediction of pore pressure from seismic data. In the Lewis Shale, the compressional velocity of the mudstone-rich portions of the highstand systems tract is moderately related to both total clay content and total carbonate content. It is poorly related to porosity. Within the transgressive systems tract, compressional velocity is unrelated to these compositional parameters but is still poorly related to porosity.

Simple seismic modeling of the third-order transgression and highstand represented by the Lewis Shale confirmed the speculation that the facies stacking pattern should produce a strong seismic reflection and significant amplitude-vs.-offset response. The basic modeling performed in this experiment used elastic rock properties and layer thicknesses (Figure 16A) observed in the Lewis Shale cores and outcrops in the study area. Shuey's (1985) approximation was used to compute angle-specific reflection coefficients for all interfaces and for angles ranging from 0 to 60°. The result was an angle gather of reflection coefficients. Each trace of this gather represents a specific reflection angle and was convolved with an angle-specific wavelet. These wavelets were derived from one extracted from data in a major exploration area and stretched or squeezed to

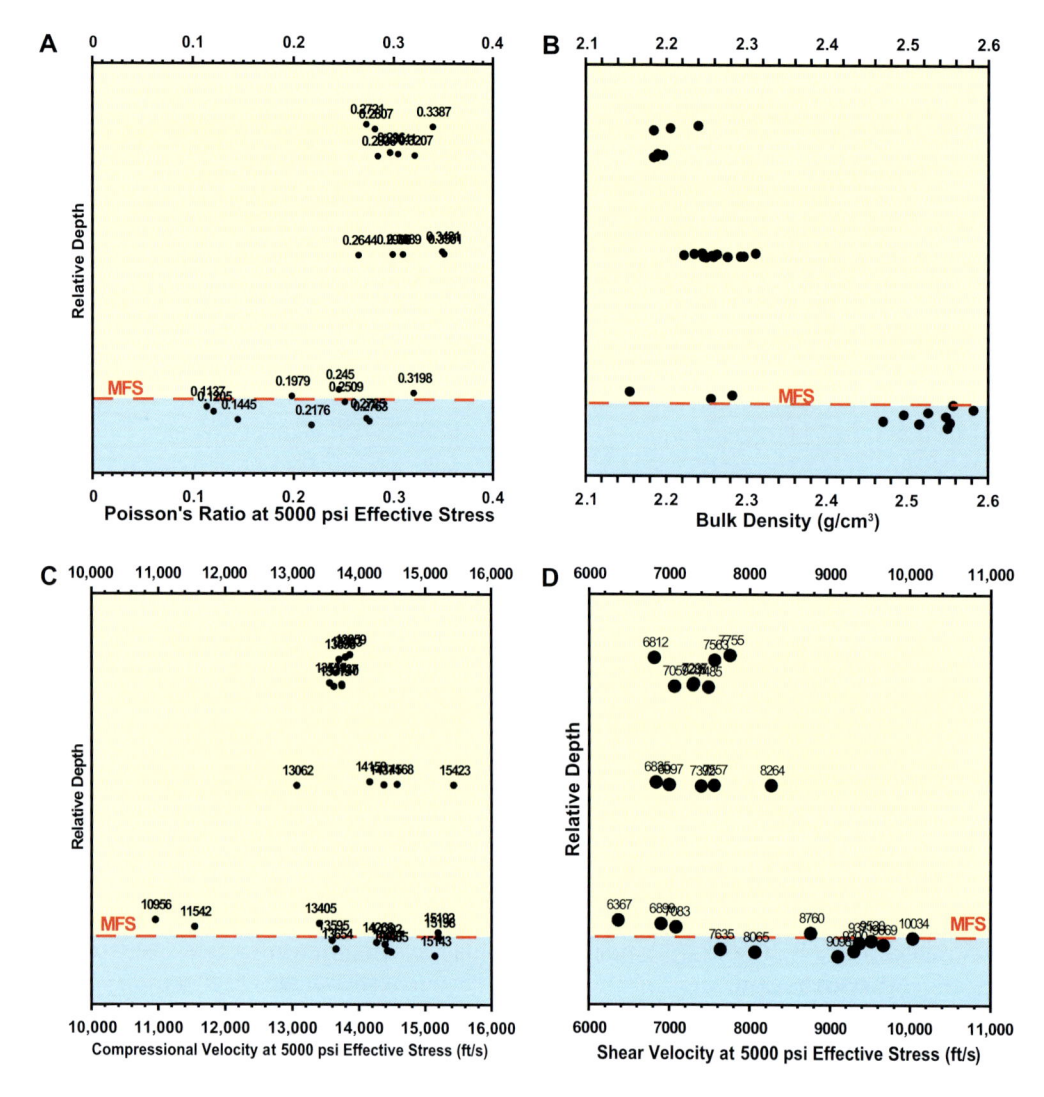

FIGURE 15. Several seismically important parameters vary in systematic ways among the various shale microfacies. (A) Poisson's ratio is generally less in the highstand systems tract (HST) shales than in the transgressive systems tract (TST) shales. (B) The TST shales are generally denser than the HST shales. In the HST, density shows a slight tendency to decrease upsection. (C) Highstand systems tract shales exhibit an overall increase in compressional velocity above the maximum flooding surface (MFS). The average compressional velocity of the TST shales is approximately equal to the maximum HST compressional velocity. (D) Highstand systems tract shales exhibit a slight general increase in shear velocity above the MFS. The average shear velocity of the TST shales significantly exceeds the maximum HST shear velocity.

simulate the effect of normal moveout. The result is a synthetic seismic angle gather that shows the amplitude-vs.-offset effect of the gather (Figure 16B).

To assess the effect of the thin, high-impedance layer (representing the condensed interval, microfacies 2), it was removed, and the calculations were repeated without it. The result shows that the effect on the seismic response is minimal (Figure 16C). If the deepest two layers, those from the transgressive systems tract (microfacies 1) and condensed interval (microfacies 2), are removed, the effect is more noticeable but still small (Figure 16D). This suggests that the properties in the low-impedance layer, representing microfacies 3, are key to the amplitude-vs.-offset response. The model response is very similar to responses seen on seismic response in some exploration areas, such as the example in Figure 17, which shows a shale horizon (strong seismic reflector) that could be misinterpreted as a hydrocarbon-saturated sandstone. Results from a well confirm the absence of sandstone and hydrocarbons.

CONCLUSIONS

- Lewis Shale strata consist of at least five argillaceous microfacies that exhibit distinctive sedimentological and petrophysical features along with significant variations in seal character.
- The uppermost transgressive and condensed shales (Lewis Shale microfacies 1 and 2) offer excellent to exceptional top seal potential. These shales occur preferentially in distal parts of marine depositional systems.
- The top seal capacity of highstand (Lewis Shale microfacies 3 and 5) and lowstand (Lewis Shale microfacies 4 and 5) intervals is reduced mainly because of elevated content (>25%) of detrital silt and disrupted fabrics (extensive bioturbation).
- Significant stratigraphic separation (several hundred feet) can exist between a lowstand sandstone reservoir and its controlling top seal horizon (i.e., overlying transgressive shale).

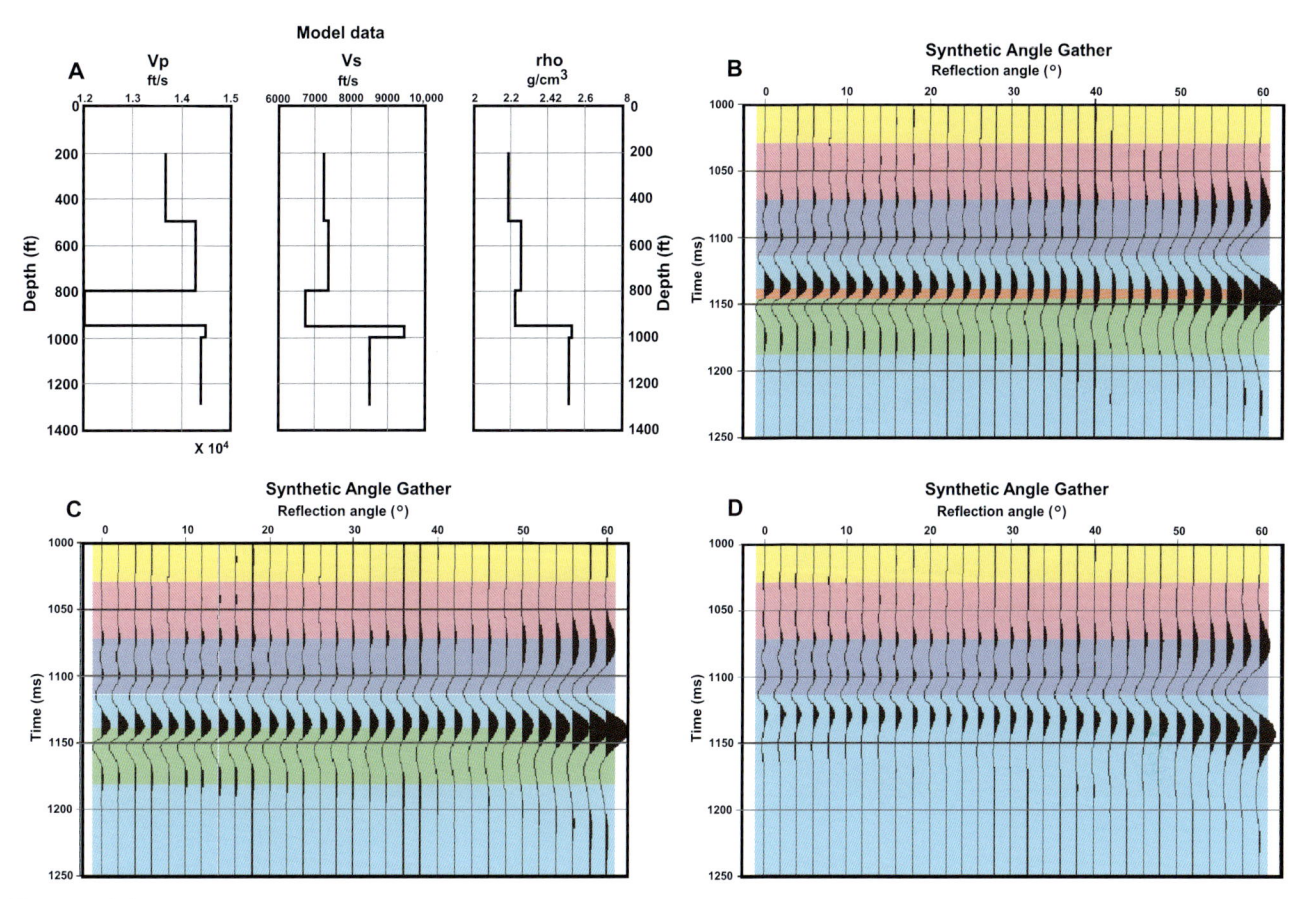

FIGURE 16. Input data (A) and results from a simple seismic model of the shale stacking pattern found in the Lewis Shale. (B) This model is a synthetic seismic angle gather that shows the amplitude-vs.-angle (AVA) effect when all shale facies layers are present. (C) The overall effect of removing the thin, high-impedance layer representing microfacies 2 is minimal. (D) Removing the thicker, high-impedance layer representing microfacies 1 reduces the AVA effect, but the difference is still small. This suggests that the properties in the low-impedance layer, representing microfacies 3, are key to the AVA response.

- Factors that tend to enhance sealing characteristics of marine shales include low content (<25%) of detrital silt; relatively slow rates of accumulation; low oxygen levels and limited bioturbation (preservation of laminar fabrics); and increasing content of Fe- and Mg-enriched minerals.
- Seismically significant parameters (e.g., density, shear velocity, Poisson's ratio, and compressional velocity) exhibit systematic variations that are consistent within the third-order sequence-stratigraphic framework of the Lewis Shale.
- Seismic modeling reveals a potential of some shales to exhibit an amplitude-vs.-offset response comparable to that exhibited by hydrocarbon-saturated sandstones.

ACKNOWLEDGMENTS

The authors thank ChevronTexaco for permission to present these data and interpretations. We are especially grateful to R. M. Slatt, D. R. Pyles, and S. M. Goolsby for sharing their knowledge concerning the Lewis Shale. C. Ward and A. Koenig made initial stratigraphic observations and collected outcrop samples. W. T. Lawrence prepared the thin sections. D. K. McCarty completed the XRD analyses, and B. J. Katz provided insight into the organic geochemical analyses. Poro-Technology (Houston, Texas) performed mercury injection capillary pressure analyses. The authors have benefited greatly from discussions concerning sequence stratigraphy with J. B. Sangree and L. M. Liro.

REFERENCES CITED

Abed, A. M., and R. Sadaqah, 1998, Role of Upper Cretaceous oyster bioherms in the deposition and accumulation of high-grade phosphorites in central Jordan: Journal Sedimentary Research, v. 68, p. 1009–1020.

Almon, W. R., Wm. C. Dawson, S. J. Sutton, F. G. Ethridge, and B. Castelblanco-Torres, 2002, Sequence stratigraphy, facies variation and petrophysical properties in deepwater shales, Upper Cretaceous Lewis Shale, south-central Wyoming: Gulf Coast Association of Geologic Societies Transactions, v. 52, p. 1041–1053.

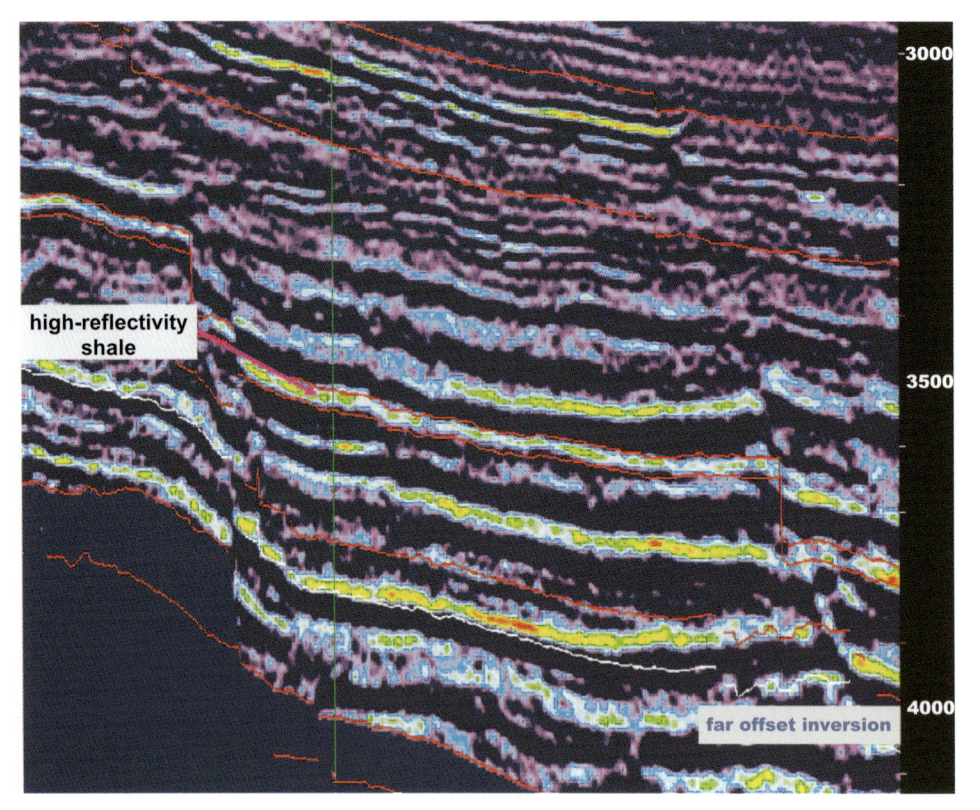

FIGURE 17. This example shows a shale horizon (strong seismic reflector) that could be misinterpreted as a hydrocarbon-saturated sandstone. Results from a well confirm the absence of sandstone and hydrocarbons.

Asquith, D. O., 1970, Depositional topography and major marine environments, Late Cretaceous, Wyoming: AAPG Bulletin, v. 54, p. 1184–1224.

Berg, R. R., 1975, Capillary pressures in stratigraphic traps: AAPG Bulletin, v. 59, p. 939–956.

Castelblanco-Torres, B., 2003, Distribution of sealing capacity within a sequence-stratigraphic framework: Upper Cretaceous Lewis Shale, south-central Wyoming: M.S. thesis, Colorado State University, Fort Collins, Colorado, 194 p.

Charvat, W. A., and R. C. Grayson, 1981, Anoxic sedimentation in the Eagle Ford Group (Upper Cretaceous) of central Texas: Gulf Coast Association of Geologic Societies Transactions, v. 31, p. 256.

Chung, F. H., 1974, Quantitative interpretation of x-ray diffraction patterns of mixtures: I. Matrix-flushing method for quantitative multicomponent analysis: Journal of Applied Crystallography, v. 7, p. 519–525.

Dawson, Wm. C., 2000, Shale microfacies: Eagle Ford Group (Cenomanian–Turonian) north-central Texas outcrops and subsurface equivalents: Gulf Coast Association of Geologic Societies Transactions, v. 50, p. 607–622.

Dawson, Wm. C., and W. R. Almon, 2002, Top seal potential of Tertiary deepwater Gulf of Mexico shales: Gulf Coast Association of Geologic Societies Transactions, v. 52, p. 167–176.

Demanison, G. J., and G. T. Moore, 1980, Anoxic environments and oil source rock genesis: AAPG Bulletin, v. 64, p. 1179–1209.

Gill, J. R., E. A. Merewether, and W. A. Cobban, 1970, Stratigraphy and nomenclature of some Upper Cretaceous and lower Tertiary rocks in south-central Wyoming: U.S. Geological Survey Professional Paper 667, 53 p.

Hale, L. A., 1961, Late Cretaceous (Montanan) stratigraphy eastern Washakie basin, Carbon County, Wyoming, in G. J. Wiloth, ed., Wyoming Geological Association Guidebook, Sixteenth Annual Field Conference, p. 129–137.

Hallam, A., 1987, Mesozoic marine organic-rich shales, in J. Brooks and A. J. Fleet, eds., Marine petroleum source rocks: Geological Society (London) Special Publication 26, p. 251–256.

Heckel, P. H., 1977, Origin of phosphatic black shale facies in Pennsylvanian cyclothems of mid-continent North America: AAPG Bulletin, v. 61, p. 1045–1068.

Ingram, G. M., and J. L. Urai, 1999, Top-seal leakage through faults and fractures: The role of mudrock properties, in A. C. Aplin, A. J. Fleet, and J. H. S. Macquaker, eds., Muds and mudstones: Physical and fluid-flow properties: Geological Society (London) Special Publication 158, p. 125–135.

International Center for Diffraction Data, 1993, Mineral powder diffraction file databook: ICDD, Swarthmore, Pennsylvania, unpaginated.

Jennings, J. J., 1987, Capillary pressure techniques: Application to exploration and development geology: AAPG Bulletin, v. 71, p. 1196–1209.

Leckie, D. A., C. Singh, F. Goodarzi, and J. H. Wall, 1990, Organic-rich, radioactive marine shales: A case study of a shallow water condensed section, Cretaceous Shaftesbury Formation, Alberta, Canada: Journal Sedimentary Petrology, v. 60, p. 101–117.

Lev, S. M., S. M. McLennan, W. J. Meyers, and G. N. Hanson, 1998, A petrographic approach for evaluating trace-element mobility in a black shale: Journal Sedimentary Petrology, v. 68, p. 970–980.

Loutit, T. S., J. Hardenbol, and P. R. Vail, 1988, Condensed sections: The key to age determination, an integrated approach: SEPM Special Publication 42, p. 188–213.

Love, J. D., and A. C. Christiansen, 1985, Geologic map of Wyoming: U.S. Geological Survey, scale 1:500,000, 1 sheet.

McMillen, K. J., and R. D. Winn, Jr., 1991, Seismic facies

of shelf slope and submarine fan environments of the Lewis Shale, Upper Cretaceous, Wyoming, *in* P. Weimer and L. G. Link, eds., Seismic facies and sedimentary processes of submarine fans and turbidite systems: New York, Springer-Verlag, p. 273–287.

O'Brian, N. R., and R. M. Slatt, 1990, Argillaceous rock atlas: New York, Springer-Verlag, 141 p.

Perman, R. C., 1990, Depositional history of the Maastrichtian Lewis Shale in south central Wyoming: Deltaic and interdeltaic, marginal marine through deep-water marine environments: AAPG Bulletin, v. 74, p. 1695–1717.

Pinous, O. V., M. A. Levchuk, and D. L. Sahagian, 2001, Regional synthesis of the productive Neocomian complex of West Siberia: Sequence-stratigraphic framework: AAPG Bulletin, v. 85, p. 1713–1730.

Pyles, D. R., 2000, A high frequency sequence-stratigraphic framework for the Lewis Shale and Fox Hills Sandstone, Great Divide and Washakie basins, Wyoming: Master's thesis, Colorado School of Mines, Golden, Colorado, 256 p.

Pyles, D. R., and R. M. Slatt, 2000a, The Upper Cretaceous Lewis Shale of south-central Wyoming: An analog to deep-water Gulf of Mexico turbidite reservoirs (abs.): AAPG Annual Meeting Program, v. 84, p. A121.

Pyles, D. R., and R. M. Slatt, 2000b, A high frequency sequence-stratigraphic framework for the Lewis Shale and Fox Hills Sandstone, Great Divide and Washakie basins, Wyoming: Central Gulf Coast SEPM Foundation 20th Annual Research Conference, December 3–6, 2000, p. 836–861.

Reynolds, M. W., 1976, Influence of recurrent Laramide structural growth on sedimentation and petroleum accumulation, Lost Soldier area, Wyoming: AAPG Bulletin, v. 60, p. 12–33.

Schieber, J., 1999, Distribution and deposition of mudstone facies in the upper Devonian Sonyea Group of New York: Journal of Sedimentary Research, v. 69, p. 909–925.

Schlanger, S. O., M. A. Arthur, H. C. Jenkyns, and P. A. Scholle, 1987, The Cenomanian–Turonian oceanic event: I. Stratigraphy and distribution of organic carbon-rich beds and the marine ^{13}C excursion, *in* J. Brooks and A. J. Fleet, eds., Marine petroleum source rocks: Geological Society (London) Special Publication 26, p. 221–241.

Schutter, S. R., 1998, Characteristics of shale deposition in relation to stratigraphic sequence systems tracts, *in* J. Schieber, W. Zimmerle, and P. S. Sethi, eds., Shales and mudstones—Basin studies, sedimentology and paleontology: Stuttgart, E. Schweizerbart'sche, p. 79–107.

Schowalter, T. T., 1979, Mechanics of secondary migration and entrapment: AAPG Bulletin, v. 63, p. 723–760.

Schowalter, T. T., 1981, Prediction of cap rock seal capacity (abs.): AAPG Bulletin, v. 65, p. 987–988.

Shuey, R. T., 1985, A simplification of the Zoeppritz equations: Geophysics, v. 50, p. 609–614.

Srodon, J., V. A. Drits, D. K. McCarty, J. C. C. Hsieh, and D. D. Eberl, 2000, Quantitative x-ray diffraction analysis of clay-bearing rocks from random preparations: Clays and Clay Minerals, v. 49, p. 514–528.

Stow, D. A. V., and A. Wetzel, 1990, Hemiturbidite: A new type of deep water sediment, *in* J. R. Concron and D. A. V. Stow, eds., Proceedings Ocean Drilling Program Scientific Results, v. 116, p. 25–34.

Stow, D. A. V., A. Y. Huc, and P. Bertrand, 2001, Depositional processes of black shales in deep water: Marine and Petroleum Geology, v. 18, p. 491–498.

Sutton, S. J., F. G. Ethridge, W. R. Almon, and Wm. C. Dawson, 2001, Variables controlling sealing capacity of Lower and Upper Cretaceous shales, Denver Basin, Colorado (abs.): AAPG Annual Meeting Program, v. 85, p. A6.

Weimer, R. J., 1960, Upper Cretaceous stratigraphy, Rocky Mountain area: AAPG Bulletin, v. 44, p. 1–20.

Wignall, P. B., and R. Newton, 2001, Black shales on the basin margin: A model based on examples from the Upper Jurassic of the Boulonnais, northern France: Sedimentary Geology, v. 144, p. 335–356.

Wilkin, R. T., and H. L. Barnes, 1997, Formation processes of framboidal pyrite: Geochimica et Cosmochimica Acta, v. 61, p. 323–339.

Winn, R. D., M. G. Bishop, and P. S. Gardner, 1985, Lewis Shale, south-central Wyoming: Shelf, delta front, and turbidite sedimentation: Wyoming Geological Association 36th Annual Field Conference Guidebook, p. 113–130.

Winn, R. D., M. G. Bishop, and P. S. Gardner, 1987, Shallow water and sub-storm-base deposition of Lewis Shale in Cretaceous Western Interior seaway, south-central Wyoming: AAPG Bulletin, v. 71, p. 859–881.

Witton-Barnes, B. M., N. F. Herley, and R. M. Slatt, 2000, Outcrop and subsurface criteria for differentiation of sheet and channel-fill strata; example from the Cretaceous Lewis Shale, Wyoming: Central Gulf Coast SEPM Foundation 20th Annual Research Conference, December 3–6, 2000, p. 1087–1104.

Using Gas Chimneys in Seal Integrity Analysis: A Discussion Based on Case Histories

Roar Heggland
Statoil ASA, Stavanger, Norway

ABSTRACT

Gas chimneys are visible in seismic data as columnar disturbances, where the continuity of reflectors is missing, and reflection amplitudes are weaker than in the surrounding areas. In this chapter, gas chimneys interpreted from three-dimensional seismic data, some of which have been confirmed by wells, have been sorted into two kinds. Type 1 chimneys are associated with faults. These chimneys commonly have a circular and limited horizontal cross section with a diameter in the order of 100 m (330 ft). The presence of gas chimneys along faults indicates that the faults are open or have been open for a time, in which case fluids can migrate through the faults. Type 2 chimneys are not associated with faults, and their lateral extent can be in the order of several hundred meters. Because open faults are not capillary barriers for hydrocarbons, as opposed to shales, type 1 chimneys can indicate hydrocarbon-migration pathways where relatively high flux rates can occur. Because type 2 chimneys are not associated with faults, capillary resistance in the shales will prevent upward movement of free gas (and oil), and the chimneys are regarded to represent gas having a very slow or no upward movement (low to zero flux rate). However, fractures beyond seismic resolution may exist, which may account for gas migration through the shales. Another explanation for the presence of gas type 2 chimneys is that gas-saturated water may release gas during upward movement caused by a drop in the pressure.

Examples from the North Sea, Gulf of Mexico, Nigeria, and the Caspian Sea show a consistency in the appearance and distribution of types 1 and 2 chimneys above hydrocarbon-charged reservoirs, as well as above dry reservoirs. Type 1 chimneys have also been observed below hydrocarbon-charged reservoirs, in which case, they indicate migration pathways into the reservoir.

Copyright ©2005 by The American Association of Petroleum Geologists.
DOI:10.1306/1060767H23170

INTRODUCTION

To find out if gas chimneys can be used to identify leaking faults and for the seal integrity analysis of cap rocks, gas chimneys have been mapped in different areas by the use of exploration three-dimensional (3-D) seismic data. It has been found useful to separate the observed chimneys into two types and, as such, help to distinguish between high-risk and low-risk hydrocarbon prospects.

- type 1, which can identify faults that are likely to be hydrocarbon-migration pathways
- type 2, which are associated with cap rocks of hydrocarbon-charged structures

METHODS

The diffuse character and common weak expression of gas chimneys in seismic data make them difficult to map, and in most cases, they are best visible in vertical seismic sections but not very visible on 3-D seismic time slices or attribute maps. To improve the identification of gas chimneys in seismic data and to map their extents and distribution in a consistent manner, a method for detection of gas chimneys in poststack 3-D seismic data was developed (Heggland et al., 2000; Meldahl et al., 2001).

The method makes use of multiattribute calculations and a neural network (de Groot, 1999a, b). The inputs are standard 3-D seismic data and other seismic attributes. To distinguish the chimneys from background, attributes that best increase the contrast between the chimneys and the background are selected as input. The different attributes that are input to the neural network are weighted according to their contribution to the enhancement of the chimneys. The attributes that give the highest contributions to the detection of gas chimneys are trace-to-trace similarity, energy (or absolute amplitude), which both generally are lower in chimneys than in the areas surrounding them, and the variance of the dip of seismic reflectors, which is higher inside chimneys than outside because of the chaotic reflection pattern in chimneys. The neural network is trained on the attributes extracted at chimney and nonchimney example locations, which are chosen by the interpreter based on chimney interpretation experience. After training, the network is applied to the entire data set to recognize chimneys from the background. In the chimney detection process, multiple vertical attribute extraction windows are used. This enables the network to distinguish between gas chimneys and objects with similar seismic characteristics but having a smaller vertical extent than the gas chimneys. In the final stage, the neural network makes a classification of the seismic data into chimney and nonchimney samples. The output samples are assigned a high value for chimneys (high probability) and a low value for nonchimneys (low probability). The resulting 3-D probability volume is called a chimney cube.

The method has since been generalized for the detection of other seismic objects, like faults and diapirs, in which case, the detection is steered along the orientation of the actual seismic object (Tingdahl et al., 2001).

RESULTS

For hydrocarbons to move through a shale, an open fault or a fracture system has to be present. This can only occur in an extensional regime as a result of overpressure in a reservoir, and the fault or fracture will be open for a time until the pressure has dropped (Bjørkum et al., 1998).

Water, as well as gas-saturated water, is not prevented by capillary forces to move through the shales. When water moves through a fault caused by overpressure in an underlying reservoir, some water is believed to move horizontally into the shale for a limited distance, i.e., in the order of 10–100 m (33–330 ft). If the water is gas saturated, gas may be released when the pressure drops. In the seismic data, this may appear to be a gas chimney. Alternatively, if gas is migrating through a fault, some of the gas may occupy fractures existing along the fault and generate what is observed in the seismic data to be a gas chimney. It is believed that if a fault is or has been open for a water flux, it will also be open for hydrocarbons (free gas or oil) to move through and, as such, represent the most likely migration pathway for hydrocarbons. Gas chimneys observed from the seismic data, which are found to be associated with faults, are here named type 1 chimneys.

During the upward movement of water, gas may be released when the pressure drops. In this case, chimneys can have a large lateral extent of several hundred meters. The gas chimneys are, in this case, believed to be free gas, which is trapped in the shales. Gas chimneys that are observed in the seismic data to have large lateral extents and are not found to be associated with faults are here named type 2 chimneys. The detection of gas chimneys on different 3-D seismic data has revealed chimneys of types 1 and 2. The detected chimneys have been displayed together with attribute maps generated from the 3-D seismic data to relate the chimneys to geological structures.

Type 1 Chimneys

Several 3-D seismic examples show the presence of chimneys along faults and fractures. This is believed to indicate that the faults are or have been working as fluid-migration pathways. Depending on the location of the faults, the presence of chimneys along the faults can indicate fluid-migration pathways between a source rock and a reservoir, or they can indicate leakage from a reservoir, either through a fault across the top of the structure or through a fault located at the flank of the structure (Figure 1a, b). If the leaking fault is located at the top of a structure, only small amounts of hydrocarbons are expected to remain in the reservoir. If the leakage is occurring through a fault at the flank of a structure, a hydrocarbon column may still be preserved in the reservoir. Figure 2 shows an example from the Danish North Sea, where gas chimneys are surrounding a deeper oil-charged reservoir (Heggland, 1998). In this example, conventional exploration 3-D seismic data show buried depressions (pockmarks) on top of the gas chimneys just below the seabed. The depressions are indications that gas has escaped through the seabed. A high-resolution, deep tow sparker profile across one of the depressions shows high reflectivity in the water column above the crater (Figure 3). This indicates that escape of gas through the seabed is still occurring. Some chimneys are grouped as lineaments, indicating that they are appearing from faults present at the flank of the deeper reservoir.

An example from the Gulf of Mexico is presented (Figure 4), in which two reservoirs contain only small amounts of hydrocarbons (shows). This is assumed to be caused by leaking faults across the top of the structure. Figure 4a shows a 3-D visualization with three mapped horizons from a 3-D seismic volume, the sea-

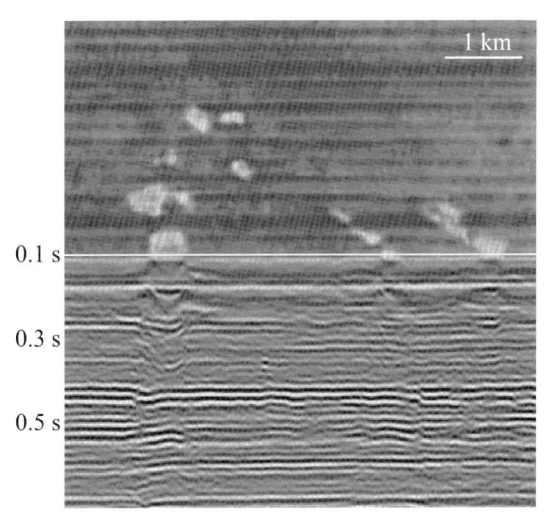

FIGURE 2. Composite display of a time slice at seabed level and a vertical section from 3-D seismic data from the Danish North Sea. A cluster of gas chimneys is surrounding a deeper, oil-charged structure. Buried depressions are present on top of the chimneys. Some of the chimneys form lineaments, indicating that the chimneys appear from faults at the flank of the deeper structural closure (cf. Figure 1b).

bed (brown), and two subseabed horizons (green and blue). Seismic high-amplitude anomalies (red), clustered at two levels between the lowest horizon and the salt diapir at the base, represent the outline of two prospects. Detected chimneys (yellow) indicate gas migration between the two prospects and the seabed. An average amplitude map centered at the green surface (Figure 4b) shows that the chimneys are located along faults across the crest of the structure, indicating a nonsealing fault. The faults are visible as low-amplitude features in black. A well through the two

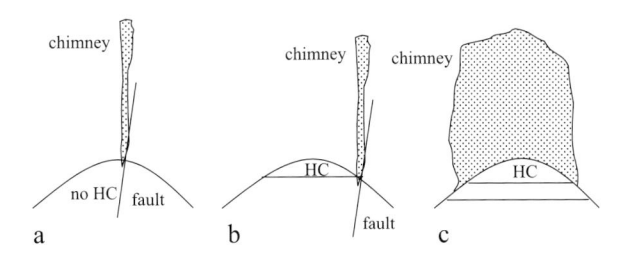

FIGURE 1. Illustrations of chimney types. (a) Type 1 chimney on top of a structural closure. The chimney is associated with a fault, which involves a risk that hydrocarbons (HC) have left the trap. (b) Type 1 chimney on the flank of a structural closure. In this case, a column of hydrocarbons may be preserved. (c) Type 2 chimney covers a large area on top of a structural closure. This may be an indication that hydrocarbons are present in the underlying structure.

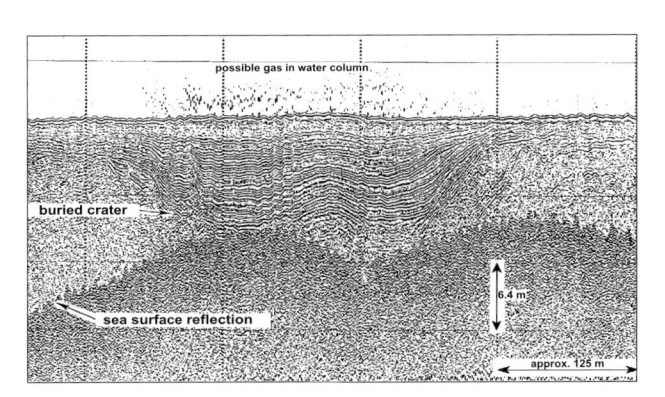

FIGURE 3. High-resolution sparker section through one of the buried depressions in Figure 2. High reflectivity in the water column across the buried depression indicates gas seepage.

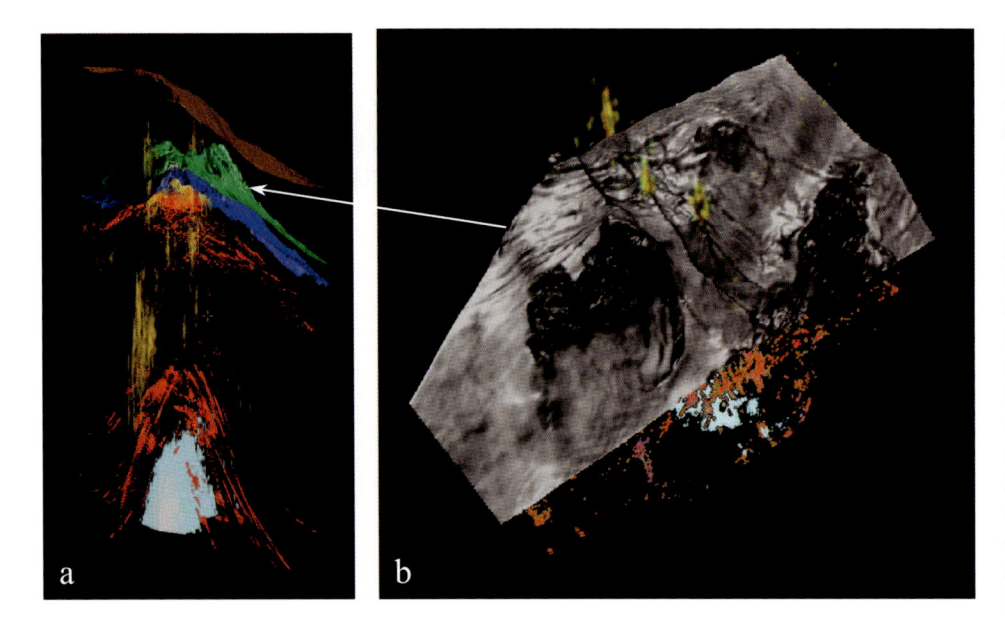

FIGURE 4. (a) High-amplitude clusters (in red) indicate the outline of two prospects in the Gulf of Mexico. Three mapped surfaces are displayed, the seabed (brown) and subseabed surfaces (green and blue). A salt diapir has been detected at the base using a modified version of the chimney-detection method. (b) Average absolute amplitude map centered at the green surface and detected chimneys. Faults are visible as low-amplitude features in black. The chimneys are located along faults across the top of the structure. This involves a risk that hydrocarbons have left the underlying structures. A well drilled through the two prospects encountered only small amounts of hydrocarbons (shows).

prospects encountered only small amounts of hydrocarbons (shows). The latter observation of chimneys indicating nonsealing faults may explain why only small amounts of hydrocarbons were present in these reservoirs.

Figure 5a shows a seabed azimuth map generated from 3-D seismic data from offshore Nigeria. Figure 5b shows the same map as in Figure 5a with detected gas chimneys (white) superimposed. Type 1 gas chimneys show a continuous presence along an east–west-extended

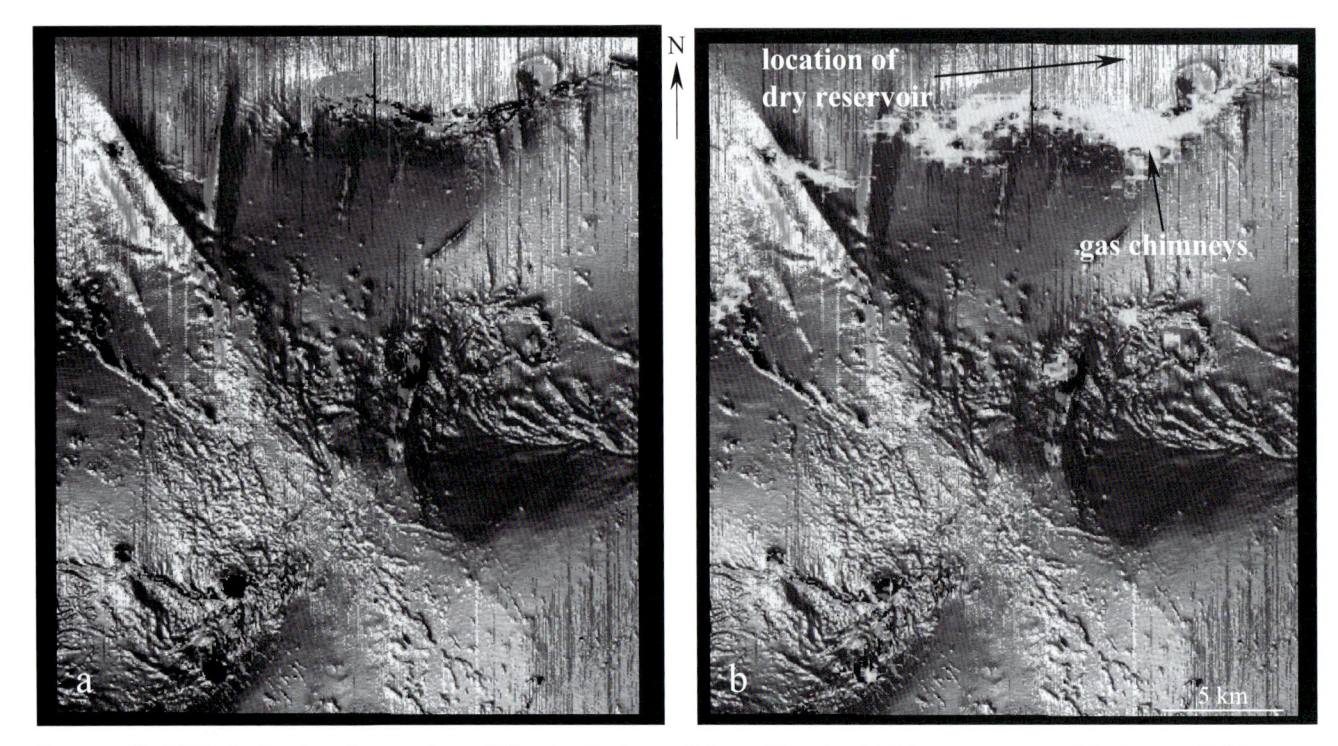

FIGURE 5. (a) Seabed azimuth map from 3-D seismic data offshore Nigeria. (b) The same map with detected chimneys (white) superimposed. Type 1 chimneys are associated with a fault. A prospect on the north side of this fault was originally believed to be sealed by the fault. A well was drilled and found to be dry. The latter observation of chimneys along this fault indicates that it was nonsealing.

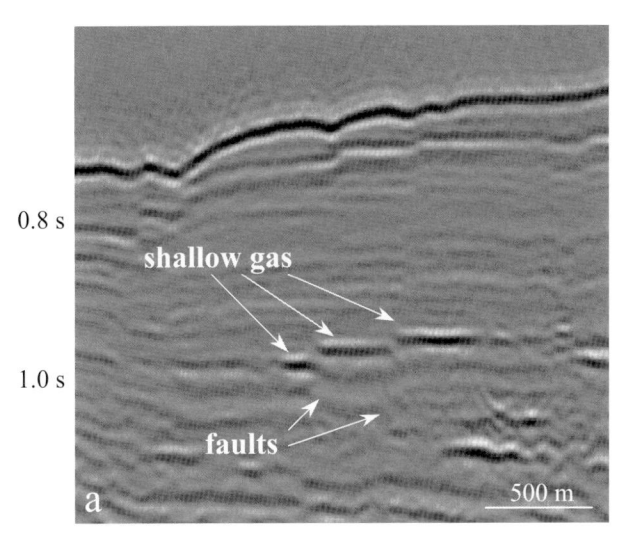

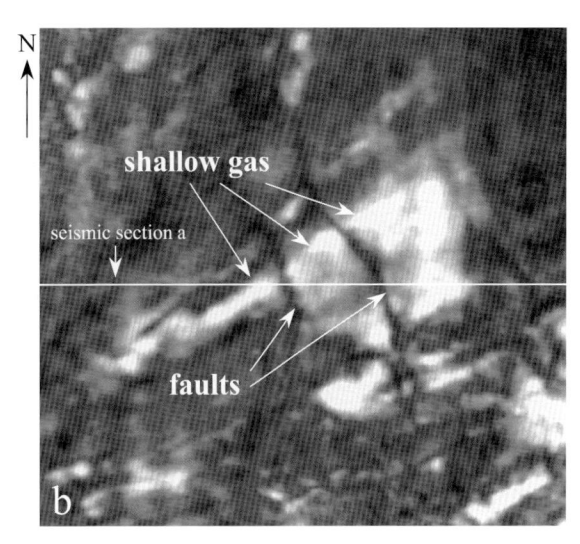

FIGURE 6. (a) This seismic section from 3-D data offshore Nigeria shows three high-amplitude reflectors. (b) An average absolute amplitude map over a time interval, including the three high-amplitude reflectors. These high-amplitude anomalies (white) are interpreted to represent a gas-charged sand. The sand is segmented by faults visible as low-amplitude features in black. Gas is believed to have migrated through the faults, charging the sand and migrating to the seabed. The presence of pockmarks along the faults at the seabed supports this interpretation (cf. Figure 7).

fault. A prospect on the north side is connected to the fault, which was originally believed to be sealing, and a well drilled into the prospect was found to be dry. The later discovery of gas chimneys along the fault indicates that the fault is nonsealing, which may explain the lack of hydrocarbons in the prospect.

In some cases, other indications suggest that leakage is (or has been) occurring through faults. High-amplitude anomalies along faults can indicate reservoirs that have been charged by gas migrating up the faults. Figure 6 shows an example from offshore Nigeria, where high-amplitude reflectors are interpreted to represent a sand that is segmented by faults and possibly charged with gas that has migrated up the faults. Pockmarks at the seabed (Figure 7) are located along the faults, indicating that gas has migrated to the seabed. Core samples from the seabed have confirmed the presence of gas at fault locations, whereas core samples taken at a distance from the faults showed no gas contents.

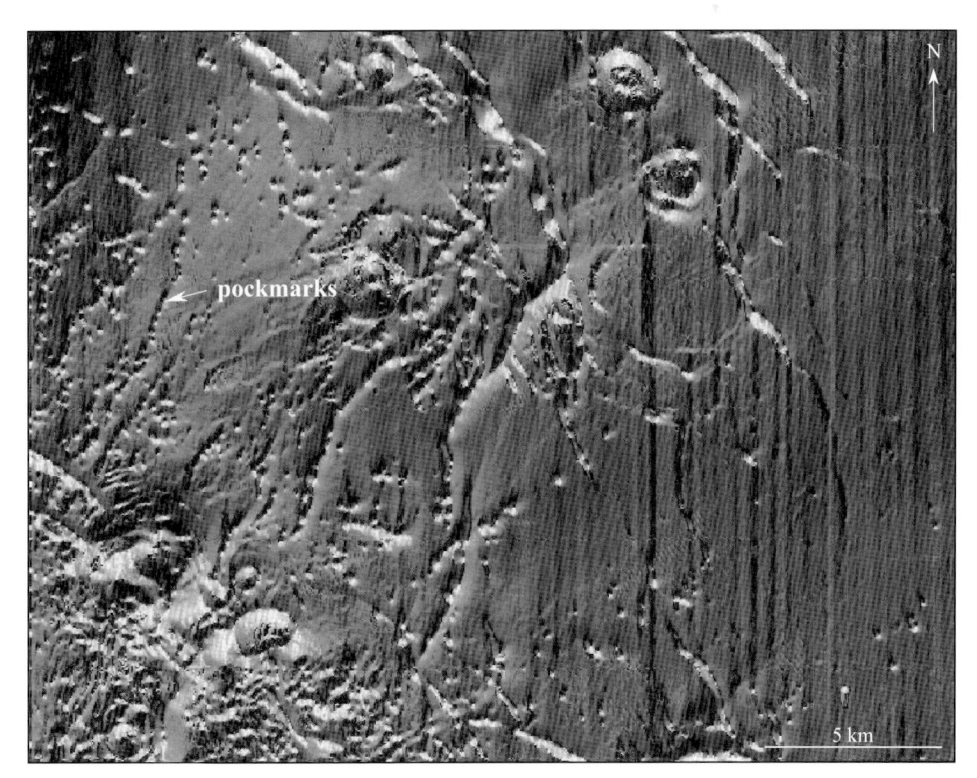

FIGURE 7. A seabed azimuth map generated from the same 3-D data as in Figure 6 shows pockmarks along faults reaching the seabed, indicating gas escape through the faults. Seabed samples taken at fault locations have showed contents of gas. Core samples taken at a distance from the faults showed no gas contents.

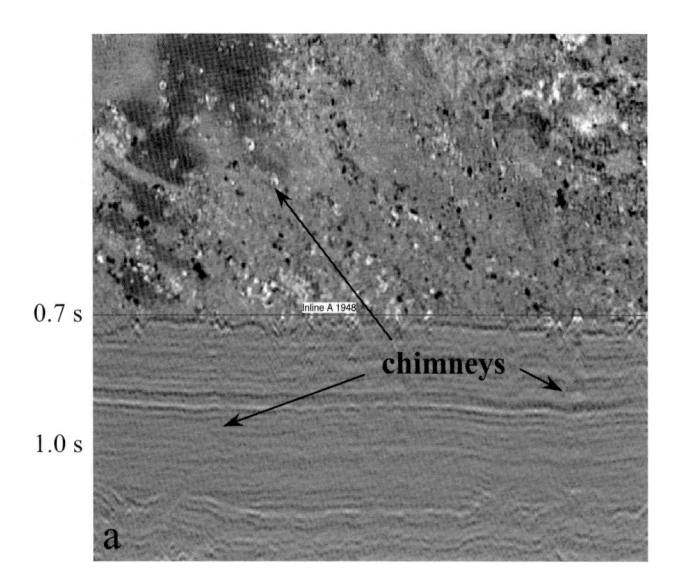

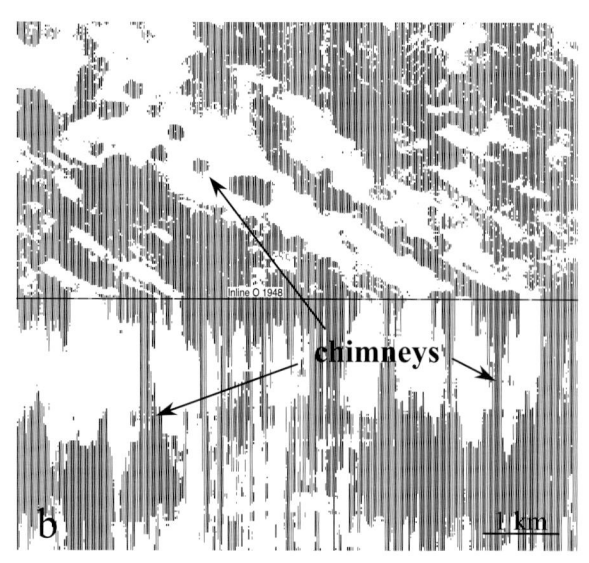

FIGURE 8. (a) Composite display of a time slice and a vertical section from Norwegian North Sea. The time slice is taken through a level of possible carbonate buildups in the late Pliocene. (b) Chimney cube version of the composite display in (a) that shows possible gas chimneys in the vertical section as well as in the time slice. The chimneys are located below the possible carbonate buildups along detected faults forming lineaments like the ones formed by the buildups. Based on this observation, carbonate buildups are believed to have been formed at gas seepage locations in the late Pliocene.

High-amplitude anomalies on an average absolute amplitude map generated from 3-D data from the Norwegian North Sea show structures in the late Pliocene that are believed to be carbonate buildups formed on top of gas seeps from faults in the late Pliocene. The structures form a pattern of straight parallel lines, indicating the presence of underlying faults. Gas chimney detection has revealed chimneys below the build-up structures (see also Heggland, 1997) and faults. The standard 3-D seismic data (Figure 8a) do not show chimneys in time slices. In the chimney cube (Figure 8b), however, chimneys are visible in time slices as circular features, and they are distributed along faults picked up by the chimney detection.

Type 1 chimneys are believed to indicate faults that are or have been open for fluid migration. Flux rates may be large enough to be a risk with regard to remaining hydrocarbons in a reservoir or could charge a reservoir from beneath with hydrocarbons. Type 1 chimneys can represent a risk or can be a positive indication with regard to the presence of hydrocarbons in a structure, depending on the location of the faults relative to the structure.

Type 2 Chimneys

Gas chimneys are also found to be present in areas where seismic data show no faults or fractures. In such a case, the chimneys occupy a much larger space (i.e., several hundred meters lateral extent) than when chimneys are associated with a fault. In some cases, such chimneys are present on top of hydrocarbon-charged reservoirs, such as in Figure 1c. An example from the North Sea (Figure 9) shows a chimney present between the top of an Eocene oil and gas reservoir and the seabed. Figure 9a shows a standard 3-D seismic section. Figure 9b shows the corresponding section after chimney detection. Figure 10a shows a time map of the top of the Eocene structure. In Figure 10b, a time slice from the chimney cube at 1-s two-way time is displayed, and it can be seen that the gas chimney almost images the outline of the structural closure. The two wells displayed on the maps encountered columns of gas and oil in the Eocene reservoir.

An example from the Caspian Sea is presented in Figure 11. In this case, type 2 chimneys are present on top of two hydrocarbon-charged reservoirs. The blue surface is the top reservoir time map. Mud volcanoes are present at the flanks of the reservoirs. The mud trajectories below the mud volcanoes have been picked up by the chimney detection, because they have similar seismic characteristics as chimneys.

Type 2 chimneys are believed to be gas that either has come out of solution from upward-moving water and got trapped in the shales (zero flux rate), or gas migrating with a relatively slow flux rate through small faults or fractures beyond seismic resolution. The amount of gas transported through the top seal is regarded to be small, in which case, type 2 chimneys are not regarded to represent a risk with respect to remaining hydrocarbons in a reservoir, but rather an indication that hydrocarbons are present in the underlying

FIGURE 9. A 3-D seismic section from the Norwegian North Sea. (a) Gas chimney present above an oil and gas reservoir in the top Eocene. The base of the actual gas chimney is believed to be at the top of the reservoir, and a shadow zone is believed to exist below it as seen in the seismic section. (b) Corresponding section from the chimney cube (detected chimneys).

structure. Type 2 chimneys have, in many cases, been observed to be present on top of oil- and gas-charged reservoirs.

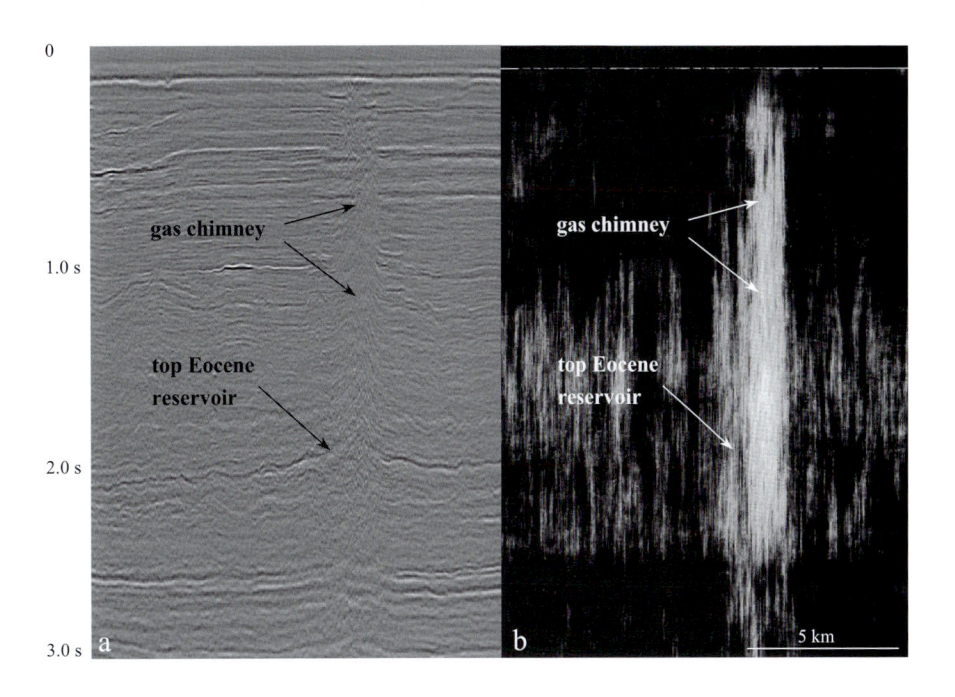

CONCLUSIONS

To evaluate sealing vs. nonsealing faults, the presence of chimneys along faults may indicate which faults are likely to be hydrocarbon-migration pathways. Cases also exist where chimneys are not present, but other seismic anomalies can indicate current or previous fluid migration, like pockmarks, high-amplitude reflectors, and buildup structures. These structural features are observed to exist at the present-day seabed and at paleoseabeds. A cap rock can be evaluated in a similar manner. If a fault is cutting across (or near) the top of a potential hydrocarbon trap, and if gas chimneys or other features indicating fluid migration are present along the fault at the structural high, the prospect can be regarded as a high risk with respect to presence of hydrocarbons.

Chimneys having a large lateral extent and that cannot be related to faults are believed to represent the absence of or very slow flux rates. If such a chimney is present on top of a prospect, it can rather be an

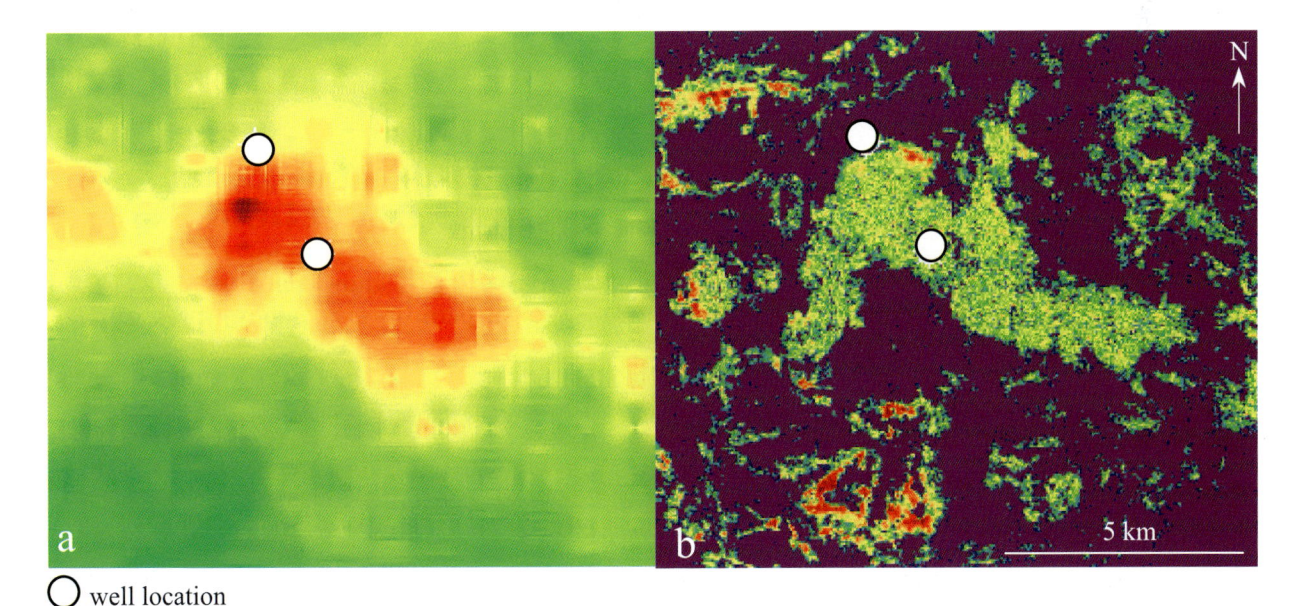

○ well location

FIGURE 10. (a) Time map at the top of the Eocene reservoir. (b) Time slice at 1-s two-way time through the chimney cube, showing that the gas chimney has a similar lateral extent and shape as the underlying oil and gas reservoir. The two wells displayed on the maps encountered columns of gas and oil in the Eocene reservoir.

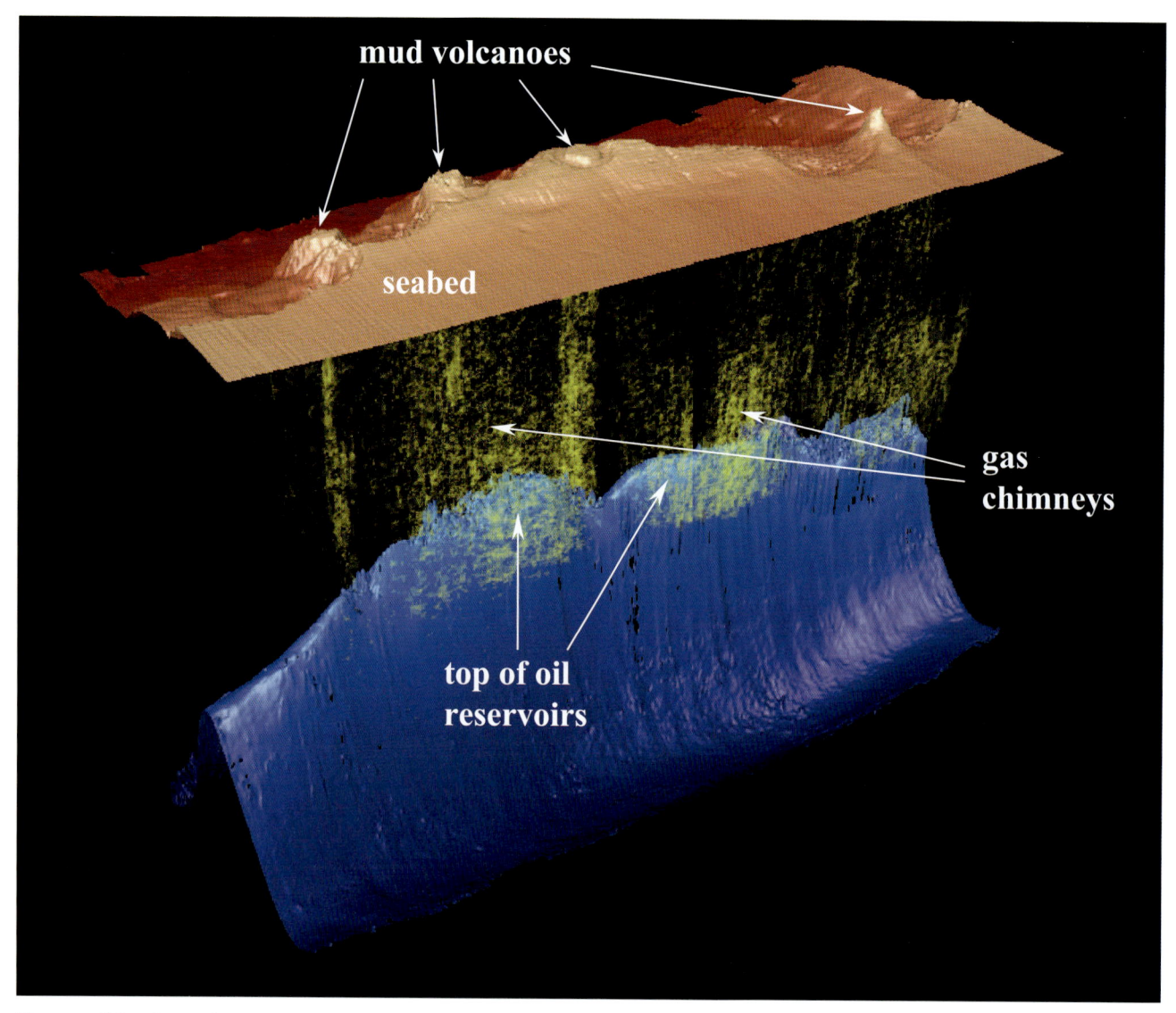

FIGURE 11. Three-dimensional visualization from the Caspian Sea. Type 2 chimneys are present on top of two oil reservoirs (blue surface). The seabed map (brown) shows the presence of mud volcanoes. The mud transport pathways have been picked up in the chimney-detection process.

indication of the presence of hydrocarbons in the underlying structure.

The chimney detection has revealed chimneys lining up not only with faults but also with features associated with gas seepage (e.g., pockmarks, carbonate buildups, and mud volcanoes), shallow gas accumulations, and deeper hydrocarbon accumulations and, as such, have revealed potential hydrocarbon-migration pathways. The chimneys indicate fluid migration between hydrocarbon source rock and reservoirs, between reservoirs (remigration), and between reservoirs and the seabed. As such, detection of gas chimneys in seismic data, as well as mapping of other features that can indicate fluid-migration pathways, have significance in fault seal and top seal integrity analysis and in the process of risking hydrocarbon prospects.

ACKNOWLEDGMENTS

Statoil ASA is acknowledged for the use of their data and the permission to publish this chapter.

REFERENCES CITED

Bjørkum, P. A., O. Walderhaug, and P. Nadeau, 1998, Physical constraints on hydrocarbon leakage and trapping revisited: Petroleum Geoscience, v. 4, p. 237–239.

de Groot, P. F. M., 1999a, Seismic reservoir characterisation using artificial neural networks: Proceedings of the 19th Mintrop Seminar, May 16–18, 1999, Munster, Germany, p. 71–85.

de Groot, P. F. M., 1999b, Volume transformation by way of neural network mapping: Proceedings of the 61st European Association of Geoscientists and Engineers Conference, June 7–11, 1999, Helsinki, p. 3–7.

Heggland, R., 1997, Detection of gas migration from a deep source by the use of exploration 3-D seismic data: Marine Geology, v. 137, p. 41–47.

Heggland, R., 1998, Gas seepage as an indicator of deeper prospective reservoirs. A study based on exploration 3-D seismic data: Marine and Petroleum Geology, v. 15, p. 1–9.

Heggland, R., P. Meldahl, P. de Groot, and F. Aminzadeh, 2000, Chimneys in the Gulf of Mexico: The American Oil and Gas Reporter, February 2000, v. 43, p. 78–83.

Meldahl, P., R. Heggland, A. H. Bril, and P. F. M. de Groot, 2001, Identifying targets like faults and chimneys using multi-attributes and neural networks: The Leading Edge, May 2001, v. 20, p. 474–482.

Tingdahl, K. M., P. de Groot, R. Heggland, and H. Ligtenberg, 2001, Semiautomated object detection in 3-D seismic data: Offshore Magazine, August 2001, v. 61, p. 124–128.

Formation Fluids in Faulted Aquifers: Examples from the Foothills of Western Canada and the North West Shelf of Australia

J. R. Underschultz
Commonwealth Scientific and Industrial Research Organization Petroleum, Bentley, Western Australia, Australia

C. J. Otto
Commonwealth Scientific and Industrial Research Organization Petroleum, Bentley, Western Australia, Australia

R. Bartlett
Hydro-Fax Resources Ltd., Calgary, Alberta, Canada

ABSTRACT

Faults and fault zones commonly represent key geological factors in determining migration fairways and assessing the retention and leakage history for hydrocarbons in the subsurface. Although formation pressure data are sparsely acquired from within fault zones themselves, hydrodynamic analysis of faulted aquifers can be used as an indirect indicator of the fault zone hydraulic properties. Case studies from the foothills of Western Canada and the North West Shelf of Australia are used to define a workflow for hydrodynamic analysis in faulted strata and to identify the manifestation of fault zone hydraulic properties on adjacent aquifer pressure systems for various tectonic settings.

Faults with significant displacement can form hydraulic barriers. In this case, fluid flow in the aquifer next to the fault is predominantly parallel to the structural grain, and a discontinuity occurs in the potentiometric surface for the aquifer being crosscut. Localized hydraulic communication (leakage), either across a fault in an aquifer or vertically along a fault zone between aquifers, tends to occur (1) where the fault zone bends out of plane from the dominant stress field; (2) where the main structural grain is crosscut by steeply dipping high-angle faults; or (3) where deformation is transferred from one fault zone to another through a relay zone or transfer fault. These are manifest by chemical or thermal anomalies and potentiometric highs or lows closed against the fault trace. Although conditions of fault zone conductivity tend to be localized, they can limit the trapping potential of structural closures by allowing the leakage and further migration of hydrocarbons.

INTRODUCTION

Faults can both be barriers and conduits to flow, and because faults form key risk factors in the capture and retention of hydrocarbons, understanding their effects on mass transport processes in sedimentary basins is essential at the geological timescale. On the reservoir production timescale, faults can compartmentalize a reservoir or act as a thief to injected fluids. This chapter examines flow systems in faulted strata under a range of tectonic settings, with the aim to identify a linkage between structural style, the hydraulic behavior of faults themselves, and more generally, the impact of faults on formation water flow systems. Understanding the fault systems that make up part of the plumbing of a sedimentary basin, in turn, aids the characterization of hydrocarbon migration and trapping.

To examine the hydrodynamic behavior of faults to fluid flow, hydrodynamic analyses are presented at various scales from two basins that represent different tectonic settings. The first case study is located on the transition between undisturbed Mississippian carbonates of the west central Alberta Basin in Canada and the outboard equivalent strata in the adjacent thrust-fold belt of the Rocky Mountains. This region has extensive data to constrain the stratigraphic and structural geometry of the rock framework, as well as formation pressure and rock property data needed to define the formation water flow system. The overall tectonic setting is that of a shortened continental margin, developed through transpressional collision of the North American continent with various tectonic elements of the Pacific plate. Mississippian strata host extensive gas accumulations in the thrust-fold belt and gas and oil reserves in the adjacent foreland basin.

The second case study is a reservoir-scale evaluation located on the North West Shelf of Australia, in a contrasting tectonic setting to that of Western Canada. The North West Shelf represents a Late Devonian to Early Carboniferous rifted passive margin, with thermal subsidence to the Late Triassic. Late Jurassic to Early Cretaceous saw regional transpression and middle Miocene to Holocene reactivation of faults, because of convergence and subduction of the northern edge of the Australian continental plate margin at the Timor trough (O'Brien et al., 1993). The North West Shelf of Australia is an active exploration region with extensive geological and hydrodynamic data. The complex tectonic history with fault reactivation make fault seal issues an important regional exploration risk.

METHODOLOGY

Pressure data are obtained from drillstem (DST), wireline (WLT), and production or static gradient tests. Initially, only data from zones of formation water are used to characterize the hydraulic head distribution, but these are supplemented by preproduction hydrocarbon pressure data extrapolated to known free-water-level elevations. Data are located in a well-defined structural stratigraphic framework.

Standard hydrodynamic approaches to characterizing flow systems in unfaulted aquifer systems include the analysis of pressure data, both in vertical profile (e.g., pressure-elevation plot), and within the plane of the aquifer after conversion to hydraulic head. Pressure data are supplemented with formation-water analysis and formation-temperature data to aid in the evaluation of the flow system. Bachu (1995), Dahlberg (1995), Otto et al. (2001), and Bachu and Michael (2002) provide an overview of hydrodynamic analysis techniques. Evaluation techniques for the culling and analysis of formation water samples are described by Hitchon and Brulotte (1994) and Underschultz et al. (2002). Techniques for the evaluation of formation temperature are described by Bachu and Burwash (1991) and Bachu et al. (1995).

The hydraulic properties of a fault should be considered separately from the impact that fault may have on the aquifer it crosscuts. A fault zone may hydraulically separate the aquifer on either side yet put one side in hydraulic communication vertically with a separate stratigraphic level. Therefore, both juxtaposition and fault zone rock properties need to be considered. Because pressure data from a fault zone itself is not typically available, inferences about the hydraulic nature of the fault need to be made by evaluating the pressure data in the aquifer near the fault. Otto et al. (2001)

describe some theoretical patterns of hydraulic head in faulted aquifers and pressure gradients on pressure-elevation plots for faults with various hydraulic properties.

Although flow systems are three dimensional in nature, they are commonly visualized in more simplistic ways first and then built into a three-dimensional model. In the plane of the aquifer, the hydraulic head can be contoured to show the fluid potentials for formation water (Bachu, 1995; Bachu and Michael, 2002), provided that no significant density variations are present. For faulted aquifers in the studies described here, the hydraulic head distribution is first characterized in unfaulted blocks of the aquifer. Then, the hydraulic head distributions in adjacent blocks are compared and built as a patchwork into a flow model that is representative of the faulted strata as a whole.

Upfault (or Downfault) Flow

If a fault is acting as a conduit, with higher permeability along the fault than the aquifer it crosscuts, and if the fault zone permeability pathway is vertically continuous to either a separate aquifer or the land surface (or seabed), then the aquifer will see the fault as either a source or a sink for fluids. The hydraulic head distribution in the aquifer will form either a closed high or low against the fault surface, indicating that formation water is either emanating from the fault zone into the aquifer or flowing from the aquifer into the fault zone, respectively (Figure 1a). This analysis can be supplemented by evidence such as thermal springs at the land surface or thermal and chemical anomalies of the formation water in the aquifer adjacent to the fault. If, for example, a fault zone is acting as a conduit recharging an aquifer at depth, it is expected that the formation water in the aquifer would be relatively fresh and have a meteoric ionic signature, such as elevated HCO_3^- and/or SO_4^{2-}. The likelihood of an aquifer being in hydraulic communication with the land surface can be deduced by comparing the hydraulic head in the aquifer with the elevation of the water table (commonly similar to the topographic elevation). If the hydraulic head in the aquifer is similar to topographic elevation, it is possible that the water table elevation is in hydraulic communication with the aquifer.

At any one location, the vertical hydraulic communication can be examined by pressure-elevation plot analysis. If two vertically separated aquifers are in hydraulic communication via a fault zone conduit, the pressure data from the two aquifers near the fault should define a near-common hydrostatic gradient on a pressure-elevation plot (Figure 1b). In this schematic example, the hydraulic head in the deeper aquifer is slightly higher than the shallower aquifer, which al-

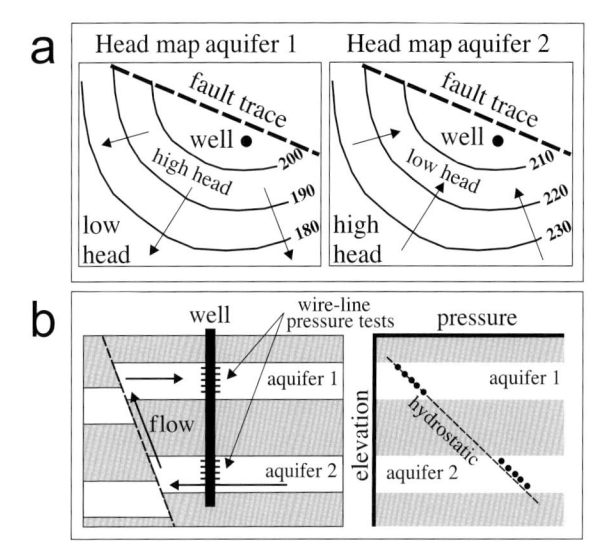

FIGURE 1. Schematic fault hydrodynamics (modified from Otto et al., 2001). (a) Hydraulic head distributions in two stacked aquifers connected by a crosscutting conduit fault; (b) cross section of the conduit fault in (a) with upfault flow, and the corresponding pressure-elevation plot of WLT data from the two aquifers at the well location.

lows for the flow up the fault zone. For this reason, the WLT pressure data in aquifer 2 fall slightly above the pressure gradient defined by the data in aquifer 1 on the pressure-elevation plot (Figure 1b).

Faults as Flow Barriers

When a fault zone has lower permeability than the aquifer it crosscuts, the flow direction in the aquifer adjacent to the fault will tend to be parallel to the plane of the fault (Figure 2a). This nature of the flow leads to a relatively abrupt contrast in hydraulic head values in the aquifer directly across the fault (a hydraulic head discontinuity). The more significant the fault is as a barrier, either because of juxtaposition or low fault rock permeability, the more severe is the hydraulic head discontinuity. In this schematic example (Figure 2a), there is a 70 m drop in hydraulic head across the fault zone. If the fault zone could be examined at a small scale, the discontinuity in hydraulic head would actually be a very steep hydraulic head gradient along the plane of the fault, thus leading to some flow across the fault plane, but the flux would be negligible compared to that moving parallel to the fault plane in the adjacent aquifer. Corroborating evidence for fault zone barriers are the accumulations of hydrocarbon on one side of the fault, and discontinuities in the formation water chemistry across the fault.

At any one location, the vertical hydraulic communication can be examined by pressure-elevation

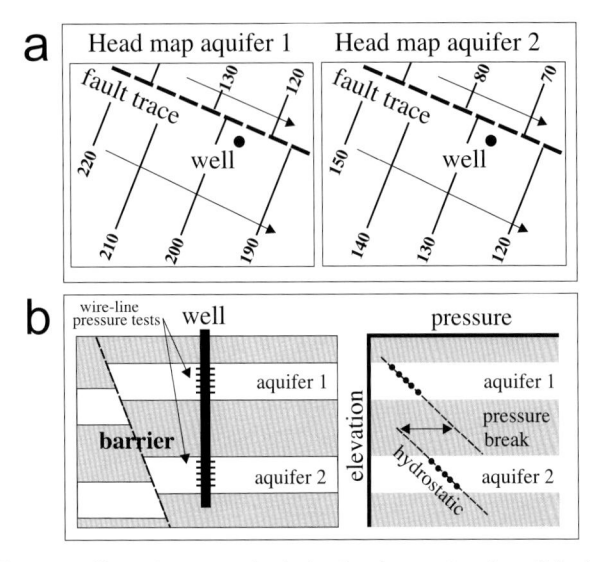

FIGURE 2. Schematic fault hydrodynamics (modified from Otto et al., 2001). (a) Hydraulic head distributions in two stacked aquifers crosscut by a barrier fault; and (b) cross section of the barrier fault in (a) with no upfault flow and the corresponding pressure-elevation plot of WLT data from the two aquifers at the well location.

plot analysis for a well near to, or crosscut by a fault (Figure 2b). Zones of low hydraulic transmissivity manifest themselves as a break in the observed pressure gradient with depth (Dahlberg, 1995).

WESTERN CANADIAN FOLD AND THRUST BELT

The western edge of the Alberta Basin is deformed by thrusting and folding in the foothills and main ranges. Further westward, allochthonous rocks make up the accreted terranes that collided with the North American craton between the Jurassic and Cretaceous. The geological history of the thrust-fold belt has been the subject of extensive investigation and is described by various authors (e.g., Price, 1994, 2001; Struik and MacIntyre, 2001; Begin and Spratt, 2002).

One of the first hydrodynamic evaluations of the foothills and main ranges of the Rocky Mountains is by Wilkinson (1995) in the Burnt Timber area. He observed abrupt changes in both hydraulic head and formation-water salinity across the main thrust faults, suggesting that they were acting as barriers to flow. Underschultz and Bartlett (1999) examined the Mississippian strata at Moose Mountain, Wildcat Hills, and Burnt Timber fields and noticed that a component of flow moved parallel to the structural grain in individual thrust sheets. Grasby and Hutcheon (2001) observed the distribution and chemistry of thermal springs in the southern Canadian Cordillera. They found springs with temperatures as much as 67°C and circulation depths of as much as about 3 km (1.8 mi).

Study Area and Structural Style

For the purpose of examining the regional characteristics of the flow system, the hydraulic head distributions for the Mississippian aquifer (Livingstone strata) have been selected (Figure 3). The area with the best data control extends from Township (Tp) 30 in the south to Tp 46 in the north, representing a distance of about 160 km (99 mi) (Figure 4). The structural trend is from southeast to northwest, so the study area was selected to be about 120 km (74 mi) wide and straddling the boundary between thrusted and undisturbed strata. At any location in the study area, moving from undisturbed strata westward, minor thrusts are encountered

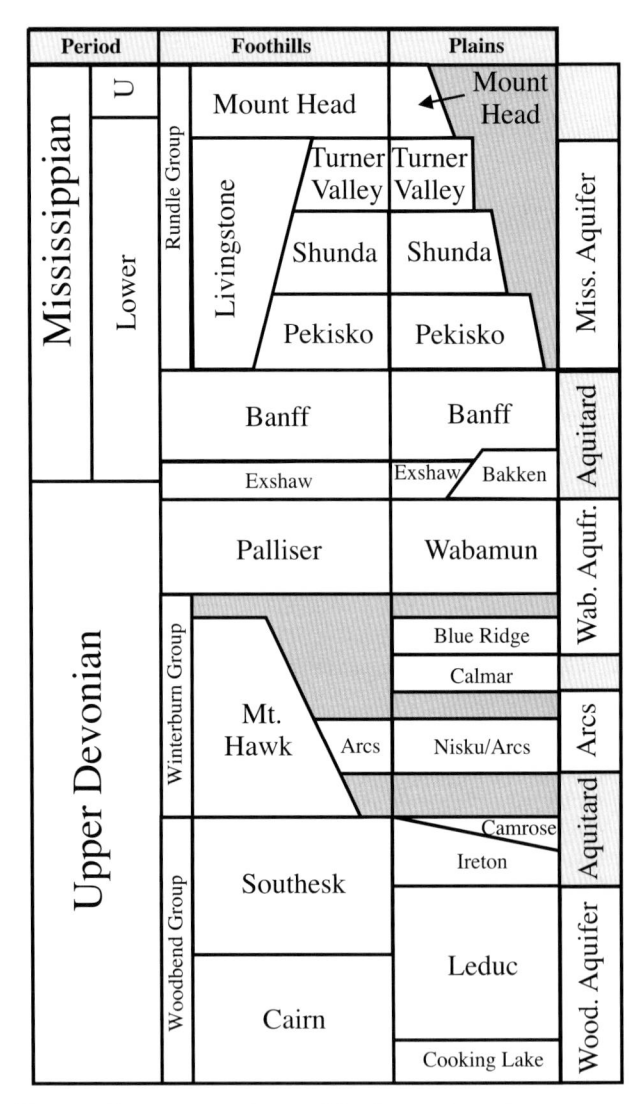

FIGURE 3. Stratigraphic and hydrostratigraphic nomenclature for the west central Alberta Basin and foothills.

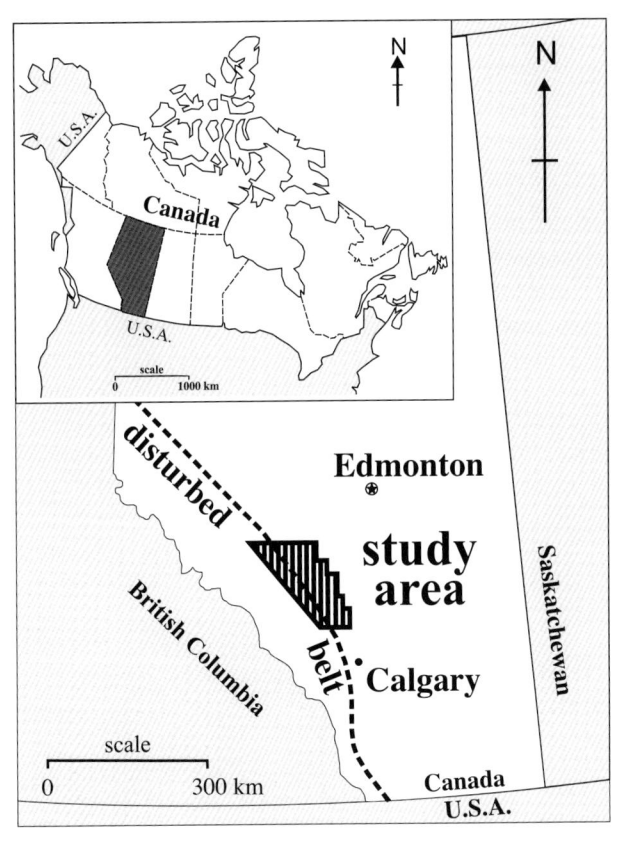

FIGURE 4. Study area for the west central Alberta Basin.

with small displacement and limited geographic extent. Typically, the maximum displacement along an individual fault is greatest in the center, decreasing to zero displacement at the ends. These minor thrust faults commonly form doubly plunging anticlines, making them excellent hydrocarbon traps. When taken in series, as displacement is lost along one thrust, it is taken up on the next subparallel thrust either to the east or west. This style of deformation is characteristic for the foothills section of the thrust-fold belt in the study area. Further west, more significant thrusts occur, such as the Bighorn and Brazeau sheets. These faults are continuous over much larger distances and also represent more significant displacement. Finally, the McConnell thrust to the west is the first regional thrust sheet, and it defines the start of the Front Ranges of the Rocky Mountains. It represents a continuous structure from one end of the study area to the other.

Hydraulic head was calculated for pressure data based on a constant density correction for formation-water salinity of 70 g/L, which is average for Paleozoic formation water in this region. More than 4000 pressure tests (DSTs, WLTs, and production and static gradient tests) and nearly 4400 formation water analyses in the Paleozoic strata of the thrust-fold belt and adjacent undeformed strata of Western Canada have been evaluated. Both the thrusted and the undeformed parts of the Alberta Basin are characterized to observe the relation between the two. Care has been taken to ensure that data used to constrain the flow model is unaffected by production, and that standard data quality assurance has been observed (Otto et al., 2001).

Mississippian Aquifer

The regional distribution of hydraulic head for Mississippian strata is shown in Figure 5. At some locations, as much as three repeat sections of Mississippian strata are present, each with its own pressure data. For display purposes, the hydraulic head distribution for the aquifer in the upper sheet (Brazeau and Bighorn) is shown in green color shade with black contours, and the aquifer repeated below is shown in blue color shade with blue contours. At North Limestone (34-11W5), a window is cut out of the green color shade of the Brazeau flow system to show the aquifer system below the Brazeau sheet. Some generalities observed in the foreland thrust belt of the Alberta Basin common to all the Paleozoic aquifers are as follows. (1) Within the same aquifer, hydraulic head generally increases westward and to structurally higher thrust sheets; (2) faults with large displacement tend to correspond to a discontinuity of hydraulic head for the aquifer displaced across the fault; and (3) in faulted aquifers, the hydraulic gradient in an unfaulted block tends to drive flow parallel to the structural grain (i.e., along strike).

In the case of the study area described here, the amount of data is much less in the thrusted part of the aquifer than in the undisturbed part of the aquifer to the east. With the additional constraint of initially contouring unfaulted blocks separately from one another, very few control points may be present in a particular contouring domain. The result is that the contour distribution is very simplistic for most parts of the thrusted strata in any individual block. With more data control, it is expected that there would be much more detail and complexity to the distribution. Nonetheless, because significant differences in hydraulic head commonly exist between one thrust sheet and the next, even a rudimentary flow model can provide significant predictive capability for the pressures and flow directions expected for individual thrust sheets.

In the undisturbed part of the Mississippian aquifer, the hydraulic head ranges from 500 to 800 m (1600 to 2600 ft). Two significant troughs of low hydraulic head extend southwestward into the foothills region. These are separated by a region (ridge) of high hydraulic head. The first extends westward to the North Limestone gas pool. In this region, a series of small thrusts form the Limestone, North Limestone, and Clearwater gas fields. The Brazeau thrust brings Paleozoic (Mississippian and Devonian) strata in a separate thrust sheet over the top of the Mississippian aquifer containing the Limestone and Clearwater pools. The

252 Underschultz et al.

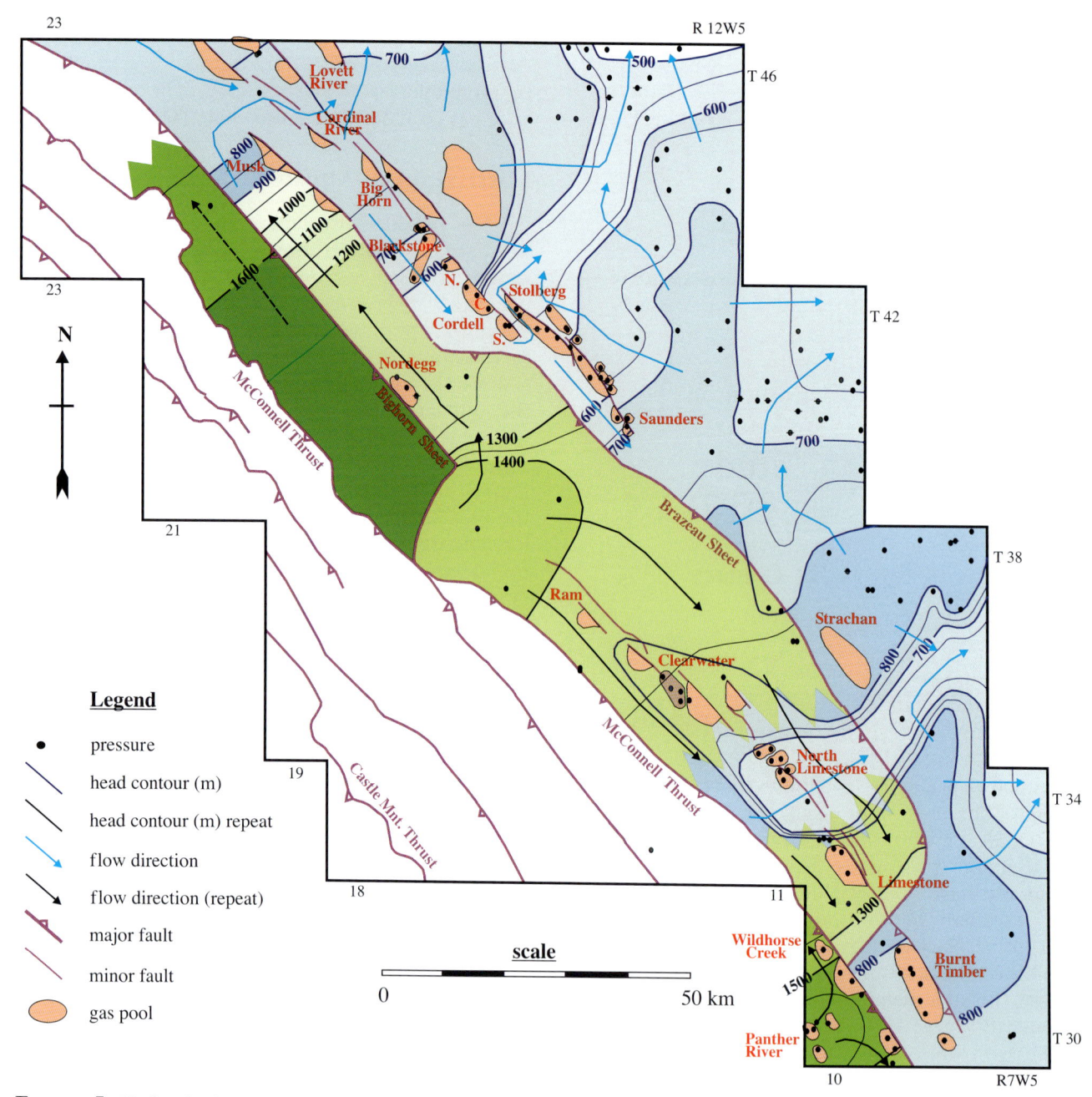

FIGURE 5. Hydraulic head (m) distribution for the Mississippian aquifer.

trough of low hydraulic head in the Mississippian aquifer below the Brazeau thrust turns parallel to the structural grain and extends northwest to the Clearwater field. The hydraulic head in this trough is between 700 and 800 m (2300 and 2600 ft), whereas the Mississippian aquifer in the Brazeau sheet above has a hydraulic head of 1300 m (4300 ft). The second trough of low hydraulic head extends southwestward to the foothills in the region of the Stolburg field. This region of low hydraulic head extends past the first small thrusts and westward at least to the edge of the Brazeau sheet. Each of these troughs of low hydraulic head channels formation-water flow like a collector system in the Mississippian aquifer, and each enters the thrust-fold belt at a location where a continuous Mississippian aquifer between small faults is present.

In the foothills and Front Ranges of the Rocky Mountains, the flow system was characterized by first examining the hydraulic head distribution in each thrust sheet separately and then combining these into a composite. For example, the Saunders to Stolburg region has northwest-directed flow between the first foothills thrusts and the leading edge of the Brazeau thrust. This flow proceeds into the trough of low hydraulic head described previously that enters the foothills belt at the north end of the Stolburg bounding fault. From Blackstone to South Cordell, a southeast-directed flow system is present between the series of first foothills faults (that bound the Blackstone and Cordell pools) and the leading edge of the Brazeau thrust. This system decreases from 700 to 600 m (2300 to 1900 ft) of hydraulic head before joining into the Stolburg flow system around the south end of the South Cordell thrust.

Flow in the southern part of the Brazeau thrust sheet is southeastward, parallel to the trend of the structure. The hydraulic head in the Mississippian aquifer of the Brazeau sheet is 1400 m (4600 ft) in the vicinity of 39-16W5. A decrease is observed in head southeastward to about 1300 m (4300 ft) in the limestone area. From the 39-16W5 region, a decrease is also observed in hydraulic head parallel to strike northwestward, until displacement on the Brazeau thrust is lost (in the region of the Musk field). Here, the Brazeau thrust flow system comes into hydraulic equilibrium with the small discontinuous thrusts that make up the first structures on the western edge of the undisturbed strata. From Musk, the flow turns east and southeastward toward the Blackstone field described previously.

Further into the thrust-fold belt, less data exist; however, the hydraulic head in the Bighorn thrust is more than 1600 m (5200 ft) (a 200-m [660-ft] jump from the hydraulic head in the adjacent Brazeau thrust). Similarly, the hydraulic head in the Panther River region at the southern end of the study area is about 1600 m (5200 ft).

Because formation water analyses are extremely subject to contamination (Hitchon and Brulotte, 1994; Underschultz et al., 2002), the number of usable salinity data points is small. For the northern part of the undisturbed Mississippian aquifer, the formation water has salinity of as much as 70 g/L (Figure 6), but with low salinity (less than 30 g/L) at the north end of Stolberg and extending northeastward into the plains to about 45-13w5, exactly the same location as the trough of low hydraulic head described earlier. Toward the southern edge of the study area, the salinity increases to 120 g/L in the undisturbed strata. Insufficient data is observed in the Mississippian aquifer below the Brazeau sheet to contour. Based on very little data, the entire foothills region east of the Bighorn thrust sheet (northern half of the study area) appears relatively fresh (less than 30 g/L). Within the Brazeau sheet, the salinity of Mississippian aquifer increases toward the southeast to about 70 g/L (Figure 6).

Flow and Faults in the Foothills and Front Ranges

For the Mississippian aquifer in the foothills of Western Canada, it is observed that when crosscut by faults with significant displacement, an abrupt change (discontinuity) generally occurs in the hydraulic head between one side of the fault and the other, resulting in a flow parallel to structural strike. As the fault loses displacement along its length, the hydraulic head in the aquifer on either side becomes similar, and at the end of the fault where displacement becomes zero, the hydraulic head in the aquifer equilibrates, and hydraulic communication is reestablished in the aquifer. However, faults with significant displacement are not exclusively sealing to the aquifers they crosscut.

Grasby and Hutcheon (2001) demonstrated, by examining thermal springs in the southern Canadian Cordillera, that gravity-driven flow systems occur in the front and main ranges of the Rocky Mountains down to more than 3 km (1.8 mi) depth. They also showed that in the overall compressive tectonic regime, at locations of complex structure, at locations where near-vertical orthogonal structures crosscut the main structural trend, and at locations of steeply dipping lateral ramps, vertical hydraulic communication between the deep subsurface and the ground level was greatly enhanced. Because of variable fault geometry, opportunities exist for localized extensional stress in an overall compressional setting. A similar scenario is described by Craw (2000) in the compressional tectonic setting of the Southern Alps in New Zealand. He observed enhanced formation water flow with meteoric signatures at the intersection of steeply dipping extensional structures and the main fault zones.

Although discharging thermal springs indicate highly permeable pathways from depth along fault zones, the flux tends to be highly focused. Heterogeneous fault zone properties are observed worldwide. For example, in the Mesozoic- and Tertiary-age strata of the Pechelbronn–Soultz subbasin, France, Otto (1992), Toth and Otto (1993), and Otto and Yassir (1997) observed vertically ascending flow of groundwater concentrated largely through the fault zones. These constitute the main pathways for cross-formational fluid flow. However, because of the discontinuous sealing nature of the faults and fault zones, rising fluids may be deflected and forced into highly permeable sections (e.g., lenses) of strata juxtaposed onto the fault plane.

The thermal springs observed by Grasby and Hutcheon (2001) represent a discharge phenomena over an exceedingly small part of the geographic area of the foothills and main ranges, occurring as focused fluid flow enabled by particular geological conditions. Similar structural scenarios of locally extensional settings could equally be located in regions of recharge, where formation water at high topographic elevation is placed in hydraulic communication with a deep aquifer. In this case, it is expected that the formation water in the deep aquifer would have high hydraulic head (approaching that of the surface elevation) and fresh salinity, because the residency time after recharge would understanding the flow and transport mechanisms in faulted aquifers, because much of the flux is directed parallel to strike, whereas interpretations of various processes are commonly interpreted and published on dip-oriented cross sections. These invariably indicate no hydraulic connection along the line of section between one sheet and the next; however, eventual communication may well be present out of the plane of the section.

NORTH WEST SHELF OF AUSTRALIA

The North West Shelf of Australia can be broadly characterized as a rift margin with thermal sag. Late-stage convergence resulted in the reactivation of some structures and basin inversion. This has proved to be an exploration challenge, particularly in the Timor Sea region, because the main period of hydrocarbon generation and trap charge occurred prior to reactivation of the structures. The late-stage reactivation has resulted in some previously filled traps leaking some, or all, of their hydrocarbons. Predicting which structures have leaked and which are likely to have retained their hydrocarbons has proven to be difficult and is the subject of significant efforts in fault seal research. It has been recognized that fault intersections where the main structural grain is crosscut at a high angle by deep-seated transfer faults are a high risk for leakage and seal breach (Cowley and O'Brien, 2000; Gartrell et al., 2002). A review of some published studies and a hydrodynamic interpretation of the Challis oil field in the Vulcan subbasin is presented here to illustrate how fault zone hydraulic properties can be deduced from aquifer hydrodynamics. Figure 9 shows a location map of Australia's North West Shelf with the main basin outlines and the Australian coastline.

Commonwealth Scientific and Industrial Research Organization Petroleum conducted the first regional hydrodynamic evaluation of the North West Shelf strata from 1998 to 2001. With the support of an industry consortium, a database was developed of more than 5000 pressure tests and 800 water chemistry analyses. Otto et al. (2001) define the regional hydrostratigraphy across the entire North West Shelf and provide a regional flow system characterization that includes pressure, temperature, and water-salinity data. Two subbasin-scale studies have been completed that examine the influence of faults on regional flow systems, one in the Vulcan subbasin (Underschultz et al., 2002) and one in the Barrow and Dampier subbasins (Underschultz et al., 2003).

Underschultz et al. (2002) show flow systems in the Plover strata of the Vulcan subbasin (Figure 9) to be similarly affected by faulting as those previously

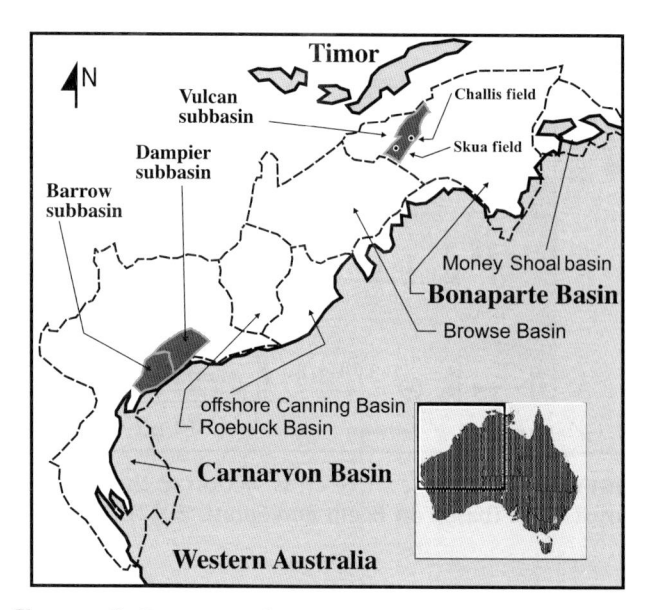

FIGURE 9. Basins on the North West Shelf of Australia modified from Geoscience Australia Map (modified after Kennard, 2004).

described for the Mississippian strata of the Alberta Basin. Outboard faults that form continuous zones of displacement separate flow systems, which is evidenced by discontinuities both in hydraulic head and formation-water salinity across the faults. In these regions, formation water migration is parallel to the structural grain. Nearer to shore, the faults are less continuous. Here, tortuous flow paths are defined, which feed regions of low hydraulic head. These drain the aquifer to the basin margins and eventual discharge.

A hydraulic head distribution for the Barrow strata (Underschultz et al., 2003) was updated from Hennig et al. (2002) for the Barrow and Dampier subbasins (Figure 9). Underschultz et al. (2003) evaluated the likelihood of oil migration from source rocks in the central part of the basin to the south and east across the basin-edge Flinders fault zone and onto the adjacent margin. The basin-edge fault pattern has a change in character north and south of 21°S latitude. To the south, the faults have a general orientation of southwest–northeast; they are closely spaced; and they are interconnected. North of 21°S latitude, the faults change to a north–south orientation (at high angle to regional stress); they become widely spaced; and few connecting structures between the main fault zones are present. Underschultz et al. (2003) conclude that north of 21°S latitude, the faults act as barriers, except for specific locations where transfer zones shift displacement from one fault to the next. Conversely, south of 21°S latitude, the complex fault orientations and interconnectedness give opportunity for the faults to contain conductive pathways.

Some field-scale examples are present where faults locally provide zones of vertical hydraulic communication. Gartrell et al. (2002) show how the intersection

Formation Fluids in Faulted Aquifers

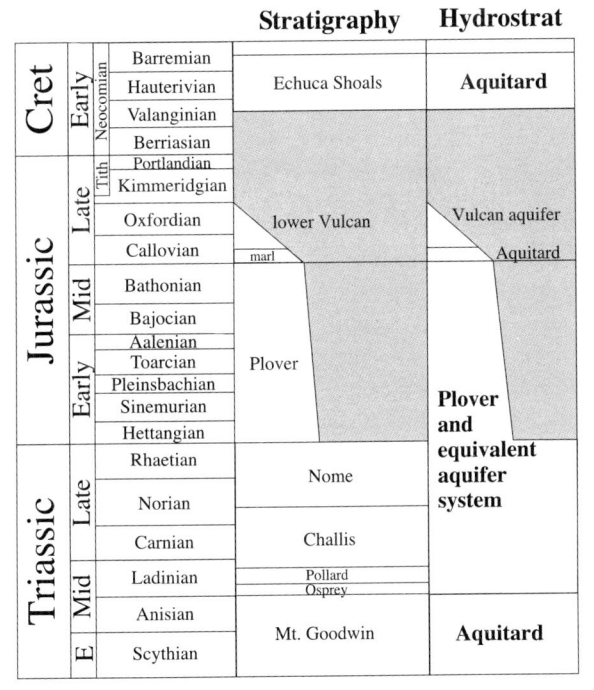

FIGURE 10. Stratigraphic and hydrostratigraphic nomenclature for the Challis Field.

taceous unconformity. The Challis-11ST1 well completion report indicates that a thin Jurassic sand exists between the base of the seal and the Triassic sands, but pressure data from the Jurassic and Triassic sands form a common gradient and are considered the same aquifer. Top seal is provided by Cretaceous claystones of the Echuca Shoals Formation (Figure 10) and is reliant on fault juxtaposition for lateral seal. The seal capacity of the top seal is excellent, with Kivior et al. (2002) showing it capable of holding more than 400 m (1300 ft) of oil column. The trap is heavily faulted and is both structurally and sedimentologically complex. With the top seal thickness of about 10 m (33 ft) and fault throws locally in excess of this amount, the structural seal integrity is the main sealing risk. Gorman (1990) and Wormald (1988) describe the structural geometry of the Challis field. The study area extends slightly southwest of the field itself to include Cassini-1 (Figure 11). Within the Challis field horst block, the Jurassic sands are mainly eroded, so the stratigraphic horizons containing the bulk of the hydrodynamic data are the Triassic Challis and Pollard formations (Figure 10). In the vicinity surrounding the Challis field where Jurassic sands occur, the Middle to Early Jurassic sands of the Plover Formation form a single aquifer system with the Triassic sands below and are jointly termed the Plover aquifer system by Underschultz et al. (2002). This is demonstrated by the pressure-elevation profile for Cypress-1, just north of the Challis field (Figure 12). Here, the Late Jurassic sands of the Vulcan Formation have slightly lower hydraulic head than the Middle to Early Jurassic and Triassic sands of the Plover aquifer system.

Of the 18 wells in the study area, 15 have pressure data either from DSTs or WLTs. Eleven of the wells with

of southwest–northeast-orientated Late Jurassic rift faults and north–south-orientated late Proterozoic basement faults may have controlled the leakage history for the Skua oil field (Figure 9). This interpretation is supported by a structural analysis and restoration, charge history analysis, and an evaluation of the hydraulic communication using hydrodynamic techniques. The intersections of steeply dipping, high-angle basement faults with basin-forming faults create zones of vertical hydraulic communication. A second field in the Vulcan subbasin that shows evidence for vertical hydraulic communication is Challis (Figure 9).

CHALLIS FIELD

The Challis oil field was discovered in 1984 when Challis-1 encountered a 29-m (95-ft) gross oil column in Triassic sandstones immediately below the base Cretaceous

FIGURE 11. Hydraulic head (m) distribution for the Triassic sands at the Challis field.

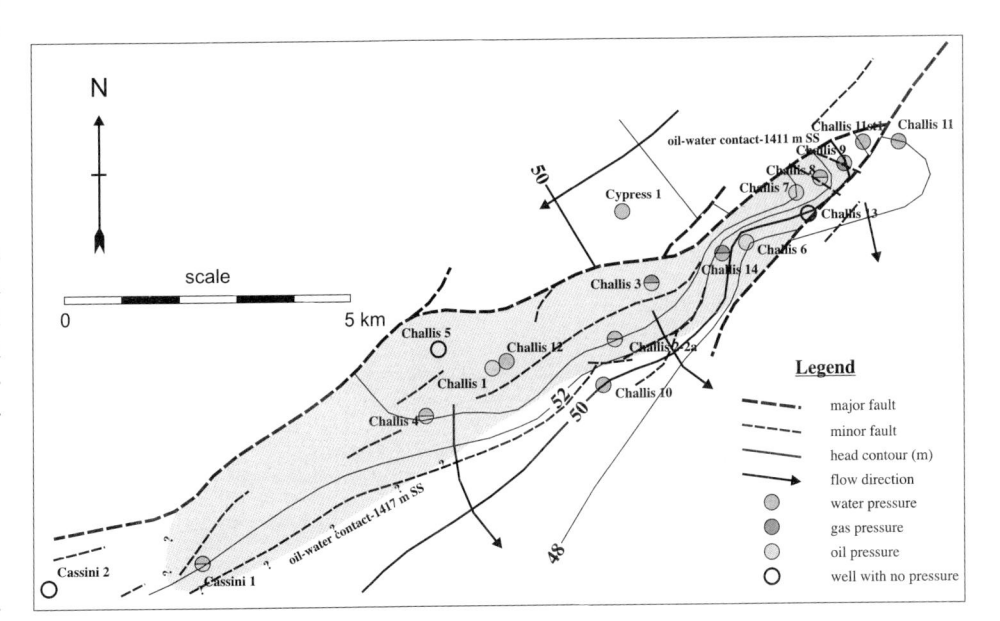

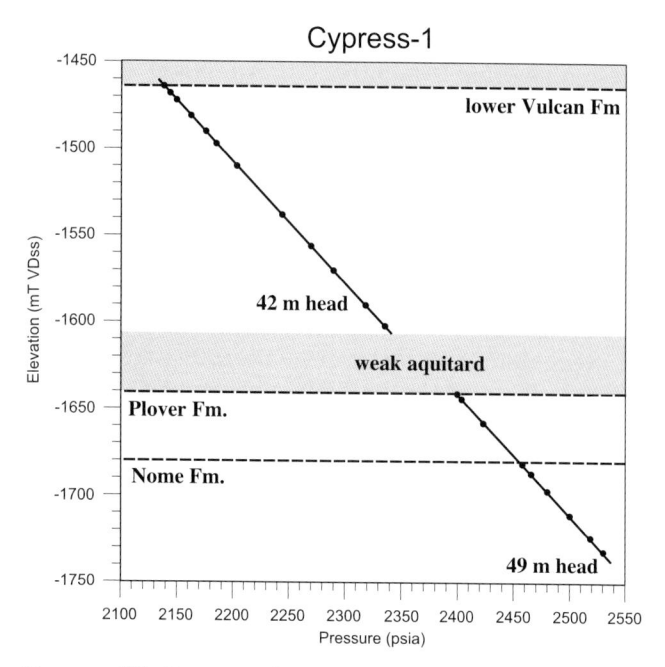

FIGURE 12. Pressure-elevation profile for Cypress-1.

pressure data record information directly on the formation water system. For the four wells with pressure measurements only in the hydrocarbon phase, the pressure was extrapolated down to the estimated hydrocarbon-water contact to gain information on the water system at these locations. To accomplish this, an estimate of the hydrocarbon water contact elevation was made from a combination of WLT data at nearby wells and log analyses from the well completion reports. Formation-water analyses were not found for any of the wells in the study area, but there are production water analyses from Cassini-1, Challis-1, and Challis-2A. Supplementing these data are log analyses values and water salinity estimated from the hydrostatic water pressure gradient recorded by WLTs (Underschultz et al., 2002) in the aquifer.

It has previously been recognized (Yassir and Otto, 1997) and confirmed in this evaluation that production from the Challis field has impacted the pressure recorded at wells drilled into the field after about 1990 (Challis 9–14).

Hydraulic Head

Hydraulic head values for the Triassic sands were calculated based on freshwater density (Figure 11). The dashed lines on the base map represent the position of faults obtained from Wormald (1988). The predominant feature of the map is the closed hydraulic head high in the aquifer beneath the Challis field, against the main Challis-bounding fault. A drop in hydraulic head is present to the north side of the fault, as defined by Cypress-1. No other data control points in the study area are present on the north side of the fault, so a hydraulic gradient on this information alone cannot be established. By examining regional hydraulic head maps from Underschultz et al. (2002), it appears that groundwater flow on the north side of the Challis-bounding fault in the Plover aquifer system is roughly parallel to strike and toward the west.

Within the aquifer, at the base of the Challis field, the fault bounding the north side of the field appears to be a hydraulic barrier. The hydraulic head distribution on the south side of the fault defines a system where flow emanates from the fault and migrates away in a radial fashion to the southwest, south, and east. This pattern is a field example of the idealized transmissive fault described by Otto et al. (2001) and shown in Figure 1a. The formation water must be flowing vertically, from above or below, along the Challis-bounding fault or a subsidiary fault and into the aquifer at the base of the Challis field. The only well that has formation water pressure data above the Challis aquifer is Cypress-1, which defines a value of 42 m (137 ft) head in the Late Jurassic sands of the Vulcan Formation. If this value of hydraulic head is representative, it would suggest that an upward hydraulic gradient is present. Therefore, the formation water traveling along the Challis-bounding fault is likely to be originating from a deeper aquifer. This is broadly consistent with a dewatering compacting basin. From the regional hydrodynamic assessment (Underschultz et al., 2002), the salinity of the Plover aquifer system is described as being 30–45 g/L in the vicinity of the Challis field. If the formation water from deeper in the stratigraphic column was migrating up the Challis fault zone and appearing in the sands at the base of the field, then the salinity would be higher than the regional values in the Plover aquifer system. The pressure-derived salinity for the Challis wells range from 48 to 63 g/L, slightly higher than would have been predicted from the regional trend. Further, at 1510-m (4954-ft) TVDss elevation, the estimated formation-water salinity is only 38 g/L at Cypress-1 on the north side of the Challis fault, fitting well with the low salinity predicted from the regional study of Underschultz et al. (2002). The limited evidence suggests that the source of the formation water entering the aquifer at the base of the Challis field is most likely to be from below. If the Challis fault zone becomes sealing as it passes through the claystone at the top of the Challis field, it would explain the maintenance of the hydrocarbon accumulation.

CONCLUSIONS

When mapping the potential energy distribution in faulted aquifers, unfaulted regions of aquifer are initially considered separately, and then unfaulted blocks are combined in a patchwork to form a regional flow model. In this way, the impact of the faults on the aquifer system can be assessed. Through the examination of hydrodynamic systems in faulted strata from various regions and tectonic regimes, there emerges some commonality to the impact of faults on hydrodynamic systems. These are as follows.

- Aquifer flow tends to be parallel to the structural grain where faults are sealing.
- Vertical hydraulic communication tends to occur where (1) fault orientation bends out of plane from the dominant stress regime; (2) fault intersections are present, particularly where one fault set is steeply dipping and at a high angle to the other fault set; and (3) at relay zones and transfer faults.
- Flow along conductive faults, instead of an entire fault plane, is likely to be focused.

Conditions of fault zone transmissivity can be identified if sufficient pressure data is available in the aquifer adjacent to the fault. Features that identify a fault zone as being of lower permeability than the reservoir horizon and thus having the potential to be sealing are

- hydraulic head discontinuity across a fault
- flow directions in the aquifer adjacent to the fault, which are parallel with the structural grain
- formation water chemistry discontinuity across a fault
- a vertical break in the pressure gradient on a pressure-elevation plot for wells adjacent to or crosscut by a fault

Features that identify a fault zone as being leaky are

- hydraulic head at depth and adjacent to a fault being similar to topographic elevation and accompanied by low-salinity formation water with elevated CO_3^{2-} or SO_4^{2-}
- hydraulic head highs or lows in the aquifer closing onto the plane of the fault and accompanied by anomalous formation-water chemistry
- a vertically continuous pressure gradient on a pressure-elevation plot for wells adjacent to or crosscut by a fault

ACKNOWLEDGMENTS

We thank Commonwealth Scientific and Industrial Research Organization Petroleum and Hydro-Fax Resources Ltd for supporting the publication of this work. Appreciation is due to Pat Ward, who instigated the initial hydrodynamic work at Moose Mountain. We are grateful for thought-provoking discussions with Allison Hennig, Dan Barson, and Kent Wilkinson. Comments from Jennifer Adams, Jenny Stedmon, and Tim Wood and technical reviews by Barb Tilley, Neil Tupper, and John Kaldi helped improve this chapter.

REFERENCES CITED

Bachu, S., 1995, Flow of variable-density formation water in deep sloping aquifers: Review of methods of representation with case studies: Journal of Hydrology, v. 164, p. 19–38.

Bachu, S., and R. A. Burwash, 1991, Regional-scale analysis of the geothermal regime in the Western Canada sedimentary basin: Geothermics, v. 20, p. 387–407.

Bachu, S., and K. Michael, 2002, Flow of variable-density formation water in deep sloping aquifers: Minimizing the error in representation and analysis when using hydraulic-head distributions: Journal of Hydrology, v. 259, p. 49–65.

Bachu, S., J. C. Ramon, M. E. Villegas, and J. R. Underschultz, 1995, Geothermal regime and thermal history of the Llanos basin, Colombia: AAPG Bulletin, v. 79, p. 116–129.

Begin, N. J., and D. A. Spratt, 2002, Role of transverse faulting in along-strike termination of Limestone Mountain culmination, Rocky Mountain thrust-and-fold belt, Alberta, Canada: Journal of Structural Geology, v. 24, p. 689–707.

Cowley, R., and G. W. O'Brien, 2000, Identification and interpretation of leaking hydrocarbons using seismic data: A comparative montage of examples from major fields in Australia's North West Shelf and Gippsland basin: Australian Petroleum Production and Exploration Association Journal, v. 40, p. 121–150.

Craw, D., 2000, Fluid flow at fault intersections in an active oblique collision zone, Southern Alps, New Zealand: Journal of Geochemical Exploration, v. 69–70, p. 523–526.

Dahlberg, E. C., 1995, Applied hydrodynamics in petroleum exploration: New York, Springer-Verlag, Inc., 295 p.

Gartrell, A., M. Lisk, and J. R. Underschultz, 2002, Controls on the trap integrity of the Skua oil field, Timor Sea, in The sedimentary basins of Western Australia: 3. Proceedings of the Petroleum Exploration Society of Australia Symposium, Perth, Western Australia, p. 389–407.

Gorman, I. G. D., 1990, The role of reservoir simulation in the development of the Challis and Cassini fields: Australian Petroleum Exploration Association Journal, v. 30, p. 212–221.

Grasby, S. E., and I. Hutcheon, 2001, Controls on the distribution of thermal springs in the southern Canadian Cordillera: Canadian Journal of Earth Sciences, v. 38, p. 427–440.

Hennig, A., J. R. Underschultz, and C. J. Otto, 2002, Hydrodynamic analysis of the Early Cretaceous aquifers in the Barrow sub-basin in relation to hydraulic continuity and fault seal, in The sedimentary basins of Western

Australia: 3. Proceedings of the Petroleum Exploration Society of Australia Symposium, Perth, Western Australia, p. 305–320.

Hitchon, B., and M. Brulotte, 1994, Culling criteria for "standard" formation water analyses: Applied Geochemistry, v. 9, p. 637–645.

Kennard, J., 2004, Geoscience Australia: http://www.agso.gov.au/oceans/projects/nwr.jsp (accessed April 2004).

Kivior, T., J. G. Kaldi, and S. C. Lang, 2002, Seal potential in Cretaceous and Late Jurassic rocks of the Vulcan sub-basin, North West Shelf, Australia: Australian Petroleum Production and Exploration Association Journal, v. 42, p. 203–224.

O'Brien, G. W., M. A. Etheridge, J. B. Willcox, M. Morse, P. Symonds, C. Norman, and D. J. Needham, 1993, The structural architecture of the Timor Sea, North-Western Australia: Implications for basin development and hydrocarbon exploration: Australian Petroleum Exploration Association Journal, v. 33, p. 258–277.

Otto, C. J., 1992, Petroleum hydrogeology of the Pechelbronn–Soultz in the Upper Rhine Graben, France—Ramifications for exploration in intermontane basins. Ph.D. Thesis, University of Alberta, Canada, 357 p.

Otto, C. J., and N. Yassir, 1997, Hydrodynamic assessment of fault seal integrity: Ramifications for exploration and production (abs.), in Contributions to the Second International Conference on Fluid Evolution, Migration and Interaction in Sedimentary Basins and Orogenic Belts, Belfast, Northern Ireland: Geofluids II Extended Abstracts, p. 129–132.

Otto, C., J. Underschultz, A. Hennig, and V. Roy, 2001, Hydrodynamic analysis of flow systems and fault seal integrity in the North West Shelf of Australia: Australian Petroleum Production and Exploration Association Journal, v. 41, p. 347–365.

Price, R. A., 1994, Cordilleran tectonics and the evolution of the Western Canada sedimentary basin, in G. D. Mossop and I. Shetsen, eds., Geological atlas of Western Canada: Calgary, Canadian Society of Petroleum Geologists/Alberta Research Council, p. 13–24.

Price, R. A., 2001, An evaluation of models for the kinematic evolution of thrust and fold belts: Structural analysis of a transverse fault zone in the Front Ranges of the Canadian Rockies north of Banff, Alberta: Journal of Structural Geology, v. 23, p. 1079–1088.

Struik, L. C., and D. G. MacIntyre, 2001, Introduction to the special issue of Canadian Journal of Earth Sciences: The Nechako NATMAP Project of the central Canadian Cordillera: Canadian Journal of Earth Sciences, v. 38, p. 485–494.

Toth, J., and C. J. Otto, 1993, Hydrogeology and oil deposits at Pechelbronn–Soultz, Upper Rhine Graben: Acta Geologica Hungarica, v. 36, no. 4, p. 375–393.

Underschultz, J. R., and R. Bartlett, 1999, Hydrodynamic controls on foothills gas pools; Mississippian strata: Canadian Society of Petroleum Geologists Reservoir, v. 26, p. 10–11.

Underschultz, J. R., G. K. Ellis, A. Hennig, E. Bekele, and C. Otto, 2002, Estimating formation water salinity from wireline pressure data: Case study in the Vulcan sub-basin, in The sedimentary basins of Western Australia: 3. Proceedings of the Petroleum Exploration Society of Australia Symposium, Perth, Western Australia, p. 285–303.

Underschultz, J. R., C. J. Otto, and T. Cruse, 2003, Hydrodynamics to assess hydrocarbon migration in faulted strata—Methodology and a case study from the North West Shelf of Australia: Journal of Geochemical Exploration, v. 78–79, p. 469–474.

Wilkinson, K., 1995, Is fluid flow in Paleozoic formations of west-central Alberta affected by the Rocky Mountain thrust belt? Master's thesis, University of Alberta, Edmonton, Alberta, Canada, 122 p.

Wormald, G. B., 1988, The geology of the Challis oil field, Timor Sea, Australia: Proceedings of Petroleum Exploration Society Australia Symposium, Perth, p. 425–437.

Yassir, N., and C. J. Otto, 1997, Hydrodynamics and fault seal assessment in the Vulcan sub-basin, Timor Sea: Australian Petroleum Production and Exploration Association Journal, v. 37, part 1, p. 380–389.

Economic Evaluation of Prospects with a Top Seal Risk

David C. Lowry
Origin Energy, Brisbane, Australia

ABSTRACT

In petroleum exploration, an important task is to estimate both the reserves likely to be discovered in a prospect and the chance of success of finding them. Published advice on estimating reward and risk can be very misleading when applied to a prospect with a risk of membrane seal failure. The literature mostly recommends assessing the reward on the basis of the mean reserve volume of the mapped closure and the risk on the chance of finding a minimal amount of petroleum. This method breaks down for prospects where the perceived chance of success of the top seal being effective varies with prospect's estimated column height. If the chance of retaining a column that will give a mean reserves volume is much less than the chance of retaining a minimal column, the value of the prospect will be grossly overestimated. This chapter presents several alternative methods. The preferred method of obtaining an expected monetary value (EMV) is to construct a curve of chance of top seal against column height and calculate a graph of the net present value and EMV for numerous levels. If the EMV is positive at any level, the prospect can be considered for drilling.

INTRODUCTION

The petroleum exploration industry continually makes decisions on whether to test a prospect with a well. This involves balancing the chance of success and its value with the risk of failure and its cost. In the last two decades, companies have developed procedures for systematic estimation of reserves and risks. A common work flow (see, for example, Rose, 2001) is

1) create a probability distribution for recoverable reserves, which is based on a mapped closure and the estimated ranges of various reservoir parameters
2) truncate the reserve distribution to remove small, noncommercial fields
3) calculate the mean of the truncated reserve distribution and estimate its net present value (NPV) with a discounted cash flow model
4) estimate the chance of discovering a minimal amount of hydrocarbons by multiplying the probabilities of the various geological parameters being favorable for a discovery (probability of geological success [Pg] or chance of success)
5) compute a chance of economic success (Pe): Pe = Pg × the proportion of reserve distribution not removed by economic truncation

6) calculate an expected monetary value: EMV = NPV × Pe − dry hole cost × (1 − Pe)

A positive EMV is commonly used as a favorable criterion to invest.

Although this method has served the industry well in many areas, it cannot be applied universally. In particular it should not be applied to large, high-relief prospects where the column height is anticipated to be limited by seal capacity instead of trap closure. If the NPV is based on the mean truncated reserves of the mapped prospect, combining it with Pe in the EMV equation will overestimate the value of the prospect. In some cases, the overestimation may be gross.

METHODOLOGIES FOR EVALUATING PROSPECTS

Over the last 20 yr, several exploration companies have allowed publication of their methodologies for commercial evaluation of prospects. The most detailed descriptions of a methodology and its implementation are those of Rose (1992, 2001). As demonstrated below, these methods have a lot in common and are referred to here for convenience as the "Common Methodology."

The probability of a prospect succeeding (chance of success) is obtained by multiplying the probability of success for each of the independent elements that are needed for success. Chance of success (COS) is the complement of risk of failure; COS = 1 − risk. These risks are commonly grouped under structure, reservoir, seal, source, and migration. A good evaluation system needs to be explicit on what constitutes success and on the definition of the risk element that the explorationist is being asked to assess.

The chance of geological success (Pg) is the chance that "A well encounters mobile reservoired hydrocarbons" (Rose, 1992, p. 71). Tests for success were that the accumulation is "at least large enough to support a flowing test" or "whether the exploratory well was completed for production." He suggested that these tests would be met by reserves of 1000–20,000 BOE (Rose, 2001, p. 33). This probability (Pg) is obtained by multiplying the various geological risk components (e.g., structure, reservoir, charge, and seal). Rose (2001, p. 36) required a sealing capability to be "...compatible with at least the P99% reserves forecast...", that is, a minimal amount of reserves.

A variation of the system advocated by Rose was published by White (1993) of Exxon. He avoided economic truncation by anchoring the reserves distribution at a minimum economic pool size (MEPS) that varied according to region. However, this approach has apparently not been widely adopted (MacKay, 1996).

Nevertheless, White (1993) agreed on the need to risk a minimal amount of reserves. He wrote (p. 2005) "One chronic cause of...overrisking is trying to relate chances of adequacy to most likely or mean values rather than to minimum values."

Earlier, Sluijk and Nederlof (1984) outlined Shell's system. It is very different from the Common Methodology, in that it models the processes of generation, migration, and retention. It combined reserves and risk assessment in a probabilistic system in which seal and charge are calibrated to a worldwide database, updated by local knowledge. This system has served Shell well (Sluijk and Parker, 1986), but it is not accessible to other exploration companies. It appears that no other companies have emulated the system, although ARCO felt the need to improve their risking in frontier areas by incorporating modeling of hydrocarbon generation and migration (Nicklin et al., 1994).

Otis and Schneidermann (1997) described the system used by Chevron. It is very similar to that advocated by Rose. In assessing geological risk, "a well is considered a geological success if a stabilized flow of hydrocarbons is obtained on test" (p. 1087). The overall probability of success is obtained by assessing geological chance of success and multiplying by the commercial chance of success derived by truncating the reserves distribution at a minimum commercial pool size.

The Australian company Santos also adopted a very similar system (Johns et al., 1998). Their definition of geological (technical) success was (p. 560) "the chance of finding mobile, reservoired hydrocarbons." The reserves distribution was truncated at a MEPS (a volume sufficient to pay back exploration as well as development costs). The probability of exceeding the MEPS was referred to as Pmeps. The value of the prospect was commonly based on the mean of the truncated reserves distribution and the probability of commercial success (Pc) obtained by Pc = Pg × Pmeps. Johns et al. (1998) reported that adoption of this and related procedures had resulted in improved finding costs and more reliable prediction of exploration success in their gas exploration program. However, the experience was based largely on the Cooper/Eromanga Basins, a mature area where only small fields remain to be found. The distortion of decision making resulting from the logical discontinuity of valuing a mean reserve and risking a minimal amount of hydrocarbons only becomes apparent with larger prospects.

BHP also adopted a similar evaluation system. Watson (1998, p. 577) defined geological risk as "our confidence that the proposed petroleum system is actually working at the particular prospect... For most practical purposes... (it) ... is the chance of making a discovery regardless of size." The reserve distribution was truncated at an economic reserves threshold to obtain a commercial risk. Watson (1998) outlined a sophisticated risking system developed by BHP, and although he did not explain precisely what was being risked, some phrases suggests that it was a minimal amount. For example, his Table 2 (p. 582) states that fault seal should be adequate "at least immediately opposite the uppermost portion of the prognosed fault-dependent hydrocarbon accumulation." This implies the trapping of a minimal column in the crest of the structure.

ALTERNATIVE METHODOLOGIES

Valuation of prospects with a large element of seal risk can be improved in several ways. The argument can be advanced best by working through an example in several ways and comparing the outcomes. Bodalla South oil field (Salomon et al., 1990) is used here in an idealized form as if it was a prospect to be considered for drilling. The field lies in the Eromanga Basin, Australia, and one of its reservoirs is the thick, fluvial Hutton Sandstone, sealed by altered, lithic sandstone (Birkhead Formation).

Method 1: Common Methodology

Volumetrics

Figure 1 is a depth map for the Bodalla South field for the top of the Birkhead Formation (Boult, 1993). For this exercise, the horizon can be taken as a reasonable approximation to the form of the top of the Hutton reservoir. The rock volumetrics are given in Table 1 and plotted in Figure 2.

The lowest closing contour (LCC) or spillpoint of the closure is shown as 1425 m (4675 ft). If filled to spill, the recoverable reserves amount to 30 million bbl. This result is obtained using single values for reservoir parameters and arbitrarily adjusting them to get a simple reserves value as a starting point.

In estimating recoverable reserves, it is usual to put ranges on the reservoir parameters and compute a reserves distribution. In this case, the following approach is used to keep things simple:

gross rock volume: triangular frequency distribution based on 1410-m (4625-ft) (minimum), 1420-m (4658-ft) (mode), and 1425-m (4675-ft) (maximum) structure contours. Corresponding rock volumes are 5.93×10^6 m³ (4800 ac ft) (minimum), 33.33×10^6 m³ (27,000 ac ft) (mode), and 42.52×10^6 m³ (34,500 ac ft) (maximum)
porosity: normal, $P(50)$ 0.208; $P(10)$ 0.25

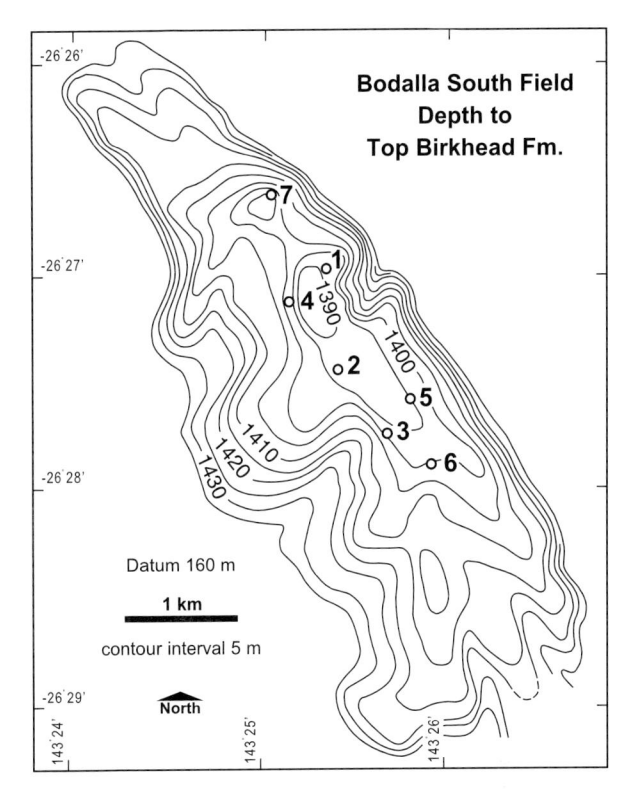

FIGURE 1. Bodalla South field structure map. From figure 5 of Boult (1993), with permission from Elsevier.

net/gross: normal, $P(50)$ 0.70; $P(10)$ 0.84, truncated at 1.0
S_w: normal, $P(50)$ 0.35; $P(10)$ 0.42
Bo: 1.1
recovery factor: normal, $P(50)$ 0.40; $P(10)$ 0.48
for simplicity, $P(10)$ values were put at 120% of the $P(50)$ values

This results in a reserves distribution of

$P(90)$ 10.1 million bbl
$P(50)$ 17.4 million bbl
$P(10)$ 28 million bbl
mean 18.4 million bbl

Calculations were performed in a reserves evaluation program using a Monte Carlo routine.

The estimates above raise two debatable issues:

1) One is the appropriate shapes for the probability distributions of the input parameters. Many workers (e.g., Rose, 2001) have argued for using lognormal distributions for all reservoir parameters on the grounds that many naturally occurring systems are log-normally distributed. However, Squire (1996) showed that fields in the Cooper/Eromanga Basins (Australia) have several parameters skewed in the opposite direction. For example, Hutton Sandstone

TABLE 1. Volumetrics for mapped closure in Figure 1.

Subsea Depth (m)	Column Height (m)	Area (km²)	Volume Above (m³ 10⁶)	
1389	0	0	0	crest
1390	1	0.2	0.07	
1395	6	0.78	2.59	
1400	11	1.55	5.93	
1405	16	2.78	10.97	
1410	21	4.38	18.05	
1415	26	5.63	25.09	
1420	31	7.65	33.33	
1425	36	9.33	42.52	LCC

porosity and hydrocarbon saturation were both positively skewed. It is assumed here that in the absence of good statistics, distributions are largely a matter of convenience. A triangular distribution is used here for gross rock volume, because it is easy to visualize and less likely than a log-normal distribution to create unintended and unrealistic high-side reserve estimates. Other parameters are given a normal distribution.

2) The other debatable topic is whether the uncertainty of the gross rock volume should be estimated with a log-normal distribution anchored at the low end to a value giving minimal reserves of a few thousand barrels (Rose, 2001, p. 131–133). However, Rose (2001, p. 134) also conceded that the $P(90)$ could be as much as 80% of the mapped closure if it was defined by dense seismic data and supported by a matching amplitude anomaly. In the case of Bodalla South, reasonable two-dimensional seismic control is present. The most likely value is based on the 1420-m (4658-ft) contour, 5 m (16 ft) above the LCC to allow for optimism in the mapping. The maximum value is taken as the LCC on the grounds that velocity variations (known to be a problem in the area) have been allowed for in the preparation of the depth map. The minimum value is based on inspection of the map and selecting the 1410-m (4625-ft) contour on the basis that it would be most unlikely that the closure could be smaller than that.

Risking

If we are to estimate the chance of success where success is defined as finding an amount of oil "at least large enough to support a flowing test" (Rose, 2001, p. 33) or at least enough to support "a stabilized flow of hydrocarbons" (Otis and Schneidermann, 1997, p. 1087), then the success criteria would easily be met by a pool filled down to the 1395-m (4576-ft) contour (about 0.5 million bbl). This requires a top seal with a seal capacity adequate to hold a column of 6 m (20 ft). A study of the seal capacity of the Birkhead Formation by mercury injection (Boult, 1993, p. 11) concluded that "the maximum oil column...that can be held by the base of the Birkhead Formation at Bodalla South is between 12.64 and 10.86 m." Thus, the top seal will have little problem holding a 6-m (20-ft) column, and the chance of seal capacity is put at 90%. To simplify the calculations, the combined COS of other risk factors (such as closure, reservoir, and charge) is put at 20%, giving an overall COS of 18%.

Economic Evaluation

Simplified (and arbitrary) economic assumptions are

dry hole cost	$1.8 million
minimum economic pool size	0.1 million bbl
oil value (discounted)	$10/bbl (above economic minimum)

With these assumptions, the NPV of the mean reserves (18.4 million bbl) can be approximated by

$$\text{NPV} = (\text{mean reserves} - \text{minimum pool size}) \times \text{discounted unit value of oil}$$
$$= (18.4 - 0.1) \times \$10$$
$$= \$183 \text{ million}$$

and the expected monetary value is

$$\text{EMV} = \$183 \times 18\% - \$1.8 \times (1 - 18\%)$$
$$= \$31.5 \text{ million}$$

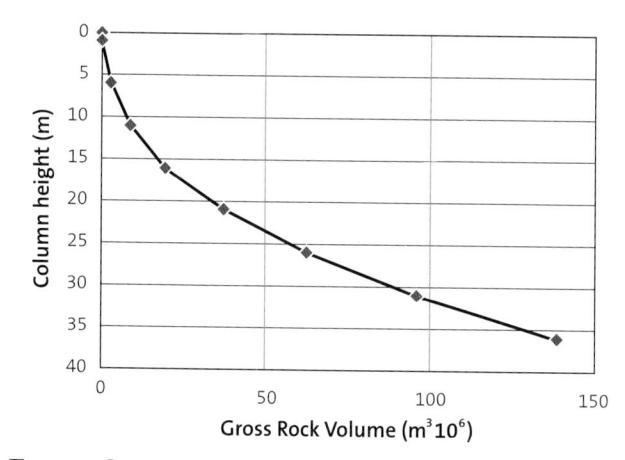

FIGURE 2. Plot of volumetrics for mapped closure.

The meaning of this EMV is that if an explorer drilled many prospects of this sort, they could expect, on average, to recover their investment ($1.8 million), plus interest at the nominated rate, plus $31.5 million: clearly a very attractive opportunity.

However, the problem is that the risking has been done on a minimal amount of oil and the reward calculated from the mapped closure. The literal application of advice published in the literature has grossly overestimated the EMV because of the limited seal capacity. To remedy this problem, either treat seal capacity as a volumetric issue and risk a model with a reduced column, or honor the mapped closure but vary the risk with column height.

Alternate Method 1: Use the Seal Capacity to Restrict the Volumetrics

The spillpoint of the structure can be ignored, and the volumetrics can be based on a column height derived from sealing capacity estimates. Boult (1993) concluded that the maximum oil column that can be retained in the Hutton Sandstone is around 11 m (36 ft). This could be interpreted as indicating that a prospect has a 50% chance of sealing at least an 11-m (36-ft) column. It may be commercially reasonable to drill the prospect on the basis of an 11-m (36-ft) column. With such a column, the chance of top seal could be put at 50%, and the overall COS would be 10%. Mean volume for an 11-m (36-ft) column is 1.84 million bbl valued at $17.4 million. The expected monetary value is given by

$$EMV = \$17.4 \times 10\% - \$1.8 \times (1 - 10\%)$$
$$= \$0.12 \text{ million}$$

This EMV suggests that the prospect is worth investing in, but it is not the bonanza indicated by the Common Methodology.

Alternate Method 2.1: Honor the Mapped Closure for Volumetrics but Risk the Mean Case

If the EMV is to be based on a mean expected volume, the chance of success should be based on that case. The mean volume (18.4 million bbl) implies a column height of around 30 m (100 ft). If the maximum column able to be retained is around 11 m (36 ft), the chance of the closure being filled with 30 m (100 ft) of oil is extremely low. The literature is largely silent on how to assign a top seal risk to a prospect with mapped closure height and an array of capillary pressure data. Kaldi and Atkinson (1997) provided perhaps the first and only published methodology. They distinguished five sealing facies for the Talang Akar Formation in the Arjuna Basin and used mercury injection capillary pressure (MICP) measurements to determine a typical seal capacity for each facies. The COS of top seal for a reservoir was assessed as the ratio of the seal capacity to closure height. If applied to Bodalla South, the COS would be 11/30 or 37%. This approach seems too optimistic in this case. For example, in materials testing, if a sample group of lifting slings was tested to destruction and had an average strength of 11 t, it would not be realistic to suggest that the next sling had a 37% chance of lifting 30 t; the value would be smaller and would depend on the frequency distribution of the sample. Extending this approach to Bodalla South, inspection of Boult's (1993) figures 9 and 10 suggests that typical Birkhead Formation has the capacity to seal a column of 11.6 m (38 ft) with a standard deviation of 3 m (10 ft) (assuming a normal distribution). With this distribution, the chance of sealing a column of 30 m (100 ft) is less than 0.1%. A value between these extremes seems likely, and the COS is arbitrarily set at 1%. Whatever the approach chosen for selecting a value of COS for sealing a 30-m (100-ft) column, the overall conclusions are unlikely to be affected. With this assumption, the overall COS is 0.2%. In this case, the expected monetary value is

$$EMV = \$183 \times 0.2\% - \$1.8 \times (1 - 0.2\%)$$
$$= -\$1.43 \text{ million}$$

The negative EMV means that, on average, the investor can expect to lose money each time they drill this sort of prospect, and that investing in the prospect would be a bad decision.

Alternate Method 2.2: Honor the Map for Volumetrics; Risk Three Cases, Calculate EMVs, and Take a Swanson's Mean

This case is developed in Table 2. Valuing a Swanson's Mean based on three reserves values is generally accepted to be more robust than using a single mean value.

The reserves values are derived by a program using a Monte Carlo routine, with the uncertainties in gross rock volume and reservoir parameters listed above.

The chance of top seal values are arbitrarily assessed for each column height; see alternate method 2.1 above for more information.

Swanson's Mean is computed as 30% of $P(90)$ + 40% of $P(50)$ + 30% of $P(10)$ values.

On this basis, the opportunity is bad but not as bad if calculated from a single value (−$0.82 million instead of −$1.43 million). This is because the column associated with the $P(90)$ reserves has a small but significantly better chance of being sealed than the $P(50)$ and $P(10)$ cases.

TABLE 2. Expected monetary value based on a three-point Swanson's Mean.

	Reserves (million bbl)	Associated Column (m)	Chance of Top Seal (%)	COS Other Risks (%)	Total COS (%)	NPV ($ million)	EMV ($ million)
P(90)	6.4	19	12.0	20.0	2.4	63.0	−0.24
P(50)	17.4	29	2.0	20.0	0.4	173.0	−1.10
P(10)	38.9	36	1.0	20.0	0.2	388.0	−1.02
						Swanson's Mean	−0.82

Alternate Method 2.3: Honor the Map for Volumetrics But Use a Variable Risk

In this method, the chance of sealing is assessed for numerous column heights (Table 3).

The COS top seal values are arbitrary and relate to MICP seal capacity estimates. A 50% chance of sealing at least an 11-m (36-ft) column exists (see above); lesser columns are given a higher chance of being sealed; and greater columns are given decreasing chances of being sealed. The method proposed here also departs from normal practice by including the trap volume uncertainty as a risk. Instead of basing the uncertainty of volumetrics on a triangular probability distribution of closure (1410 m [4625 ft] minimum, 1420 m [4658 ft] mode, and 1425 m [4675 ft] maximum), the uncertainty is treated as a risk in column 5 (100% chance of being closed down to 1410 m [4625 ft]; 50% chance of being closed down to 1420 m [4658 ft], and no chance of being closed beyond 1425 m [4675 ft]).

The values are graphed in Figure 3.

Note that for small column heights, EMV is low because oil reserves are small, and at greater column heights, the EMV is low because of the very low COS. The EMV is least negative for intermediate column heights of 10–20 m (33–66 ft). The decision to invest should be based on the maximum EMV. The maximum computed value is $0.1 million, or marginally higher if an appropriate curve is fitted to the data.

DISCUSSION

Method 2.3 recommends honoring the map volumetrics, using a variable risk, and taking the maximum EMV of the prospect. It might be argued that taking an average of the EMV profile is a more appropriate methodology. Consider a hypothetical prospect where a well will test a small, low-risk dip closure, which is part of a medium-size, medium-risk fault closure, which is part of a large, high-risk stratigraphic trap. The dip-closed model may be too small to have a positive EMV, and the stratigraphic trap may be too risky to have a positive EMV, but the fault trap may have just the right balance of reward and risk to be EMV positive, and the prospect should be drilled on that basis. The hypothetical example can be regarded as three different prospects, only one of which is economically attractive. If one model is attractive, it should be drilled; there is no need to average the unattractive aspects of the other two models. Figure 3 can be considered as a continuum of prospect models with a different model for every column height, and if any column height is EMV positive, it should be drilled.

The problem of valuing a prospect for which the chance of success varies with the reserve distribution is not a common one; it occurs mainly when seal capacity is limited. It could also occur if a large closure has a small drainage area or other charge limit, and it

TABLE 3. Expected monetary value using mapped closure, variable top seal risk based around a column height of 11 m (36 ft).

Structure Contour	Column Height (m)	Mean Reserves (million bbl)	COS Top Seal (%)	COS Structure (%)	COS Other Risks (%)	Total COS (%)	NPV ($ million)	EMV ($ million)
1389	0							
1390	1	0.02	95.0	100.0	20.0	19.0	−0.8	−2.0
1395	6	0.55	90.0	100.0	20.0	18.0	4.5	−0.7
1400	11	1.81	50.0	100.0	20.0	10.0	17.1	0.1
1405	16	4.16	20.0	100.0	20.0	4.0	40.6	−0.1
1410	21	8.3	10.0	100.0	20.0	2.0	82.0	−0.1
1415	26	13.4	2.0	75.0	20.0	0.3	133.0	−1.4
1420	31	20.6	1.0	50.0	20.0	0.1	205.0	−1.6
1425	36	29.8	1.0	0.0	20.0	0.0	297.0	−1.8

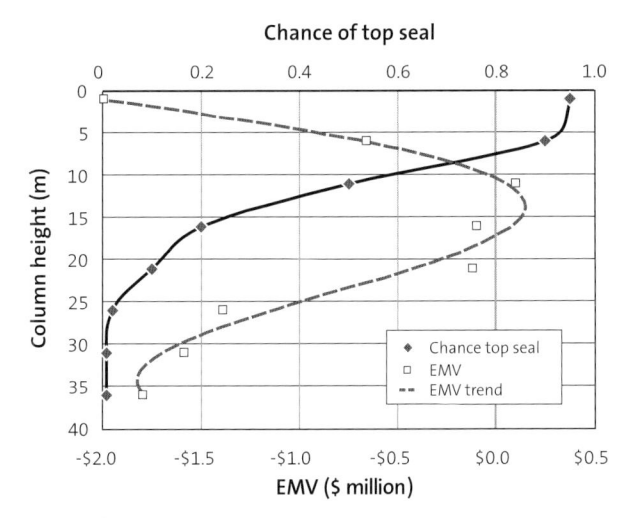

FIGURE 3. Method 2.3; variation in COS and EMV with column height.

might be an issue where water washing and biodegradation has reduced the volume of the oil initially present in the reservoir.

A "Fill Factor" could be used in an area where structures do not seem to be full to spill. In the past, a Fill Factor has been used on occasions by management to counter the natural exuberance of interpreters to close contours, but with computer-contoured, high-quality 3-D seismic data, using a Fill Factor will give distorted results. It could be used in the case of shrinkage by biodegradation, but it is not appropriate for seal capacity problems. Fill Factor is commonly applied to the gross rock volume, not the vertical closure, and in seal capacity issues, it is the column height that is constrained and not the volume. If chance of top seal is a continuously varying function of column height, using a single Fill Factor becomes a blunt instrument likely to lead to a distorted evaluation.

Method 2.2 uses Swanson's Mean, because it has been commonly used in the industry in evaluating prospects. The calculation assumes that the variable has a log-normal probability distribution, and whereas the reserves are roughly log-normally distributed, the EMV is not (see Figure 3). So the mathematics are probably technically invalid, and a judgement has to be made whether this distortion outweighs the additional robustness of using three values instead of the single value relating to the mean reserves case. In any case, this method demands that judgements have to be made on seal risk for three column heights, and it is little more trouble to use method 2.3 and get a much more reliable value.

Implementing Method 2.3 raises at least three problems:

1) Some uncertainties exist in the technology of taking MICP pressures and calculating the hydrocarbon column heights that the samples would seal (see, for example, Boult, 1993).

2) The literature is largely silent on how to take the column height data from a series of MICP measurements and estimate the column height that can be sealed at a particular prospect. Downey (1984) favored the weakest link approach and proposed using the sample with the lowest capacity to characterize the top seal. At the other extreme, it might be argued that in an unfaulted sequence of laterally extensive beds (e.g., turbidite siltstones and shales), seal capacity will be controlled by the bed with the highest seal capacity. The issue is whether the top seal is controlled by the weakest link or strongest layer or something in between.

3) A third difficulty is finding techniques for estimating COS of top seal for a prospect for various column heights. Kaldi and Atkinson's (1997) contribution is noted above, and the present chapter suggests an alternate technique (see method 2.3) that gives very different results. This area needs further attention by industry.

This chapter considers a simple prospect model where seal is constrained by pore entry pressure. In real prospect evaluation, other aspects of sealing such as thickness, brittleness, and areal extent will also have to be considered (Kaldi and Atkinson, 1997).

When can the Common Methodology be used, and when is a more sophisticated approach needed? The key question is whether the volumetrics of the reserves are partly dependent on one or more of the risk elements. If they are independent, then the Common Methodology is fine; it does not matter whether the explorationist has in mind a minimal amount, a $P(90)$, a $P(50)$, or a $P(10)$ volume when estimating the risks. If, however, doubts arise about whether the prospect will be filled to spill because of concerns about seal capacity or charge adequacy, then it is time to consider a more sophisticated approach as offered by methodology 2.3. A common early warning sign is when a group of explorationists focusing on assigning risk to a prospect start to ask "What are we risking?"

CONCLUSIONS

Most published methods of prospect evaluation recommend estimating the chance of encountering a minimal amount of hydrocarbons and valuing the mean reserves based on mapped closure. This approach is invalid if the COS is a function of column height. For example, the published methodology will overestimate the EMV of a prospect if the mapped closure is greater than the seal capacity of the top seal as determined by MICP.

An improved method is to develop a graph of COS against column height and use it to develop a graph of EMV against column height. The value of the prospect is given by the maximum EMV.

Methods for taking MICP data and estimating chance of top seal for various column heights are speculative and need further study.

ACKNOWLEDGMENTS

Elsevier gave permission to copy the map in Figure 1.

The ideas presented here were developed while working for Origin Energy, and I thank staff, particularly R. Suttill, for stimulating discussions, and Peter Boult, for advice on seal capacity. Reviewers of the chapter proposed substantial improvements. Origin Energy has allowed publication of the chapter, but responsibility for the ideas are mine.

REFERENCES CITED

Boult, P. J., 1993, Membrane seal and tertiary migration pathways in the Bodalla South oilfield, Eromanga Basin, Australia: Marine and Petroleum Geology, v. 10, p. 3–13.

Downey, M. W., 1984, Evaluating seals for hydrocarbon accumulations: AAPG Bulletin, v. 68, p. 1752–1763.

Johns, D. R., S. G. Squire, and M. J. Ryan, 1998, Measuring exploration performance and improving exploration predictions—With examples from Santos' exploration program: Australian Petroleum Production and Exploration Association Journal, v. 1998, p. 559–569.

Kaldi, J. G., and C. D. Atkinson, 1997, Evaluating seal potential: Example from the Talang Akar Formation, offshore northwest Java, Indonesia, in R. C. Surdam, ed., Seals, traps and the petroleum system: AAPG Memoir 67, p. 85–101.

MacKay, J. A., 1996, Risk management in international petroleum ventures; ideas from a Hedberg Conference: AAPG Bulletin, v. 80, p. 1845–1849.

Nicklin, D. F., C. R. K. Moore, R. A. Noble, and K. Touysinhthiphonexay, 1994, Reducing exploration risk by improving pre-drill estimates (abs.): AAPG Annual Meeting Program, v. 78, p. 224.

Otis, R. M., and N. Schneidermann, 1997, A process for evaluating exploration prospects: AAPG Bulletin, v. 81, p. 1087–1109.

Rose, P. R., 1992, Chance of success and its use in petroleum exploration, in R. Steinmetz, ed., The business of petroleum exploration: AAPG Treatise of Petroleum Geology, Handbook of Petroleum Geology, chapter 7, p. 71–86.

Rose, P. R., 2001, Risk analysis and management of petroleum exploration ventures: AAPG Methods in Exploration Series, no. 12, 164 p.

Salomon, J. A., S. L. Keenihan, and A. P. Calcraft, 1990, Bodalla South field—Australia, Queensland, Eromanga Basin, in E. A. Beaumont and N. A. Foster, eds., Structural traps 1: Tectonic fold traps in treatise of petroleum geology, atlas of oil and gas fields: AAPG, p. 129–155.

Sluijk, D., and M. R. Nederlof, 1984, Worldwide geological experience as a systematic basis for prospect appraisal, in G. Demaison and R. J. Murris, eds., Petroleum geochemistry and basin evaluation: AAPG Memoir 35, p. 15–26.

Sluijk, D., and J. R. Parker, 1986, Comparison of pre-drilling predictions with postdrilling outcomes, using Shell's prospect appraisal system, in D. D. Rice, ed., Oil and gas assessment—Methods and applications: AAPG Studies in Geology 21, p. 55–58.

Squire, S. G., 1996, Reservoir and pool parameter distributions from the Cooper/Eromanga Basin: Society of Petroleum Engineers, SPE Paper 37365, 9 p.

Watson, P., 1998, A process for estimating geological risk of petroleum exploration prospects: Australian Petroleum Production and Exploration Association Journal, v. 1998, p. 577–582.

White, D. A., 1993, Geologic risking guide for prospects and plays: AAPG Bulletin, v. 77, p. 2048–2061.